MULTIVARIABLE CALCULUS

MULTIVARIABLE CALCULUS

Produced by the Consortium based at Harvard and funded by a National Science Foundation Grant. All proceeds from the sale of this work are used to support the work of the Consortium.

William G. McCallum
University of Arizona

Deborah Hughes-Hallett
Harvard University

Daniel Flath
University of South Alabama

Douglas Quinney
University of Keele

Andrew M. Gleason
Harvard University

Wayne Raskind
University of Southern California

Sheldon P. Gordon
Suffolk County Community College

Jeff Tecosky-Feldman
Haverford College

David Mumford
Harvard University

Joe B. Thrash
University of Southern Mississippi

Brad G. Osgood
Stanford University

Thomas W. Tucker
Colgate University

John Wiley & Sons, Inc.
New York   Chichester   Brisbane   Toronto   Singapore

Dedicated to Amy, Nell, Abby, and Sally.

Cover Photo by Greg Pease

This material is based upon work supported by the National Science Foundation under Grant No. DUE-9352905. All royalties from the sale of this book will go toward the furtherance of the project.

ISBN: 0-471-12256-4

PREFACE

Calculus is one of the greatest achievements of the human intellect. Inspired by problems in astronomy, Newton and Leibniz developed the ideas of calculus 300 years ago. Since then, each century has demonstrated the power of calculus to illuminate questions in mathematics, the physical sciences, engineering, and the social and biological sciences.

Calculus has been so successful because of its extraordinary power to reduce complicated problems to simple rules and procedures. Therein lies the danger in teaching calculus: it is possible to teach the subject as nothing but the rules and procedures – thereby losing sight of both the mathematics and of its practical value. With the generous support of the National Science Foundation, our group set out to create a new calculus curriculum that would restore that insight. This book is the second stage in that endeavor. The first stage is our single variable text.

Basic Principles

The two principles that guided our efforts in developing the single variable book remain valid. The first is our prescription for restoring the mathematical content to calculus:

> **The Rule of Three:** *Every topic should be presented geometrically, numerically and algebraically.*

We continually encourage students to think and write about the geometrical and numerical meaning of what they are doing. It is not our intention to undermine the purely algebraic aspect of calculus, but rather to reinforce it by giving meaning to the symbols. In the homework problems dealing with applications, we continually ask students what their answers mean in practical terms.

The second principle, inspired by Archimedes, is our prescription for restoring practical understanding:

> **The Way of Archimedes:** *Formal definitions and procedures evolve from the investigation of practical problems.*

Archimedes believed that insight into mathematical problems is gained by investigating mechanical or physical problems first.[1] For the same reason, our text is problem driven. Whenever possible, we start with

---

[1] ... I thought fit to write out for you and explain in detail ... the peculiarity of a certain method, by which it will be possible for you to get a start to enable you to investigate some of the problems in mathematics by means of mechanics. This procedure is, I am persuaded, no less useful even for the proof of the theorems themselves; for certain things first became clear to me by a mechanical method, although they had to be demonstrated by geometry afterwards because their investigation by the said method did not furnish an actual demonstration. But it is of course easier, when we have previously acquired, by the method, some knowledge of the questions, to supply the proof than it is to find it without any previous knowledge. From *The Method*, in *The Works of Archimedes* edited and translated by Sir Thomas L. Heath (Dover, NY)

a practical problem and derive the general results from it. By practical problems we usually, but not always, mean real world applications. These two principles have led to a dramatically new curriculum – more so than a cursory glance at the table of contents might indicate.

Technology

In multivariable calculus, even more so than in single variable calculus, computer technology can be put to great advantage to help students learn to think mathematically. For example, looking at surface graphs and contour diagrams is enormously helpful in understanding functions of many variables. Furthermore, the ability to use technology effectively as a tool in itself is of the greatest importance. Students are expected to use their judgment to determine where technology is useful.

However, the book does not require any specific software or technology, and we have accommodated those without access to sufficiently powerful technology by providing supplementary master copies for overhead slides, showing surface graphs, contour diagrams, parametrized curves, and vector fields. Ideally, students should have access to technology with the ability to draw surface graphs, contour diagrams, and vector fields, and to calculate multiple integrals and line integrals numerically. Failing that, however, the combination of hand held graphing calculators and the overhead transparencies is quite satisfactory, and has been used successfully by test sites.

What Student Background is Expected?

Students using this book should have successfully completed a course in single variable calculus. It is not necessary for them to have used the single variable book from the same consortium in order for them to learn from this book.

The book is thought-provoking for well-prepared students while still accessible to students with weaker backgrounds. Providing numerical and graphical approaches as well as the algebraic gives students another way of mastering the material. This approach encourages students to persist, thereby lowering failure rates.

Content

Our approach to designing this curriculum was the same as the one we took in our single variable book: we started with a clean slate, and compiled a list of topics that we thought were fundamental to the subject, after discussions with mathematicians, engineers, physicists, chemists, biologists, and economists. In order to meet individual needs or course requirements, topics can easily be added or deleted, or the order changed.

Chapter 11: Functions of Many Variables

We introduce functions of many variables from several points of view, using surface graphs, contour diagrams, and tables. This chapter is as crucial for this course as Chapter 1 is for the single variable course; it gives students the skills to read graphs and contour diagrams and think graphically, to read tables and think numerically, and to apply these skills, along with their algebraic skills, to modeling the real world. We pay particular attention to the idea of a section of a function, obtained by varying one variable independently of the others. It is important that the student thoroughly digest this notion from both a graphical and a numerical point of view, before being exposed to the ideas of partial derivatives and gradients. We study linear functions in detail from all points of view, in preparation for the notion of local linearity.

Chapter 12: A Fundamental Tool: Vectors

We define vectors as geometric objects having direction and magnitude, with displacement vectors as the model, and then show how to resolve vectors into components. We define the dot and cross product of two vectors purely in terms of their direction and magnitude, and then give the formulas in terms of components. We continue this approach to vectors throughout the book; the geometric definition first, and the formula in terms of components immediately afterward.

Chapter 13: Differentiating Functions of Many Variables

We introduce the basic notions of partial derivative, directional derivative, gradient, and differential. In keeping with the spirit of the single variable book, we put all the different notions of derivative in the framework of local linearity. We also use local linearity as the basis for the multivariable chain rule. We discuss higher order partial derivatives, their interpretation in partial differential equations, and their application to quadratic Taylor approximations.

Chapter 14: Optimization

We apply the ideas of the previous chapter to optimization problems, both constrained and unconstrained. We derive the second derivative test for local extrema by first considering the case of quadratic polynomials, and then appealing to the quadratic Taylor approximation. We discuss the existence of global extrema for continuous functions on closed and bounded regions. In the section on constrained optimization, we discuss Lagrange multipliers, equality and inequality constraints, problems with more than one constraint, and the Lagrangian.

Chapter 15: Integrating Functions of Many Variables

We motivate the multivariable definite integral graphically by considering the problem of estimating total population from a contour diagram for population density, using finer and finer grids. We continue with numerical examples using tables, and then give two methods of calculating multiple integrals: analytically, by means of iterated integrals, and numerically, by the Monte Carlo method. We discuss both double and triple integrals in Cartesian, polar, spherical, and cylindrical coordinates. We also discuss applications to multivariate probability.

Chapter 16: Parametric Curves and Surfaces

We start with the problem of representing motion in space. This leads in two different directions, each discussed in a separate section: the problem of representing curves parametrically, and the study of velocity and acceleration of moving particles. In keeping with our approach to vectors, we define velocity and acceleration geometrically, then give the formulas in terms of components. We conclude with a section parameterizing surfaces. The chapter leads to an appendix with one of the original, and still the most inspiring, applications of calculus; the derivation of Kepler's laws of motion from Newton's laws.

Chapter 17: Vector Fields

In this brief chapter we introduce vector-valued functions of many variables, or vector fields. This chapter lays the foundation for the geometric approach in the next three chapters to line integrals, flux integrals, divergence, and curl. We start with physical examples, such as velocity vector fields and force fields, and include many sketches of vector fields to help build geometric intuition. We also discuss flow lines of vector fields and their relation with systems of differential equations.

Chapter 18: Line Integrals

We present the concept of integrating a vector field along a path with a coordinate-free definition. We spend some time building intuition using sketches of vector fields with paths superimposed, before introducing the method of calculating line integrals using parametrizations. We then discuss conservative fields, gradient fields, and the Fundamental Theorem of Calculus for Line Integrals. We conclude with a section on non-conservative vector fields and Green's Theorem.

Chapter 19: Flux Integrals

We introduce the flux integral of a vector field through a parameterized surface in the same way as we introduced line integrals. First we give a coordinate-free definition, then we discuss examples where the flux integral (or at least its sign) can be calculated geometrically. Then we show how to calculate flux integrals, first over surface graphs, and then over arbitrary parameterized surfaces.

Chapter 20: Calculus of Vector Fields

We introduce divergence and curl in a coordinate-free way; the divergence in terms of flux density, and curl in terms of circulation density. We then give the formulas in Cartesian coordinates. In the single variable book we derived the Fundamental Theorem of Calculus by pointing out that the integral of the rate of change is the total change. In much the same way, we derive the divergence theorem by showing that the integral of flux density over a volume is the total flux out of the volume, and Stokes' theorem by showing that the integral of circulation density over a surface is the total circulation around its boundary.

Changes in This Edition

We have incorporated suggestions from users of the Draft Version into this new Preliminary Edition. These changes include the following:

- *Chapter 11.* We have rewritten the material in 11.6 on the relation between surface graphs and level surfaces for greater clarity.

- *Chapter 12.* We have reorganized the first two sections: Components of vectors are now in 12.1, and vectors other than displacement vectors are now in 12.2. We have moved the material on the area vector to Chapter 18, where it is first used.

- *Chapter 13.* We have reorganized and rewritten parts of Chapter 13 to make the unifying theme of local linearity clearer. Differentials and local linearity are now together in 13.3, which now also includes a discussion of differentiability. We give a different treatment of the gradient, using local linearity rather than the previous geometric argument. Partial differential equations now come directly after higher order partial derivatives, and quadratic Taylor approximations have been moved to the end of the chapter.

- *Chapter 14.* We have added material on inequality constraints, multiple constraints, Lagrangians, and the existence of global extrema for continuous functions on closed and bounded regions.

- *Chapter 15.* We have added a section on applications of multiple integration to multivariate probability.

- *Chapter 16.* We have extensively reorganized this chapter, to make clearer the distinction between parametric equations used to describe motion in space, and parametric equations used to represent curves. We have merged the sections on velocity and acceleration, and have removed the material on curvature. Parameterized surfaces now come at the end of the chapter. We have moved the material on area vectors to Chapter 18.

- *Chapters 17-20*. The old Chapters 17 and 18 have been split into four chapters, as follows:
    - Chapter 17: Vector Fields
    - Chapter 18: Line Integrals
    - Chapter 19: Flux Integrals
    - Chapter 20: Calculus of Vector Fields

Our purpose was to make the text more flexible for a one-semester course, and to make it easier to pick out a fast track through the material. For example, instructors have the following two choices: They can stop at the end of Chapter 18, allowing time for a thorough treatment of parameterized curves and surfaces, line integrals, and Green's theorem; or they can continue to Chapter 20, covering only the earlier sections in Chapters 16, 17, 18, and 19, which will yield a brief treatment of line and flux integrals from a geometric point of view and give students enough background to understand the divergence theorem and Stokes' theorem.

- *Chapter 17*. This is the first two sections of the old Chapter 17.
- *Chapter 18*. We have reorganized for clarity the material on conservative fields, gradient fields, and circulation, and we have added material on nonconservative vector fields, the two-dimensional curl criterion, and Green's theorem.
- *Chapter 19*. This is the material on flux integrals first two sections of the old Chapter 18. We have split the old section on flux integrals over parameterized surfaces into two sections, on integrals over surface graphs and over general parameterized surfaces.
- *Chapter 20*. We have moved the formulas for divergence and curl in Cartesian coordinates closer to their definitions, and moved the more detailed arguments justifying these formulas to a new section at the end of the chapter.

Supplementary Materials

- **Instructor's Manual** with teaching tips, calculator programs, some overhead transparency masters and sample exams and quizzes.
- **Instructor's Solution Manual** with complete solutions to all problems.
- **Student's Solution Manual** with complete solutions to every other odd-numbered problem.
- **Answer Manual** with brief answers to all odd-numbered problems.
- **MultiGraph** for Windows based surface plotting software.

Our Experiences

In the process of developing the ideas incorporated in this book, we have been conscious of the need to test the materials thoroughly in a wide variety of institutions serving many different types of students. Consortium members have used previous versions of the book at a broad range of institutions. During the 1994–1995 academic year we were assisted by colleagues at over 100 schools who class-tested the Draft Version and reported their experiences and those of their students. This diverse group of schools used the book in semester and quarter systems, in computer labs, small groups, and traditional settings, and with a number of different technologies. We appreciate the valuable suggestions they made, which we have tried to incorporate into this Preliminary Edition of the text.

Acknowledgements

Thanks to Ed Alexander, Carole Anderson, Kevin Anderson, Ralph Baierlein, Roxann Batiste, Jerrie Beiberstein, Shelina Bhojani, Paul Blanchard, Melkana Brakalova, Otto Bretscher, John Brillhart, Ruvim Breydo, Chris Bowman, David Bressoud, Will Brockman, Theresa Broderick, Edward Chandler, Phil Cheifetz, C. K. Cheung, Dave Chen, Dave Chua, Robert Condon, Eric Connally, Radu Constantinescu, Josh Cowley, Greg Crow, Jie Cui, Caspar Curjel, John Drabicki, Bill Dunn, Pavel Etingof, Bill Faris, Paul Feehan, George Fennemore, Hermann Flaschka, Katy Flint, Leonid Friedlander, Deborah Gaines, Avijit Gangopadhyay, Liwei Gao, Scott Gilbert, Nikki Grant, David Grazer, Marty Greenlee, John Hagood, Robert Hanson, Angus Hendrick, Tricia Hersh, Randy Ho, Greg Holmberg, Sharon Hurst, David Hurtubise, Luke Hunsberger, Brady Hunsaker, Robert Indik, Utith Inprasit, Adrian Iovita, Jack Jackson, Jerry Johnson, Millie Johnson, Calvin Jongsma, Georgia Kamvosoulis, Joe Kanapka, Alex Kasman, Matthias Kawski, Misha Kazhdan, Charlie Kerr, Mike Klucznik, Dmitri Kountourgiannis, Matt Kruse, Robert Kuhn, Kam Kwong, Ted Laetsch, Sylvain Laroche, Janny Leung, Dave Levermore, Lei Li, Weiye Li, Li Liu, Carlos Lizzaraga, Patti Frazer Lock, John Lucas, Alex Mallozzi, James Mark, Ricardo Martinez, Mark McConnell, Dan McGee, Andrew Metrick, Karen Millstone, Michal Mlejnek, Kathy Mosher, Marshall Mundt, Don Myers, Jeff Nelson, Alan Newell, Huy Nguyen, John Olson, Myriam Oviedo, James Osterburg, Ed Park, Howard Penn, Tony Phillips, Jessica Polito, Steve Prothero, Amy Rabb-Liu, Fred Richman, Renee Robles, David Royster, W. R. Salzman, Bill Schultz, Barbara Shipman, Michael Stringer, Noah Syroid, Mike Tabor, Sulian Tay, Tepache, Denise Todd, Jose Torres, Elias Toubassi, Jerry Uhl, Doug Ulmer, Steve Uurtamo, Bill Velez, Faye Villalobos, Alice Wang, Joseph Watkins, Eric Wepsic, Steve Wheaton, Maciej Wojtkowski, Xianbao Xu, and Bruce Yoshiwara.

William G. McCallum   Sheldon P. Gordon   Wayne Raskind
Deborah Hughes-Hallett   David Mumford   Jeff Tecosky-Feldman
Daniel E. Flath   Brad G. Osgood   Joe B. Thrash
Andrew M. Gleason   Douglas Quinney   Thomas W. Tucker

To Students: How to Learn from this Book

- This book may be different from other math textbooks that you have used, so it may be helpful to know about some of the differences in advance. This book emphasizes at every stage the *meaning* (in practical, graphical or numerical terms) of the symbols you are using. There is much less emphasis on "plug-and-chug" and using formulas, and much more emphasis on the interpretation of these formulas than you may expect. You will often be asked to explain your ideas in words or to explain an answer using graphs.

- The book contains the main ideas of multivariable calculus in plain English. Your success in using this book will depend on your reading, questioning, and thinking hard about the ideas presented. Although you may not have done this with other books, you should plan on reading the text in detail, not just the worked examples.

- There are very few examples in the text that are exactly like the homework problems. This means that you can't just look at a homework problem and search for a similar–looking "worked out" example. Success with the homework will come by grappling with the ideas of calculus.

- Many of the problems that we have included in the book are open-ended. This means that there may be more than one approach and more than one solution, depending on your analysis. Many times, solving a problem relies on common sense ideas that are not stated in the problem but which you will know from everyday life.

- This book assumes that you have access to a graphing calculator or computer; preferably one that can draw surface graphs, contour diagrams, and vector fields, and can compute multivariable integrals and line integrals numerically. There are many situations where you may not be able to find an exact solution to a problem, but you can use a calculator or computer to get a reasonable approximation. An answer obtained this way is usually just as useful as an exact one. However, the problem does not always state that a calculator is required, so use your judgement.

- This book attempts to give equal weight to three methods for describing functions: graphical (a picture), numerical (a table of values) and algebraic (a formula). Sometimes you may find it easier to translate a problem given in one form into another. For example, if you have to find the maximum of a function, you might use a contour diagram to estimate its approximate position, use its formula to find equations that give the exact position, then use a numerical method to solve the equations. The best idea is to be flexible about your approach: if one way of looking at a problem doesn't work, try another.

- Students using this book have found discussing these problems in small groups very helpful. There are a great many problems which are not cut-and-dried; it can help to attack them with the other perspectives your colleagues can provide. If group work is not feasible, see if your instructor can organize a discussion session in which additional problems can be worked on.

- You are probably wondering what you'll get from the book. The answer is, if you put in a solid effort, you will get a real understanding of one of the most important accomplishments of the millennium – calculus – as well as a real sense of how mathematics is used in the age of technology.

CONTENTS

11 FUNCTIONS OF MANY VARIABLES
1

11.1 FUNCTIONS OF TWO VARIABLES 2

11.2 A TOUR OF THREE-DIMENSIONAL SPACE 10

11.3 GRAPHS OF FUNCTIONS OF TWO VARIABLES 15

11.4 CONTOUR DIAGRAMS 26

11.5 LINEAR FUNCTIONS 43

11.6 FUNCTIONS OF MORE THAN TWO VARIABLES 51

REVIEW PROBLEMS 59

12 A FUNDAMENTAL TOOL: VECTORS
65

12.1 DISPLACEMENT VECTORS 66

12.2 VECTORS IN GENERAL 78

12.3 THE DOT PRODUCT 86

12.4 THE CROSS PRODUCT 96

REVIEW PROBLEMS 105

13 DIFFERENTIATING FUNCTIONS OF MANY VARIABLES
109

13.1 THE PARTIAL DERIVATIVE 110

13.2 COMPUTING PARTIAL DERIVATIVES ALGEBRAICALLY 121

13.3 LOCAL LINEARITY AND THE DIFFERENTIAL 126

13.4 DIRECTIONAL DERIVATIVES 137

13.5 THE GRADIENT 143

13.6 THE CHAIN RULE 151

13.7 SECOND-ORDER PARTIAL DERIVATIVES 157

13.8 PARTIAL DIFFERENTIAL EQUATIONS 162

13.9 NOTES ON QUADRATIC APPROXIMATIONS 171

REVIEW PROBLEMS 175

14 OPTIMIZATION   183

14.1   LOCAL AND GLOBAL EXTREMA   184

14.2   UNCONSTRAINED OPTIMIZATION   196

14.3   CONSTRAINED OPTIMIZATION   206

REVIEW PROBLEMS   219

15 INTEGRATING FUNCTIONS OF MANY VARIABLES   221

15.1   THE DEFINITE INTEGRAL OF A FUNCTION OF TWO VARIABLES   222

15.2   ITERATED INTEGRALS   232

15.3   THREE-VARIABLE INTEGRALS   244

15.4   NUMERICAL INTEGRATION: THE MONTE CARLO METHOD   247

15.5   TWO-VARIABLE INTEGRALS IN POLAR COORDINATES   251

15.6   INTEGRALS IN CYLINDRICAL AND SPHERICAL COORDINATES   256

15.7   APPLICATIONS OF INTEGRATION TO PROBABILITY   263

15.8   NOTES ON CHANGE OF VARIABLES IN A MULTIPLE INTEGRAL   273

REVIEW PROBLEMS   277

16 PARAMETRIC CURVES AND SURFACES   281

16.1   MOTION IN SPACE   282

16.2   PARAMETERIZED CURVES   288

16.3   VELOCITY AND ACCELERATION VECTORS   295

16.4   PARAMETERIZED SURFACES   306

REVIEW PROBLEMS   319

17 VECTOR FIELDS   323

17.1   VECTOR FIELDS   324

17.2   THE FLOW OF A VECTOR FIELD   331

REVIEW PROBLEMS   338

18 LINE INTEGRALS   341

18.1 THE IDEA OF A LINE INTEGRAL   342
18.2 COMPUTING LINE INTEGRALS OVER PARAMETERIZED CURVES   352
18.3 GRADIENT FIELDS AND CONSERVATIVE FIELDS   360
18.4 NONCONSERVATIVE FIELDS AND GREEN'S THEOREM   371
REVIEW PROBLEMS   377

19 FLUX INTEGRALS   379

19.1 THE IDEA OF A FLUX INTEGRAL   380
19.2 CALCULATING FLUX INTEGRALS   393
19.3 NOTES ON FLUX INTEGRALS OVER PARAMETERIZED SURFACES   400
REVIEW PROBLEMS   406

20 CALCULUS OF VECTOR FIELDS   409

20.1 THE DIVERGENCE OF A VECTOR FIELD   410
20.2 THE DIVERGENCE THEOREM   418
20.3 THE CURL OF A VECTOR FIELD   425
20.4 STOKES' THEOREM   434
20.5 NOTES ON THE DIVERGENCE AND CURL   443
REVIEW PROBLEMS   449

APPENDICES   455

A REVIEW OF LOCAL LINEARITY FOR ONE VARIABLE   456
B MAXIMA AND MINIMA OF FUNCTIONS OF ONE VARIABLE   457
C DETERMINANTS   458
D REVIEW OF ONE-VARIABLE INTEGRATION   459

E   TABLE OF INTEGRALS   467

F   REVIEW OF DENSITY FUNCTIONS AND PROBABILITY   469

G   REVIEW OF POLAR COORDINATES   479

H   CHANGE OF VARIABLES   481

I   THE IMPLICIT FUNCTION THEOREM   484

J   NOTES ON KEPLER, NEWTON AND PLANETARY MOTION   489

CHAPTER ELEVEN

FUNCTIONS OF
MANY VARIABLES

Many quantities depend on more than one variable: the amount of food grown depends on the amount of rain and the amount of fertilizer used; the rate of a chemical reaction depends on the temperature and the pressure of the environment in which it proceeds; the strength of the gravitational attraction between two bodies depends on their masses and their distance apart; and the rate of fallout from a volcanic explosion depends on the distance from the volcano and the time since the explosion. In this chapter we will see the many different ways of looking at functions of many variables.

11.1 FUNCTIONS OF TWO VARIABLES

Function Notation

Suppose you are planning to take out a five-year loan to buy a car and you need to calculate what your monthly payment will be; this depends on both the amount of money you borrow and the interest rate. These quantities can vary separately: the loan amount can change while the interest rate remains the same, or the interest rate can change while the loan amount remains the same. To calculate your monthly payment you need to know both. If the monthly payment is m, the loan amount is L, and the interest rate is $r\%$, then we express the fact that m is a function of L and r by writing:

$$m = f(L, r).$$

This is just like the function notation of one-variable calculus. The variable m is called the dependent variable, and the variables L and r are called the independent variables. The letter f stands for the *function* or rule that gives the value of m corresponding to given values of L and r.

A function of two variables can be represented pictorially, numerically by a table of values, or algebraically by a formula. In this section we will give examples of each of these three ways of viewing a function.

Graphical Example: A Weather Map

Figure 11.1 shows a weather map from a newspaper. What information does it convey? It is displaying the predicted high temperature, T, in degrees Fahrenheit (°F), at any point in the US on that day. The curving lines on the map, called *isotherms*, separate the country into zones, according to whether T is in the 60s, 70s, 80s, 90s, or 100s. (*iso* means same and *therm* means heat.) Notice that the isotherm separating the 80s and 90s zones connects all the points where the temperature is exactly 90°F.

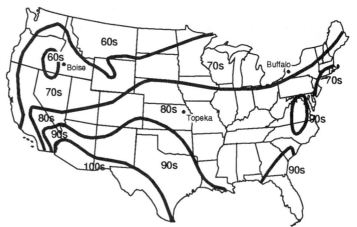

Figure 11.1: Weather map showing predicted high temperatures, T, for June 30, 1992

Example 1 Estimate the value of T in Boise, Idaho; Topeka, Kansas; and Buffalo, New York.

Solution Boise and Buffalo are in the 70s region, and Topeka is in the 80s region. Thus, the temperature in Boise and Buffalo is between 70 and 80; the temperature in Topeka is between 80 and 90.

In fact, we can say more. Although both Boise and Buffalo are in the 70s, Boise is quite close to the $T = 70$ isotherm, whereas Buffalo is quite close to the $T = 80$ isotherm. So we estimate that the temperature is in the low 70s in Boise, and the high 70s in Buffalo. Topeka is more or less halfway in between the $T = 80$ isotherm and the $T = 90$ isotherm. Thus, we guess that the temperature in Topeka is in the mid 80s. In fact, the high temperatures for that day were 71°F for Boise, 79°F for Buffalo, and 86°F for Topeka.

The predicted high temperature, T, illustrated by the weather map is a function of (that is, depends on) location. The location of a point in the US is given by two variables, often longitude and latitude, or miles east-west and miles north-south of a fixed point, say, Topeka. Thus, T is a function of two variables and the weather map is one way of visualizing that function. The weather map in Figure 11.1 is called a *contour map* or *contour diagram*. Section 11.3 shows another way of visualizing functions of two variables using surfaces; Section 11.4 looks at contour maps in detail.

Numerical Example: Beef Consumption

Suppose you are a beef producer and you want to know how much beef people will buy. This depends on how much money people have and on the price of beef. Thus, the consumption of beef, C (in pounds per week per household) is a function of household income, I (in thousands of dollars per year), and the price of beef, p (in dollars per pound). In function notation, we write:

$$C = f(I, p).$$

Table 11.1 contains values of this function. Values of p are shown across the top, values of I are down the left side, and corresponding values of $f(I, p)$ are given in the table.[1] For example, to find the value of $f(40, 3.50)$, we look in the row corresponding to $I = 40$ under $p = 3.50$, where we find the number 4.05. Thus,

$$f(40, 3.50) = 4.05.$$

This means that, on average, if a household's income is $40,000 a year and the price of beef is $3.50/lb, the family will buy 4.05 lbs of beef per week.

TABLE 11.1: *Quantity of beef bought (pounds/household/week)*

		Price of beef, p ($/lb)			
		3.00	3.50	4.00	4.50
Household income per year, I ($1000)	20	2.65	2.59	2.51	2.43
	40	4.14	4.05	3.94	3.88
	60	5.11	5.00	4.97	4.84
	80	5.35	5.29	5.19	5.07
	100	5.79	5.77	5.60	5.53

Notice how this differs from the table of values of a one-variable function, where one row or one column is enough to list the values of the function. Here many rows and columns are needed because the function has a value for every *pair* of values of the independent variables.

Algebraic Examples: Formulas

In both the weather map and beef consumption examples, there is no formula for the underlying function. That is usually the case for functions representing real-life data. On the other hand, for many idealized models in physics, engineering, or economics, there are exact formulas.

---

[1] Adapted from Richard G. Lipsey, *An Introduction to Positive Economics 3rd Ed.*, Weidenfeld and Nicolson, London, 1971

Example 2   Give a formula for the function $M = f(B, t)$ where M is the amount of money in a bank account t years after an initial investment of B dollars, if interest is accrued at a rate of 5% per year compounded (a)   annually   (b)   continuously.

Solution   (a)   Annual compounding means that M increases by a factor of 1.05 every year, so

$$M = f(B, t) = B(1.05)^t.$$

(b)   Continuous compounding means that M grows according to the function e^{kt}, with $k = 0.05$, so

$$M = f(B, t) = Be^{0.05t}.$$

Example 3   A cylinder with closed ends has a radius r and a height h. If its volume is V and surface area is A, find formulas for the functions $V = f(r, h)$ and $A = g(r, h)$.

Solution   Since the area of the circular base is πr^2, we have

$$V = f(r, h) = \text{Area of base} \cdot \text{Height} = \pi r^2 h.$$

The surface area of the side, is the circumference of the bottom, $2\pi r$, times the height h, or $2\pi rh$. Thus,

$$A = g(r, h) = 2 \cdot \text{Area of base} + \text{Area of side} = 2\pi r^2 + 2\pi rh.$$

Strategy to Investigate Functions of Two Variables: Vary One Variable at a Time

We can learn a great deal about functions of two or more variables by letting one variable vary at a time while holding the others fixed, thus obtaining a function of one variable.

The Wave

Suppose you are in a stadium where the audience is doing the wave. This is a ritual in which members of the audience stand up and down in such a way as to create a wave that moves around the stadium. Normally a single wave travels all the way around the stadium, but we will assume there is a continuous sequence of waves. What sort of function will describe the motion of the audience? To keep things simple, we will consider just one row of spectators. We consider the function which describes the motion of each individual in the row. This is a function of two variables: x (the seat number) and t (the time in seconds). For each value of x and t, we write $h(x, t)$ for the height above the ground of the head of the spectator in seat number x at time t seconds. Suppose we are told that

$$h(x, t) = 5 + \cos(0.5x - t).$$

Example 4   (a)   Explain the significance of $h(x, 5)$ in terms of the wave. Find the period of $h(x, 5)$. What does this period represent?
(b)   Explain the significance of $h(2, t)$ in terms of the wave. Find the period of $h(2, t)$. What does this period represent?

Solution   (a)   Holding t fixed at 5 means we are taking a particular moment in time; letting x vary means we are looking along the whole row at that instant. Thus, the function $h(x, 5) = 5 + \cos(0.5x - 5)$ gives the heights along the row at the instant $t = 5$. Figure 11.2 contains the graph of $h(x, 5)$ which is a snapshot of the row at $t = 5$. The heights form a wave of period 4π, or about 12.6 seats. This period tells us that the length of the wave is about 13 seats long.

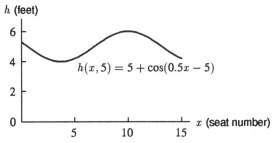

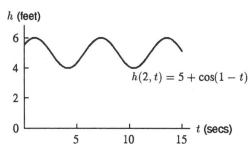

Figure 11.2: The function $h(x,5)$ shows the shape of the wave at time $t=5$

Figure 11.3: The function $h(2,t)$ shows the motion of the spectator in seat number 2

(b) Holding x fixed at 2 means we are concentrating on the spectator in seat number 2; letting t vary means we are watching the motion of that spectator as time passes. Thus, the function $h(2,t)$ describes the motion of the spectator in seat 2 as a function of time. Figure 11.3 shows the graph of $h(2,t) = 5 + \cos(1 - t)$. Notice that the value of h varies between 4 feet and 6 feet as the spectator sits down and stands up. The period is 2π, or about 6.3 seconds. This period represents the time it takes for a spectator to stand up and sit down once.

In general, the one-variable function $h(a,t)$ gives the motion of the spectator in seat a. Figure 11.3 shows the motion of the person in seat number 2. If we pick someone in another seat, we get a similar function, except that the graph may be shifted to the right or to the left.

Example 5   Show that the graph of $h(7,t)$ has the same shape as the graph of $h(2,t)$.

Solution   The motion of the person in seat 7 is described by

$$h(7,t) = 5 + \cos(0.5(7) - t) = 5 + \cos(3.5 - t).$$

Since $h(2,t) = 5 + \cos(1 - t)$, we can rewrite $h(7,t)$ as

$$h(7,t) = 5 + \cos(1 + 2.5 - t) = 5 + \cos(1 - (t - 2.5)) = h(2, t - 2.5).$$

We see that $h(7,t)$ is the same function as $h(2,t)$, except that the t has been replaced by $t - 2.5$. Thus, the graph of $h(7,t)$ is the graph of $h(2,t)$ shifted 2.5 seconds to the right, which means 2.5 seconds later. This means the spectator in seat 7 stands up 2.5 seconds later than the person in seat 2. (See Figure 11.4.) This lag is what makes the wave travel around the stadium. If all the spectators stood up and down at the same time, the wave would not move.

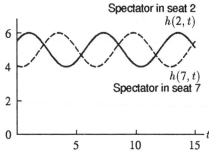

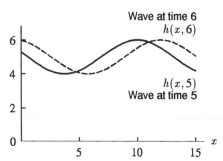

Figure 11.4: Comparison of the motion of spectators in seats 2 and 7

Figure 11.5: The shape of the wave at $t = 5$ and $t = 6$

Example 6   Use the result of Example 5 to find the speed of the wave.

Solution   Since the spectator in seat 7 does the same thing as the spectator in seat 2 but 2.5 seconds later, the wave has moved 5 seats in 2.5 seconds. Thus, the speed is 2 seats per second.

Example 7   Use the functions $h(x, 5)$ and $h(x, 6)$ to show that the speed of the wave is 2 seats per second.

Solution   Figure 11.5 shows graphs of $h(x, 5) = 5 + \cos(0.5x - 5)$ and $h(x, 6) = 5 + \cos(0.5x - 6)$, that is, snapshots of the wave at $t = 5$ and $t = 6$. The shape of the wave at $t = 6$ is the same as the shape at $t = 5$, only shifted to the right by about two seats. Thus, the wave is moving at a rate of about 2 seats per second. To confirm that the speed is exactly 2 seats per second, we must use algebra. When $t = 5$, the equation of the wave is

$$h(x, 5) = 5 + \cos(0.5x - 5),$$

which has a peak where

$$0.5x - 5 = 0,$$

so, at

$$x = 10,$$

that is, in the 10th seat. One second later, at $t = 6$, the equation of the wave is

$$h = 5 + \cos(0.5x - 6),$$

which has a peak where

$$0.5x - 6 = 0,$$

so, at

$$x = 12,$$

that is, in the 12th seat. Thus, the wave moved 2 seats in one second.

The Beef Data

For a function given by a table of values, such as the beef consumption data, we allow one variable to vary at a time by looking at one row or one column. For example, to hold the income, I, fixed at 40, we look at the row $I = 40$. This row gives the values of the function $f(40, p)$ and shows how beef consumption varies as the price varies. See Table 11.2.

TABLE 11.2: *Beef consumption by households making $40,000*

p	3.00	3.50	4.00	4.50
$f(40, p)$	4.14	4.05	3.94	3.88

Since I is fixed, we now have a function of one variable that shows how much beef is bought at various prices by people who earn $40,000 a year. Table 11.2 shows that $f(40, p)$ decreases as p increases. The other rows tell the same story; for each income, I, the consumption of beef goes down as the price, p, increases.

Weather Map

What happens on the weather map in Figure 11.1 when we allow only one variable at a time to vary? For example, suppose we moved along the east-west line through Topeka. Suppose x represents miles east-west of Topeka and y represents miles north-south. We keep y fixed at 0 and let x vary.

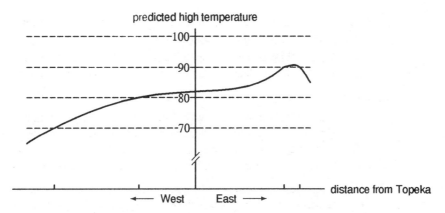

Figure 11.6: Predicted high temperature on an east-west line through Topeka

Along this line, the high temperature T goes from the 60s along the west coast, to the 70s in Nevada and Utah, to the 80s in Topeka, to the 90s just before the east coast, then returns to the 80s. A possible graph is shown in Figure 11.6. Other graphs are possible because we don't know for sure how the temperature varies between contours.

Problems for Section 11.1

Problems 1–3 refer to the weather map in Figure 11.1 on page 2.

1. Give the range of daily high temperatures on June 30, 1992 for
   (a) Pennsylvania   (b) North Dakota   (c) California.

2. Sketch the graph of the predicted high temperature T on a line north-south through Topeka.

3. Sketch the graph of the predicted high temperature on a north-south and an east-west line through Boise.

For Problems 4–8 refer to Table 11.1 on page 3.

4. Make a table showing the amount of money, M, that each household spends on beef (in dollars per household per week) as a function of the price of beef and household income.

5. Give tables for beef consumption as a function of p, with I fixed at $I = 20$ and $I = 100$. Give similar tables for $p = 3.00$ and $p = 4.00$. Comment on what you see in the tables.

6. How does beef consumption vary as a function of household income if the price of beef is held constant?

7. Make a table of the proportion, P, of household income spent on beef per week as a function of price and income.

8. Express P, the proportion of household income spent on beef per week, in terms of the original function $f(I, p)$ which gave consumption as a function of p and I.

9. Sketch the graph of the bank account functions f in Example 2 (a) on page 4, holding B fixed at three different values and letting only t vary. Then sketch the graph of f holding t fixed at three different values and letting only B vary.

10. Consider the acceleration due to gravity, g, at a height h above the surface of a planet of mass m.
    (a) If m is held constant, is g an increasing or decreasing function of h? Why?
    (b) If h is held constant, is g an increasing or decreasing function of m? Why?

11. Consider a function giving the number, n, of new cars sold in a year as a function of the price of new cars, c, and of the average price of gas, g.

     (a) If c is held constant, is n an increasing or decreasing function of g? Why?
     (b) If g is held constant, is n an increasing or decreasing function of c? Why?

12. You are planning a long driving trip and your principal cost will be gasoline.

     (a) Make a table showing how the daily fuel cost varies as a function of the price of gasoline (in dollars per gallon) and the number of gallons you buy each day.
     (b) If your car goes 30 miles on each gallon of gasoline, make a table showing how your daily fuel cost varies as a function of your travel distance and the price of gas.

13. Table 11.3 shows the wind-chill factor as a function of wind speed and temperature. The wind-chill factor is a temperature which tells you how cold it feels, as a result of the combination of wind and temperature.

TABLE 11.3: *Wind-chill factor (°F)*

		35	30	25	20	15	10	5	0
	5	33	27	21	16	12	7	0	-5
Wind	10	22	16	10	3	-3	-9	-15	-22
speed	15	16	9	2	-5	-11	-18	-25	-31
(mph)	20	12	4	-3	-10	-17	-24	-31	-39
	25	8	1	-7	-15	-22	-29	-36	-44

Temperature (°F)

     (a) If the temperature is 0°F and the wind speed is 15 mph, how cold does it feel?
     (b) If the temperature is 35°F, what wind speed makes it feel like 22°F?
     (c) If the temperature is 25°F, what wind speed makes it feel like 20°F?
     (d) If the wind is blowing at 15 mph, what temperature feels like 0°F?

14. Using Table 11.3, make tables of the windchill factor as a function of wind speed for temperatures of 20°F and 0°F.

15. Using Table 11.3, make tables of the windchill factor as a function of temperature for wind speeds of 5 mph and 20 mph.

16. Table 11.4 shows the heat index as a function of temperature and humidity. The heat index is a temperature which tells you how hot it feels as a result of the combination of the two. Heat exhaustion is likely to occur when the heat index reaches 105.

TABLE 11.4: *Heat index (°F)*

		70	75	80	85	90	95	100	105	110	115
	0	64	69	73	78	83	87	91	95	99	103
	10	65	70	75	80	85	90	95	100	105	111
Relative	20	66	72	77	82	87	93	99	105	112	120
humidity	30	67	73	78	84	90	96	104	113	123	135
(%)	40	68	74	79	86	93	101	110	123	137	151
	50	69	75	81	88	96	107	120	135	150	
	60	70	76	82	90	100	114	132	149		

Temperature (°F)

(a)   If the temperature is 80° F and the humidity is 50%, how hot does it feel?

(b)   At what humidity does 90° F feel like 90° F?

(c)   Make a table showing the approximate temperature at which heat exhaustion becomes a danger, as a function of humidity.

(d)   Explain why the heat index is sometimes above the actual temperature and sometimes below it.

17.   Using Table 11.4, graph the heat index function with temperature fixed at 70° F and at 100° F. Explain the features of each graph and the difference between them in common sense terms.

18.   Suppose the function for the stadium wave on page 4 was given by $h(x, t) = 5 + \cos(x - 2t)$. How does this wave compare with the original wave? What is the speed of this wave (in seats per second)?

19.   Suppose the stadium wave on page 4 was moving in the opposite direction, right-to-left instead of left-to-right. Give a possible formula for h.

Problems 20–23 concern a vibrating guitar string.   Suppose you pluck a guitar string and watch it vibrate. If you take snapshots of the guitar string at millisecond intervals, you might get something like Figure 11.7.

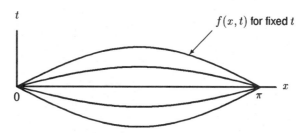

Figure 11.7: A vibrating guitar string: $f(x, t) = \cos t \sin x$

We can analyze the motion of the guitar string using a function of two variables. Think of the guitar string stretched tight along the x-axis from $x = 0$ to $x = \pi$. Each point on the string has an x value, $0 \le x \le \pi$. As the string vibrates, each point on the string moves back and forth on either side of the x-axis. The ends of the string at $x = 0$ and $x = \pi$ remain stationary, while the point at the middle of the string moves the most. Let $f(x, t)$ be the displacement at time t of the point on the string located x units from the left end. Then a possible formula for $f(x, t)$ is

$$f(x, t) = \cos t \sin x, \quad 0 \le x \le \pi, \quad t \text{ in milliseconds.}$$

20.   (a)   Sketch graphs of f versus x for fixed t values, $t = 0, \pi/4, \pi/2, 3\pi/4, \pi$.

(b)   Use your graphs to explain why f represents a vibrating guitar string.

21.   Explain what the functions $f(x, 0)$ and $f(x, 1)$ represent in terms of the vibrating string.

22.   Explain what the functions $f(0, t)$ and $f(1, t)$ represent in terms of the vibrating string.

23.   Describe the motion of the guitar strings whose displacements are given by the following:

(a) $g(x, t) = \cos 2t \sin x$ (b) $h(x, t) = \cos t \sin 2x$

11.2 A TOUR OF THREE-DIMENSIONAL SPACE

Cartesian Coordinates in Three-Space

The way we describe points in the plane by giving x- and y-coordinates can be extended to three-dimensional space. Imagine three coordinate axes meeting at the *origin*: a vertical axis, and two horizontal axes at right angles to each other. (See Figure 11.8.) Think of the xy-plane as being horizontal, while the z-axis extends vertically above and below the plane. The labels x, y, and z show which part of each axis is positive; the other side is negative. We specify a point in 3-space by giving its coordinates (x, y, z) with respect to these axes. Think of the coordinates as instructions telling you how to get to the point; starting at the origin, going x units in the direction parallel to the x-axis, then y units in the direction parallel to the y-axis, and finally z units in the direction parallel to the z-axis. The coordinates can be positive, zero or negative; a zero coordinate means "don't move in this direction," and a negative coordinate means "go in the negative direction along this axis." For example, the origin has coordinates $(0, 0, 0)$, since you get there from the origin by doing nothing at all.

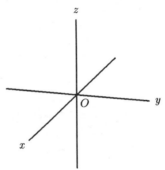

Figure 11.8: Coordinate axes in
three-dimensional space

Example 1   Describe the position of the points with coordinates $(1, 2, 3)$ and $(0, 0, -1)$.

Solution   We get to the point $(1, 2, 3)$ by starting at the origin, going 1 unit along the x-axis, 2 units in the direction parallel to the y-axis, and 3 units up in the direction parallel to the z-axis. (See Figure 11.9.)

To get to $(0, 0, -1)$, we don't move at all in the x and y directions, and we move 1 unit in the negative z direction. So the point is on the negative z-axis. (See Figure 11.9.)

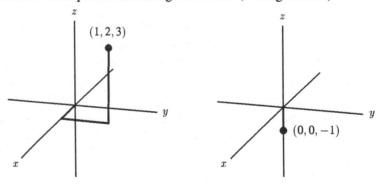

Figure 11.9: Coordinates of points in three-space

Example 2   You start at the origin, go along the y-axis a distance of 2 units in the positive direction, and then move vertically upward a distance of 1 unit. What are your coordinates?

Solution   You started at the point $(0, 0, 0)$. When you went along the y-axis your y-coordinate increased to 2, and when you went vertically your z-coordinate increased to 1; your x-coordinate didn't change because you did not move in the x direction. So your coordinates are $(0, 2, 1)$.

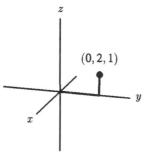

Figure 11.10: The point
$(0, 2, 1)$

It is often helpful to picture a three dimensional coordinate system in terms of a room you are in. The origin is a corner at floor level where two walls meet the floor. The vertical or z-axis is the intersection of the two walls; the x- and y-axes are the intersections of each wall with the floor. Points with negative coordinates lie behind a wall in the next room or below the floor.

Graphing Equations in Three-Dimensional Space

We can graph equations involving the variables x, y, and z in three-dimensional space.

Example 3   What do the graphs of the equations $z = 0$, $z = 3$, and $z = -1$ look like?

Solution   Graphing an equation means drawing the set of points in space whose coordinates satisfy the equation. So to graph $z = 0$ we need to visualize the set of points whose z-coordinate is zero. If your z-coordinate is 0, then you must be at the same vertical level as the origin, that is, you are in the horizontal plane containing the origin. So the graph of $z = 0$ is the middle plane in Figure 11.11. The graph of $z = 3$ is a plane parallel to the graph of $z = 0$, but three units above it. Similarly, the graph of $z = -1$ is a plane parallel to the graph of $z = 0$, but one unit below it.

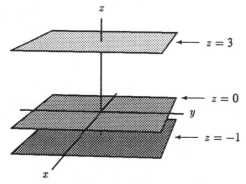

Figure 11.11: The planes $z = -1$, $z = 0$, and
$z = 3$

The plane $z = 0$ contains the x- and y-coordinate axes, and hence is called the xy-coordinate plane, or xy-plane for short. There are two other coordinate planes. The yz-plane contains both the y- and the z-axes, and the xz-plane contains the x- and z-axes. See Figure 11.12.

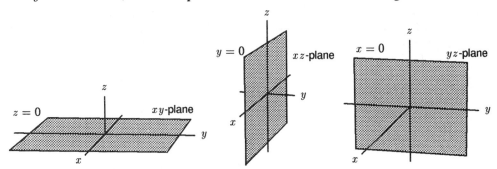

Figure 11.12: The three coordinate planes

Example 4   Which of the points $A = (1, -1, 0)$, $B = (0, 3, 4)$, $C = (2, 2, 1)$, and $D = (0, -4, 0)$ lies closest to the xz-plane? Which point lies on the y-axis?

Solution   The absolute value of the y-coordinate gives the distance to the xz-plane. The point A lies closest to that plane, because it has the smallest y-coordinate in absolute value. (It actually lies to the left of the plane, since its y-coordinate is negative.) To get to a point on the y-axis, you move along the y-axis, but you don't move at all in the x or z directions; thus, a point on the y-axis has both its x- and z-coordinates equal to zero. The only point of the four that satisfies this is D, so it is the only one on the y-axis. Figure 11.13 shows the points.

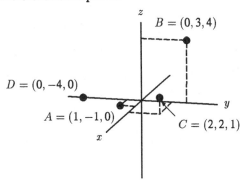

Figure 11.13: Which point lies closest to the xz-plane? Which point lies on the y-axis?

In general, if a point has one of its coordinates equal to zero, it lies in one of the coordinate planes. If a point has two of its coordinates equal to zero, it lies on one of the coordinate axes.

Example 5   You are 2 units below the xy-plane and in the yz-plane. What are your coordinates?

Solution   Since you are 2 units below the xy-plane, your z-coordinate is -2. Since you are in the yz-plane, your x-coordinate is 0. Your y-coordinate can be anything. Thus, you are at the point $(0, y, -2)$, where y can be anything. The set of all such points forms a line parallel to the y-axis, 2 units below the xy-plane, and in the yz-plane. See Figure 11.14.

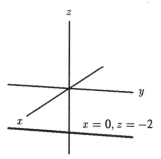

Figure 11.14: The line $x = 0$,
$z = -2$

Example 6   You are standing at the point $(4, 5, 2)$, looking at the point $(0.5, 0, 3)$. Are you looking up or down?

Solution   The point you are standing at has z-coordinate 2, whereas the point you are looking at has z-coordinate 3; hence you are looking up.

Example 7   Imagine that the yz-plane in Figure 11.12 on page 12 is a page of this book. Describe the region behind the page.

Solution   The positive part of the x-axis pokes out of the page; moving in the positive x direction brings you out in front of the page. The region behind the page corresponds to negative values of x, and so it is the set of all points in three-dimensional space satisfying the inequality $x < 0$. Think of it as all the pages behind the one you are looking at.

Distance

In 2-space, the formula for the distance between two points (x, y) and (a, b) is given by

$$\text{Distance} = \sqrt{(x - a)^2 + (y - b)^2}.$$

The distance between two points (x, y, z) and (a, b, c) in 3-space is represented by PG in Figure 11.15.

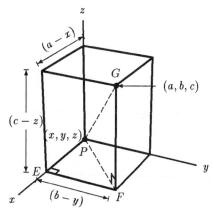

Figure 11.15: Distance between the points
(x, y, z) and (a, b, c)

Using Pythagoras' theorem twice gives

$$PG = \sqrt{(PF)^2 + (FG)^2} = \sqrt{(PE)^2 + (EF)^2 + (FG)^2} = \sqrt{(x - a)^2 + (y - b)^2 + (z - c)^2}.$$

Thus, we have a formula for the distance between the points (x, y, z) and (a, b, c) in 3-space

$$\text{Distance} = \sqrt{(x - a)^2 + (y - b)^2 + (z - c)^2}.$$

Example 8   Find the distance between $(1, 2, 1)$ and $(-3, 1, 2)$.

Solution   The formula gives a distance of
$$\sqrt{(-3 - 1)^2 + (1 - 2)^2 + (2 - 1)^2} = \sqrt{18} = 4.24.$$

Example 9   Find an expression for the distance from the origin to the point (x, y, z).

Solution   The origin has coordinates $(0, 0, 0)$, so the distance from the origin to a point (x, y, z) is
$$\sqrt{(x - 0)^2 + (y - 0)^2 + (z - 0)^2} = \sqrt{x^2 + y^2 + z^2}.$$

Example 10   Find the equation for a sphere of radius 1 with center at the origin.

Solution   The sphere consists of all points (x, y, z) whose distance from the origin is 1, that is, which satisfy the equation
$$\sqrt{x^2 + y^2 + z^2} = 1.$$
This is an equation for the sphere. If we square both sides we get the equation in the more pleasant looking form
$$x^2 + y^2 + z^2 = 1.$$
Note that this equation represents the surface of the sphere. The solid ball enclosed by the sphere is represented by the inequality $x^2 + y^2 + z^2 < 1$.

Problems for Section 11.2

1. Which of the points $A = (1.3, -2.7, 0)$, $B = (0.9, 0, 3.2)$, $C = (2.5, 0.1, -0.3)$ is closest to the yz-plane? Which one lies on the xz-plane? Which one is farthest from the xy-plane?

2. Which of the points $A = (23, 92, 48)$, $B = (-60, 0, 0)$, $C = (60, 1, -92)$ is closest to the yz-plane? Which one lies on the xz-plane? Which one is farthest from the xy-plane?

3. You are at the point $(-1, -3, -3)$ facing the yz-plane. Assume the axes are oriented as in Figure 11.8 on page 10, and that your head is pointing upwards. You walk 2 units forward, turn left, and walk for another 2 units. Are you in front of or behind the yz-plane? Are you to the left or to the right of the xz-plane? Are you above or below the xy-plane?

4. You are at the point $(3, 1, 1)$ facing the yz-plane. Assume the axes are oriented as in Figure 11.8 on page 10, and that your head is pointing upwards. You walk 2 units forward, turn left, and walk for another 2 units. Are you in front of or behind the yz-plane? Are you to the left of or to the right of the xz-plane? Are you above or below the xy-plane?

Sketch graphs of the equations in Problems 5–7 in 3-space.

5. $x = -3$ 6. $y = 1$ 7. $z = 2$ and $y = 4$

8. Find a formula for the shortest distance between a point (a, b, c) and the y-axis.

9. Describe the set of points whose distance from the x-axis is 2.

10. Describe the set of points whose distance from the x-axis equals the distance from the yz-plane.

11. Which of the points $P = (1, 2, 1)$ and $Q = (2, 0, 0)$ is closest to the origin?

12. Which two of the three points $(1, 2, 3)$, $(3, 2, 1)$ and $(1, 1, 0)$ are closest to each other?

13. A cube is located such that its top four corners have the coordinates of $(-1, -2, 2)$, $(-1, 3, 2)$, $(4, -2, 2)$ and $(4, 3, 2)$. Give the coordinates of the center of the cube.

14. A rectangular solid lies with its length parallel to the y-axis, and its top and bottom faces parallel to the plane $z = 0$. If the center of the object is at $(1, 1, -2)$ and it has a length of 13, a height of 5 and a width of 6, give the coordinates of all eight corners and draw the figure labeling the eight corners.

15. Which of the points $(-3, 2, 15)$, $(0, -10, 0)$, $(-6, 5, 3)$ and $(-4, 2, 7)$ is closest to $(6, 0, 4)$?

16. You are standing on the point $(3, 4, -2)$. North is in the negative x-direction, east is in the positive y-direction, and up in the positive z-direction.

    (a) What coordinates will you be at after you move north 3 units, up 5 units, west 7 units, down 2 units, east 7 units, and south 2 units?

    (b) After completing part (a), will you be looking up or down to the point $(6, 7, 7)$?

17. On a set of x, y, and z axes oriented as in Figure 11.8 on page 10, draw a straight line through the origin, lying in the xz-plane and such that if you move along the line with your x-coordinate increasing, your z-coordinate is decreasing.

18. On a set of x, y and z axes oriented as in Figure 11.8 on page 10, draw a straight line through the origin, lying in the yz-plane and such that if you move along the line with your y-coordinate increasing, your z-coordinate is increasing.

19. Find the equation of the sphere of radius 5 centered at the origin.

20. Find the equation of the sphere of radius 5 centered at $(1, 2, 3)$.

21. Given the sphere
$$(x - 1)^2 + (y + 3)^2 + (z - 2)^2 = 4,$$

    (a) Find the equations of the circles (if any) where the sphere intersects each coordinate plane.

    (b) Find the points (if any) where the sphere intersects each coordinate axis.

11.3 GRAPHS OF FUNCTIONS OF TWO VARIABLES

How do You Visualize a Function of Two Variables?

The weather map on page 2 is one way of visualizing a function of two variables. In this section we see how to visualize a function of two variables in another way, using a surface in 3-space.

The Graph of a Function and How to Plot One

For a function of one variable, $y = f(x)$, the graph of f is the set of all points (x, y) in 2-space such that $y = f(x)$. In general, these points lie on a curve in the plane. When a computer or calculator graphs f, it plots points in the xy-plane and joins consecutive points by line segments. This doesn't give the exact graph, because the exact graph does not go straight from one plotted point to the next. However, the more points, the better the approximation.

Now consider a function of two variables.

> The **graph** of a function of two variables, f, is the set of all points (x, y, z) such that $z = f(x, y)$. In general, the graph of a function of two variables is a surface in 3-space.

Plotting the Graph of the Function $f(x, y) = x^2 + y^2$

To sketch the graph of f we connect points just as we did for a function of one variable. We first make a table of values of f, such as in Table 11.5.

TABLE 11.5: *Table of values of* $f(x, y) = x^2 + y^2$

x \ y	-3	-2	-1	0	1	2	3
-3	18	13	10	9	10	13	18
-2	13	8	5	4	5	8	13
-1	10	5	2	1	2	5	10
0	9	4	1	0	1	4	9
1	10	5	2	1	2	5	10
2	13	8	5	4	5	8	13
3	18	13	10	9	10	13	18

Now we plot points. For example, we plot $(1, 2, 5)$ because $f(1, 2) = 5$, and we plot $(0, 2, 4)$ because $f(0, 2) = 4$. Then, we connect the points corresponding to the rows and columns in the table. The result is called a *wire-frame* picture of the graph. Filling in between the wires gives a surface. That is the way a computer drew the graph of $f(x, y) = x^2 + y^2$ in Figure 11.16. Figure 11.17 shows a more accurate sketch obtained by plotting more points in each direction.

You should check to see if the sketches make sense. Notice that the graph goes through the origin since $(x, y, z) = (0, 0, 0)$ satisfies $z = x^2 + y^2$. Observe that if x is held fixed and y is allowed to vary, the graph dips down and then goes back up, just like the rows of Table 11.5. Similarly, if y is held fixed and x is allowed to vary, the graph dips down and then goes back up, just like the columns of Table 11.5.

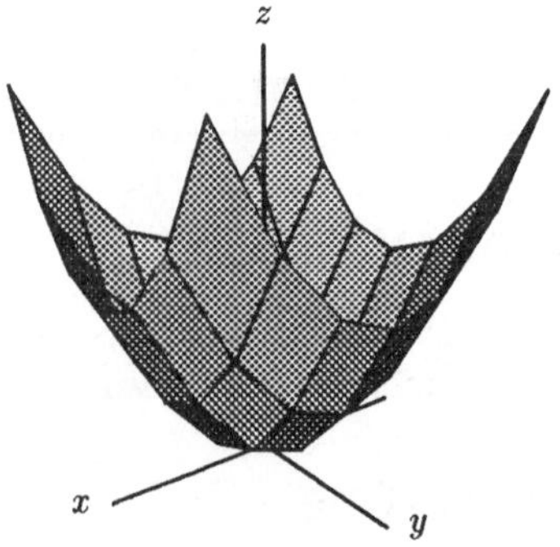

Figure 11.16: Graph of $f(x, y) = x^2 + y^2$ for $-3 \le x \le 3$, $-3 \le y \le 3$

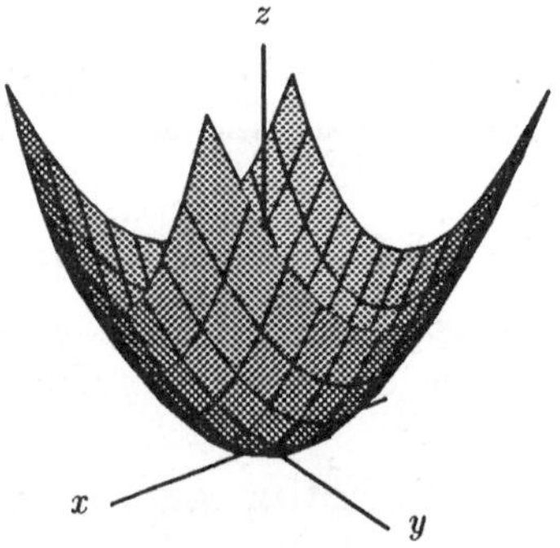

Figure 11.17: Graph of $f(x, y) = x^2 + y^2$ with more points plotted

New Graphs from Old

We can use the graph of a function to visualize the graphs of related functions.

Example 1   Let $f(x, y) = x^2 + y^2$. Describe the graphs of the following functions:

(a) $g(x, y) = x^2 + y^2 + 3$,   (b) $h(x, y) = 5 - x^2 - y^2$,   (c) $k(x, y) = x^2 + (y - 1)^2$.

Solution   We already know from Figure 11.17 that the graph of f looks like a bowl with its vertex at the origin. From this we can work out what the graphs of g, h, and k will look like.

(a)   The function $g(x, y) = x^2 + y^2 + 3 = f(x, y) + 3$, so the graph of g looks like the graph of f, but raised by 3 units. See Figure 11.18.

(b)   Since $-x^2 - y^2$ is the negative of $x^2 + y^2$, the graph of $-x^2 - y^2$ is an upside down bowl. Thus, the graph of $h(x, y) = 5 - x^2 - y^2 = 5 - f(x, y)$ looks like an upside down bowl with vertex at $(0, 0, 5)$, as in Figure 11.19.

(c)   The graph of $k(x, y) = x^2 + (y - 1)^2 = f(x, y - 1)$ is a bowl with vertex at $x = 0$, $y = 1$, since that is where $k(x, y) = 0$, as in Figure 11.20.

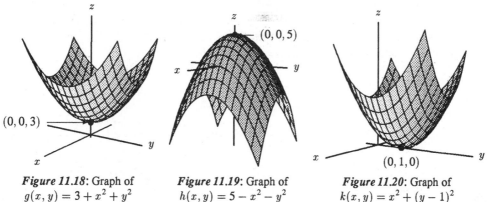

Figure 11.18: Graph of
$g(x, y) = 3 + x^2 + y^2$

Figure 11.19: Graph of
$h(x, y) = 5 - x^2 - y^2$

Figure 11.20: Graph of
$k(x, y) = x^2 + (y - 1)^2$

Example 2   Describe the graph of $G(x, y) = e^{-(x^2 + y^2)}$. What symmetry does it have?

Solution   Since the exponential function is always positive, the graph lies entirely above the xy-plane. From the graph of $x^2 + y^2$ we see that $x^2 + y^2$ is zero at the origin and gets larger as you move farther from the origin in any direction. Thus, $e^{-(x^2 + y^2)}$ is 1 at the origin, and gets smaller as you move from the origin in any direction. It can't go below the xy-plane; instead it flattens out, getting closer and closer to the plane. We say the surface is *asymptotic* to the xy-plane. See Figure 11.21.

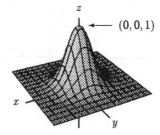

Figure 11.21: Graph of
$G(x, y) = e^{-(x^2 + y^2)}$

Now consider a point (x, y) on the circle $x^2 + y^2 = r^2$. Since

$$G(x, y) = e^{-(x^2+y^2)} = e^{-r^2},$$

the value of the function G is the same at all points on this circle. Thus, we say the graph of G has *circular symmetry*.

Cross-Sections and the Graph of a Function

We have seen that a good way to analyze a function of two variables is to let one variable vary at a time while the other is kept fixed.

> For a function $f(x, y)$, the function we get by holding x fixed and letting y vary is called a **section** of f with x fixed. The graph of the section of $f(x, y)$ with $x = c$ is the curve, or cross-section, we get by intersecting the graph of f with the plane $x = c$. We define a section of f with y fixed similarly.

For example, the section of $f(x, y) = x^2 + y^2$ with $x = 2$ is $f(2, y) = 4 + y^2$. The graph of this section is the curve we get by intersecting the graph of f with the plane perpendicular to the x-axis at $x = 2$. (See Figure 11.22.)

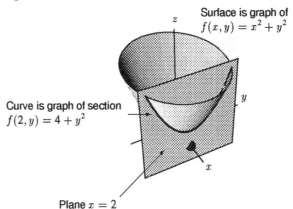

Figure 11.22: Cross-section of the the surface $z = f(x, y)$ by the plane $x = 2$

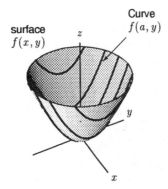

Figure 11.23: The curves $z = f(a, y)$ with a constant: cross-sections with x fixed

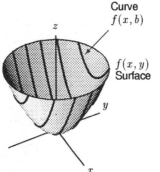

Figure 11.24: The curves $z = f(x, b)$ with b constant: cross-sections with y fixed

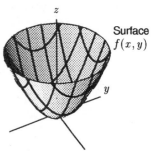

Figure 11.25: Wire-frame representation of the surface $z = f(x, y)$

Figure 11.23 shows graphs of other sections of f with x fixed; Figure 11.24 shows graphs of sections with y fixed. Together these sections generate the wire-frame surface in Figure 11.25.

Example 3   Describe the sections of the function $g(x, y) = x^2 - y^2$ with y fixed and then with x fixed. Use these sections to describe the shape of the graph of g.

Solution   The sections with y fixed at $y = a$ are given by

$$z = g(x, a) = x^2 - a^2.$$

Thus, each section with y fixed gives a parabola opening upwards, with minimum $z = -a^2$. The sections with x fixed are of the form

$$z = g(a, y) = a^2 - y^2,$$

which are parabolas opening downwards with a maximum of $z = a^2$. Figure 11.26 shows the graphs of sections with y fixed all drawn on one xz-plane; Figure 11.27 shows the graphs of sections with x fixed.

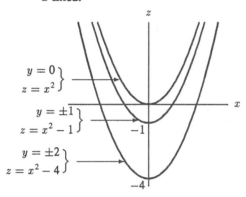

Figure 11.26: Sections of $g(x, y) = x^2 - y^2$ with y fixed

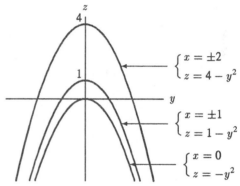

Figure 11.27: Sections of $g(x, y) = x^2 - y^2$ with x fixed

The graph of g is shown in Figure 11.28. Notice the upward opening parabolas in the x-direction and the downward opening parabolas in the y-direction. We say that the surface is *saddle-shaped*.

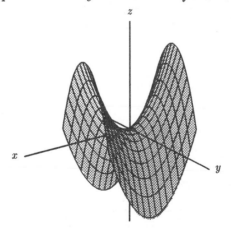

Figure 11.28: Graph of $g(x, y) = x^2 - y^2$

Example 4   Suppose the graph in Figure 11.29 shows your happiness as a function of love and money. Describe in words your happiness:
(a)   As a function of money, with love fixed.
(b)   As a function of love, with money fixed.
(c)   Draw the graphs of two different sections with love fixed and two different sections with money fixed.

Figure 11.29: Happiness as a function of
love and money

Solution   (a)   As money increases, with love fixed, your happiness goes up, reaches a maximum and then goes back down. Evidently, there is such a thing as too much money.
(b)   On the other hand, as love increases, with money fixed, your happiness keeps going up.
(c)   A section with love fixed will show your happiness as money increases; the curve goes up to a maximum then back down, as in Figure 11.30. The higher section, showing more overall happiness, corresponds to a larger amount of love, because as love increases so does happiness. Figure 11.31 shows two sections with money fixed. Happiness increases as love increases. We cannot say, however, which section corresponds to a larger fixed amount of money, because as money increases happiness can either increase or decrease.

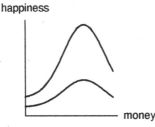

Figure 11.30: Sections with love
fixed

Figure 11.31: Sections with
money fixed

Linear Functions

You may be able to guess the shape of the graph of a linear function of two variables. (It's a plane.) Let's look at an example.

Example 5   Describe the graph of $f(x, y) = 1 + x - y$.

Solution   The plane $x = a$ is vertical and parallel to the yz-plane. Thus, the section with $x = a$ is the line $z = 1 + a - y$ which slopes downward in the y-direction. Similarly, the plane $y = b$ is parallel to the xz-plane. Thus, the section with $y = b$ is the line $z = 1 + x - b$ which slopes upward in the x-direction. Since all the sections are lines, you might expect the graph to be a flat plane, sloping down in the y-direction and up in the x-direction. This is indeed the case. See Figure 11.32.

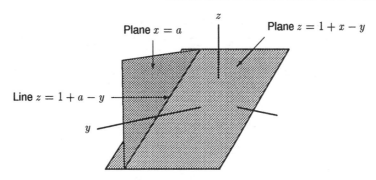

Figure 11.32: Graph of the plane $z = 1 + x - y$ showing section with $x = a$

When One Variable is Missing: Cylinders

Suppose we graph an equation like $z = x^2$ which has one variable missing. What does the surface look like? Since y is missing from the equation, the cross-sections with y fixed are all the same parabola, namely $z = x^2$. Letting y vary up and down the y-axis, this parabola sweeps out the trough-shaped surface shown in Figure 11.33. The cross-sections with x fixed are horizontal lines, obtained by cutting the surface by a plane perpendicular to the x-axis.

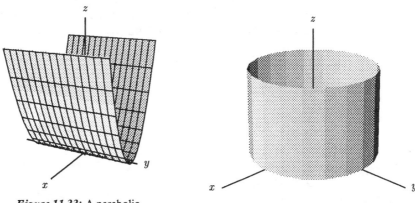

Figure 11.33: A parabolic cylinder $z = x^2$

Figure 11.34: Circular cylinder $x^2 + y^2 = 1$

This surface is called a *parabolic cylinder*, because it is formed from a parabola in the same way that an ordinary cylinder is formed from a circle; it has a parabolic cross-section instead of a circular one.

Example 6   Graph the equation $x^2 + y^2 = 1$ in 3-space.

Solution   Although the equation $x^2 + y^2 = 1$ does not represent a function, the surface representing it can be graphed by the method used for $z = x^2$. The graph of $x^2 + y^2 = 1$ in the xy-plane is a circle. Since z does not appear in the equation, the intersection of the surface with any horizontal plane, $z = c$, will be the same circle $x^2 + y^2 = 1$. Thus, the surface is the vertical circular cylinder shown in Figure 11.34.

Problems for Section 11.3

1. The surface in Figure 11.35 is the graph of the function $z = f(x, y)$ for positive x and y.
   (a) Suppose y is fixed and positive. Does z increase or decrease as x increases? Sketch a graph of z against x.
   (b) Suppose x is fixed and positive. Does z increase or decrease as y increases? Sketch a graph of z against y.

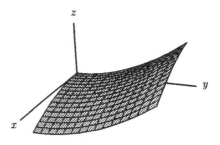

Figure 11.35

2. Match the following descriptions of a company's success with the graphs in Figure 11.36.
   (a) Although we aren't always totally successful, it seems that the amount of money invested doesn't matter. As long as we put hard work into the company our success will increase.
   (b) No matter how much money or hard work we put into the company, we just couldn't make a go of it.
   (c) Our success is measured in dollars, plain and simple. More hard work won't hurt, but it also won't help.
   (d) The company seemed to take off by itself.

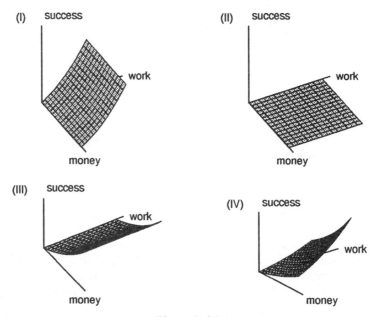

Figure 11.36

For Problems 3–6, use a computer to sketch the graph of a function with the given shapes. Include the axes and the equation used to generate it in your sketch.

3.   A cone of circular cross-section opening downward and with its vertex at the origin.

4.   A bowl which opens upward and has its vertex at 5 on the z-axis.

5.   A plane which has its x, y, and z intercepts all positive.

6.   A parabolic cylinder opening upward from along the line $y = x$ in the xy-plane.

7.   For each of the following functions, decide whether it could be a bowl, a plate, or neither. Consider a plate to be any fairly flat surface and a bowl to be anything that could hold water, assuming the positive z-axis is up.

    (a) $z = x^2 + y^2$ (b) $z = 1 - x^2 - y^2$ (c) $x + y + z = 1$

    (d) $z = -\sqrt{5 - x^2 - y^2}$ (e) $z = 3$

8.   For each function in Problem 7 sketch:

    (a) Sections with x fixed at $x = 0$ and $x = 1$.

    (b) Sections with y fixed at $y = 0$ and $y = 1$.

9.   Match the following functions with their graphs in Figure 11.37.

    (a) $z = \dfrac{1}{x^2 + y^2}$ (b) $z = -e^{-x^2 - y^2}$ (c) $z = x + 2y + 3$

    (d) $z = -y^2$ (e) $z = x^3 - \sin y.$

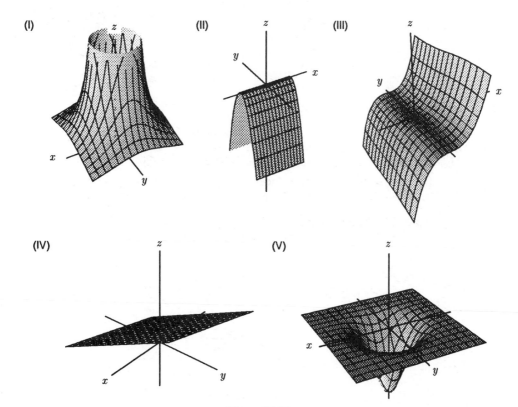

Figure 11.37

10. You like pizza and you like cola. Which of the graphs in Figure 11.38 represents your happiness as a function of how many pizzas and how much cola you have if

    (a) There is no such thing as too many pizzas and too much cola?
    (b) There is such a thing as too many pizzas or too much cola?
    (c) There is such a thing as too much cola but no such thing as too many pizzas?

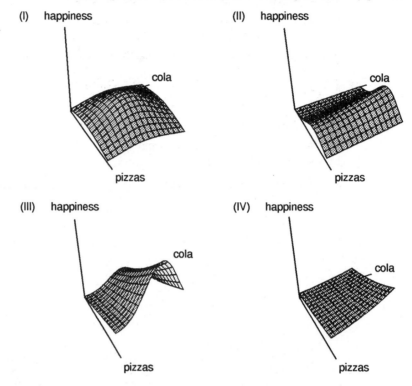

Figure 11.38

11. For each of the graphs I-IV in Problem 10 draw:

    (a) two sections with pizza fixed
    (b) two sections with cola fixed.

12. Figure 11.39 contains graphs of $z = f(x, b)$ for $b = -2, -1, 0, 1, 2$. Which of the graphs of $z = f(x, y)$ in Figure 11.40 best fits this information?

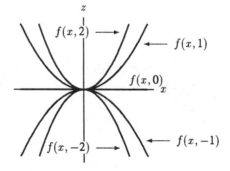

Figure 11.39

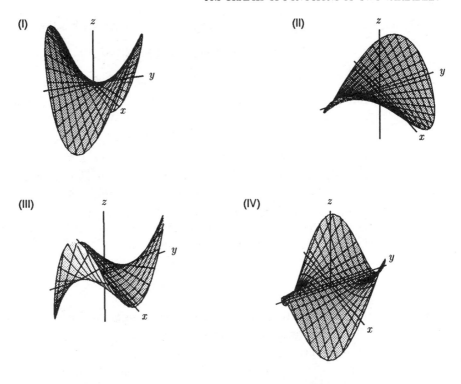

Figure 11.40

13. Imagine a single wave traveling along a canal. Suppose x is the distance from the middle of the canal, t is the time, and z is the height of the water above the equilibrium level. The graph of z as a function of x and t is shown in Figure 11.41.

   (a) Draw the profile of the wave for $t = -1, 0, 1, 2$. (Show the x-axis to the right and the z-axis vertically.)
   (b) Is the wave traveling in the direction of increasing or decreasing x?
   (c) Sketch a surface representing a wave traveling in the opposite direction.

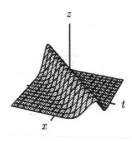

Figure 11.41

14. Use a computer to draw the graph of the vibrating string function:

$$g(x, t) = \cos t \sin 2x, \quad 0 \le x \le \pi, \quad 0 \le t \le 2\pi.$$

Explain the shape of the graph using sections with t fixed and sections with x fixed.

15. Use a computer to draw the graph of the traveling wave function:

$$h(x, t) = 3 + \cos(x - 0.5t), \quad 0 \le x \le 2\pi, \quad 0 \le t \le 2\pi.$$

Explain the shape of the graph using sections with t fixed and sections with x fixed.

16. Describe the sections with t fixed and the sections with x fixed of the vibrating guitar string function

$$f(x, t) = \cos t \sin x, \quad 0 \le x \le \pi,$$

on page 9. Explain the relation of these sections to the graph of f.

17. Consider the function f given by $f(x, y) = y^3 + xy$. Draw graphs of sections with:
    (a) x fixed at $x = -1$, $x = 0$, and $x = 1$.   (b) y fixed at $y = -1$, $y = 0$, and $y = 1$.

18. A swinging pendulum consists of a mass at the end of a string. At one moment the string makes an angle x with the vertical and the mass has speed y. At that time, the energy, E, of the pendulum, is given by the expression[2]

$$E = 1 - \cos x + \frac{y^2}{2}.$$

(a) Consider the surface representing the energy. Sketch a cross-section of the surface:
   (i) Perpendicular to the x-axis at $x = c$.
   (ii) Perpendicular to the y-axis at $y = c$.

(b) For each of the graphs in Figures 11.42 and 11.43 use your answer to part (a) to decide which is the x-axis and which is the y-axis and to put reasonable units on each one.

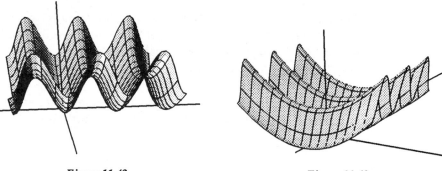

Figure 11.42   *Figure 11.43*

11.4 CONTOUR DIAGRAMS

The surface which represents a function of two variables often gives a good idea of the function's general behavior—for example, whether it is increasing or decreasing. However it is difficult to read numerical values off a surface and it can be hard to see all of the function's behavior from a surface. Thus, functions of the two variables are often represented by contour diagrams like the weather map on page 2. Contour diagrams have the additional advantage that they are readily extended to functions of three variables.

[2]Adapted from *Calculus in Context*, by James Callahan, Kenneth Hoffman, (New York: W.H. Freeman, 1995)

Topographical Maps

One of the most common examples of a contour diagram is a topographical map like that shown in Figure 11.44. It gives the elevation in the region and is a good way of getting an overall picture of the terrain: where the mountains are, where the flat areas are. Such topographical maps are frequently colored green at the lower elevations and brown, red, or even white at the higher elevations.

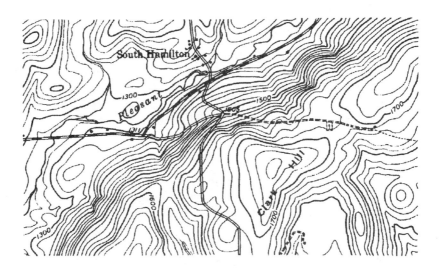

Figure 11.44: A topographical map showing the region around South Hamilton, NY

The curves on a topographical map that separate lower elevations from higher elevations are called *contour lines* because they outline the contour or shape of the land.[3] Because every point along the same contour has the same elevation, contour lines are also called *level curves* or *level sets*. The more closely spaced the contours, the steeper the terrain; the more widely spaced the contours, the flatter the terrain. (Provided, of course, that the elevation between contours varies by a constant amount.) Certain features have distinctive characteristics. A mountain peak is typically surrounded by contour lines like those in Figure 11.45. A pass in a range of mountains may have contours that look like Figure 11.46. A long valley has parallel contour lines indicating the rising elevations on both sides of the valley (see Figure 11.47); a long ridge of mountains has the same type of contour lines, only the elevations decrease on both sides of the ridge. Notice that the elevation numbers on the contour lines are just as important as the curves themselves.

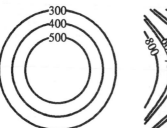

Figure 11.45: Mountain peak

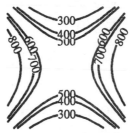

Figure 11.46: Pass between two mountains

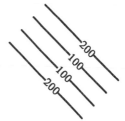

Figure 11.47: Long valley

---

[3]In fact they are usually not straight lines, but curves.

Figure 11.48: Impossible
contour lines

There are some things contour lines cannot do. Two contours corresponding to different elevations cannot cross each other as shown in Figure 11.48. If they did, the point of intersection of the two curves would have two different elevations, which is impossible (assuming the terrain has no overhangs). We will usually follow the convention of drawing contours for equally spaced values of z. Thus, if a diagram has contours for $z = 1, 2$, and 4, then it will usually (but not always) have a contour for $z = 3$ as well.

Corn Production

Contour maps can also be useful to display information about a function of two variables without reference to a surface. Consider how to represent the effect of different weather conditions on US corn production. What would happen if the average temperature were to increase (due to global warming, for example)? What would happen if the rainfall were to decrease (due to a drought)? One way of estimating the effect of these climatic changes is to use Figure 11.49. This is a contour diagram giving the corn production $f(R, T)$ in the United States as a function of the total rainfall, R, in inches and average temperature, T, in degrees Fahrenheit, during the growing season[4]. Suppose at the present time, $R = 15$ inches and $T = 76°$. Production is measured as a percentage of the present production; thus, the contour through $R = 15, T = 76$ has value 100, that is, $f(15, 76) = 100$.

Example 1   Use Figure 11.49 to evaluate $f(18, 78)$ and $f(12, 76)$ and explain the answer in terms of corn production.

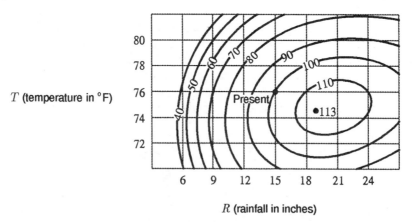

T (temperature in °F)

R (rainfall in inches)

Figure 11.49: Corn production as a function of rainfall and temperature

Solution   The point with R-coordinate 18 and T-coordinate 78 is on the contour $C = 100$, so $f(18, 78) = 100$. This means that if the annual rainfall were 18 inches and the temperature were 78° F, the country

---

[4]Adapted from S. Beaty and R. Healy, *The Future of American Agriculture*, Scientific American, Vol. 248, No.2, February, 1983

would produce about the same amount of corn at present, although it would be wetter and warmer than it is now.

The point with R-coordinate 12 and T-coordinate 76 is about halfway between the $C = 80$ contour and the $C = 90$ contour, so $f(12, 76) \approx 85$. This means that if the rainfall dropped to 12 inches and the temperature stayed at 76°, then corn production would drop to about 85% of what it is now.

Example 2 Describe in words the sections with T and R constant through the point representing present conditions. Give a common sense explanation of your answer.

Solution To see what happens to corn production if the temperature stays fixed at the present value of 76 but the rainfall changes look along the horizontal line $T = 76$. As we stay on the $T = 76$ line and move to the left, the values on the contour lines go down. In other words, if there is a drought, corn production decreases. Conversely, if rainfall increases, that is, as we move to the right from $R = 15, T = 76$, corn production increases, reaching a maximum of more than 110% when $R = 21$, and then decreases (too much rainfall floods the fields).

If, instead, rainfall remains at the present value and temperature increases, we move up the vertical line $R = 15$. Under these circumstances corn production decreases; a 2° increase causes a 10% drop in production. This makes sense since hotter temperatures lead to greater evaporation and hence drier conditions, even with rainfall constant at 15 inches. Similarly, a decrease in temperature leads to a very slight increase in production, reaching a maximum of around 102% when $T = 74$, followed by a decrease (the corn won't grow if it is too cold).

Contour Diagrams and Graphs

Contour diagrams and graphs are two different ways of representing a function of two variables. How do you go from one to the other? In the case of the topographical map, the contour diagram was created by joining all the points at the same height on the surface and dropping the curve into the xy-plane.

How do we go the other way? Suppose we wanted to plot the surface representing the corn production function $C = f(R, T)$ given by the contour diagram in Figure 11.49. Along each contour the function has a constant value; if we take each contour and lift it above the plane to a height equal to this value, we get the surface in Figure 11.50.

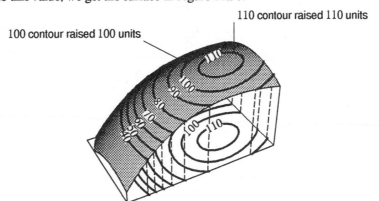

100 contour raised 100 units

110 contour raised 110 units

Figure 11.50: Getting the graph of the corn yield function from the contour diagram

In general, we have the following result:

> Contour lines, or level curves, are obtained from a surface by slicing it by horizontal planes.

Finding Contours Algebraically

Algebraic equations for the contours of a function f are easy to find if we have a formula for $f(x, y)$. Suppose the surface has equation

$$z = f(x, y).$$

A contour is obtained by slicing the surface with a horizontal plane with equation $z = c$. Thus, the equation for the contour at height c is given by:

$$f(x, y) = c.$$

Example 3   Find equations for the contours of $f(x, y) = x^2 + y^2$ and draw a contour diagram for f. Relate the contour diagram to the graph of f.

Solution   The contour at height c is given by

$$f(x, y) = x^2 + y^2 = c.$$

This is the equation of a circle of radius $\sqrt{c}$. Thus, the contours at an elevation of $c = 1, 2, 3, 4, \ldots$ are all circles centered at the origin of radius 1, $\sqrt{2}$, $\sqrt{3}$, 2, $\ldots$. The contour diagram is shown in Figure 11.51. The bowl-shaped graph of f is shown in Figure 11.52. Notice that the graph of f gets steeper and steeper as we move further away from the origin. This is reflected in the fact that the contours become more closely packed as we move further from the origin; for example, the contours for $c = 6$ and $c = 8$ are closer together than the contours for $c = 2$ and $c = 4$.

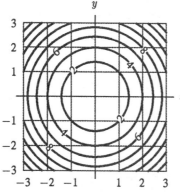

Figure 11.51: Contour diagram for $f(x, y) = x^2 + y^2$ (Even c values only)

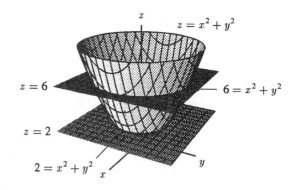

Figure 11.52: The graph of $f(x, y) = x^2 + y^2$

Example 4   Draw a contour diagram for $f(x, y) = \sqrt{x^2 + y^2}$ and relate it to the graph of f.

Solution   The contour at level c is given by

$$f(x, y) = \sqrt{x^2 + y^2} = c$$

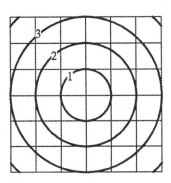

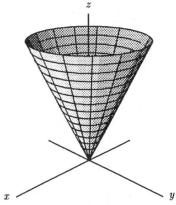

Figure 11.53: A contour diagram
for $f(x, y) = \sqrt{x^2 + y^2}$

Figure 11.54: The graph of
$f(x, y) = \sqrt{x^2 + y^2}$

This is a circle, just as in the previous example, but here the radius is c instead of $\sqrt{c}$. Thus, if the level c increases by 1, the radius of the contour increases by 1. This means the contours are equally spaced concentric circles (see Figure 11.53) which do not become more closely packed further from the origin. Thus, the graph of f has the same constant slope as we move away from the origin (see Figure 11.54), making it a cone rather than a bowl as in the previous example.

In both of the previous examples the level curves are concentric circles because the surfaces have circular symmetry. Any function of two variables which depends only on the quantity $(x^2 + y^2)$ has such symmetry: for example, $G(x, y) = e^{-(x^2 + y^2)}$ or $H(x, y) = \sin(\sqrt{x^2 + y^2})$.

Example 5   Draw a contour diagram for $f(x, y) = 2x + 3y + 1$.

Solution   The contour at level c has equation $2x + 3y + 1 = c$. Rewriting this as $y = -(2/3)x + (c - 1)/3$, we see that the contours are parallel lines with slope $-2/3$. The y-intercept for the contour at level c is $(c - 1)/3$; each time c increases by 3, the y-intercept moves up by 1. The contour diagram is shown in Figure 11.55.

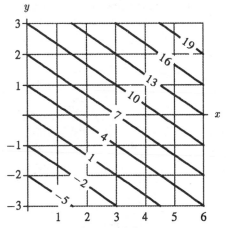

Figure 11.55: A contour diagram for
$f(x, y) = 2x + 3y + 1$

Contour Diagrams and Tables

Sometimes you can get an idea of what the contour map of a function looks like from its table.

Example 6　Relate the values of $f(x, y) = x^2 - y^2$ in Table 11.6 to its contour diagram in Figure 11.56.

TABLE 11.6: *Table of values of* $f(x, y) = x^2 - y^2$

3	0	-5	-8	-9	-8	-5	0
2	5	0	-3	-4	-3	0	5
1	8	3	0	-1	0	3	8
y　0	9	4	1	0	1	4	9
-1	8	3	0	-1	0	3	8
-2	5	0	-3	-4	-3	0	5
-3	0	-5	-8	-9	-8	-5	0
	-3	-2	-1	0	1	2	3

x

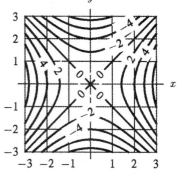

Figure 11.56: Contour map of $f(x, y) = x^2 - y^2$

Solution　One striking feature of the values in Table 11.6 are the zeros along the diagonals. This occurs because $x^2 - y^2 = 0$ along the lines $y = x$ and $y = -x$. So the $z = 0$ contour consists of these two lines. In the triangular regions of the table that lie to the right and left of both diagonals, the entries are positive. Thus, in the contour diagram, the positive contours will lie in the triangular regions to the right and left of the lines $y = x$ and $y = -x$. Further, the table shows that the numbers on the left are the same as the numbers on the right; thus, each contour will have two pieces, one on the left and one on the right. See Figure 11.56. As we move away from the origin, along the x-axis, we cross contours corresponding to successively larger values. On the saddle-shaped graph of $f(x, y) = x^2 - y^2$ shown in Figure 11.57, this corresponds to climbing out of the saddle along one of the ridges. Similarly, the negative contours occur in pairs in the top and bottom triangular regions and get more and more negative as we go out along the y-axis. This corresponds to descending from the saddle along the valleys that are submerged below the xy-plane in Figure 11.57. Notice that we could get the contour diagram instead by graphing the family of hyperbolas $x^2 - y^2 = 0, \pm 2, \pm 4, \ldots$.

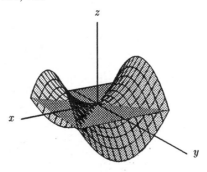

Figure 11.57: Graph of $f(x, y) = x^2 - y^2$
showing plane $z = 0$

Using Contour Diagrams: The Cobb-Douglas Production Function

Suppose you are running a small printing business, and decide to expand because you have more orders than you can handle. How should you expand? Should you start a night-shift and hire more workers? Should you buy more expensive but faster computers which will enable the current staff to keep up with the work? Or should you do some combination of the two?

Obviously, the way such a decision is made in practice involves many other considerations — such as whether you could get a suitably trained night shift, or whether there are any faster computers available that your current staff could use. Nevertheless, you might model the quantity of work produced by your business as a function of two variables: your total number, N, of workers, and the total value, V, of your equipment.

How would you expect such a production function to behave? In general, having more equipment and more workers enables you to produce more. However, increasing equipment without increasing the number of workers will increase production a bit, but not beyond a point. (If equipment is already lying idle, having more of it won't help.) Similarly, increasing the number of workers without increasing equipment will increase production, but not past the point where the equipment is fully utilized, as any new workers would have no equipment available to them.

Example 7   Explain why the contour diagram in Figure 11.58 does not model the behavior expected of the production function, whereas the contour diagram in Figure 11.59 does.

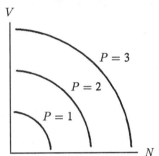

Figure 11.58: Incorrect contours for printing production

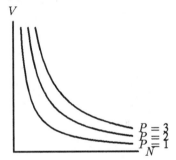

Figure 11.59: Correct contours for printing production

Solution   Look at the contour diagram in Figure 11.58. Fixing V at a particular value and letting N increase means moving to the right on the contour diagram. As you do so, you cross contours with larger and larger P values, meaning that production increases indefinitely. On the other hand, in Figure 11.59, as you move in the same direction you find yourself moving nearly parallel to the contours, crossing them less and less frequently. Therefore, production increases more and more slowly as N increases while V is held fixed. Similarly, if you hold N fixed and let V increase, the contour diagram in Figure 11.58 shows production increasing at a steady rate, whereas Figure 11.59 shows production increasing, but at a decreasing rate. Thus, Figure 11.59 fits the expected behavior of the production function best.

Formula for a Production Function

Production functions with the quantitative behavior we want are often approximated by formulas of the form

$$P = f(N, V) = cN^\alpha V^\beta$$

where P is the total quantity produced and c, α, and β are positive constants with $0 < \alpha < 1$ and $0 < \beta < 1$.

Example 8   Show that the contours of the function $P = cN^\alpha V^\beta$ have approximately the shape of the contours in Figure 11.59.

Solution   The contours are the curves where P is equal to a constant value, say P_0, that is, where

$$cN^\alpha V^\beta = P_0.$$

Solving for V we get

$$V = \left(\frac{P_0}{c}\right)^{1/\beta} N^{-\alpha/\beta}$$

Thus, V is a power function of N with a negative exponent, and hence its graph has the shape shown in Figure 11.59.

The Cobb-Douglas Production Model

In 1928, Cobb and Douglas used a similar function to model the production of the entire US economy in the first quarter of this century. Using government estimates of P, the total yearly production between 1899 and 1922, and of K, the total capital investment over the same period and of L, the total labor force, they found that P was well approximated by

---

The Cobb-Douglas Production Formula

$$P = 1.01 L^{0.75} K^{0.25}$$

This function turned out to model the US economy surprisingly well, both for the period on which it was based, and for some time afterwards.

Problems for Section 11.4

For the functions in Problems 1–15, sketch a contour diagram with at least four labeled contours. Describe in words the contours and how they are spaced.

1. $f(x, y) = x + y$
2. $f(x, y) = 2x - y$
3. $f(x, y) = xy$

4. $f(x, y) = x^2 + y^2$
5. $f(x, y) = x + y + 1$
6. $f(x, y) = 3x + 3y$

7. $f(x, y) = -x - y$
8. $f(x, y) = x^2 + y^2 - 1$
9. $f(x, y) = -x^2 - y^2 + 1$

10. $f(x, y) = x^2 + 2y^2$
11. $f(x, y) = \sqrt{x^2 + 2y^2}$
12. $f(x, y) = y - x^2$

13. $f(x, y) = 2y - 2x^2$
14. $f(x, y) = e^{-x^2 - y^2}$
15. $f(x, y) = \cos(\sqrt{x^2 + y^2})$

16. Figure 11.60 shows the monthly payment you must make on a 5-year car loan as a function of the interest rate and the amount you borrow. Suppose the interest rate is 13% and that you decide to borrow $6,000.

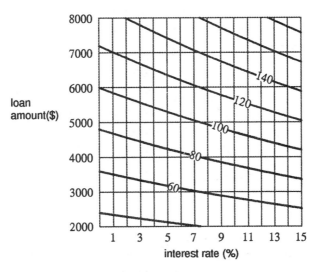

Figure 11.60

(a) What is your monthly payment?
(b) If interest rates drop to 11%, how much more can you borrow without increasing your monthly payment?
(c) Make a table of how much you can borrow without increasing your monthly payment, as a function of the interest rate.

17. Figure 11.61 shows a contour map of a hill with two paths, A and B.

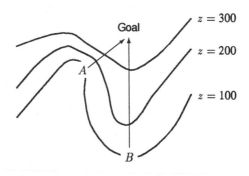

Figure 11.61

(a) On which path, A or B, will you have to climb more steeply?
(b) On which path, A or B, will you probably have a better view of the surrounding countryside? (Assuming trees do not block your view.)
(c) Near which path is there more likely to be a river?

18. Each of the contour diagrams in Figure 11.62 shows population density in a certain region of a city. Choose the contour diagram that best corresponds to each of the following situations. Many different matchings are possible. Pick any reasonable one and justify your choice.
    (a) The middle contour line is a highway.
    (b) The middle contour line is an open sewage canal.
    (c) The middle contour line is a railroad line.

(I) 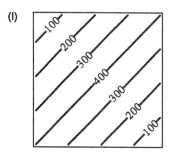 (II) 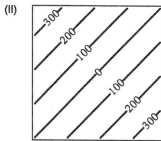 (III)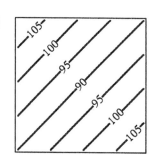

Figure 11.62

19. Each of the contour diagrams in Figure 11.63 shows population density in a certain region. Choose the contour diagram that best corresponds to each of the following situations. Many different matchings are possible. Pick any reasonable one and justify your choice.
    (a) The center of the diagram is a city.
    (b) The center of the diagram is a lake.
    (c) The center of the diagram is a power plant.

(I) 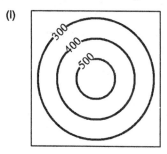 (II) 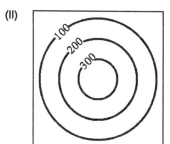 (III)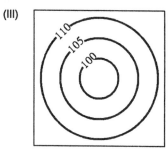

Figure 11.63

For each of the surfaces in Problems 20–22, draw a possible contour diagram, marked with reasonable z values. (Note: There are many possible answers.)

20.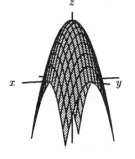

21.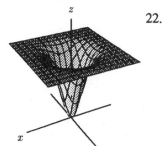

22.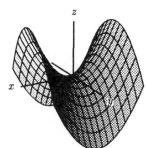

Figure 11.64

Figure 11.65

Figure 11.66

23. Figure 11.67 shows the density of the fox population P (in foxes per square mile) for southern England. Draw two different sections along a north-south line and two different sections along an east-west line of the population density P.

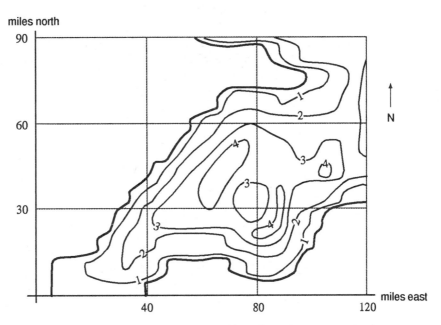

Figure 11.67: Population density of foxes in southwestern England

24. Use a computer to sketch a contour diagram for the vibrating string function

$$f(x,t) = \cos t \sin 2x, \qquad 0 \le x \le \pi, \ 0 \le t \le \pi.$$

Use $c = -2/3, -1/3, 0, 1/3, 2/3$. (You will not be able to do this algebraically.)

25. On page 4 we introduced the traveling wave function

$$h(x,t) = 5 + \cos(0.5x - t).$$

Draw a contour diagram using $c = 4, 4.5, 5, 5.5, 6$ for this function. Explain how your contour diagram relates to the sections of h discussed on page 4. Where are the contours most closely spaced? Most widely spaced?

26. Draw contour diagrams for each of the pizza-cola-happiness graphs given in Problem 10 on page 24.

27. A manufacturer sells two goods, one at a price of $3000 a unit and the other at a price of $12000 a unit. Suppose a quantity q_1 of the first good and q_2 of the second good are sold at a total fixed cost of $4000 to the manufacturer.

    (a) Express the manufacturer's profit, π, as a function of q_1 and q_2.
    (b) Sketch curves of constant profit in the q_1q_2-plane for $\pi = 10000$, $\pi = 20000$, and $\pi = 30000$ and the break-even curve $\pi = 0$.

28. The cornea is the front surface of the eye. Corneal specialists use a TMS, or Topographical Modeling System, to produce a "map" of the curvature of the eye's surface. A computer analyzes light reflected off the eye and draws level curves joining points of constant curvature. The regions between these curves are colored different colors. The first two pictures in Figure 11.68 are cross-sections of eyes with constant curvature, the smaller being about 38 units and the larger about 50 units. For contrast, the third eye has varying curvature.

   (a) Describe in words how the TMS map of an eye of constant curvature will look.

   (b) Draw the TMS map of an eye with the cross-section in Figure 11.69. Assume the eye is circular when viewed from the front, and the cross-section is the same in every direction. Put reasonable numeric labels on your level curves.

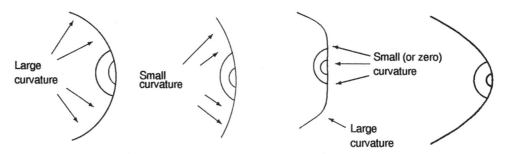

Figure 11.68: Pictures of eyes with different curvature   **Figure 11.69**

29. Match the tables (a) - (d) with the contour diagrams (I) - (IV) in Figure 11.70.

(a)

		x		
		-1	0	1
	-1	2	1	2
y	0	1	0	1
	1	2	1	2

(b)

		x		
		-1	0	1
	-1	0	1	0
y	0	1	2	1
	1	0	1	0

(c)

		x		
		-1	0	1
	-1	2	0	2
y	0	2	0	2
	1	2	0	2

(d)

		x		
		-1	0	1
	-1	2	2	2
y	0	0	0	0
	1	2	2	2

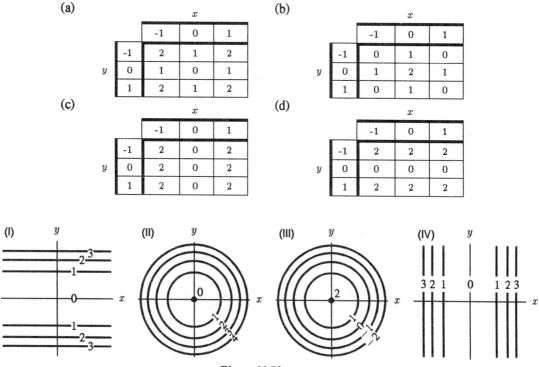

Figure 11.70

30. Match the surfaces (a)–(e) in Figure 11.71 with the contour diagrams (I)–(V) in Figure 11.72.

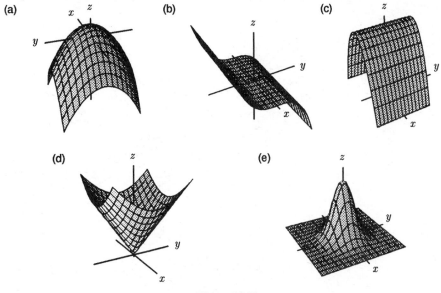

Figure 11.71

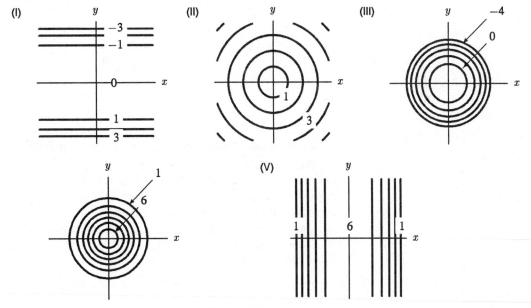

Figure 11.72

31. A city on an island has a large rectangular central park. Draw a possible contour diagram showing light intensity at night as a function of position. Label your contours with values between 0 and 1, where 0 represents total darkness and 1 represents maximum artificial illumination.

32.   The map in Figure 11.73 is from the undergraduate senior thesis of Professor Robert Cook, Director of Harvard's Arnold Arboretum. It shows level curves of the function giving the species density of breeding birds at each point in the US, Canada, and Mexico.

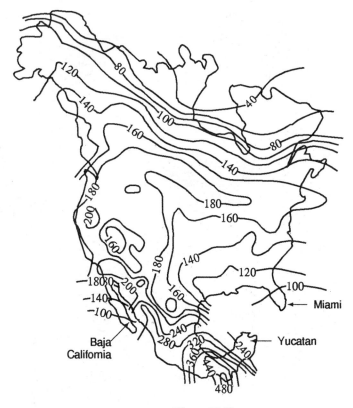

Figure 11.73

Looking at the map, are the following statements true or false? Explain your answers.
(a)   Moving from south to north across Canada, the species density increases.
(b)   The species density in the area around Miami is over 100.
(c)   In general, peninsulas (for example, Florida, Baja California, the Yucatan) have lower species densities than the areas around them.
(d)   The greatest rate of change in species density with distance is in Mexico. If you think this is true, mark the point and direction which give the maximum and explain why you picked the point and direction you did.

33.   The temperature T (in $°C$) at any point in the region $-10 \le x \le 10, -10 \le y \le 10$ is given by the function
$$T(x, y) = 100 - x^2 - y^2.$$

(a)   Sketch isothermal curves for $T = 100°C$, $T = 75°C$, $T = 50°C$, $T = 25°C$, and $T = 0°C$.
(b)   Suppose a heat-seeking bug is put down at any point on the xy-plane. In which direction should it move to increase its temperature fastest? How is that direction related to the level curve through that point?

34. Identify the contour diagrams (I)–(V) in Figure 11.74 and the surfaces (F)–(K) in Figure 11.75 corresponding to the following equations (a)–(e). Assume that each contour diagram is shown in a square window.

(a) $z = \sin x$ (b) $z = xy$ (c) $z = e^{-(x^2+y^2)}$ (d) $z = 1 - 2x - y$

(e) $z = x^2 + 4y^2$

(I)
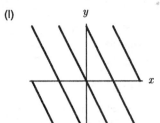

(II)
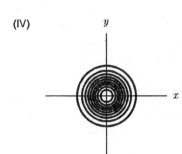

(III)
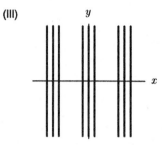

(IV)
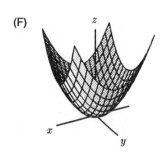

(V)
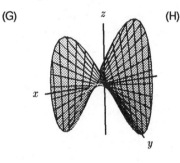

Figure 11.74

(F)
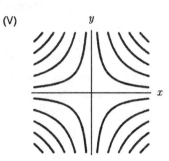

(G)
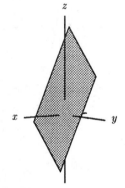

(H)
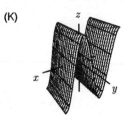

(J)
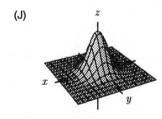

(K)

Figure 11.75

35.

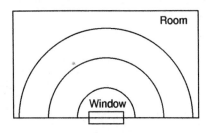

Figure 11.76

Figure 11.76 shows the level curves of the temperature H in a room near a recently opened window. Label the three level curves with reasonable values of H if the house is in the following locations.

(a)   Minnesota in winter (where winters are harsh).
(b)   San Francisco in winter (where winters are mild).
(c)   Houston in summer (where summers are hot).
(d)   Oregon in summer (where summers are mild).

36.   Match each Cobb-Douglas production function with the correct graph and the correct statements. (Note: A graph or statement can be used more than once.)

(a) $F(L, K) = L^{0.25} K^{0.25}$ (D)   Tripling each input triples output.
(b) $F(L, K) = L^{0.5} K^{0.5}$ (E)   Quadrupling each input doubles output.
(c) $F(L, K) = L^{0.75} K^{0.75}$ (G)   Doubling each input almost triples output.

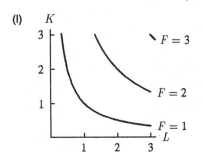 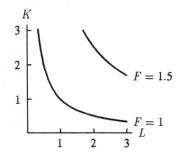 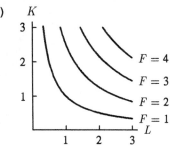

Figure 11.77

37.   Consider the Cobb-Douglas production function on page 34. What is the effect on production of doubling both labor and capital?

38.   A general Cobb-Douglas production function has the form

$$P = cL^{\alpha} K^{\beta}$$

An important economic question is what happens to production if labor and capital are both "scaled up"? For example, does production double if both labor and capital are doubled?
    Economists talk about

- *increasing returns to scale* if doubling L and K more than doubles P,
- *constant returns to scale* if doubling L and K exactly doubles P,
- *decreasing returns to scale* if doubling L and K less than doubles P.

What conditions on α and β lead to increasing, constant, or decreasing returns to scale?

39. Figure 11.78 shows the contours of the temperature along one wall of a heated room through one winter day, with time indicated as on a 24-hour clock. The room has a heater located at the left-most corner of the wall, and one window in the wall. The heater is controlled by a thermostat a couple of feet from the window.

    (a) Where is the window?
    (b) When is the window open?
    (c) When is the heat on?
    (d) Draw graphs of the temperature along the wall of the room at 6 am, at 11 am, at 3 pm (15 hours) and at 5 pm (17 hours).
    (e) Draw a graph of the temperature as a function of time at the heater, at the window and midway between them.
    (f) Can you explain why the temperature at the window at 5 pm (17 hours) is less than at 11 am (11 hours)?
    (g) To what temperature do you think the thermostat is set? How do you know?
    (h) Where is the thermostat?

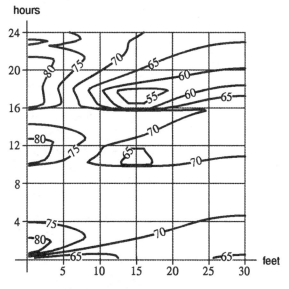

Figure 11.78

11.5 LINEAR FUNCTIONS

What is a Linear Function of Two Variables?

Linear functions played a central role in one-variable calculus because most one-variable functions are locally linear. (That is, their graphs look like a line when we zoom in.) In two-variable calculus, *linear functions* are those whose graph is a plane. In Chapter 13, we see that the graphs of most two-variable functions look like planes when we zoom in.

What Makes a Plane Flat?

What is it about a function $f(x, y)$ that makes its graph $z = f(x, y)$ a plane? Linear functions of *one* variable have straight line graphs because they have constant slope. No matter where we are on the graph, the slope is the same: a given increase in the x-coordinate always gives the same increase in the y-coordinate. In other words, the ratio $\Delta y / \Delta x$ is constant.

On a plane, the situation is a bit more complicated. If we walk around on a tilted plane, the slope is not always the same: it depends on the direction we walk in. However, it is still true that at every point on the plane, the slope is the same as long as we choose the same direction. If we walk in the direction parallel to the x-axis, we always find ourselves walking up or down with the same slope; the same thing is true if we walk in the direction parallel to the y-axis. In other words, the slope ratios $\Delta z / \Delta x$ (with y fixed) and $\Delta z / \Delta y$ (with x fixed) are both constant.

Example 1   We are on a plane that cuts the z-axis at $z = 5$, has slope 2 in the x direction and slope -1 in the y direction. What is the equation of the plane?

Solution   Finding the equation of the plane means constructing a formula for the z-coordinate in terms of the x- and y-coordinates. What is the z-coordinate of the point on the plane directly above the point (x, y) in the xy-plane? To get to that point starting from the point above the origin, first we walk x units in the x direction. Since the slope in the x direction is 2, this means that our height increases by $2x$. Then we walk y units in the y direction; since the slope in the y direction is -1, our height decreases by y units. In total, our height has changed by $2x - y$ units. Since we started at $z = 5$ on the z-axis, our height is now $5 + 2x - y$. So our z-coordinate is $5 + 2x - y$. Thus, the equation for the plane is

$$z = 5 + 2x - y.$$

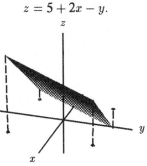

Figure 11.79: The plane $z = 5 + 2x - y$

For any linear function, if we know its value at a point (x_0, y_0), its slope in the x direction, and its slope in the y direction, then we can write the equation of the function. This is just like the equations of lines in the one-variable case, except that there are two slopes instead of one.

> If a plane has slope m in the x direction, slope n in the y direction, and passes through the point (x_0, y_0, z_0), then its equation is
>
> $$z = z_0 + m(x - x_0) + n(y - y_0).$$
>
> This plane is the graph of the linear function
>
> $$f(x, y) = z_0 + m(x - x_0) + n(y - y_0).$$
>
> If we write $c = z_0 - mx_0 - ny_0$, then we can write $f(x, y)$ in the form
>
> $$f(x, y) = c + mx + ny.$$

Just as in 2-dimensional space a line is determined by 2 points, so in 3-dimensional space a plane is determined by 3 points; see the following example.

Example 2 Find the equation of the plane that passes through the three points $(1, 0, 1)$, $(1, -1, 3)$, and $(3, 0, -1)$.

Solution The first two points have the same x-coordinate, so we can use them to find the slope of the plane in the y-direction. As the y-coordinate changes from 0 to -1, the z-coordinate changes from 1 to 3, so the slope in the y-direction is $\Delta z/\Delta y = (3 - 1)/(-1 - 0) = -2$. The first and third points have the same y-coordinate, so we can use them to find the slope in the x-direction; it is $\Delta z/\Delta x = (-1 - 1)/(3 - 1) = -1$. Because the plane passes through $(1, 0, 1)$, it has equation

$$z = 1 - (x - 1) - 2(y - 0)$$
$$= 2 - x - 2y.$$

You should check that this equation is also satisfied by the points $(1, -1, 3)$ and $(3, 0, -1)$.

The previous example was made easier by the fact that two of the points had the same x-coordinate and two of them had the same y-coordinate. An alternative method to solve this problem is to substitute the x, y, and z values for each of the three given points into the equation $z = c + mx + ny$. The resulting three equations in c, m, n can then be solved simultaneously. See Problem 22 on page 50.

Linear Functions from a Numerical Point of View

To avoid flying planes with too many empty seats, airlines sell some tickets at full price and some at a discount. Table 11.7 shows an airline's revenue in dollars from tickets sold on a particular route, as a function of the number of full-price tickets sold and the number of discount tickets sold.

TABLE 11.7: *Revenue from ticket sales (dollars)*

		Full-price tickets			
		100	200	300	400
	200	39,700	63,600	87,500	111,400
	400	55,500	79,400	103,300	127,200
Discount tickets	600	71,300	95,200	119,100	143,000
	800	87,100	111,000	134,900	158,800
	1000	102,900	126,800	150,700	174,600

Looking down any column, we see that the revenue jumps by \$15,800 for each extra 200 discount tickets. Thus, the column is a linear function of the number of discount tickets sold. Furthermore, every column has the same slope, which is $15,800/200 = 79$ dollars/ticket. This is the price of a discount ticket. Similarly, each row is a linear function and all the rows have the same slope, 239, which is the price of a full-fare ticket. Thus, the formula for the total revenue R from f full fares and d discount tickets is:

$$R = 239f + 79d.$$

Thus, R is a linear function of f and d. We have the following general result:

A linear function can be recognized from its table by the following features:
- Each row and each column is linear.
- All the rows have the same slope.
- All the columns have the same slope (although the slope of the rows and the slope of the columns are generally different).

Example 3   Table 11.8 gives some values of a linear function. Fill in the blank and give a formula for the function.

TABLE 11.8: *A linear function*

		x	
		2	3
y	1.5	0.5	-0.5
	2.0	1.5	?

Solution   In the first row the function decreases by 1 (from 0.5 to -0.5) as x goes from 2 to 3. Since the function is linear, it must decrease by the same amount in the second row. So the missing entry must be $1.5 - 1 = 0.5$. The slope of the function in the x-direction is -1 and the slope in the y-direction is 2, since in each column the function increases by 1 when y increases by 0.5. Thus, the formula is

$$f(x, y) = c - x + 2y.$$

From the table we get $f(2, 1.5) = 0.5$. Substituting this into the equation gives

$$0.5 = c - 2 + 2 \cdot 1.5 = c + 1,$$

so $c = -0.5$. Therefore, the formula is

$$f(x, y) = -0.5 - x + 2y.$$

What Does the Contour Diagram of a Linear Function Look Like?

Consider the function for airline revenue given in Table 11.7. Its formula is

$$R = 239f + 79d,$$

where f is the number of full-fares and d is the number of discount fares sold. Figure 11.80 gives the contour diagram for this function.

Notice that the contours are parallel straight lines. What is the practical significance of the slope? Consider the contour $R = 100,000$; that means we are looking at combinations of ticket sales that yield $100,000 in revenue. If we move down and to the right on the contour, the f-coordinate increases and the d-coordinate decreases, so we sell more full-fares and fewer discount fares. This makes sense, because to receive a fixed revenue of $100,000, we must sell more full-fares if we sell fewer discount fares. The exact trade-off depends on the slope of the contour; the diagram shows that each contour has a slope of about -3. This means that for a fixed revenue, we must sell three discount fares to replace one full-fare. This can also be seen by comparing prices. Each full fare

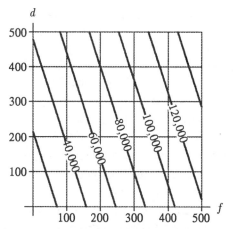

Figure 11.80: Revenue as a function of the
number of full and discount fares sold

brings in \$239; to earn the same amount in discount fares we need to sell $239/79 \approx 3.03 \approx 3$ fares. Since the price ratio is independent of how many of each type of fare we sell, this slope remains constant over the whole contour map; thus, the contours are all parallel straight lines.

You should also observe that the contours are evenly spaced. This means that no matter which contour we are on, a fixed increase in one of the variables will cause the same increase in the value of the function. In terms of revenue, this says that no matter how many fares we have sold, an extra fare, whether full or discount, will bring the same revenue as it did before.

Example 4   Find the equation of the linear function whose contour diagram is in Figure 11.81.

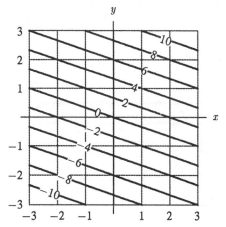

Figure 11.81: Contour map of linear function
$f(x, y)$

Solution   The contours correspond to values of the function that are 2 units apart. Moving two units in the y direction we cross three contours; that is, a change of 2 in y changes the function by 6, so the slope in the y direction is $\Delta z/\Delta y = 6/2 = 3$. Similarly, a move of 2 in the x-direction crosses one contour line and changes the function by 2; so the slope in the x direction is $\Delta z/\Delta x = 2/2 = 1$. Hence, $f(x, y) = c + x + 3y$. We see from the diagram that $f(0, 0) = 0$, so $c = 0$. Therefore, $f(x, y) = x + 3y$.

Problems for Section 11.5

1. Suppose that z is a linear function with slope of 2 in the x direction and slope of 3 in the y direction. A change of 0.5 in x and -0.2 in y produces what change in z? If $z = 2$ when $x = 5$ and $y = 7$, what is the value of z when $x = 4.9$ and $y = 7.2$?

2. Find the equation of the linear function $z = c + mx + ny$ whose graph contains the points $(0, 0, 0)$, $(0, 2, -1)$, and $(-3, 0, -4)$.

3. Find the equation of the plane through the points $(4, 0, 0)$, $(0, 3, 0)$ and $(0, 0, 2)$.

4. Find an equation for the plane containing the line in the xy-plane where $y = 1$, and the line in the xz-plane where $z = 2$.

5. Find the equation of the linear function $z = c + mx + ny$ whose graph intersects the xz-plane in the line $z = 3x + 4$ and intersects the yz-plane in the line $z = y + 4$.

6. Find the equation of the linear function $z = c + mx + ny$ whose graph intersects the xy-plane in the line $y = 3x + 4$ and contains the point $(0, 0, 5)$.

7. Is the function represented in Table 11.9 linear? Give reasons for your answer.

TABLE 11.9

		v			
		1.1	1.2	1.3	1.4
	3.2	11.06	12.06	13.07	14.07
	3.4	11.75	12.82	13.89	14.95
u	3.6	12.44	13.57	14.70	15.83
	3.8	13.13	14.33	15.52	16.71
	4.0	13.82	15.08	16.34	17.59

For Problems 8–10, find equations for the linear functions with the given tables.

8.

		y			
		-1	0	1	2
	0	1.5	1	0.5	0
x	1	3.5	3	2.5	2
	2	5.5	5	4.5	4
	3	7.5	7	6.5	6

9.

		y			
		1	1.1	1.2	1.3
	2.2	7.9	7.5	7.1	6.7
x	2.4	8.2	7.8	7.4	7.0
	2.6	8.5	8.1	7.7	7.3
	2.8	8.8	8.4	8.0	7.6

10.

		y			
		10	20	30	40
	100	3	6	9	12
x	200	2	5	8	11
	300	1	4	7	10
	400	0	3	6	9

Problems 11–13 each contain a partial table of values for a linear function. Fill in the blanks.

11.

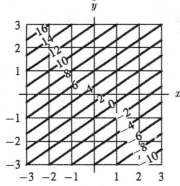

	y	
x	0.0	1.0
0.0		1.0
2.0	3.0	5.0

12.

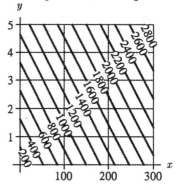

	y		
x	3.0	4.0	5.0
1.0	5.0		7.0
2.0		9.0	
3.0			

13.

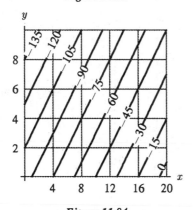

	y		
x	-1.0	0.0	1.0
2.0	4.0		
3.0		3.0	5.0

For Problems 14–16, find equations for the linear functions with the given contour diagrams.

14.

Figure 11.82

15.

Figure 11.83

16.

Figure 11.84

17. Let f be the linear function $f(x, y) = c + mx + ny$, where c, m, n are constants and $m, n \neq 0$.
    (a) Show that all the contours of f are lines of slope $-m/n$, if $n \neq 0$.
    (b) Show that
    $$f(x + n, y - m) = f(x, y),$$
    for all x and y.
    (c) Explain the relation between parts (a) and (b).

It is difficult to sketch a plane that is the graph of a linear function by hand. One method that works if the x, y, and z-intercepts of the graph are positive is to plot the intercepts and join them by a triangle as shown in Figure 11.85; this shows the part of the graph in the octant where $x \geq 0$, $y \geq 0$, $z \geq 0$. If the intercepts are not all positive, the same method works if the x, y, and z-axes are drawn from a different perspective. Use this method to sketch the graphs of the linear functions in Problems 18–20.

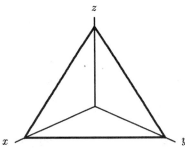

Figure 11.85: The graph of a linear function with positive x, y, and z-intercepts.

18. $z = 6 - 2x - 3y$ 19. $z = 4 + x - 2y$ 20. $z = 2 - 2x + y$

21.   Figure 11.86 shows the revenue of a music store as a function of the number t of tapes and the number c of compact discs that it sells. What is the price of tapes? What is the price of compact discs?

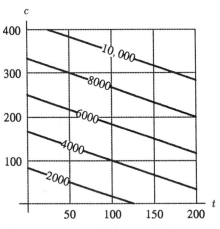

Figure 11.86

22.   Find the equation of the plane that passes through the points $(1, 1, 3)$, $(-1, 2, 2)$, and $(0, 3, 3)$. [Hint: Write the general form for a linear function, substitute the given points into it, and solve for the coefficients.]

23.   A manufacturer makes two products out of two raw materials. Let q_1, q_2 be the quantities sold of the two products, p_1, p_2 their prices, and m_1, m_2 the quantities purchased of the two raw materials. Which of the following functions do you expect to be linear, and why? In each case, assume that all variables except the ones mentioned are held fixed.

(a) Expenditure as a function of m_1 and m_2.

(b) Revenue as a function of q_1 and q_2.

(c) Revenue as a function of p_1 and q_1.

24. Give an example of a nonlinear function $f(x, y)$ such that all the sections with x fixed and all the sections with y fixed are straight lines.

25. A college admissions office uses the following linear equation to predict the grade point average of an incoming student:

$$z = 0.003x + 0.8y - 4$$

where z is the predicted college GPA on a scale of 0 to 4.3, and x is the sum of the student's SAT Math and SAT Verbal on a scale of 400 to 1600, and y is the student's high school GPA on a scale of 0 to 4.3. The college admits students whose predicted GPA is at least 2.3.

(a) Will a student with SAT's of 1050 and high school GPA of 3.0 be admitted?

(b) Will every student with SAT's of 1600 be admitted?

(c) Will every student with a high school GPA of 4.3 be admitted?

(d) Draw a contour diagram for predicted GPA z with $400 \leq x \leq 1600$ and $0 \leq y \leq 4.3$. Shade the points corresponding to students who will be admitted.

(e) Which is more important, an extra 100 points on the SAT or an extra 0.5 of high school GPA?

11.6 FUNCTIONS OF MORE THAN TWO VARIABLES

In applications of calculus, functions of any number of variables can arise. The density of matter in the universe is a function of three variables, since it takes three variables to specify a point in space. If we want to study the evolution of this density over time, we need to add the fourth variable, time. Thus we need to be able to apply calculus to functions of arbitrarily many variables.

The main problem with functions of more than two variables is that it is hard to visualize their graphs. The graph of a function of one variable is a curve in 2-space, the graph of a function of two variables is a surface in 3-space, and so the graph of a function of three variables would be a solid in 4-space. We can't easily visualize 4-space, or any higher dimensional space, and so we won't use the graphs of functions of three or more variables.

On the other hand, the idea of sectioning a function by keeping one variable fixed allows us to picture a function of three variables. It is also possible to give contour diagrams for these functions, only now the contours are surfaces in 3-space.

Representing a Function of Three Variables Using Contour Diagrams

One way to analyze a function of three variables, $f(x, y, z)$, is to look at sections with one variable fixed. Suppose we keep z fixed and view f as a function of the remaining two variables, x and y. For each fixed value, $z = c$, we represent the two-variable function $f(x, y, c)$ by a contour diagram. As c varies, we get a whole collection of contour diagrams.

Example 1   A pond is 30 feet deep in the middle and 200 feet across. The pond is infested with algae. Suppose the density of algae at a point z feet below, x feet east, and y feet north of the center of the surface of the pond is approximated by the formula

$$f(x, y, z) = \frac{1}{10}(50 + \sqrt{x^2 + y^2})(30 - z),$$

where density is measured in ounces per cubic feet. Draw contour diagrams for f at the surface of the pond and at a depth of 10 feet. Describe in words how the density of algae varies with depth and distance from the center of the pond.

Solution   At the surface of the pond, $z = 0$, so the formula for the $z = 0$ section is

$$f(x, y, 0) = \frac{1}{10}(50 + \sqrt{x^2 + y^2})(30 - 0) = 150 + 3\sqrt{x^2 + y^2}.$$

Notice that we now have a function of two variables. The contours of this function are circles, since $f(x, y, 0)$ is constant when $\sqrt{x^2 + y^2}$ is constant. The contour diagram is shown in Figure 11.87. At a depth of 10 feet, $z = 10$, and we have

$$f(x, y, 10) = \frac{1}{10}(50 + \sqrt{x^2 + y^2})(30 - 10) = 100 + 2\sqrt{x^2 + y^2}.$$

The contour diagram for the $z = 10$ section is shown in Figure 11.88. The contour diagrams show that there is more algae near the surface than deeper down in the pond and that there is more algae near the shoreline than near the center of the pond.

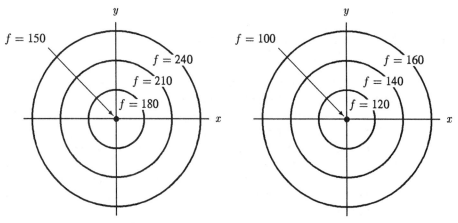

Figure 11.87: Density of algae in the pond at the surface: Contour diagram for $z = 0$ *Figure 11.88*: Density of algae in the pond 10 feet down: Contour diagram for $z = 10$

For weather forecasting, it is important to know the barometric pressure and temperature not only at the surface of the earth but also aloft in the atmosphere above the earth. On TV, you may have heard about a "surface" low pressure system or an "upper level" low pressure system (upper level lows influence the jet stream). The surface and upper level maps studied by meteorologists are contour diagrams at different altitudes of the pressure or temperature function. These maps are similar to the contour diagrams of algae density of the previous example.

Representing a Function of Three Variables Using Tables

In the previous example, we viewed the sections of a three-variable function $f(x, y, z)$ as contour diagrams of a two-variable function. We can also view the sections as tables of values.

Example 2   In Section 11.1, the consumption of beef as a function of the price, p, of beef and household income, I, was given in a table of values. The consumption, C, of beef is also a function of the price q of a competing meat, say chicken, as well as the price of beef and household income. That is, $C = f(I, p, q)$. This function can be given by a collection of tables, one for each different value of the price, q, of chicken. Tables 11.10 and 11.11 show two different sections of the consumption function f with q held fixed. Explain how the two tables are related.

TABLE 11.10: *Beef consumption when price of chicken $q = 1.50$*

		p		
		3.00	3.50	4.00
	20	2.65	2.59	2.51
I	60	5.11	5.00	4.97
	100	5.80	5.77	5.60

TABLE 11.11: *Beef consumption when price of chicken $q = 2.00$*

		p		
		3.00	3.50	4.00
	20	2.75	2.75	2.71
I	60	5.21	5.05	5.11
	100	5.80	5.77	5.60

Solution   Comparing tables shows that for households with large incomes (say $I = 100$), changes in the price of chicken have little effect on the amount of beef consumed (because price is not a factor in their buying). However for small incomes (say $I = 20$), an increase in the price of chicken (from $q = 1.50$ to $q = 2.00$) causes families to spend more on beef.

Representing a Function of Three Variables Using a Family of Level Surfaces

A function of two variables, $f(x, y)$, is represented by a *family* of level curves of the form $f(x, y) = c$ for various values of the constant, c.

> A function of three variables, $f(x, y, z)$, is represented by a *family* of surfaces of the form $f(x, y, z) = c$, each one of which is called a *level surface*.

Example 3   Suppose the temperature, in °F, at a point (x, y, z) is given by $T = f(x, y, z) = x^2 + y^2 + z^2$. What do the level surfaces of the function f look like and what do they mean in terms of temperature?

Solution   The level surface corresponding to $T = 100$ is the set of all points where the temperature is 100°F. That is, where $f(x, y, z) = 100$, so

$$x^2 + y^2 + z^2 = 100.$$

This is the equation of a sphere of radius 10, with center at the origin. Similarly, the level surface corresponding to $T = 200$ is the sphere with radius $\sqrt{200}$. The other level surfaces will be concentric spheres. The temperature is constant on each sphere. We may view the temperature distribution

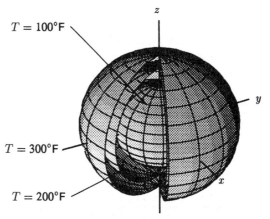

Figure 11.89: Level surfaces of
$T = f(x, y, z) = x^2 + y^2 + z^2$, each one having a
constant temperature

as a set of nested spheres, like concentric layers of an onion, each one labeled with a different temperature, starting from low temperatures in the middle and getting hotter as we go out from the center. (See Figure 11.89.) The level surfaces become more closely spaced as we move farther from the origin because the temperature increases more rapidly the farther we get from the origin.

In general, the level surfaces of a function are nested together in some way, so it is often difficult to draw them. As in Figure 11.89, we generally use cut-outs to show the inner surfaces.

Example 4 What do the level surfaces of $f(x, y, z) = x^2 + y^2 - z^2$ look like?

Solution From Section 11.4, you may recall that the two-variable quadratic function $g(x, y) = x^2 - y^2$ has a saddle-shaped graph and three types of contours. The contour equation $x^2 - y^2 = c$ gives a hyperbola opening right-left when $c > 0$, a hyperbola opening up-down when $c < 0$, and a pair of intersecting lines when $c = 0$. Similarly, the three-variable quadratic function $f(x, y, z) = x^2 + y^2 - z^2$ has three types of level surfaces depending on the value of c in the equation $x^2 + y^2 - z^2 = c$.

Suppose that $c > 0$, say $c = 1$. Rewrite the equation as $x^2 + y^2 = z^2 + 1$ and think of what happens as we section the surface perpendicular to the z axis by holding z fixed. The result is a circle, $x^2 + y^2 = $ constant, of radius at least 1 (since the constant $z^2 + 1 \geq 1$). The circles get larger as z gets larger. If we took the $x = 0$ section instead we would get the hyperbola $y^2 - z^2 = 1$. The result is shown in Figure 11.90(a).

Suppose instead $c < 0$, say $c = -1$. Then the horizontal sections, of $x^2 + y^2 = z^2 - 1$ are again circles except that the radii shrink to 0 at $z = \pm 1$ and between $z = -1$ and $z = 1$ there are no sections at all. The result is shown in Figure 11.90(b).

When $c = 0$, we get the equation $x^2 + y^2 = z^2$. Again the horizontal sections are circles, this time with the radius shrinking down to exactly 0 when $z = 0$. The resulting surface, shown in Figure 11.90(c), is the cone $z = \sqrt{x^2 + y^2}$ studied in Section 11.4, together with the lower cone $z = -\sqrt{x^2 + y^2}$. Notice that the surfaces in Figure 11.90(a) will be outside the cone in Figure 11.90(c), whereas the surfaces in Figure 11.90(b) will be inside the cone.

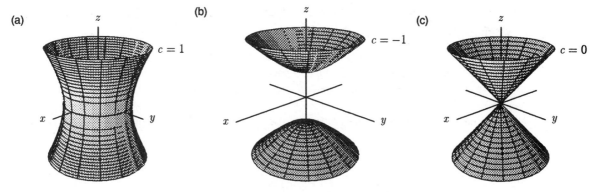

Figure 11.90: Level surfaces for $f(x, y, z) = x^2 + y^2 - z^2$

Example 5   What do the level surfaces of $f(x, y, z) = x^2 + y^2$ and $g(x, y, z) = z - y$ look like?

Solution   The level surface of f corresponding to the constant c is the surface consisting of all points satisfying the equation

$$x^2 + y^2 = c.$$

Since there is no z-coordinate in the equation, z can take any value whatsoever, so this is a circular cylinder of radius $\sqrt{c}$ around the z-axis. The level surfaces are concentric cylinders; the narrow ones near the z-axis are the ones where f has small values, and the wider ones are where it has larger values. See Figure 11.91.

The level surface of g corresponding to the constant c is the surface

$$z - y = c.$$

This time there is no x variable, so this surface is the one you get by taking each point on the straight line $z - y = c$ in the yz-plane and letting x roam back and forth. You get a plane which cuts the yz-plane diagonally; the x-axis is parallel to this plane. See Figure 11.92.

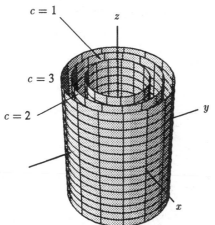

Figure 11.91: Level surfaces of
$f(x, y, z) = x^2 + y^2$

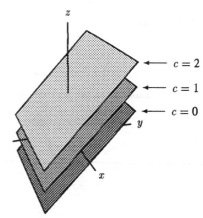

Figure 11.92: Level surfaces of
$g(x, y, z) = z - y$

How Surfaces Can Represent Functions of Two Variables and Functions of Three Variables

You may have noticed that we have used surfaces to represent functions in two different ways. First, we used a *single* surface to represent a two-variable function $z = f(x, y)$. Second, we used a *family* of level surfaces to represent a three-variable function $w = F(x, y, z)$. These level surfaces have equation $F(x, y, z) = c$.

What is the relation between these two uses of surfaces? For example, consider the function

$$z = x^2 + y^2 + 3.$$

Define

$$G(x, y, z) = x^2 + y^2 + 3 - z$$

Then the surface $z = x^2 + y^2 + 3$ is the level surface $G = 0$ of the function $G(x, y, z)$, since

$$G(x, y, z) = x^2 + y^2 + 3 - z = 0.$$

Thus, we have the following result:

A single surface representing a two-variable function $z = f(x, y)$ can always be thought of as one member of the family of level surfaces representing a three-variable function $G(x, y, z) = f(x, y) - z$. The graph of $z = f(x, y)$ is the level surface $G = 0$.

Conversely, a single member of a family of level surfaces can be regarded as the graph of a function of the form $z = f(x, y)$ if it is possible to solve for z. For example, if $F(x, y, z) = x^2 + y^2 + z^2$, then one member of the family of level surfaces is the sphere

$$x^2 + y^2 + z^2 = 1.$$

This is an implicit function which can be solved to give two functions

$$z = \sqrt{1 - x^2 - y^2} \quad \text{and} \quad z = -\sqrt{1 - x^2 - y^2}.$$

The graph of the first function is the top half of the sphere and the graph of the second function is the bottom half.

A Catalog of Surfaces

For later reference, here is a small catalog of the surfaces we have encountered.

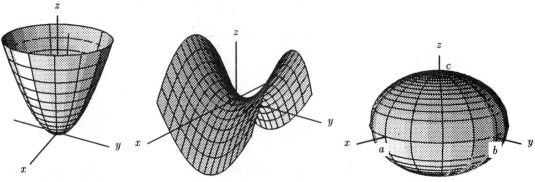

Figure 11.93: Elliptical paraboloid $z = \frac{x^2}{a^2} + \frac{y^2}{b^2}$

Figure 11.94: Hyperbolic paraboloid $z = -\frac{x^2}{a^2} + \frac{y^2}{b^2}$

Figure 11.95: Ellipsoid $\frac{x^2}{a^2} + \frac{y^2}{b^2} + \frac{z^2}{c^2} = 1$

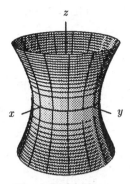

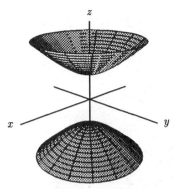

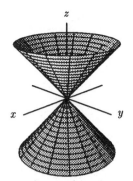

Figure 11.96: Hyperboloid of one sheet $\frac{x^2}{a^2} + \frac{y^2}{b^2} - \frac{z^2}{c^2} = 1$

Figure 11.97: Hyperboloid of two sheets $\frac{x^2}{a^2} + \frac{y^2}{b^2} - \frac{z^2}{c^2} = -1$

Figure 11.98: Cone $\frac{x^2}{a^2} + \frac{y^2}{b^2} - \frac{z^2}{c^2} = 0$

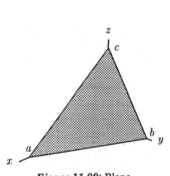

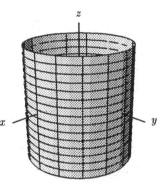

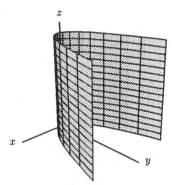

Figure 11.99: Plane $\frac{x}{a} + \frac{y}{b} + \frac{z}{c} = 1$

Figure 11.100: Cylindrical surface $x^2 + y^2 = a^2$

Figure 11.101: Parabolic cylinder $y = ax^2$

(These are viewed as equations in three variables x, y, and z)

Problems for Section 11.6

1. Hot water is entering a rectangular swimming pool at the surface of the pool in one corner. Sketch a possible contour diagram for the temperature of the pool at the surface and three feet below the surface.

2. Draw contour diagrams for three different sections with t fixed, of the function

$$f(x, y, t) = \cos t \cos \sqrt{x^2 + y^2}, \quad 0 \le \sqrt{x^2 + y^2} \le \pi/2.$$

3. Describe in words the level surfaces of $f(x, y, z) = \sin(x + y + z)$.

4. Describe in words the level surfaces of $g(x, y, z) = e^{-(x^2+y^2+z^2)}$.

5. The height (in feet) of the water above the bottom of a pond at time t is given by the function $h(x, y, t) = 20 + \sin(x + y - t)$, where x and y are Cartesian coordinates arranged so that the positive y-axis is north and the positive x-axis is east, and where t is in seconds. By considering contour diagrams for different t sections (snapshots) of h, describe the motion of the water surface in the pond.

6. Match the following functions with the level surfaces in Figure 11.102.

    (a) $f(x, y, z) = y^2 + z^2$ (b) $h(x, y, z) = x^2 + z^2$.

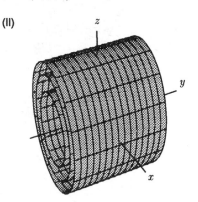

Figure 11.102

7. Figure 11.103 shows contour diagrams of temperature in degrees Fahrenheit in a room at three different times. Describe the heat flow in the room. What could be causing this?

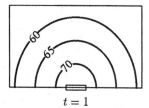   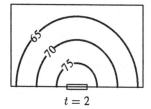   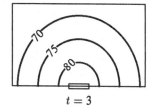

Figure 11.103

In Problems 8–9, give the equation of the linear function $f(x, y, z) = ax + by + cz + d$ that has the given values for the sections with $z = 1$ and $z = 4$.

8.   *Section with $z = 1$*   *Section with $z = 4$*

		y	
		3	5
x	0	4	8
	1		

		y	
		3	5
x	0		14
	1		9

9.   *Section with $z = 1$*   *Section with $z = 4$*

		y	
		3	5
x	0	4	
	1		8

		y	
		3	5
x	0		
	1	14	9

10. In Problems 8–9, suppose the two tables had only three values filled in. Could you determine f? Suppose the two tables had five values filled in. Can you always determine f?

11. Find the linear function $f(x, y, z) = ax + by + cz + d$ that has the contour diagrams for the sections with $z = 3$ and $z = 4$ shown in Figures 11.104 and 11.105.

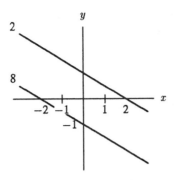

Figure 11.104: Section with $z = 3$ **Figure 11.105**: Section with $z = 4$

12. What do the level surfaces of $f(x, y, z) = x^2 - y^2 + z^2$ look like? [Hint: Use sections with y constant instead of sections with z constant.]

Use the catalog of surfaces to identify the surfaces in Problems 13–22.

13. $-x^2 - y^2 + z^2 = 1$ 14. $x^2 - y^2 - z^2 = 1$ 15. $-x^2 + y^2 - z^2 = 0$

16. $x^2 + y^2 - z = 0$ 17. $x^2 + z^2 = 1$ 18. $x^2 - y^2 + z = 0$

19. $x^2 + y^2/4 + z^2 = 1$ 20. $x + y = 1$ 21. $(x - 1)^2 + y^2 + z^2 = 1$

22. $x + y + z = 1$

23. Describe the surface $x^2 + y^2 = (2 + \sin z)^2$. In general, if $f(z) \geq 0$ for all z, describe the surface $x^2 + y^2 = (f(z))^2$.

REVIEW PROBLEMS FOR CHAPTER ELEVEN

1. A point has coordinates (a, b, c) where a, b, c are all positive. Interpret $a, b,$ and c as distances from the coordinate planes. Illustrate your answer with a picture.

2. Sketch the graph of the cylinder volume function $f(r, h) = \pi r^2 h$ first by keeping h fixed, then by keeping r fixed.

3. Describe the set of points whose x coordinate is 2 and whose y coordinate is 1.

4. Find the center and radius of the sphere with equation

$$x^2 + 4x + y^2 - 6y + z^2 + 12z = 0.$$

5. Find the equation of the plane through the points $(0, 0, 2), (0, 3, 0), (5, 0, 0)$.

6. Find an equation for the plane containing the line in the xy-plane defined by $y = 3x + 1$, and the point $(1, 0, 3)$.

7. Find the equation of the plane that intersects the xy-plane in the line $y = 2x + 2$ and contains the point $(1, 2, 2)$.

Decide if the statements in Problems 8–12 must be true, might be true, or could not be true. The function $z = f(x, y)$ is defined everywhere.

8. The level curves corresponding to $z = 1$ and $z = -1$ cross at the origin.

9. The level curve $z = 1$ consists of the circle $x^2 + y^2 = 2$ and the circle $x^2 + y^2 = 3$, but no other points.

10. The level curve $z = 1$ consists of two lines which intersect at the origin.

11. If $z = e^{-(x^2+y^2)}$, there is a level curve for every value of z.

12. If $z = e^{-(x^2+y^2)}$, there is a level curve through every point (x, y).

For each of the functions in Problems 13–16, make a contour plot in the region $-2 < x < 2$ and $-2 < y < 2$. Describe the shape of the contour lines.

13. $z = \sin y$ 14. $z = 3x - 5y + 1$ 15. $z = 2x^2 + y^2$ 16. $z = e^{-2x^2-y^2}$

17. Draw the contour diagrams for the functions $f(x, y) = (x - y)$ and $g(x, y) = (x - y)^2$ for $-3 \leq x \leq 3$, $-3 \leq y \leq 3$. Where does each function attain its maximum value on this region?

18. Suppose you are in a room 30 feet long with a heater at one end. In the morning the room is $65°F$. You turn on the heater, which quickly warms up to $85°F$. Let $H(x, t)$ be the temperature x feet from the heater, t minutes after the heater is turned on. Figure 11.106 shows the contour diagram for H. How warm is it 10 feet from the heater 5 minutes after it was turned on? 10 minutes after it was turned on?

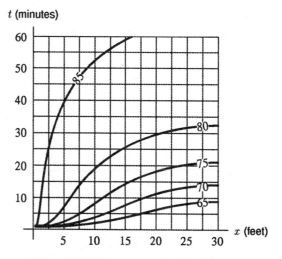

Figure 11.106: Temperature in a heated room.

19. Using the contour diagram in Figure 11.106, sketch the graphs of the one-variable functions $H(x, 5)$ and $H(x, 20)$. Interpret the two graphs in practical terms, and explain the difference between them.

Find equations for the linear functions with the contour diagrams in Problems 20–21

20.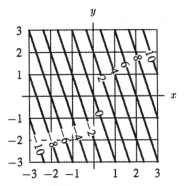

Figure 11.107: A contour map of the linear function $g(x, y)$

21.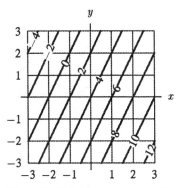

Figure 11.108: A contour map of the linear function $h(x, y)$

22. You are an anthropologist observing a native ritual. Sixteen people arrange themselves along a bench, with all but the three on the far left side seated. The first person is standing with her hands at her side, the second is standing with his hands raised and the third is standing with her hands at her side. At some unseen signal, the first one sits down, and everyone else copies what his neighbor to the left was doing one second earlier. Every second that passes, this behavior is repeated until all are once again seated.

    (a) Draw graphs at several different times showing how the height depends upon the distance along the bench.
    (b) Graph the location of the raised hands as a function of time.
    (c) What US ritual is most closely related to what you have observed?

23. Figure 11.109 shows the contour diagram for the vibrating string function from page 9:

$$f(x, t) = \cos t \sin x, \quad 0 \le x \le \pi.$$

Using the diagram, describe in words the sections of f with t fixed and the sections of f with x fixed. Explain what you see in terms of the behavior of the string.

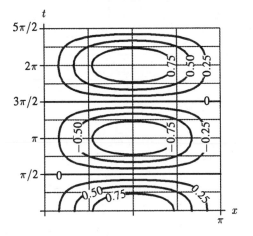

Figure 11.109

24. Consider the function $z = \cos\sqrt{x^2 + y^2}$.
    (a) Sketch the level curves of this function.
    (b) Sketch a cross-section through the surface $z = \cos\sqrt{x^2 + y^2}$ in the plane containing the x- and z-axes. Put units on your axes.
    (c) Sketch the cross-section through the surface $z = \cos\sqrt{x^2 + y^2}$ in the plane containing the z-axis and the line $y = x$ in the xy-plane.

25. Table 11.12 shows the predictions of one simple model on how average yearly ultra-violet (UV) exposure might vary with the year and the latitude.

TABLE 11.12: *Ultra-violet exposure.*

		Year								
		1970	1975	1980	1985	1990	1995	2000	2005	2010
	90	0.00	0.00	0.00	0.00	0.00	0.00	0.00	0.00	0.00
	80	0.00	0.00	0.00	0.00	0.00	3.03	5.79	6.37	6.52
	70	0.00	0.00	0.00	0.00	0.00	0.02	3.14	6.80	8.21
	60	0.01	0.01	0.01	0.01	0.01	0.01	0.12	1.78	3.41
	50	0.08	0.08	0.08	0.08	0.08	0.08	0.08	0.08	0.33
	40	0.25	0.25	0.25	0.25	0.25	0.25	0.25	0.25	0.25
	30	0.49	0.49	0.49	0.49	0.49	0.49	0.49	0.49	0.49
	20	0.74	0.74	0.74	0.74	0.74	0.74	0.74	0.74	0.74
	10	0.93	0.93	0.93	0.93	0.93	0.93	0.93	0.93	0.93
Latitude	0	1.00	1.00	1.00	1.00	1.00	1.00	1.00	1.00	1.00
	-10	0.93	0.93	0.93	0.93	0.93	0.93	0.93	0.93	0.93
	-20	0.74	0.74	0.74	0.74	0.74	0.74	0.74	0.74	0.74
	-30	0.49	0.49	0.49	0.49	0.49	0.49	0.49	0.49	0.49
	-40	0.25	0.25	0.25	0.25	0.25	0.25	0.25	1.39	2.31
	-50	0.08	0.08	0.08	0.08	0.08	0.08	2.41	5.68	7.27
	-60	0.01	0.01	0.01	0.01	0.01	1.41	7.64	10.82	11.97
	-70	0.00	0.00	0.00	0.00	0.08	6.36	10.36	11.46	11.80
	-80	0.00	0.00	0.00	0.00	3.69	6.32	6.71	6.80	6.82
	-90	0.00	0.00	0.00	0.00	0.00	0.00	0.00	0.00	0.00

    (a) Graph the sections of UV exposure for the years 1970, 1990, and 2000.
    (b) Produce a table showing what latitude has the most severe exposure to UV, as a function of the year.
    (c) What do you notice in your answer to part (b)? What is a possible explanation for this phenomenon?

26. Give sections of the linear function $f(x, y, z) = 2x - y + 3z + 4$ with z fixed, $z = 1$, $z = 2$, $z = 3$, first in terms of tables, then in terms of contour diagrams.

27. Figure 11.110 shows the contours of light intensity as a function of location and time in a microscopic wave-guide.

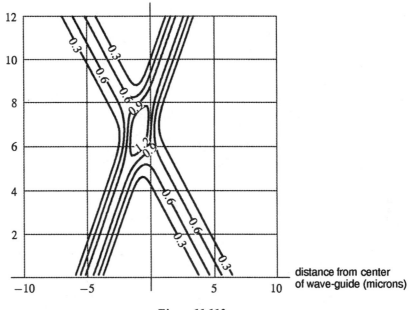

Figure 11.110

(a) Draw graphs showing intensity as a function of location at times 0, 2, 4, 6, 8, and 10 nanoseconds.

(b) If you could create an animation showing how the graph of intensity as a function of location varies as time passes, what would it look like?

(c) Draw a graph of intensity as a function of time at locations −5, 0 and 5 microns from center of wave-guide.

(d) Describe what the light beams are doing in this wave-guide.

CHAPTER TWELVE

A FUNDAMENTAL
TOOL: VECTORS

In one-variable calculus we were able to represent quantities such as velocity by numbers. Once we enter the world of more than one variable, we can no longer do this. To specify the velocity of a moving object, we need to say how fast it is moving, and also what direction it is moving in. In this chapter we will learn how to use *vectors* to represent quantities that have direction as well as magnitude.

12.1 DISPLACEMENT VECTORS

Suppose you are a pilot planning a flight from Dallas to Pittsburgh. There are two things you must know: The distance to be traveled (so you have enough fuel to make it) and in what direction to go (so you don't miss Pittsburgh). Both these quantities together specify the displacement or *displacement vector* between the two cities.

> The **displacement vector** from one point to another is an arrow with its tail at the first point and its tip at the second. The **magnitude** (or length) of the displacement vector is the distance between the points, and is represented by the length of the arrow. The **direction** of the displacement vector is the direction of the arrow.

Figure 12.1 shows the displacement vectors from Dallas to Pittsburgh, from Albuquerque to Oshkosh and from Los Angeles to Buffalo, SD. These displacement vectors have the same length and the same direction. We say that the displacement vectors between the corresponding cities are the same, even though they do not coincide. In other words:

> Displacement vectors which point in the same direction and have the same magnitude are considered to be the same, even if they do not coincide.

Notation and Terminology

The displacement vector is our first example of a vector. Vectors have both magnitude and direction; in comparison, a quantity specified only by a number, but no direction, is called a *scalar*.[1] For

Figure 12.1: Displacement vectors between cities

---

[1]So named by W. R. Hamilton because they are merely numbers on the *scale* from $-\infty$ to ∞.

instance, the weight of the plane at take-off and the time taken by the flight from Dallas to Pittsburgh are both scalar quantities.

In this book, vectors are written with an arrow over them, $\vec{v}$, to distinguish them from scalars. Other books use a bold **v** to denote a vector. We will also use the notation $\overrightarrow{PQ}$ to denote the displacement vector from a point P to a point Q. The magnitude, or length, of a vector $\vec{v}$ is written $\|\vec{v}\|$.

Addition and Subtraction of Displacement Vectors

Suppose NASA is controlling a robot on Mars. NASA commands the robot to move 75 meters in a certain direction, and then commands it to move 50 meters in a second direction. See Figure 12.2. Where does the robot end up as a result of these two commands? Each movement corresponds to a displacement vector, first $\vec{v}$ and then $\vec{w}$. The combined displacement is the sum of these two displacement vectors, $\vec{v} + \vec{w}$.

The **sum**, $\vec{v} + \vec{w}$, of two vectors $\vec{v}$ and $\vec{w}$ is the combined displacement resulting from first applying $\vec{v}$ and then $\vec{w}$. It is represented by an arrow connecting the tail of $\vec{v}$ to the tip of $\vec{w}$ in Figure 12.3.

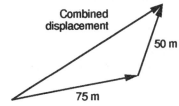

Figure 12.2: Sum of displacements of robots on Mars

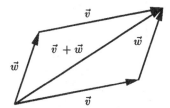

Figure 12.3: The sum $\vec{v} + \vec{w}$

Notice that the sum of $\vec{w}$ and $\vec{v}$, $\vec{w} + \vec{v}$, gives the same displacement.

Suppose now that NASA is controlling two different robots, starting out from the same location. One moves along a displacement vector $\vec{v}$ and the second along a displacement vector $\vec{w}$. What is the displacement vector from the first robot to the second? We draw the displacement vector and call it $\vec{x}$, as shown in Figure 12.4. Notice that $\vec{v} + \vec{x} = \vec{w}$, so $\vec{x}$ is the difference between $\vec{w}$ and $\vec{v}$: $\vec{x} = \vec{w} - \vec{v}$. Note that $\vec{w} - \vec{v}$ points from the tip of $\vec{v}$ to the tip of $\vec{w}$.

The **difference**, $\vec{w} - \vec{v}$, is the displacement vector which when added to $\vec{v}$ gives $\vec{w}$, that is, such that $\vec{w} = \vec{v} + (\vec{w} - \vec{v})$. See Figure 12.4.

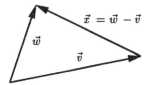

Figure 12.4: The difference $\vec{w} - \vec{v}$

It is possible that after a series of movements the robot ends up where it started. Then its total displacement vector is the *zero vector*.

The **zero vector** $\vec{0}$ is a displacement vector with zero length.

We don't need to specify the direction of the zero vector.

Scalar Multiplication of Displacement Vectors

If $\vec{v}$ represents a displacement vector, the vector $2\vec{v}$ represents a displacement of twice the magnitude in the same direction as $\vec{v}$. Similarly, $-2\vec{v}$ represents a displacement of twice the magnitude in the opposite direction, and $0.5\vec{v}$ represents a displacement in the same direction as $\vec{v}$ but half the magnitude. See Figure 12.5.

Definition of Multiplication by a Scalar

If λ is a scalar and $\vec{v}$ is a displacement vector, the **scalar multiple of $\vec{v}$ by λ**, written $\lambda\vec{v}$, is the displacement vector with the following properties:

- The displacement vector $\lambda\vec{v}$ is parallel to $\vec{v}$, pointing in the same direction if $\lambda > 0$, and in the opposite direction if $\lambda < 0$.
- The magnitude of $\lambda\vec{v}$ is $|\lambda|$ times the magnitude of $\vec{v}$, i.e.,

$$\|\lambda\vec{v}\| = |\lambda|\,\|\vec{v}\|.$$

Note that $|\lambda|$ represents the absolute value of the scalar λ while $\|\lambda\vec{v}\|$ represents the magnitude of the vector $\lambda\vec{v}$.

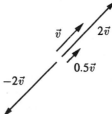

Figure 12.5: Scalar multiples of the vector $\vec{v}$

Example 1   Explain why $\vec{w} - \vec{v} = \vec{w} + (-1)\vec{v}$.

Solution   According to the definition, $(-1)\vec{v}$ is the vector with the same magnitude as $\vec{v}$, but pointing in the opposite direction. Figure 12.6 shows that $\vec{w} + (-1)\vec{v}$ is the same as $\vec{w} - \vec{v}$.

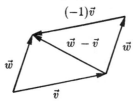

Figure 12.6: $\vec{w} - \vec{v} = \vec{w} + (-1)\vec{v}$

Components of Displacement Vectors

Suppose you live in a city with equally spaced streets running East-West and North-South, and you want to tell someone how to get from one place to another. Rather than giving them the distance and direction, you would tell them how many blocks east or west and how many blocks north or south to go. For example, to get from point P to point Q in Figure 12.7, you would go 4 blocks east and 1 block south. If $\vec{i}$ represents a displacement of one block east and $\vec{j}$ represents a displacement of one block north, then the displacement vector from P to Q is $4\vec{i} - \vec{j}$. This suggests a way of representing displacement vectors using a coordinate system.

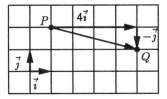

Figure 12.7: The displacement
vector from P to Q is $4\vec{i} - \vec{j}$

The Displacement Vectors $\vec{i}, \vec{j}$, and $\vec{k}$

The three displacement vectors of length 1 shown in Figure 12.8 are: The vector $\vec{i}$, which points along the positive x-axis, the vector $\vec{j}$, which points along the positive y-axis, and the vector $\vec{k}$, which points along the positive z-axis. Figure 12.9 shows the vectors $\vec{i}$ and $\vec{j}$ in two dimensions.

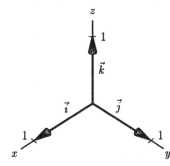

Figure 12.8: The vectors $\vec{i}, \vec{j}$ and
$\vec{k}$ in 3-space

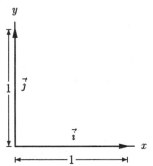

Figure 12.9: The vectors $\vec{i}$ and
$\vec{j}$ in the plane

Writing Displacement Vectors Using $\vec{i}, \vec{j}, \vec{k}$

Any displacement in 3-space or in the plane can be expressed as a combination of displacements in the coordinate directions. For example, Figure 12.10 shows that the displacement vector $\vec{v}$ from the origin to the point $P = (3, 2)$ can be written as a sum of displacement vectors along the x and y-axes:

$$\vec{v} = 3\vec{i} + 2\vec{j}.$$

If
$$\vec{v} = v_1\vec{i} + v_2\vec{j} + v_3\vec{k}$$
then $v_1\vec{i}$, $v_2\vec{j}$, and $v_3\vec{k}$ are the **components** of $\vec{v}$, and we say that we have **resolved** $\vec{v}$ into components.

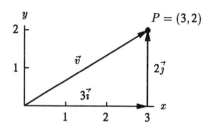

Figure 12.10: We write $\vec{v} = 3\vec{i} + 2\vec{j}$

Example 2   Resolve the displacement vector from the point $P_1 = (2, 4, 10)$ to the point $P_2 = (3, 7, 6)$ into components.

Solution   To get from P_1 to P_2, you move 1 unit in the positive x-direction, 3 units in the positive y-direction, and 4 units in the negative z-direction. Hence $\vec{v} = \vec{i} + 3\vec{j} - 4\vec{k}$.

In general, we have the following rule (see Figure 12.11):

Components of Displacement Vectors

The components of the displacement vector from the point $P_1 = (x_1, y_1, z_1)$ to the point $P_2 = (x_2, y_2, z_2)$ are given by

$$\overrightarrow{P_1P_2} = (x_2 - x_1)\vec{i} + (y_2 - y_1)\vec{j} + (z_2 - z_1)\vec{k}.$$

Position Vectors: the Displacement of a Point from the Origin

If the initial point of a displacement vector is the origin, then the displacement vector is called the position vector. Thus the point (x_0, y_0, z_0) in space has position vector $\vec{r}_0 = x_0\vec{i} + y_0\vec{j} + z_0\vec{k}$. See Figure 12.12. In general, the position vector gives the position of the point with respect to the origin. A displacement vector, on the other hand, gives the change in coordinates from one point to another.

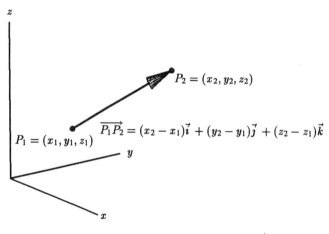

Figure 12.11: The displacement vector $\overrightarrow{P_1P_2}$

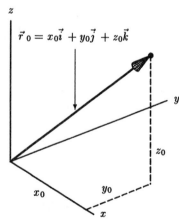

Figure 12.12: The position vector $\vec{r}_0$

The Components of the Zero Vector

Recall that the zero displacement vector has zero magnitude and is written $\vec{0}$. So

$$\vec{0} = 0\vec{i} + 0\vec{j} + 0\vec{k}.$$

Working with Components

How Do We Find the Magnitude of a Vector Given in Components?

For a vector $\vec{v}$ represented by an arrow,

> Magnitude of $\vec{v}$ = $\|\vec{v}\|$ = Length of the arrow.

If $\vec{v} = v_1\vec{i} + v_2\vec{j} + v_3\vec{k}$, then

$$\|\vec{v}\| = \sqrt{v_1^2 + v_2^2 + v_3^2}.$$

For instance, if $\vec{v} = 3\vec{i} - 4\vec{j} + 5\vec{k}$, then $\|\vec{v}\| = \sqrt{3^2 + (-4)^2 + 5^2} = \sqrt{50}$. Figure 12.13 illustrates this formula for a two-dimensional vector (so $v_3 = 0$).

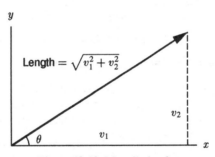

Figure 12.13: Magnitude of a two-dimensional vector

Addition and Scalar Multiplication of Vectors Using Components

Suppose two vectors $\vec{v}$ and $\vec{w}$ are given in components:

$$\vec{v} = v_1\vec{i} + v_2\vec{j} + v_3\vec{k},$$
$$\vec{w} = w_1\vec{i} + w_2\vec{j} + w_3\vec{k}.$$

Then

$$\vec{v} + \vec{w} = (v_1 + w_1)\vec{i} + (v_2 + w_2)\vec{j} + (v_3 + w_3)\vec{k},$$

and

$$\lambda\vec{v} = \lambda v_1\vec{i} + \lambda v_2\vec{j} + \lambda v_3\vec{k}.$$

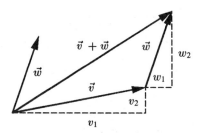

Figure 12.14: Sum $\vec{v} + \vec{w}$ in components

Figures 12.14 and 12.15 illustrate these properties in two dimensions. Since

$$\vec{v} - \vec{w} = \vec{v} + (-1)\vec{w},$$

we can write

$$\vec{v} - \vec{w} = (v_1 - w_1)\vec{i} + (v_2 - w_2)\vec{j} + (v_3 - w_3)\vec{k}.$$

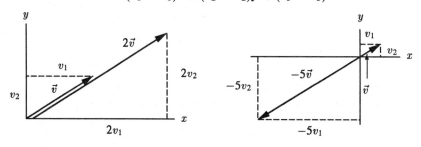

Figure 12.15: Scalar multiples of vectors showing $\vec{v}$, $2\vec{v}$, and $-5\vec{v}$

How to Resolve a Vector into Components

You may wonder how to find the components of a two-dimensional vector, given its length and direction. Suppose the vector $\vec{v}$ has length v and makes an angle of θ with the x-axis, as in Figure 12.16. If $\vec{v} = v_1\vec{i} + v_2\vec{j}$, the triangle in Figure 12.16 shows

$$v_1 = v \cos \theta$$
$$v_2 = v \sin \theta.$$

Thus,

$$\vec{v} = (v \cos \theta)\vec{i} + (v \sin \theta)\vec{j}.$$

For example, a vector of length 2 making an angle of $\pi/6$ with the x-axis has components $\sqrt{3}\,\vec{i}$ and $\vec{j}$, since

$$\vec{v} = 2\cos(\pi/6)\vec{i} + 2\sin(\pi/6)\vec{j} = 2(\sqrt{3}/2)\,\vec{i} + 2(1/2)\vec{j} = \sqrt{3}\vec{i} + \vec{j}.$$

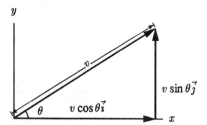

Figure 12.16: Resolving a vector:
$\vec{v} = (v \cos \theta)\vec{i} + (v \sin \theta)\vec{j}$

Unit Vectors

A *unit vector* is a vector whose magnitude is 1. The vectors $\vec{i}$, $\vec{j}$, and $\vec{k}$ are unit vectors in the direction of the coordinate axes. It is often helpful to find a unit vector in the same direction as a given vector $\vec{v}$. Suppose that $\|\vec{v}\| = 10$; a unit vector in the same direction as $\vec{v}$ is $\frac{1}{10}\vec{v}$. In general, a unit vector in the direction of any nonzero vector $\vec{v}$ is

$$\vec{u} = \frac{\vec{v}}{\|\vec{v}\|}.$$

Example 3 Find a unit vector, $\vec{u}$, in the direction of the vector $\vec{v} = \vec{i} + 3\vec{j}$.

Solution If $\vec{v} = \vec{i} + 3\vec{j}$, then $\|\vec{v}\| = \sqrt{1^2 + 3^2} = \sqrt{10}$.
Thus,

$$\vec{u} = \frac{\vec{v}}{\sqrt{10}} = \frac{1}{\sqrt{10}}(\vec{i} + 3\vec{j}) = \frac{1}{\sqrt{10}}\vec{i} + \frac{3}{\sqrt{10}}\vec{j} \approx 0.32\vec{i} + 0.95\vec{j}.$$

Example 4 Find a unit vector at the point (x, y, z) that points away from the origin.

Solution The position vector

$$\vec{r} = x\vec{i} + y\vec{j} + z\vec{k}$$

points from the origin to (x, y, z). Thus, if we put its tail at (x, y, z) it will point away from the origin. Its magnitude is

$$\|\vec{r}\| = \sqrt{x^2 + y^2 + z^2},$$

so a unit vector pointing in the same direction is

$$\frac{1}{\|\vec{r}\|}\vec{r} = \frac{1}{\sqrt{x^2 + y^2 + z^2}}(x\vec{i} + y\vec{j} + z\vec{k})$$

$$= \frac{x}{\sqrt{x^2 + y^2 + z^2}}\vec{i} + \frac{y}{\sqrt{x^2 + y^2 + z^2}}\vec{j} + \frac{z}{\sqrt{x^2 + y^2 + z^2}}\vec{k}.$$

Problems for Section 12.1

1. The vectors $\vec{w}$ and $\vec{u}$ are shown in Figure 12.17. Match the vectors $\vec{p}$, $\vec{q}$, $\vec{r}$, $\vec{s}$, $\vec{t}$ with five of the following vectors: $\vec{u} + \vec{w}$, $\vec{u} - \vec{w}$, $\vec{w} - \vec{u}$, $2\vec{w} - \vec{u}$, $\vec{u} - 2\vec{w}$, $2\vec{w}$, $-2\vec{w}$, $2\vec{u}$, $-2\vec{u}$, $-\vec{w}$, $-\vec{u}$.

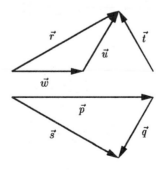

Figure 12.17

2. Given the displacement vectors $\vec{v}$ and $\vec{w}$ in Figure 12.18, draw the following vectors:
    (a) $\vec{v} + \vec{w}$ (b) $\vec{v} - \vec{w}$ (c) $2\vec{v}$ (d) $2\vec{v} + \vec{w}$ (e) $\vec{v} - 2\vec{w}$.

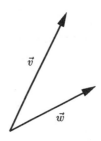

Figure 12.18

3. On a sketch of Figure 12.19, draw:
    (a)   The vector $\vec{v} = 4\vec{i} + \vec{j}$ with initial point at the origin, and
    (b)   The vector $\vec{v} = 4\vec{i} + \vec{j}$ with initial point $(3, 2)$.

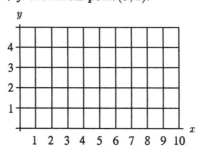

Figure 12.19

Resolve the vectors in Problems 4 – 7 into components.

4. The displacement vector from the point $(-1, 1)$ to the point $(-3, 0)$.

5. The displacement vector from the point $(1.5, 3.2)$ to the point $(0.5, 3.3)$.

6. A vector starting at the point $P = (1, 2)$ and ending at the point $Q = (4, 6)$.

7. A vector starting at the point $Q = (4, 6)$ and ending at the point $P = (1, 2)$.

8. Figure 12.20 shows a molecule with atoms at O, A, B and C. Verify that every atom in the molecule is 2 units away from every other atom.

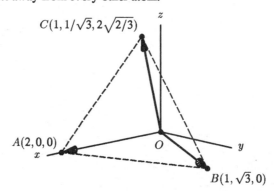

Figure 12.20

For Problems 9–10, consider the map in Figure 11.1 on page 2.

9. If you leave Topeka along the following vectors, does the temperature increase or decrease?
    (a) $\vec{u} = 3\vec{i} + 2\vec{j}$ (b) $\vec{v} = -\vec{i} - \vec{j}$ (c) $\vec{w} = -5\vec{i} - 5\vec{j}$.

10. Starting in Buffalo, sketch a vector pointing in the direction in which the temperature is increasing most rapidly. Starting in Boise, sketch a vector pointing in the direction in which the temperature is decreasing most rapidly.

11. Resolve the vector shown in Figure 12.21 into components.

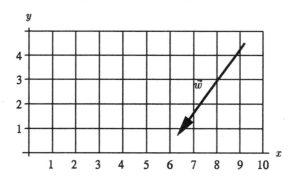

Figure 12.21: Scale: 1 grid length = 0.25 inches

A cat is sitting on the ground at the point $(1, 4, 0)$ watching a squirrel at the top of a tree. The tree is one unit high and its base is at the point $(2, 4, 0)$. In Problems 12-15, find the given displacement vectors.

12. From the origin to the cat.

13. From the bottom of the tree to the squirrel.

14. From the bottom of the tree to the cat.

15. From the cat to the squirrel.

16. Resolve the two-dimensional vectors $\vec{a}$, $\vec{b}$, $\vec{v}$, $\vec{w}$ in Figure 12.22 into components.

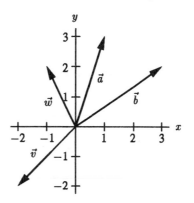

Figure 12.22

17. A car is traveling due north at 30 mph approaching a crossroad. On a perpendicular road a police car is traveling west towards the intersection at 40 mph. Suppose that both cars will reach the crossroad in exactly one hour. Find the vector representing the displacement of the car with respect to the police car.

18. Resolve into components the following two-dimensional vectors $\vec{a}$, $\vec{b}$, $\vec{c}$, $\vec{d}$, and $\vec{e}$ in Figure 12.23.

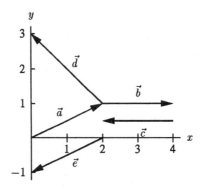

Figure 12.23

19. Resolve into components and find the length of each of the vectors shown in Figure 12.24.

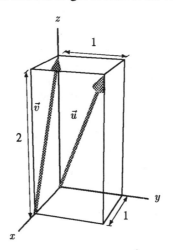

Figure 12.24

20. Resolve into components each of the vectors in Figure 12.25.

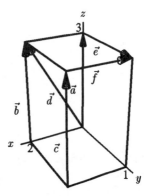

Figure 12.25

21. Which of the following vectors are parallel?

$$\vec{u} = 2\vec{i} + 4\vec{j} - 2\vec{k} \qquad \vec{v} = \vec{i} - \vec{j} + 3\vec{k} \qquad \vec{w} = -\vec{i} - 2\vec{j} + \vec{k}$$

$$\vec{p} = \vec{i} + \vec{j} + \vec{k} \qquad \vec{q} = 4\vec{i} - 4\vec{j} + 12\vec{k} \qquad \vec{r} = \vec{i} - \vec{j} + \vec{k}.$$

For Problems 22–29 perform the indicated computation.

22. $(0.6\vec{i} + 0.2\vec{j} - \vec{k}) + (0.3\vec{i} + 0.3\vec{k})$

23. $(\vec{i} + 2\vec{j}) + (-3)(2\vec{i} + \vec{j})$

24. $(3\vec{i} - 4\vec{j} + 2\vec{k}) - (6\vec{i} + 8\vec{j} - \vec{k})$

25. $(4\vec{i} + 2\vec{j}) - (3\vec{i} - \vec{j})$

26. $2(0.45\vec{i} - 0.9\vec{j} - 0.01\vec{k}) - 0.5(1.2\vec{i} - 0.1\vec{k})$

27. $(4\vec{i} - 3\vec{j} + 7\vec{k}) - 2(5\vec{i} + \vec{j} - 2\vec{k})$

28. $-4(\vec{i} - 2\vec{j}) - 0.5(\vec{i} - \vec{k})$

29. $\frac{1}{2}(2\vec{i} - \vec{j} + 3\vec{k}) + 3(\vec{i} - \frac{1}{6}\vec{j} + \frac{1}{2}\vec{k})$

For Problems 30–39, perform the following operations on the given vectors:

$$\vec{a} = 2\vec{j} + \vec{k}, \qquad \vec{b} = -3\vec{i} + 5\vec{j} + 4\vec{k} \qquad \vec{c} = \vec{i} + 6\vec{j}$$

$$\vec{x} = -2\vec{i} + 9\vec{j} \qquad \vec{y} = 4\vec{i} - 7\vec{j} \qquad \vec{z} = \vec{i} - 3\vec{j} - \vec{k}$$

30. $\|\vec{z}\|$

31. $\vec{a} + \vec{z}$

32. $5\vec{b}$

33. $2\vec{c} + \vec{x}$

34. $4\vec{z}$

35. $\|\vec{y}\|$

36. $5\vec{a} + 2\vec{b}$

37. $\|\vec{y} - \vec{x}\|$

38. $3\vec{a}$

39. $2\vec{a} + 7\vec{b} - 5\vec{z}$

40. Resolve the following vectors into components:
    (a) The vector in the plane of length 2 pointing up and to the right at an angle of $\pi/4$ with the x-axis.
    (b) The vector in 3-space of length 1 pointing upward in the xz plane at an angle of $\pi/6$ with the xy plane.

41. Find a vector that points in the same direction as $\vec{i} - \vec{j} + 2\vec{k}$ but has length 2.

42. (a) Find a unit vector starting at the point $P = (1, 2)$ and pointing toward the point $Q = (4, 6)$.

    (b) Find a vector of length 10 pointing in the same direction.

Find the length of the vectors in Problems 43–46.

43. $\vec{v} = \vec{i} - \vec{j} + 3\vec{k}$

44. $\vec{v} = \vec{i} - \vec{j} + 2\vec{k}$

45. $\vec{v} = 1.2\vec{i} - 3.6\vec{j} + 4.1\vec{k}$

46. $\vec{v} = 7.2\vec{i} - 1.5\vec{j} + 2.1\vec{k}$

47. Show that the medians of a triangle intersect at a point 1/3 of the way along each median from the side it bisects.

48. Show that the lines joining the centroid (the intersection point of the medians) of a face of the tetrahedron and the opposite vertex meet at a point.

12.2 VECTORS IN GENERAL

There are many quantities other than displacement that have both magnitude and direction, and that can be added together and scaled in the same way that displacements can. Such a quantity is called a *vector*, and we can represent it by an arrow in the same way we represent displacements. The length of the arrow is the *magnitude* of the vector, and the direction of the arrow is the direction of the vector.

Velocity versus Speed

The speed of a moving body tells us how fast it is moving, say 50 mph. The speed is just a number; it is therefore a scalar. The velocity, on the other hand, tells us both how fast the body is moving and the direction of motion; it is a vector. For instance, if a car is heading northeast at 50 mph, then its velocity is a vector of length 50 pointing northeast.

> The **velocity vector** of a moving object is a vector whose magnitude is the speed of the object and whose direction is the direction of its motion.

Example 1   A car is traveling north at a speed of 60 mph, while a plane above is flying horizontally south-west at a speed of 300 mph. Draw the velocity vectors of the car and the plane.

Solution   Figure 12.26 shows the velocity vectors. The plane's velocity vector is five times as long as the car's, because its speed is five times greater ($300 = 5 \cdot 60$), and speed is the magnitude of velocity.

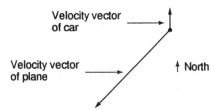

Velocity vector of car

Velocity vector of plane

↑ North

Figure 12.26: Velocity vectors of the car and the plane

The next example illustrates that the velocity vectors for two motions add together to give the velocity vector for the combined motion, just as displacements do.

Example 2   A riverboat is moving with a velocity, $\vec{v}$, of 8 km/hr relative to the water. In addition, there is a current, $\vec{c}$, of 1 km/hr at an angle to $\vec{v}$. See Figure 12.27. What is the physical significance of $\vec{v} + \vec{c}$?

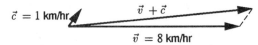

$\vec{c} = 1$ km/hr

$\vec{v} + \vec{c}$

$\vec{v} = 8$ km/hr

Figure 12.27: Sum of boat's velocities

Solution   The vector $\vec{v}$ shows how the boat is moving relative to the water, while $\vec{c}$ shows how the water is moving relative to the riverbed. Thus, $\vec{v} + \vec{c}$ shows how the boat is moving relative to the riverbed.

Scalar multiplication also makes sense for velocity vectors: For example, if $\vec{v}$ is a velocity vector, then $-2\vec{v}$ represents a velocity of twice the magnitude in the opposite direction.

Example 3   A particle is moving with velocity $\vec{v}$, when it hits a barrier at a right angle and bounces straight back, with its speed reduced by 20%. Express its new velocity in terms of the old one.

Solution   The new velocity is $-0.8\vec{v}$, where the negative sign expresses the fact that the new velocity is in the direction opposite to the old.

We can represent velocity vectors in components in the same way as for displacement vectors.

Example 4   Represent the velocity vectors of the car and the plane in Example 1 using components. Use a coordinate system in which north is the positive y-axis, east is the positive x-axis, and the positive z-axis is up.

Solution   The car is traveling north at 60 mph, so the component of its velocity in the y-direction is $60\vec{j}$ and the component in the x-direction is $0\vec{i}$. Since it is traveling horizontally, the component in the z-direction is $0\vec{k}$. So its velocity vector is given by

$$\vec{v} = 0\vec{i} + 60\vec{j} + 0\vec{k} = 60\vec{j}.$$

The plane's velocity vector also has $\vec{k}$ component equal to zero. Since it is traveling southwest, its $\vec{i}$ and $\vec{j}$ components have negative coefficients (north and east are positive). To find the components, note that the plane travels 300 miles in an hour. This represents a distance of $300/\sqrt{2} \approx 212$ miles in the westerly direction and the same distance in the southerly direction. See Figure 12.28.

Thus, the velocity vector of the plane is given by

$$\vec{v} = -\frac{300}{\sqrt{2}}\vec{i} - \frac{300}{\sqrt{2}}\vec{j} \approx -212\vec{i} - 212\vec{j}.$$

Of course if the car were climbing a steep hill or if the plane were descending for a landing, then the $\vec{k}$ component would not be zero.

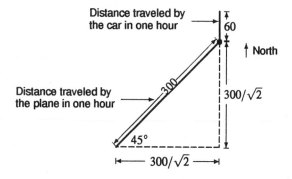

Figure 12.28: Distance traveled by the plane and car in one hour

Acceleration

Another important example of a vector quantity is acceleration. Acceleration, like velocity, is specified by both a magnitude and a direction — for example, the acceleration due to gravity at the surface of the Earth is 9.81 m/sec² *downward*.

Force

Force is another example of a vector quantity. Suppose you push on an open door. The result depends not just on how hard you push but also in what direction. Thus, to specify a force we must give its magnitude (or strength) and the direction in which it is acting. For example, the gravitational force exerted on a planet by a star is a vector pointing from the planet in the direction of the star; its magnitude is the strength of the gravitational attraction, which depends on the masses of the two bodies and the distance between them. (Newton's Law of Gravitation gives the exact relation.)

Example 5   Kepler's first two laws of planetary motion are:
Kepler's First Law: Planets follow elliptical orbits with the sun at one focus.
Kepler's Second Law: The line joining a planet to the sun sweeps out equal areas in equal times.
(a)   Sketch force vectors representing the gravitational force of the sun on the earth at two different positions in its orbit.
(b)   Sketch the velocity vector of the earth at two points of its orbit.

Solution   (a)   Figure 12.29 shows the earth orbiting the sun, with the gravitational force vectors at two different positions. Note that the force vector is larger when the earth is closer to the sun because the gravitational force there is greater. (In fact, the real orbit looks much more like a circle than we have shown here.)

(b)   The velocity vector always points in the direction of motion of the earth. Thus, the velocity vector must be tangent to the ellipse. See Figure 12.30. Furthermore, the velocity vector is longer at points of the orbit where the planet is moving quickly, and shorter at points where it is moving slowly, because the magnitude of the velocity vector represents the speed. Kepler's Second Law enables us to determine when the earth is moving quickly and when it is moving slowly. The meaning of the second law is illustrated in Figure 12.30. Over a fixed period of time, say one month, the line joining the earth to the sun sweeps out a sector having a certain area. Figure 12.30 shows two sectors swept out in two different one-month time-intervals. The second law says that the areas of the two sectors are the same. For the areas to be the same, the earth must move farther in a month when it is close to the sun than when it is far from the sun. Thus, the earth moves faster when it is closer to the sun and slower when it is further away.

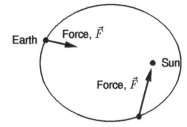

Figure 12.29: Gravitational force, $\vec{F}$, exerted by the sun on the earth

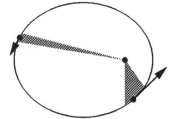

Figure 12.30: The velocity vector, $\vec{v}$, of the earth as determined by Kepler's Second Law

Properties of Addition and Scalar Multiplication

In general, vectors add, subtract, and scale in the same way as displacement vectors. Thus, they have many of the same properties. For any vectors $\vec{u}$, $\vec{v}$, and $\vec{w}$ and any scalars α and β, we have the following results.

Commutativity

1. $\vec{u} + \vec{v} = \vec{v} + \vec{u}$

Distributivity

4. $(\alpha + \beta)\vec{v} = \alpha\vec{v} + \beta\vec{v}$

5. $\alpha(\vec{v} + \vec{w}) = \alpha\vec{v} + \alpha\vec{w}$

Associativity

2. $(\vec{u} + \vec{v}) + \vec{w} = \vec{u} + (\vec{v} + \vec{w})$

3. $\alpha(\beta\vec{v}) = (\alpha\beta)\vec{v}$

Identity

6. $\vec{v} + \vec{0} = \vec{v}$

7. $1 \cdot \vec{v} = \vec{v}$

Problems 19–25 at the end of this section ask for a justification of these results, in terms of displacement vectors.

Using Components

Example 6   The velocity, $\vec{c}$, of the current in Example 2 on page 78 is of magnitude 1 and at an angle of about $53°$ to the x-axis. Verify that $\vec{c} \approx 0.6\vec{i} + 0.8\vec{j}$.

Solution   We explained how to resolve a displacement vector into components on page 72. In general, vectors can be resolved into components in the same way. Since $\|\vec{c}\| = 1$,

$$\vec{c} = (1 \cdot \cos 53°)\vec{i} + (1 \cdot \sin 53°)\vec{j} \approx 0.6\vec{i} + 0.8\vec{j}.$$

Example 7   Suppose the boat in Example 2 on page 78 is moving with a velocity $\vec{v}$, relative to the water, given in km/hr by

$$\vec{v} = 8\vec{i},$$

and that the velocity of the current in km/hr is

$$\vec{c} = 0.6\vec{i} + 0.8\vec{j}.$$

(a)   What is the speed of the boat relative to the riverbed?

(b)   What angle does the velocity of the boat relative to the riverbed make with the vector $\vec{v}$? What does this angle tell you in practical terms?

Solution   (a)   We start by finding the velocity, $\vec{w}$, of the boat relative to the riverbed. It is given by

$$\vec{w} = \vec{v} + \vec{c} = 8\vec{i} + (0.6\vec{i} + 0.8\vec{j}) = 8.6\vec{i} + 0.8\vec{j}.$$

The speed of the boat relative to the riverbed is the magnitude of $\vec{w}$:

$$\text{Speed} = \|\vec{w}\| = \sqrt{8.6^2 + 0.8^2} \approx 8.64 \text{ km/hr}.$$

(b)   Since the velocity, $\vec{v} = 8\vec{i}$, is parallel to the x-axis, we want to find the angle between $\vec{w}$ and the x-axis. From Figure 12.31, this angle is

$$\theta = \arctan\left(\frac{0.8}{8.6}\right) \approx 0.093 \text{ radians} \approx 5.3°.$$

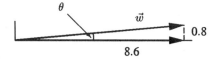

Figure 12.31: Angle between $\vec{v}$ and $\vec{w}$

In practical terms, this angle tells you that if you set your boat parallel to the x-axis at 8 km/hr, the current will take you about 5° off course.

An Alternative Notation

Many people write

$$\vec{v} = (v_1, v_2, v_3) \quad \text{instead of} \quad \vec{v} = v_1\vec{i} + v_2\vec{j} + v_3\vec{k}.$$

Since the first notation can be confused with a point and the second cannot, we will mostly use the second form.

Vectors in n Dimensions

Using the alternative notation for components, we can define a vector in n dimensions as a string of n numbers, or n components. Thus, a vector in n dimensions can be represented by an expression of the form

$$\vec{c} = (c_1, c_2, \ldots, c_n).$$

Addition and scalar multiplication are defined by the formulas

$$\vec{v} + \vec{w} = (v_1, v_2, \ldots, v_n) + (w_1, w_2, \ldots, w_n) = (v_1 + w_1, v_2 + w_2, \ldots, v_n + w_n)$$

and

$$\lambda\vec{v} = \lambda(v_1, v_2, \ldots, v_n) = (\lambda v_1, \lambda v_2, \ldots, \lambda v_n).$$

Why Would We Want Vectors in n Dimensions?

Vectors in two and three dimensions can be used to model displacement, velocities, or forces. But what about vectors in n dimensions? There is another interpretation of n-dimensional vectors (or n-vectors) which is useful: they can be thought of as listing n different quantities. Even a 3-vector is also a way of keeping three quantities organized — for example, the displacements parallel to the x, y, and z axes. Similarly

$$\vec{c} = (c_1, c_2, \ldots, c_n)$$

is a way of keeping n different quantities organized. For example, a *consumption* vector ,

$$\vec{q} = (q_1, q_2, \ldots, q_n)$$

shows the quantities $q_1, q_2, \ldots, q_n$ consumed of each of n different goods. A *price* vector

$$\vec{p} = (p_1, p_2, \ldots, p_n)$$

contains the prices of n different commodities. A *population* vector $\vec{N}$ might show the number of children and adults in a population:

$$\vec{N} = (\text{number of children, number of adults}),$$

or, if we are interested in a more detailed breakdown of ages, we might give the number in each ten-year age bracket in the population (up to age 110) in the form

$$\vec{N} = (N_1, N_2, N_3, N_4, \ldots, N_{10}, N_{11}),$$

where N_1 is the population aged 0–9, and N_2 is the population aged 10–19, and so on.

In 1907-8, Hermann Minkowski used vectors with four components when he introduced *space-time coordinates*, whereby each event is assigned a vector position $\vec{v}$ with four coordinates, three for its position in space and one for time:

$$\vec{v} = (x, y, z, t).$$

Example 8   The vector $\vec{I}$ represents the number of copies, in thousands, made by each of four copy centers in the month of December, and $\vec{J}$ represents the number of copies made at the same four copy centers during the previous eleven months (the "year-to-date"). If $\vec{I} = (25, 211, 818, 642)$, and $\vec{J} = (331, 3227, 1377, 2570)$, compute $\vec{I} + \vec{J}$. What does this sum represent?

Solution   The sum is

$$\vec{I} + \vec{J} = (25 + 331, 211 + 3227, 818 + 1377, 642 + 2570) = (356, 3438, 2195, 3212).$$

Each term in $\vec{I} + \vec{J}$ represents the sum of the number of copies made in December plus those in the previous eleven months, that is, the total number of copies made during the entire year at each copy center.

Example 9   The price vector $\vec{p} = (p_1, p_2, p_3)$ represents the prices in dollars of three goods. Write a vector which gives the prices of the same goods in cents.

Solution   The prices in cents are $100p_1$, $100p_2$, and $100p_3$ respectively, so the new price vector is

$$(100p_1, 100p_2, 100p_3) = 100\vec{p}.$$

Problems for Section 12.2 ━━━━━━━━━━━━━━━━━━━━━━━━━━━━━━━

In Problems 1-4, say whether the given quantity is a vector or a scalar.

1. The distance from Seattle to St. Louis.

2. The population of the US.

3. The magnetic field at a point on the earth's surface.

4. The temperature at a point on the earth's surface.

5. A car drives clockwise around the course in Figure 12.32, slowing down at the curves and speeding up along the straight portions of the track. Sketch velocity vectors at the points P, Q, and R.

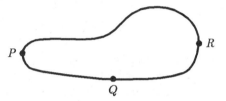

Figure 12.32

6. A racing car drives clockwise around the track shown in Figure 12.32 at a constant speed. At what point on the track does the car have the longest acceleration vector, and in roughly what direction is it pointing? Remember that acceleration is the rate of change of velocity.

7. There are five students in a class. Their scores on the midterm (out of 100) are given by the score vector $\vec{v} = (73, 80, 91, 65, 84)$. Their scores on the final (out of 100) are given by $\vec{w} = (82, 79, 88, 70, 92)$. If the final counts twice as much as the midterm, find a vector for the total scores of the students (as a percentage).

8. A car is traveling at a speed of 50 km/hr. Assume the positive y-axis is north and the positive x-axis is east. Resolve into components of its velocity vector (in two-space) if it is traveling in each of the following directions:
   (a) East   (b) South   (c) Southeast   (d) Northwest.

9. Which is traveling faster, a car whose velocity vector is $21\vec{i} + 35\vec{j}$, or a car whose velocity vector is $40\vec{i}$, assuming that the units are the same for both directions?

10. Shortly after taking off, a plane is climbing northwest through still air at an airspeed of 200 mph, and rising at a rate of 1000 ft/min. Resolve into components its velocity vector in a coordinate system in which east is the x-axis, north is the y-axis, and up is the z-axis.

11. A plane is flying due east and climbing at the rate of 50 mph. If its airspeed is 300 mph and there is a wind blowing 70 mph to the northeast, what is the ground speed of the plane?

12. An object is moving counterclockwise at a constant speed around the circle $x^2 + y^2 = 1$, where x and y are measured in meters. It completes one revolution every minute.
   (a) What is its speed?
   (b) What is its velocity vector 30 seconds after it passes the point $(1, 0)$? Does it change your answer if the object is moving clockwise? Explain.

13. A moving object has velocity vector $50\vec{i} + 20\vec{j}$ in feet per second. Express the velocity in miles per hour.

14. Suppose the current in Example 7 on page 81 is twice as fast and in the opposite direction. What is the speed of the boat with respect to the riverbed?

15. An airplane is heading northeast at an airspeed of 700 km/hr, but there is a wind blowing from the west at 60 km/hr. In what direction does the plane end up flying? What is its speed relative to the ground?

16. An airplane is flying at an airspeed of 600 km/hr in a cross-wind that is blowing from the northeast at a speed of 50 km/hr. In what direction should the plane head to end up going due east?

17. A particle moving with speed v hits a barrier at an angle of $60°$ and bounces off at an angle of $60°$ in the opposite direction with speed reduced by 20 percent, as shown in Figure 12.33.

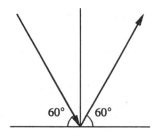

Figure 12.33

Find the velocity vector of the object after impact.

18. In the game of laser tag, you shoot a harmless laser gun and try to hit a target worn at the waist by other players. Suppose you are standing at the origin of a three dimensional coordinate system and that the xy-plane is the floor. Suppose that waist-high is 3 feet above floor level and that eye level is 5 feet above the floor. Three of your friends are your opponents. One is standing so that his target is 30 feet along the x-axis, the other lying down so that his target is at the point $x = 20, y = 15$, and the third lying in ambush so that his target is at a point 8 feet above the point $x = 12, y = 30$.

    (a) If you aim carefully with your gun at eye level, find the vector from your gun to each of the three targets.

    (b) If you try a quick draw and shoot approach (shoot from waist height, with your gun one foot to the right of the center of your body as you face along the x-axis), find the vector from your gun to each of the three targets.

Use the geometric definition of addition, subtraction, and scalar multiplication to explain each of the properties in Problems 19–25.

19. $\vec{u} + \vec{v} = \vec{v} + \vec{u}$

20. $(\alpha + \beta)\vec{u} = \alpha\vec{u} + \beta\vec{u}$

21. $\alpha(\vec{u} + \vec{v}) = \alpha\vec{u} + \alpha\vec{v}$

22. $(\vec{u} + \vec{v}) + \vec{w} = \vec{u} + (\vec{v} + \vec{w})$

23. $\alpha(\beta\vec{u}) = (\alpha\beta)\vec{u}$

24. $\vec{v} + \vec{0} = \vec{v}$

25. $1 \cdot \vec{v} = \vec{v}$

12.3 THE DOT PRODUCT

We have seen how to add vectors; can we multiply them? In the next two sections we will see two different ways of doing so: the *scalar product* (or *dot product*) which produces a scalar, and the *vector product* (or *cross product*) which produces a vector.

Definition of the Dot Product

In general, for any two vectors $\vec{v}$ and $\vec{w}$:

> The **dot product**, or **scalar product**, $\vec{v} \cdot \vec{w}$, is the scalar defined as follows:
> $$\vec{v} \cdot \vec{w} = \|\vec{v}\|\|\vec{w}\| \cos\theta.$$
> where θ is the angle between $\vec{v}$ and $\vec{w}$ and $0 \le \theta \le \pi$.

Example 1   Suppose the vector $\vec{b}$ is fixed and has length 2; the vector $\vec{a}$ is free to rotate and has length 3. What are the maximum and minimum values of the dot product $\vec{a} \cdot \vec{b}$ as the vector $\vec{a}$ rotates through all possible positions? What positions of $\vec{a}$ and $\vec{b}$ lead to these values?

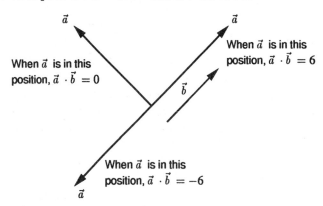

When $\vec{a}$ is in this position, $\vec{a} \cdot \vec{b} = 0$

When $\vec{a}$ is in this position, $\vec{a} \cdot \vec{b} = 6$

$\vec{b}$

When $\vec{a}$ is in this position, $\vec{a} \cdot \vec{b} = -6$

Figure 12.34: Fixed vector $\vec{b}$ of length 2 and rotating vector $\vec{a}$ of length 3

Solution   Since $\vec{a} \cdot \vec{b} = \|\vec{a}\|\|\vec{b}\| \cos\theta = 3 \cdot 2 \cos\theta = 6 \cos\theta$, the maximum value of $\vec{a} \cdot \vec{b}$ occurs when $\cos\theta = 1$ so $\theta = 0$, that is, when $\vec{a}$ and $\vec{b}$ point in the same direction. The maximum value of $\vec{a} \cdot \vec{b}$ is therefore 6. The minimum value of $\vec{a} \cdot \vec{b}$ occurs when $\cos\theta = -1$ so $\theta = \pi$, that is, when $\vec{a}$ and $\vec{b}$ point in opposite directions, so that the minimum value of $\vec{a} \cdot \vec{b}$ is therefore -6. See Figure 12.34.

A Physical Interpretation of the Dot Product: Work

In physics, the word "work" has a slightly different meaning from its everyday meaning. In physics, when a force of magnitude F acts on an object through a distance d, we say the work, W, done by the force is

$$W = Fd,$$

provided the force and the displacement are in the same direction. For example, if a 1 kg body falls 10 meters under the force of gravity, which is 9.8 newtons, then the work done is

$$W = (9.8 \text{ newtons}) \cdot (10 \text{ meters}) = 98 \text{ joules}.$$

What if the force and the displacement are not in the same direction? If a force $\vec{F}$ acts on an object as it moves along a displacement vector $\vec{d}$, then, using Figure 12.35, we see that we can write $\vec{F}$ as a sum of two forces,

$$\vec{F} = \vec{F}_{\text{parallel}} + \vec{F}_{\text{perp}},$$

where $\vec{F}_{\text{parallel}}$ is parallel to $\vec{d}$ and $\vec{F}_{\text{perp}}$ is perpendicular to $\vec{d}$. Suppose that $0 \leq \theta \leq \pi/2$, where θ is the angle between $\vec{F}$ and $\vec{d}$. Then the work done by $\vec{F}$ is defined to be

$$W = \|\vec{F}_{\text{parallel}}\| \, \|\vec{d}\|.$$

In particular, if the force and the displacement are perpendicular, then no work is done in the technical definition of the word, since $\vec{F}_{\text{parallel}} = \vec{0}$. For example, if you carry a box across the room at the same horizontal height, no work is done by gravity because the force of gravity is vertical but the motion is horizontal.

We can interpret this definition in terms of the dot product. Figure 12.35 shows that $\vec{F}_{\text{parallel}}$ has magnitude $\|\vec{F}\| \cos \theta$. So

$$W = (\|\vec{F}\| \cos \theta) \|\vec{d}\| = \|\vec{F}\| \|\vec{d}\| \cos \theta = \vec{F} \cdot \vec{d}.$$

Thus we have the following result:

The **work** done by a force $\vec{F}$ acting on an object through a displacement $\vec{d}$ is the dot product, $\vec{F} \cdot \vec{d}$.

Notice that when the vectors are parallel and in the same direction, $\cos \theta = \cos 0 = 1$, so $W = \|\vec{F}\| \|\vec{d}\| = Fd$, which is the original definition. When the vectors are perpendicular, $\cos \theta = \cos \frac{\pi}{2} = 0$, and so $W = 0$. In fact, this formula works when $\pi/2 < \theta \leq \pi$ also. In that case the dot product is negative and the object is moving against the force. Thus, the work done by the force is negative.

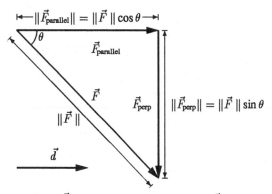

Figure 12.35: Replacing force $\vec{F}$ by two forces, one parallel to $\vec{d}$, one perpendicular to $\vec{d}$

Properties of the Dot Product

The dot product has the following properties in common with ordinary multiplication of scalars:

> For vectors $\vec{u}$, $\vec{v}$, and $\vec{w}$ and a scalar λ,
> 1. $\vec{v} \cdot (\lambda \vec{w}) = \lambda(\vec{v} \cdot \vec{w}) = (\lambda \vec{v}) \cdot \vec{w}$
> 2. $\vec{v} \cdot \vec{w} = \vec{w} \cdot \vec{v}$
> 3. $(\vec{v} + \vec{w}) \cdot \vec{u} = \vec{v} \cdot \vec{u} + \vec{w} \cdot \vec{u}$.

Property 1 says that the dot product is commutative, i.e., you can calculate the dot product of two vectors in any order.

Property 2 says that multiplying one of the vectors by a scalar simply multiplies the dot product by the same scalar. If $\lambda > 0$, then when one vector is multiplied by λ, the angle between the vectors does not change, but the length of one vector, and hence the dot product, is multiplied by λ. For the case when $\lambda < 0$, see Problem 34 on page 94.

Property 3 says that the dot product is distributive over vector addition. For a justification, see Problem 35 on page 94.

Perpendicularity, Magnitude, and Dot Products

Two vectors are perpendicular if the angle between them is $\pi/2$ or $90°$. Since $\cos(\pi/2) = 0$, if $\vec{v}$ and $\vec{w}$ are perpendicular, then $\vec{v} \cdot \vec{w} = 0$. Conversely, provided $\vec{v}$ and $\vec{w}$ do not have magnitude 0, the dot product $\vec{v} \cdot \vec{w}$ is zero when $\cos \theta = 0$, so $\theta = \pi/2$ and the vectors are perpendicular. Thus, we have the following result:

> Two nonzero vectors $\vec{v}$ and $\vec{w}$ are **perpendicular** if and only if
> $$\vec{v} \cdot \vec{w} = 0.$$

For example: $\vec{i} \cdot \vec{j} = 0, \vec{j} \cdot \vec{k} = 0, \vec{i} \cdot \vec{k} = 0$.

If we take the dot product of a vector with itself, then $\theta = 0$ and $\cos \theta = 1$. For any vector $\vec{v}$:

$$\vec{v} \cdot \vec{v} = \|\vec{v}\|^2.$$

For example: $\vec{i} \cdot \vec{i} = 1, \vec{j} \cdot \vec{j} = 1, \vec{k} \cdot \vec{k} = 1$.

Calculation of the Dot Product Using Components

How do we compute the dot product of two vectors $\vec{v}$ and $\vec{w}$ that are resolved into components? If $\vec{v} = v_1 \vec{i} + v_2 \vec{j}$ and $\vec{w} = w_1 \vec{i} + w_2 \vec{j}$ are both vectors in the plane, then using the properties of the dot product:

$$\begin{aligned}
\vec{v} \cdot \vec{w} &= (v_1 \vec{i} + v_2 \vec{j}) \cdot (w_1 \vec{i} + w_2 \vec{j}) \\
&= v_1 w_1 \vec{i} \cdot \vec{i} + v_1 w_2 \vec{i} \cdot \vec{j} + v_2 w_1 \vec{j} \cdot \vec{i} + v_2 w_2 \vec{j} \cdot \vec{j} \\
&= v_1 w_1(1) + v_1 w_2(0) + v_2 w_1(0) + v_2 w_2(1) \\
&= v_1 w_1 + v_2 w_2.
\end{aligned}$$

For example, if $\vec{v} = 3\vec{i} + 4\vec{j}$ and $\vec{w} = 2\vec{i} - \vec{j}$, then $\vec{v} \cdot \vec{w} = 3 \cdot 2 + 4(-1) = 2$.

A similar calculation gives the following result in 3-dimensions:

If $\vec{v} = v_1\vec{i} + v_2\vec{j} + v_3\vec{k}$ and $\vec{w} = w_1\vec{i} + w_2\vec{j} + w_3\vec{k}$, then
$$\vec{v} \cdot \vec{w} = v_1w_1 + v_2w_2 + v_3w_3.$$

Example 2 Which pairs from the following list of 3-dimensional vectors are perpendicular to one another?

$$\vec{u} = \vec{i} + \sqrt{3}\,\vec{k}, \quad \vec{v} = \vec{i} + \sqrt{3}\,\vec{j}, \quad \vec{w} = \sqrt{3}\,\vec{i} + \vec{j} - \vec{k}.$$

Solution Two vectors are perpendicular if and only if their dot product is zero. So we calculate dot products:
$$\vec{v} \cdot \vec{u} = (\vec{i} + \sqrt{3}\,\vec{j} + 0\vec{k}) \cdot (\vec{i} + 0\vec{j} + \sqrt{3}\,\vec{k}) = 1 \cdot 1 + \sqrt{3} \cdot 0 + 0 \cdot \sqrt{3} = 1,$$
$$\vec{v} \cdot \vec{w} = (\vec{i} + \sqrt{3}\,\vec{j} + 0\vec{k}) \cdot (\sqrt{3}\,\vec{i} + \vec{j} - \vec{k}) = 1 \cdot \sqrt{3} + \sqrt{3} \cdot 1 + 0(-1) = 2\sqrt{3},$$
$$\vec{w} \cdot \vec{u} = (\sqrt{3}\,\vec{i} + \vec{j} - \vec{k}) \cdot (\vec{i} + 0\vec{j} + \sqrt{3}\,\vec{k}) = \sqrt{3} \cdot 1 + 1 \cdot 0 + (-1) \cdot \sqrt{3} = 0.$$

So the only two vectors which are perpendicular are $\vec{w}$ and $\vec{u}$.

Example 3 A video store sells videos, tapes, CD's, and computer games. We define the quantity vector $\vec{q} = (q_1, q_2, q_3, q_4)$, where q_1, q_2, q_3, q_4 denote the quantities sold of each of the goods, and the price vector $\vec{p} = (p_1, p_2, p_3, p_4)$, where p_1, p_2, p_3, p_4 denote the price per unit of each good. What does the dot product $\vec{p} \cdot \vec{q}$ represent?

Solution The dot product is $\vec{p} \cdot \vec{q} = p_1q_1 + p_2q_2 + p_3q_3 + p_4q_4$. The quantity p_1q_1 represents the revenue received by the store for the videos, p_2q_2 represents the revenue for the tapes, and so on. The dot product represents the total revenue received by the store for the sale of these four goods.

Uses of the Dot Product

Normal Vectors and the Equation of a Plane

A *normal vector* to a plane is a vector that is perpendicular to the plane. In Section 11.5 we saw how to write the equation of a plane, given its x-slope, y-slope and z-intercept. Now we will see how to do it using normal vectors and dot products. Let $\vec{n} = a\vec{i} + b\vec{j} + c\vec{k}$ be a normal vector, let (x_0, y_0, z_0) be a fixed point P_0 in the plane, and let (x, y, z) be any other point P in the plane. Then $\overrightarrow{P_0P} = (x - x_0)\vec{i} + (y - y_0)\vec{j} + (z - z_0)\vec{k}$ is a vector whose head and tail both lie in the plane. (See Figure 12.36.) Thus, the vectors $\vec{n}$ and $\overrightarrow{P_0P}$ are perpendicular, so $\vec{n} \cdot \overrightarrow{P_0P} = 0$. Therefore,

$$\vec{n} \cdot \overrightarrow{P_0P} = (a\vec{i} + b\vec{j} + c\vec{k}) \cdot ((x - x_0)\vec{i} + (y - y_0)\vec{j} + (z - z_0)\vec{k}) = 0.$$

Thus, we obtain the following result:

The **equation of the plane** with normal vector $\vec{n} = a\vec{i} + b\vec{j} + c\vec{k}$ and containing the point (x_0, y_0, z_0) is
$$\vec{n} \cdot \overrightarrow{P_0P} = a(x - x_0) + b(y - y_0) + c(z - z_0) = 0.$$

Letting $d = ax_0 + by_0 + cz_0$ (a constant), we can write the equation of the plane in the form
$$ax + by + cz = d.$$

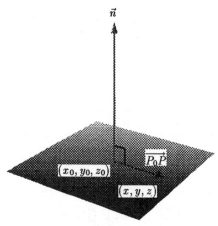

Figure 12.36: Plane with normal $\vec{n}$ and
containing a fixed point (x_0, y_0, z_0)

Notice that the coefficients of x, y, z are the same as those of $\vec{i}$, $\vec{j}$, $\vec{k}$.

Example 4   Find the equation of the plane perpendicular to $-\vec{i} + 3\vec{j} + 2\vec{k}$ and passing through the point $(1, 0, 4)$.

Solution   The equation is

$$-(x - 1) + 3(y - 0) + 2(z - 4) = 0,$$

which simplifies to

$$-x + 3y + 2z = 7.$$

Example 5   Find a normal vector to the plane with equation (a) $\quad x - y + 2z = 5$ $\qquad$ (b) $\quad z = 0.5x + 1.2y$.

Solution   (a)   Since the coefficients of $\vec{i}$, $\vec{j}$, and $\vec{k}$ in a normal vector are the coefficients of x, y, z in the
equation of the plane, a normal vector is $\vec{n} = \vec{i} - \vec{j} + 2\vec{k}$.

(b)   Before we can find a normal vector, we rewrite the equation of the plane in the form

$$0.5x + 1.2y - z = 0.$$

Then a normal vector is $\vec{n} = 0.5\vec{i} + 1.2\vec{j} - \vec{k}$.

Projections of Vectors: Components in Arbitrary Directions

When we defined the work, W, done by a force $\vec{F}$ over a displacement $\vec{d}$, we used the
decomposition:

$$\vec{F} = \vec{F}_{\text{perp}} + \vec{F}_{\text{parallel}},$$

where $\vec{F}_{\text{perp}}$ is perpendicular to $\vec{d}$ and $\vec{F}_{\text{parallel}}$ is parallel to $\vec{d}$. (See Figure 12.37.) It is useful to
be able to decompose $\vec{F}$ this way when $\vec{F}$ and $\vec{d}$ are any vectors, and $\vec{d} \neq \vec{0}$. So now we express
$\vec{F}_{\text{perp}}$ and $\vec{F}_{\text{parallel}}$ in terms of dot products. Suppose $0 \leq \theta \leq \pi/2$, as in Figure 12.37. Then

$$\|\vec{F}_{\text{parallel}}\| = \|\vec{F}\| \cos\theta.$$

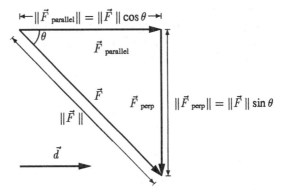

Figure 12.37: Expressing a vector $\vec{F}$ as a sum of two vectors, one parallel to $\vec{d}$, one perpendicular to $\vec{d}$

If $\vec{d} \neq \vec{0}$, we know that $\vec{d}/\|\vec{d}\|$ is a unit vector parallel to $\vec{d}$, so we have

$$\vec{F}_{\text{parallel}} = \left(\|\vec{F}\| \cos\theta\right) \frac{\vec{d}}{\|\vec{d}\|},$$

since $\vec{F}_{\text{parallel}}$ is a vector of length $\|\vec{F}\| \cos\theta$ in the direction of $\vec{d}$. Multiplying top and bottom by $\|\vec{d}\|$, we get

$$\vec{F}_{\text{parallel}} = \left(\frac{\|\vec{F}\|\|\vec{d}\| \cos\theta}{\|\vec{d}\|}\right) \frac{\vec{d}}{\|\vec{d}\|} = \left(\frac{\vec{F} \cdot \vec{d}}{\|\vec{d}\|}\right) \frac{\vec{d}}{\|\vec{d}\|} = \left(\frac{\vec{F} \cdot \vec{d}}{\|\vec{d}\|^2}\right) \vec{d}.$$

Thus, we have the following result:

Given two vectors $\vec{F}$ and $\vec{d}$, where $\vec{d} \neq \vec{0}$, we can write

$$\vec{F} = \vec{F}_{\text{parallel}} + \vec{F}_{\text{perp}},$$

where $\vec{F}_{\text{parallel}}$ is parallel to $\vec{d}$ and $\vec{F}_{\text{perp}}$ is perpendicular to $\vec{d}$. We have

$$\vec{F}_{\text{parallel}} = \left(\frac{\vec{F} \cdot \vec{d}}{\|\vec{d}\|^2}\right) \vec{d}$$

and

$$\vec{F}_{\text{perp}} = \vec{F} - \vec{F}_{\text{parallel}}.$$

We call $\vec{F}_{\text{parallel}}$ the *component of $\vec{F}$ parallel to $\vec{d}$*, and $\vec{F}_{\text{perp}}$ the *component of $\vec{F}$ perpendicular to $\vec{d}$*. In fact, the formulas for $\vec{F}_{\text{parallel}}$ and $\vec{F}_{\text{perp}}$ are valid when $\pi/2 < \theta \leq \pi$ also. In that case the dot product $\vec{F} \cdot \vec{d}$ is negative and $\vec{F}_{\text{parallel}}$ points in the opposite direction to $\vec{d}$.

Example 6   Figure 12.38 shows the force the wind exerts on the sail of a sailboat. Find the component of the force in the direction in which the sailboat is traveling.

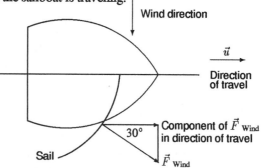

Figure 12.38: Wind moving a sailboat

Solution   Let $\vec{u}$ be a unit vector in the direction of travel. The force of the wind on the sail makes an angle of 30° with the heading of the boat. Thus, the component of this force along the heading is

$$\left(\frac{\vec{F} \cdot \vec{u}}{\|\vec{u}\|^2}\right)\vec{u} = \|F\|(\cos 30°)\vec{u} = 0.87\|\vec{F}\|\vec{u}.$$

Thus, the boat is being pushed forward with about 87% of the total force due to the wind. (In fact, the interaction of wind and sail is much more complex than our simple model suggests.)

Problems for Section 12.3

For Problems 1–3, perform the following operations on the given 3-dimensional vectors.

$$\vec{a} = 2\vec{j} + \vec{k} \qquad \vec{b} = -3\vec{i} + 5\vec{j} + 4\vec{k} \qquad \vec{c} = \vec{i} + 6\vec{j} \qquad \vec{y} = 4\vec{i} - 7\vec{j} \qquad \vec{z} = \vec{i} - 3\vec{j} - \vec{k}$$

1. $\vec{c} \cdot \vec{y}$ 2. $\vec{a} \cdot \vec{z}$ 3. $\vec{a} \cdot \vec{b}$

4. Compute the angle between the vectors $\vec{i} + \vec{j} + \vec{k}$ and $\vec{i} - \vec{j} - \vec{k}$.

5. Which pairs of the vectors $\sqrt{3}\vec{i} + \vec{j}$, $3\vec{i} + \sqrt{3}\vec{j}$, $\vec{i} - \sqrt{3}\vec{j}$ are parallel and which are perpendicular?

6. For what values of t are $\vec{u} = t\vec{i} - \vec{j} + \vec{k}$ and $\vec{v} = t\vec{i} + t\vec{j} - 2\vec{k}$ perpendicular? Are there values of t for which $\vec{u}$ and $\vec{v}$ are parallel?

In Problems 7–12, find a normal vector to the given plane.

7. $\pi(x - 1) = (1 - \pi)(y - z) + \pi$ 8. $2x + y - z = 23$

9. $1.5x + 3.2y + z = 0$ 10. $z = 3x + 4y - 7$

11. $2(x - z) = 3(x + y)$ 12. $2x + y - z = 5$

In Problems 13–17, find the equation of a plane that satisfies the given conditions.

13. Perpendicular to $-\vec{i} + 2\vec{j} + \vec{k}$ and passing through the point $(1, 0, 2)$.

14. Perpendicular to $5\vec{i} + \vec{j} - 2\vec{k}$ and passing through the point $(0, 1, -1)$.

15. Perpendicular to the vector $2\vec{i} - 3\vec{j} + 7\vec{k}$ and passing through the point $(1, -1, 2)$.

16. Parallel to $2x + 4y - 3z = 1$ and through the point $(1, 0, -1)$.

17. Through the point $(-2, 3, 2)$ and parallel to $3x + y + z = 4$.

18. (a) Find a vector perpendicular to the plane $z = 2 + 3x - y$.
    (b) Find a vector parallel to the plane $z = 2 + 3x - y$.

19. A plane wants to head due east. In the following situations the plane's airspeed (speed relative to the air) and the wind speed are given. For each case, find the direction in which the plane should head and its speed relative to the ground.

    (a) The airspeed is 300 mph; wind is 70 mph blowing in the direction of 30° north of east.
    (b) The airspeed is 400 mph; wind is 100 mph blowing in the direction of 45° north of west.

20. A plane is headed due east flying at 400 mph relative to the air. The wind is blowing as follows. In each case, find the resulting direction of the plane and its speed relative to the ground.

    (a) 60 mph blowing towards the north.
    (b) 100 mph blowing in the direction of 60° south of east.

21. Let S be the triangle with vertices $A = (2, 2, 2)$, $B = (4, 2, 1)$, and $C = (2, 3, 1)$.

    (a) Find the length of the shortest side of S.
    (b) Find the cosine of the angle BAC at vertex A.

22. The points $(5, 0, 0)$, $(0, -3, 0)$, and $(0, 0, 2)$ form a triangle. Find the lengths of the sides of the triangle and each of its angles.

23. Write $\vec{a} = 3\vec{i} + 2\vec{j} - 6\vec{k}$ as the sum of two vectors, one parallel to $\vec{d} = 2\vec{i} - 4\vec{j} + \vec{k}$ and the other perpendicular to $\vec{d}$.

24. Find the points where the plane
$$z = 5x - 4y + 3$$
intersects each of the coordinate axes. Then find the lengths of the sides and the angles of the triangle formed by these points.

25. Find the angle between the planes $5(x-1)+3(y+2)+2z = 0$, and $x+3(y-1)+2(z+4) = 0$.

26. Show that the vectors $(\vec{b} \cdot \vec{c})\vec{a} - (\vec{a} \cdot \vec{c})\vec{b}$ and $\vec{c}$ are perpendicular.

27. The cosine law, for a triangle with side lengths a, b, and c, and with angle C opposite side c, says
$$c^2 = a^2 + b^2 - 2ab\cos C.$$
Use the properties of the dot product to prove this law. [Hint: Let $\vec{u}$ and $\vec{v}$ be the displacement vectors from C to the other two vertices, and express c^2 in terms of $\vec{u}$ and $\vec{v}$.]

28. (a) If a vector $\vec{v}$ of magnitude v makes an angle α with the positive x-axis, angle β with the positive y-axis, and angle γ with the positive z-axis, then show that $\vec{v} = v\cos\alpha\vec{i} + v\cos\beta\vec{j} + v\cos\gamma\vec{k}$.
    (b) Show that $\cos^2\alpha + \cos^2\beta + \cos^2\gamma = 1$.

29. Find the vector $\vec{v}$ with all of the following properties:

    (i) magnitude 10
    (ii) angle of 45° with positive x-axis
    (iii) angle of 75° with positive y-axis
    (iv) positive $\vec{k}$-component.

30. Show that if $\vec{u}$ and $\vec{v}$ are two vectors such that

$$\vec{u} \cdot \vec{w} = \vec{v} \cdot \vec{w}$$

for every vector $\vec{w}$, then

$$\vec{u} = \vec{v}.$$

31. A consumption vector of three goods is defined by $\vec{x} = (x_1, x_2, x_3)$, where x_1, x_2 and x_3 are the quantities consumed of the three goods. Consider a budget constraint represented by the equation $\vec{p} \cdot \vec{x} = k$, where $\vec{p}$ is the price vector of the three goods and k is a constant. Show that the difference between two consumption vectors corresponding to points satisfying the same budget constraint are perpendicular to the price vector $\vec{p}$.

32. A 100-meter dash is run on a track in the direction of the vector $\vec{v} = 2\vec{i} + 5\vec{j}$. The wind velocity $\vec{w}$ is $4\vec{i} + \vec{j}$ mph. The rules say that a legal wind speed measured in the direction of the dash must not exceed 3 mph. Will the race results be disqualified due to an illegal wind? Justify your answer.

33. A room is 10 feet high, 12 feet wide, and 20 feet long. Two strings, one from each of the two corners of the ceiling at one end of the room are stretched to the diagonally opposite corners on the floor. What is the cosine of the angle made by the strings as they cross?

34. (a) Using

$$\vec{u} \cdot \vec{v} = \|\vec{u}\| \|\vec{v}\| \cos\theta,$$

show that

$$\vec{u} \cdot (-\vec{v}) = -(\vec{u} \cdot \vec{v}).$$

(Hint: What happens to the angle when you multiply $\vec{v}$ by -1?)

(b) Show for any negative scalar λ that

$$\vec{u} \cdot (\lambda\vec{v}) = \lambda(\vec{u} \cdot \vec{v})$$
$$(\lambda\vec{u}) \cdot \vec{v} = \lambda(\vec{u} \cdot \vec{v}).$$

35. Figure 12.39 shows that, given three vectors $\vec{u}$, $\vec{v}$, and $\vec{w}$, the sum of the components of $\vec{v}$ and $\vec{w}$ in the direction of $\vec{u}$ is the component of $\vec{v} + \vec{w}$ in the direction of $\vec{u}$. (Although the figure is drawn in two dimensions, this result is also true in three dimensions.)

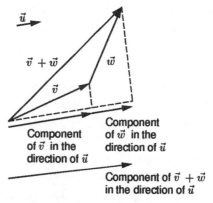

Figure 12.39: The component of $\vec{v} + \vec{w}$ in the direction of $\vec{u}$ is the sum of the components of $\vec{v}$ and $\vec{w}$ in that direction

Use this fact to explain why $(\vec{v} + \vec{w}) \cdot \vec{u} = \vec{v} \cdot \vec{u} + \vec{w} \cdot \vec{u}$.

36. Recall that in 2 or 3 dimensions, the dot product is

$$\vec{v} \cdot \vec{w} = \|\vec{v}\|\|\vec{w}\| \cos\theta,$$

where θ is the angle between $\vec{v}$ and $\vec{w}$. The result of Problem 37 enables us to use this relationship to define the angle between two vectors in n dimensions. If $\vec{v}$, $\vec{w}$ are n-vectors, then

$$\vec{v} \cdot \vec{w} = v_1 w_1 + v_2 w_2 + \cdots + v_n w_n$$

and we define the angle θ by

$$\cos\theta = \frac{\vec{v} \cdot \vec{w}}{\|\vec{v}\|\|\vec{w}\|} \quad \text{provided } \|\vec{v}\|, \|\vec{w}\| \neq 0.$$

Problem 37 shows that the right hand side of this equation is between -1 and 1, so this makes sense. We now use this idea of angle to measure how close two populations are to one another genetically. The following table shows the relative frequencies of four alleles (variants of a gene) in four populations.

TABLE 12.1

Allele	Eskimo	Bantu	English	Korean
A_1	0.29	0.10	0.20	0.22
A_2	0.00	0.08	0.06	0.00
B	0.03	0.12	0.06	0.20
O	0.67	0.69	0.66	0.57

Let $\vec{a}_1$ be the 4-vector showing relative frequencies in the Eskimo population;

$\vec{a}_2$ be the 4-vector showing relative frequencies in the Bantu population;

$\vec{a}_3$ be the 4-vector showing relative frequencies in the English population;

$\vec{a}_4$ be the 4-vector showing relative frequencies in the Korean population.

The genetic distance between two populations is defined as the angle between the corresponding vectors. Using this definition, is the English population closer genetically to the Bantus or to the Koreans? Explain.[2]

37. Suppose that $\vec{v}$ and $\vec{w}$ are n-dimensional vectors. Consider the following function of t:

$$q(t) = (\vec{v} + t\vec{w}) \cdot (\vec{v} + t\vec{w})$$

(a) Explain why $q(t) \geq 0$ for all real t.
(b) Expand $q(t)$ as a quadratic polynomial in t.
(c) Using the discriminant of the quadratic, show that,

$$|\vec{v} \cdot \vec{w}| \leq \|\vec{v}\|\|\vec{w}\|.$$

---

[2]Adapted from Cavalli-Sforza and Edwards, "Models and Estimation Procedures," Am J. Hum. Genet., Vol. 19 (1967), pp. 223-57.

12.4 THE CROSS PRODUCT

In this section we will see how to multiply two vectors $\vec{v}$ and $\vec{w}$ to produce a third vector, $\vec{v} \times \vec{w}$, called the **cross product**, or **vector product**, of $\vec{v}$ and $\vec{w}$.

Definition of the Cross Product

We will define the cross product, $\vec{v} \times \vec{w}$, of two vectors $\vec{v}$ and $\vec{w}$ to be perpendicular to both of them. First we will specify the direction. If, for example, $\vec{v}$ and $\vec{w}$ are in the xy-plane, there are two possible perpendicular directions—vertically upward and vertically downward. To choose the correct direction we use the following rule.

> **The right-hand rule:** Place $\vec{v}$ and $\vec{w}$ so that their tails coincide and curl the fingers of your right hand in an arc from $\vec{v}$ to $\vec{w}$; your thumb points in the direction of $\vec{v} \times \vec{w}$.

Now we define the cross product as follows.

> **The cross product, $\vec{v} \times \vec{w}$**, is the vector whose magnitude is
>
> $$\|\vec{v} \times \vec{w}\| = \|\vec{v}\|\|\vec{w}\| \sin\theta,$$
>
> where $0 \leq \theta \leq \pi$ is the angle between $\vec{v}$ and $\vec{w}$, and which points in the direction perpendicular to $\vec{v}$ and $\vec{w}$ determined by the right-hand rule. Notice that if $\vec{n}$ is the unit vector in the direction of $\vec{v} \times \vec{w}$, then
>
> $$\vec{v} \times \vec{w} = \|\vec{v}\|\|\vec{w}\| \sin\theta\, \vec{n}.$$

See Figure 12.40. Note that the cross product of $\vec{v}$ and $\vec{w}$ is a vector, while the dot product $\vec{v} \cdot \vec{w}$ is a scalar. Unlike the dot product, the cross product is only defined for three-dimensional vectors.

Example 1 Find $\vec{i} \times \vec{j}$ and $\vec{j} \times \vec{i}$.

Solution The vectors $\vec{i}$ and $\vec{j}$ both have magnitude 1 and the angle between them is $\pi/2$. By the right-hand rule, the direction of $\vec{i} \times \vec{j}$ is $\vec{k}$, so we have

$$\vec{i} \times \vec{j} = \left(\|\vec{i}\|\|\vec{j}\| \sin\frac{\pi}{2}\right) \vec{k} = \vec{k}.$$

Similarly, the right-hand rule says that the direction of $\vec{j} \times \vec{i}$ is $-\vec{k}$, so

$$\vec{j} \times \vec{i} = (\|\vec{j}\|\|\vec{i}\| \sin(\frac{\pi}{2}))(-\vec{k}) = -\vec{k}.$$

Example 2 For any vector $\vec{v}$, find $\vec{v} \times \vec{v}$.

Solution Since the angle between $\vec{v}$ and $\vec{v}$ is 0 and $\sin 0 = 0$, we have

$$\vec{v} \times \vec{v} = \vec{0}.$$

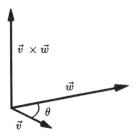

Figure 12.40: The cross
product $\vec{v} \times \vec{w}$

Example 3   Suppose $\vec{a}$ is a fixed vector of length 3 in the direction of the positive x-axis and the vector $\vec{b}$
of length 2 is free to rotate in the xy-plane. What are the maximum and minimum values of the
magnitude of $\vec{a} \times \vec{b}$? In what direction is $\vec{a} \times \vec{b}$ as $\vec{b}$ rotates?

Solution   The magnitude of $\vec{a} \times \vec{b}$ is given by

$$\|\vec{a} \times \vec{b}\| = \|\vec{a}\|\|\vec{b}\| \sin\theta = 3 \cdot 2 \sin\theta = 6 \sin\theta.$$

Therefore, the maximum possible value of $\|\vec{a} \times \vec{b}\|$ occurs when $\sin\theta = 1$ so $\theta = \pi/2$; that is, when
$\vec{a}$ and $\vec{b}$ are perpendicular. The maximum value of $\|\vec{a} \times \vec{b}\|$ is 6. The minimum value of $\|\vec{a} \times \vec{b}\|$
occurs when $\sin\theta = 0$ so $\theta = 0$ or π; that is, when $\vec{a}$ and $\vec{b}$ are parallel. Then $\|\vec{a} \times \vec{b}\|$ is 0.

   The direction of $\vec{a} \times \vec{b}$ will be along the positive z-axis when $\vec{b}$ is in the first or second quadrant
and along the negative z-axis when $\vec{b}$ is in the third or fourth quadrant. See Figure 12.41.

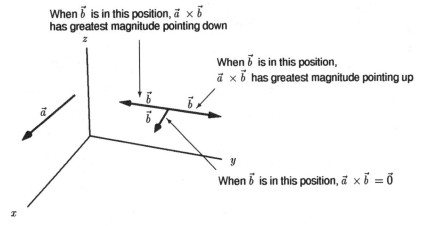

Figure 12.41: A fixed vector $\vec{a}$ and a rotating vector $\vec{b}$

A Physical Interpretation of the Cross Product: Torque

Consider the difference between hammering a nail and twisting a screw into the top of a table. Both
involve applying a force, but in two different directions. In the case of the hammer, the force and
the movement of the nail are both vertical. In the case of the screw, the force exerted is horizontal

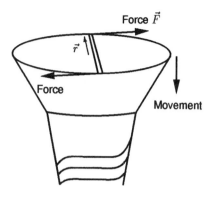

Figure 12.42: Motion of screw is perpendicular to both $\vec{F}$ and $\vec{r}$

but the movement of the screw is perpendicular to both the force, $\vec{F}$, applied to the screw (see Figure 12.42) and to the radius vector $\vec{r}$, that is, it is parallel to $\vec{F} \times \vec{r}$.

In general, a force acting on a body can have both a pushing tendency and a tendency to rotate the body. If a force $\vec{F}$ acts at a point with position vector $\vec{r}$, then the rotational tendency of the force about the origin is measured by the cross product $\vec{F} \times \vec{r}$, which is called the *torque* or *moment* of the force about the origin. The torque is a vector whose magnitude measures the rotational tendency of the force and whose direction is the axis of rotation. Notice that the magnitude of the torque grows larger when the point is further away from the origin. This makes sense; the same force applied at a greater distance has more effect. For example, a pipe wrench is effective in turning very stiff pipes because it has a long handle, which enables the plumber to apply a force at a distance from the axis of rotation (in this case, the pipe).

Properties of the Cross Product

The right-hand rule shows that $\vec{v} \times \vec{w}$ and $\vec{w} \times \vec{v}$ point in opposite directions. The magnitudes of $\vec{v} \times \vec{w}$ and $\vec{w} \times \vec{v}$ are the same, so $\vec{w} \times \vec{v} = -\vec{v} \times \vec{w}$ (see Figure 12.43).

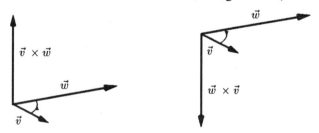

Figure 12.43: Diagram showing $\vec{v} \times \vec{w} = -\vec{w} \times \vec{v}$

This explains the second of the properties listed below. The other two are derived in Problems 22 and 25 at the end of this section.

> For vectors $\vec{u}$, $\vec{v}$, $\vec{w}$ and scalar λ
> 1. $(\lambda \vec{v}) \times \vec{w} = \lambda(\vec{v} \times \vec{w}) = \vec{v} \times (\lambda \vec{w})$
> 2. $\vec{w} \times \vec{v} = -\vec{v} \times \vec{w}$
> 3. $\vec{u} \times (\vec{v} + \vec{w}) = \vec{u} \times \vec{v} + \vec{u} \times \vec{w}$.

Calculation of the Cross Product Using Components

In order to use the cross product in practice, we need to be able to calculate $\vec{v} \times \vec{w}$ when $\vec{v}$ and $\vec{w}$ are given in components. We have already shown that $\vec{i} \times \vec{j} = \vec{k}$ and $\vec{i} \times \vec{i} = \vec{0}$. By similar arguments, we can show $\vec{j} \times \vec{i} = -\vec{k}$, $\vec{k} \times \vec{i} = \vec{j}$, $\vec{j} \times \vec{k} = \vec{i}$, $\vec{j} \times \vec{j} = \vec{0}$, and so on. Thus, we have

$$
\begin{aligned}
\vec{v} \times \vec{w} &= (v_1\vec{i} + v_2\vec{j} + v_3\vec{k}) \times (w_1\vec{i} + w_2\vec{j} + w_3\vec{k}) \\
&= v_1 w_1 \vec{i} \times \vec{i} + v_1 w_2 \vec{i} \times \vec{j} + v_1 w_3 \vec{i} \times \vec{k} \\
&\quad + v_2 w_1 \vec{j} \times \vec{i} + v_2 w_2 \vec{j} \times \vec{j} + v_2 w_3 \vec{j} \times \vec{k} \\
&\quad + v_3 w_1 \vec{k} \times \vec{i} + v_3 w_2 \vec{k} \times \vec{j} + v_3 w_3 \vec{k} \times \vec{k} \\
&= \vec{0} + v_1 w_2 \vec{k} + v_1 w_3(-\vec{j}) + v_2 w_1(-\vec{k}) + \vec{0} + v_2 w_3 \vec{i} + v_3 w_1 \vec{j} + v_3 w_2(-\vec{i}) + \vec{0} \\
&= (v_2 w_3 - v_3 w_2)\vec{i} + (v_3 w_1 - v_1 w_3)\vec{j} + (v_1 w_2 - v_2 w_1)\vec{k}.
\end{aligned}
$$

This expression can be more easily remembered by writing it as a determinant:

$$
\vec{v} \times \vec{w} = \begin{vmatrix} \vec{i} & \vec{j} & \vec{k} \\ v_1 & v_2 & v_3 \\ w_1 & w_2 & w_3 \end{vmatrix} = (v_2 w_3 - v_3 w_2)\vec{i} - (v_1 w_3 - v_3 w_1)\vec{j} + (v_1 w_2 - v_2 w_1)\vec{k}.
$$

The properties of determinants are outlined in Appendix C on page 458.

Example 4   Find the cross product of $\vec{v} = 2\vec{i} + \vec{j} - 2\vec{k}$ and $\vec{w} = 3\vec{i} + \vec{k}$ and check that it is perpendicular to both $\vec{v}$ and $\vec{w}$.

Solution

$$
\vec{v} \times \vec{w} = \begin{vmatrix} \vec{i} & \vec{j} & \vec{k} \\ 2 & 1 & -2 \\ 3 & 0 & 1 \end{vmatrix}.
$$

Writing $\vec{v} \times \vec{w}$ as a determinant and expanding it into three two-by-two determinants, we have

$$
\vec{v} \times \vec{w} = \begin{vmatrix} \vec{i} & \vec{j} & \vec{k} \\ 2 & 1 & -2 \\ 3 & 0 & 1 \end{vmatrix} = \vec{i} \begin{vmatrix} 1 & -2 \\ 0 & 1 \end{vmatrix} - \vec{j} \begin{vmatrix} 2 & -2 \\ 3 & 1 \end{vmatrix} + \vec{k} \begin{vmatrix} 2 & 1 \\ 3 & 0 \end{vmatrix}
$$

$$
= \vec{i}\,((1)(1) - 0(-2)) - \vec{j}\,(2(1) - 3(-2)) + \vec{k}\,(2(0) - 3(1))
$$

$$
= \vec{i} - 8\vec{j} - 3\vec{k}.
$$

To check that $\vec{v} \times \vec{w}$ is perpendicular to $\vec{v}$, we compute the dot product:

$$
\vec{v} \cdot (\vec{v} \times \vec{w}) = (2\vec{i} + \vec{j} - 2\vec{k}) \cdot (\vec{i} - 8\vec{j} - 3\vec{k}) = 2 - 8 + 6 = 0.
$$

Similarly,

$$
\vec{w} \cdot (\vec{v} \times \vec{w}) = (3\vec{i} + 0\vec{j} + \vec{k}) \cdot (\vec{i} - 8\vec{j} - 3\vec{k}) = 3 + 0 - 3 = 0.
$$

Uses of the Cross Product

The Equation of a Plane Through Three Points

The equation of a plane is determined by a point P_0 on it and a normal vector $\vec{n}$. If $P_0 = (x_0, y_0, z_0)$ and $\vec{n} = a\vec{i} + b\vec{j} + c\vec{k}$, we have seen that the equation is

$$a(x - x_0) + b(y - y_0) + c(z - z_0) = 0.$$

However a plane can also be determined by three points on it (provided they do not lie on the same line). In that case we find the equation of the plane by first determining two vectors in the plane and then finding a normal vector using the cross product, as in the following example.

Example 5   Find the equation of the plane containing the points $P = (1, 3, 0), Q = (3, 4, -3)$ and $R = (3, 6, 2)$.

Solution   Since the points P and Q are in the plane, the displacement vector between them, $\vec{v}$, is in the plane:

$$\vec{v} = \overrightarrow{PQ} = (3 - 1)\vec{i} + (4 - 3)\vec{j} + (-3 - 0)\vec{k} = 2\vec{i} + \vec{j} - 3\vec{k}.$$

The vector displacement $\vec{w} = \overrightarrow{PR}$ is also in the plane:

$$\vec{w} = \overrightarrow{PR} = (3 - 1)\vec{i} + (6 - 3)\vec{j} + (2 - 0)\vec{k} = 2\vec{i} + 3\vec{j} + 2\vec{k}.$$

Thus, a normal vector, $\vec{n}$, to the plane is given by

$$\vec{n} = \vec{v} \times \vec{w} = \begin{vmatrix} \vec{i} & \vec{j} & \vec{k} \\ 2 & 1 & -3 \\ 2 & 3 & 2 \end{vmatrix} = 11\vec{i} - 10\vec{j} + 4\vec{k}.$$

Since the point $(1, 3, 0)$ is on the plane, the equation of the plane is

$$11(x - 1) - 10(y - 3) + 4(z - 0) = 0,$$

which simplifies to

$$11x - 10y + 4z = -19.$$

You should check that P, Q, and R are actually on the plane.

Area and the Cross Product

Consider the parallelogram formed by the vectors $\vec{v}$ and $\vec{w}$, shown in Figure 12.44. The area of the parallelogram is given by

$$\text{Area} = \text{Base} \cdot \text{Height} = \|\vec{v}\| \cdot \|\vec{w}\| \sin\theta.$$

Thus, as shown in Figure 12.45, we have

$$\text{Area of the parallelogram} = \|\vec{v} \times \vec{w}\|.$$

We conclude that the cross product can be visualized as follows:

The cross product $\vec{v} \times \vec{w}$ is a vector with
- Magnitude equal to the area of the parallelogram determined by $\vec{v}$ and $\vec{w}$.
- Direction perpendicular to $\vec{v}$ and $\vec{w}$ and determined by the right-hand rule.

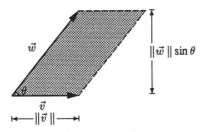

Figure 12.44: The parallelogram determined by $\vec{v}$ and $\vec{w}$

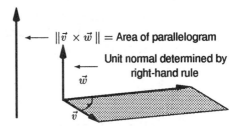

Figure 12.45: Area vector of parallelogram $= \vec{v} \times \vec{w}$

Example 6   Find the area of the parallelogram determined by $\vec{v} = 2\vec{i} + \vec{j} - 3\vec{k}$ and $\vec{w} = \vec{i} + 3\vec{j} + 2\vec{k}$.

Solution   We calculate the cross product:

$$\vec{v} \times \vec{w} = \begin{vmatrix} \vec{i} & \vec{j} & \vec{k} \\ 2 & 1 & -3 \\ 1 & 3 & 2 \end{vmatrix} = (2+9)\vec{i} - (4+3)\vec{j} + (6-1)\vec{k}$$

$$= 11\vec{i} - 7\vec{j} + 5\vec{k}.$$

The area of the parallelogram determined by $\vec{v}$ and $\vec{w}$ is the magnitude of the vector $\vec{v} \times \vec{w}$:

$$\text{Area} = \|\vec{v} \times \vec{w}\| = \sqrt{11^2 + (-7)^2 + 5^2} = \sqrt{195}.$$

Example 7   Show that $\left| \vec{a} \cdot (\vec{b} \times \vec{c}) \right|$ is the volume of the parallelepiped with sides formed by $\vec{a}$, $\vec{b}$, and $\vec{c}$.

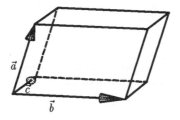

Figure 12.46

Solution   If we think of the base as formed by the vectors $\vec{b}$ and $\vec{c}$, we have

$$\text{Area of base} = \|\vec{b} \times \vec{c}\|.$$

The vectors $\vec{a}$, $\vec{b}$, and $\vec{c}$ can be arranged either as shown in Figure 12.47 (in which $\vec{b} \times \vec{c}$ and $\vec{a}$ have an angle of less than $\pi/2$ between them) or as in Figure 12.48 (where the angle is more than $\pi/2$). In either case,

$$\text{Height} = \|\vec{a}\| \cos \theta.$$

In the case shown in Figure 12.47 (in which $\vec{a}$, $\vec{b}$, and $\vec{c}$ are called a *right-handed* set), the triple product, $\vec{a} \cdot (\vec{b} \times \vec{c})$ is positive because the angle, θ, between $\vec{a}$ and $\vec{b} \times \vec{c}$ is less than $\pi/2$. Thus, in this case

$$\text{Volume} = \text{Base} \cdot \text{Height} = \|\vec{b} \times \vec{c}\| \cdot \|\vec{a}\| \cos \theta = (\vec{b} \times \vec{c}) \cdot \vec{a}.$$

For the case shown in Figure 12.48 (in which $\vec{a}$, $\vec{b}$, and $\vec{c}$ are called a *left-handed set*), the triple product $\vec{a} \cdot (\vec{b} \times \vec{c})$ is negative because the angle, $\pi - \theta$, between $\vec{a}$ and $\vec{b} \times \vec{c}$ is more than $\pi/2$. Thus, in this case we have

$$\text{Volume} = \text{Base} \cdot \text{Height} = \|\vec{b} \times \vec{c}\| \cdot \|\vec{a}\| \cos\theta = -\|\vec{b} \times \vec{c}\| \cdot \|\vec{a}\| \cos(\pi - \theta)$$
$$= -(\vec{b} \times \vec{c}) \cdot \vec{a} = \left| (\vec{b} \times \vec{c}) \cdot \vec{a} \right|.$$

Therefore, in both cases we have

$$\text{Volume} = \left| \vec{a} \cdot (\vec{b} \times \vec{c}) \right|.$$

Notice that the volume is given by the absolute value of the determinant:

$$\begin{vmatrix} a_1 & a_2 & a_3 \\ b_1 & b_2 & b_3 \\ c_1 & c_2 & c_3 \end{vmatrix}.$$

(See Problem 27 on page 104.)

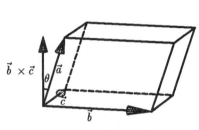

Figure 12.47: The vectors $\vec{a}$, $\vec{b}$, $\vec{c}$ are right-handed

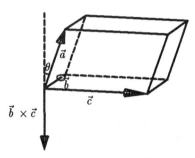

Figure 12.48: The vectors $\vec{a}$, $\vec{b}$, $\vec{c}$ are left-handed

Problems for Section 12.4

1. Find $\vec{k} \times \vec{j}$.

2. Does $\vec{i} \times \vec{i} = \vec{i} \cdot \vec{i}$? Explain your answer.

In Problems 3–6, find $\vec{a} \times \vec{b}$.

3. $\vec{a} = \vec{i} + \vec{k}$ and $\vec{b} = \vec{i} + \vec{j}$.

4. $\vec{a} = -\vec{i}$ and $\vec{b} = \vec{j} + \vec{k}$.

5. $\vec{a} = \vec{i} + \vec{j} + \vec{k}$ and $\vec{b} = \vec{i} + \vec{j} + -\vec{k}$.

6. $\vec{a} = 2\vec{i} - 3\vec{j} + \vec{k}$ and $\vec{b} = \vec{i} + 2\vec{j} - \vec{k}$.

7. You are using a jet pilot dogfight simulator. Your monitor tells you that two missiles have homed in on your plane along the directions $3\vec{i} + 5\vec{j} + 2\vec{k}$ and $\vec{i} - 3\vec{j} - 2\vec{k}$. In what direction should you turn to have the maximum choice of avoiding both missiles?

8. Show that if $\vec{r} = x\vec{i} + y\vec{j} + z\vec{k}$, then $\vec{k} \times \vec{r} = -y\vec{i} + x\vec{j}$.

9. Suppose $\vec{n}$ is a unit vector which is perpendicular to the vector $\vec{r} = x\vec{i} + y\vec{j} + z\vec{k}$, and $\|\vec{r}\| = a$. If $\vec{H} = -y\vec{i} + x\vec{j}$, show that

$$\vec{n} \cdot (\vec{r} \times \vec{H}) = a^2 \vec{n} \cdot \vec{k}.$$

10. Given $\vec{a} = 3\vec{i} + \vec{j} - \vec{k}$ and $\vec{b} = \vec{i} - 4\vec{j} + 2\vec{k}$, find $\vec{a} \times \vec{b}$ and check that $\vec{a} \times \vec{b}$ is perpendicular to both $\vec{a}$ and $\vec{b}$.

11. If $\vec{v} \times \vec{w} = 2\vec{i} - 3\vec{j} + 5\vec{k}$, and $\vec{v} \cdot \vec{w} = 3$, find $\tan \theta$ where θ is the angle between $\vec{v}$ and $\vec{w}$.

Find an equation of the plane through the points in Problems 12–13.

12. $(1, 0, 0), (0, 1, 0), (0, 0, 1)$.

13. $(3, 4, 2), (-2, 1, 0), (0, 2, 1)$.

14. Given the points $P = (0, 1, 0)$, $Q = (-1, 1, 2)$, $R = (2, 1, -1)$, find:
    (a) The area of the triangle PQR.
    (b) The equation for a plane that contains P, Q, and R.

15. Find a vector parallel to the intersection of the planes $2x - 3y + 5z = 2$ and $4x + y - 3z = 7$.

16. Find the equation of the plane through the origin which is perpendicular to the line of intersection of the planes in Problem 15.

17. Find the equation of the plane through the point $(4, 5, 6)$ which is perpendicular to the line of intersection of the planes in Problem 15.

18. Use the formula for the cross product in components to check that

$$\vec{a} \times (\vec{b} + \vec{c}) = (\vec{a} \times \vec{b}) + (\vec{a} \times \vec{c}).$$

19. In this problem, we approach the idea of finding the cross product algebraically. Let $\vec{a} = a_1\vec{i} + a_2\vec{j} + a_3\vec{k}$ and $\vec{b} = b_1\vec{i} + b_2\vec{j} + b_3\vec{k}$. We seek a vector $\vec{v} = x\vec{i} + y\vec{j} + z\vec{k}$ which is perpendicular to both $\vec{a}$ and $\vec{b}$. Use this requirement to construct two equations for x, y, and z. Eliminate x and solve for y in terms of z. Then eliminate y and solve for x in terms of z. Since z can be any value whatsoever (the direction of $\vec{v}$ is unaffected), select the value for z which eliminates the denominator in the equation you obtained. How does the resulting expression for $\vec{v}$ compare to the formula we derived on page 99?

20. Suppose $\vec{a}$ and $\vec{b}$ are vectors in the xy-plane, such that $\vec{a} = a_1\vec{i} + a_2\vec{j}$ and $\vec{b} = b_1\vec{i} + b_2\vec{j}$ with $0 < a_2 < a_1$ and $0 < b_1 < b_2$.
    (a) Sketch $\vec{a}$ and $\vec{b}$ and the vector $\vec{c} = -a_2\vec{i} + a_1\vec{j}$. Shade the parallelogram formed by $\vec{a}$ and $\vec{b}$.
    (b) What is the relation between $\vec{a}$ and $\vec{c}$? [Hint: Find $\vec{c} \cdot \vec{a}$ and $\vec{c} \cdot \vec{c}$.]
    (c) Find $\vec{c} \cdot \vec{b}$.
    (d) Explain why $\vec{c} \cdot \vec{b}$ gives the area of the parallelogram formed by $\vec{a}$ and $\vec{b}$.
    (e) Verify that in this case $\vec{a} \times \vec{b} = (a_1 b_2 - a_2 b_1)\vec{k}$.

21. If $\vec{a} + \vec{b} + \vec{c} = \vec{0}$, show that

$$\vec{a} \times \vec{b} = \vec{b} \times \vec{c} = \vec{c} \times \vec{a}.$$

Geometrically what does this imply about $\vec{a}, \vec{b}$, and $\vec{c}$?

22. If $\vec{v}$ and $\vec{w}$ are nonzero vectors, use the geometric definition of the cross product to explain why

$$(\lambda \vec{v}) \times \vec{w} = \lambda(\vec{v} \times \vec{w}) = \vec{v} \times (\lambda \vec{w}).$$

Consider the cases $\lambda > 0$, $\lambda = 0$, and $\lambda < 0$ separately.

23. Use Example 7 on page 101 to show that $\vec{a} \cdot (\vec{b} \times \vec{c}) = (\vec{a} \times \vec{b}) \cdot \vec{c}$ for any vectors $\vec{a}$, $\vec{b}$, and $\vec{c}$.

24. Suppose that $\vec{a}$, $\vec{b}$, $\vec{c}$ is a right-handed set of vectors in 3 space. Show that $\vec{b}$, $\vec{a}$, $\vec{c}$ is a left-handed set, and that $\vec{c}$, $\vec{a}$, $\vec{b}$ is right-handed set.

25. Use the result of Problem 23 to show that the cross product distributes over addition, that is, show that

$$(\vec{a} + \vec{b}) \times \vec{c} = (\vec{a} \times \vec{c}) + (\vec{b} \times \vec{c}).$$

First, use distributivity for the dot product to show that for any vector $\vec{d}$,

$$[(\vec{a} + \vec{b}) \times \vec{c}] \cdot \vec{d} = [(\vec{a} \times \vec{c}) + (\vec{b} \times \vec{c})] \cdot \vec{d}.$$

Next, show that for any vector $\vec{d}$,

$$[((\vec{a} + \vec{b}) \times \vec{c}) - (\vec{a} \times \vec{c}) - (\vec{b} \times \vec{c})] \cdot \vec{d} = 0.$$

Finally, since the above equation is true for all vectors $\vec{d}$, explain why you can conclude that

$$(\vec{a} + \vec{b}) \times \vec{c} = (\vec{a} \times \vec{c}) + (\vec{b} \times \vec{c}).$$

26. Consider the tetrahedron determined by three vectors $\vec{a}$, $\vec{b}$, $\vec{c}$. The *area vector* of a face is a vector perpendicular to the face, pointing outward, whose magnitude is the area of the face. Show that the sum of the four outward pointing area vectors of the faces equals the zero vector.

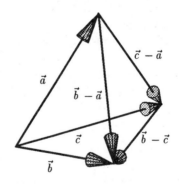

Figure 12.49

27. If $\vec{a} = a_1\vec{i} + a_2\vec{j} + a_3\vec{k}$, $\vec{b} = b_1\vec{i} + b_2\vec{j} + b_3\vec{k}$ and $\vec{c} = c_1\vec{i} + c_2\vec{j} + c_3\vec{k}$ are any three vectors in space, show that

$$\vec{a} \cdot (\vec{b} \times \vec{c}) = \begin{vmatrix} a_1 & a_2 & a_3 \\ b_1 & b_2 & b_3 \\ c_1 & c_2 & c_3 \end{vmatrix}.$$

REVIEW PROBLEMS FOR CHAPTER TWELVE

1. Given the displacement vectors $\vec{u}$ and $\vec{v}$ in Figure 12.50, draw the following vectors:
   (a) $(\vec{u} + \vec{v}) + \vec{u}$ (b) $\vec{v} + (\vec{v} + \vec{u})$ (c) $(\vec{u} + \vec{u}) + \vec{u}$.

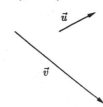

Figure 12.50

2. Figure 12.51 shows five points A, B, C, D and E.

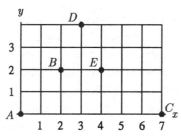

Figure 12.51

   (a) Read off the coordinates of the five points and thus resolve into components the following two vectors:
$$\vec{u} = (2.5)\overrightarrow{AB} + (-0.8)\overrightarrow{CD}, \quad \vec{v} = (2.5)\overrightarrow{BA} - (-0.8)\overrightarrow{CD}$$
   (b) What is the relation between $\vec{u}$ and $\vec{v}$? Why was this to be expected?

3. Find the components of a vector $\vec{p}$ which has the same direction as $\overrightarrow{EA}$ in Figure 12.51 and whose length equals two units.

4. For each of the four statements below, answer the following questions: Does the statement make sense? If yes, is it true for all possible choices of $\vec{a}$ and $\vec{b}$? If no, why not? Use complete sentences for your answers.
   (a) $\vec{a} + \vec{b} = \vec{b} + \vec{a}$ (b) $\vec{a} + \|\vec{b}\| = \|\vec{a} + \vec{b}\|$ (c) $\|\vec{b} + \vec{a}\| = \|\vec{a} + \vec{b}\|$
   (d) $\|\vec{a} + \vec{b}\| = \|\vec{a}\| + \|\vec{b}\|$.

5. Two adjacent sides of a regular hexagon are given as the vectors $\vec{u}$ and $\vec{v}$ in Figure 12.52. Label the remaining sides in terms of $\vec{u}$ and $\vec{v}$.

Figure 12.52

6. Figure 12.53 shows seven vectors $\vec{a}$, $\vec{b}$, $\vec{c}$, $\vec{d}$, $\vec{e}$, $\vec{f}$ and $\vec{g}$.

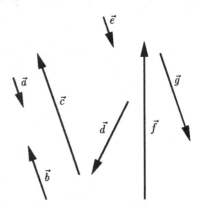

Figure 12.53

(a) Are any of these two vectors equal? Write down all equal pairs.
(b) Can you find a scalar x so that $\vec{a} = x\vec{g}$? If yes, find such an x; if no, explain why not.
(c) Same question as in part b), but for the equation $\vec{b} = x\vec{d}$.
(d) Can you solve the equation $\vec{f} = u\vec{c} + v\vec{d}$ for the scalars u and v? If yes, find them; if no, explain why not.

7. For what values of t are the following pairs of vectors parallel?
    (a) $2\vec{i} + (t^2 + \frac{2}{3}t + 1)\vec{j} + t\vec{k}$, $6\vec{i} + 8\vec{j} + 3\vec{k}$
    (b) $t\vec{i} + \vec{j} + (t-1)\vec{k}$, $2\vec{i} - 4\vec{j} + \vec{k}$
    (c) $2t\vec{i} + t\vec{j} + t\vec{k}$, $6\vec{i} + 3\vec{j} + 3\vec{k}$.

8. If north is the direction of the positive y-axis and east is the direction of the positive x-axis, give the unit vector pointing northwest.

9. Find an equation of the line perpendicular to the vector $3\vec{i} + 5\vec{j}$ and passing through the point $(1, 2)$.

10. Find the equation of the line perpendicular to the vector $3\vec{i} + 2\vec{j}$ and passing through the point $(5, 7)$.

11. For each of the following lines, find a 2-vector parallel to the line and find a 2-vector perpendicular to the line: (a) $3x + 2y = 7$ (b) $6x + 4y = 1$ (c) $y = 4$
    (d) $x = 5$ (e) $y = 5x - 7$ (f) $x = 2y$

Use the definition to compute the cross products in Problems 12–13.

12. $2\vec{i} \times (\vec{i} + \vec{j})$

13. $(\vec{i} + \vec{j}) \times (\vec{i} - \vec{j})$

Compute the cross products for Problem 14–15.

14. $[(\vec{i} + \vec{j}) \times \vec{i}] \times \vec{j}$

15. $(\vec{i} + \vec{j}) \times (\vec{i} \times \vec{j})$

16. True or false? $\vec{a} \times \vec{b} = -(\vec{b} \times \vec{a})$ for all $\vec{a}$ and $\vec{b}$. Explain your answer.

17. Show that if $\vec{r}$ is a vector of magnitude a and $\vec{n}$ is a unit vector perpendicular to $\vec{r}$, then

$$\vec{r} \times (\vec{n} \times \vec{r}) = a^2 \vec{n}.$$

18. Find the area of the triangle with vertices $(-2, 2, 0)$, $(1, 3, -1)$, and $(-4, 2, 1)$ using the cross product.

19. An object is attached by an inelastic string to a fixed point and rotates 30 times per minute in a horizontal plane. Show that the speed of the object is constant but the velocity is not. What does this imply about the acceleration?

20. A man wishes to row across a river which is flowing at 4 mph from the east.
    (a) If he wishes to travel the shortest possible distance from the north and can row at 5 mph, in which direction should he steer?
    (b) If there is a wind of 10 mph from the southwest, how will this affect his steering?

21. A large ship is being towed by two tugs. The larger tug exerts a force which is 25% greater than the smaller tug and at an angle of 30 degrees north of east. Which direction must the smaller tug pull to ensure that the ship travels due east?

22. An airport is at the point $(200, 10, 0)$ and an approaching plane is at the point $(550, 60, 4)$. Assume that the xy-plane is horizontal, with the x-axis pointing eastward and the y-axis pointing northward. Also assume that the z-axis is upward and that all distances are measured in miles. The plane flies due west at a constant altitude at a speed of 500 mph for half an hour. It then descends at 200 mph, heading straight for the airport.
    (a) Find a vector representation for the velocity of the plane while it is flying at constant altitude.
    (b) Find the coordinates of the point at which the plane starts to descend.
    (c) Find a vector representing the velocity of the plane when it is descending.

23. Is the collection of populations of each of the 50 states a vector or scalar quantity?

24. Find the equation of the plane through the origin which is parallel to $z = 4x - 3y + 8$.

25. Find $\|\vec{z}\|$ where $\vec{z} = \vec{i} - 3\vec{j} - \vec{k}$.

26. Find a vector normal to the plane $4(x - 1) + 6(z + 3) = 12$.

27. Consider the plane $5x - y + 7z = 21$.
    (a) Find a point on the x-axis on this plane.
    (b) Find two other points on the plane.
    (c) Find a vector perpendicular to the plane.
    (d) Find a vector parallel to the plane.

28. Given the points $P = (1, 2, 3)$, $Q = (3, 5, 7)$, and $R = (2, 5, 3)$, find:
    (a) A unit vector perpendicular to a plane containing P, Q, R.
    (b) The angle between PQ and PR.
    (c) The area of triangle PQR.
    (d) The distance from R to the line through P and Q.

29. Find all vectors $\vec{v}$ in the plane such that $\|\vec{v}\| = 1$ and $\|\vec{v} + \vec{i}\| = 1$.

30. Find all vectors $\vec{w}$ in 3-space such that $\|\vec{w}\| = 1$ and $\|\vec{w} + \vec{i}\| = 1$. Describe this set geometrically.

31. The price vector of beans, rice, and tofu is $(0.30, 0.20, 0.50)$ in dollars per pound. Express it in dollars per ounce.

32. Three people are trying to hold a ferocious lion still for the veterinarian. The lion, in the center, is wearing a collar with three ropes attached to it and each person has hold of a rope. Charlie is pulling in the direction 62° West of North with a force of 350 pounds and Sam is pulling in the direction 43° East of North with a force of 400 pounds. What is the direction and magnitude of the force which must be exerted by Alice on the third rope to counterbalance Sam and Charlie? (Draw a diagram showing all variables that you use.)

33. Using vectors, show that the perpendicular bisectors of a triangle intersect at a point.

34. Find the distance from the point $P = (2, -1, 3)$ to the plane $2x + 4y - z = -1$.

35. Find an equation of the plane passing through the three points $(1, 1, 1), (1, 4, 5), (-3, -2, 0)$. Find the distance from the origin to the plane.

36. Two lines in space are skew if they are not parallel and do not intersect. Determine the minimum distance between two such lines.

CHAPTER THIRTEEN

DIFFERENTIATING FUNCTIONS OF MANY VARIABLES

For a function of one variable, $y = f(x)$, the derivative $dy/dx = f'(x)$ may be thought of as the rate of change of y with respect to x. For a function of two variables $z = f(x, y)$, there is no such thing as *the* rate of change, since each of the independent variables x and y can be held fixed while the other varies. However, we can still consider the rate of change with respect to one of the independent variables. In this chapter we will see how to compute and interpret these partial derivatives, and the various ways that they can be combined to give a total picture of the way the function varies.

13.1   THE PARTIAL DERIVATIVE

In one-variable calculus you learned how the derivative measures the rate of change of a function. First we will quickly review this idea.

Temperature in a Metal Rod: a One Variable Problem

Let $m(x)$ be the temperature (in °F) of an unevenly heated metal rod, x feet from its left end. Table 13.1 gives some values of $m(x)$, and Figure 13.1 shows a graph of m. From the table and the graph you can see that the temperature increases as you move along the rod, reaching its maximum at around 4 feet from the left end, after which it starts to decrease.

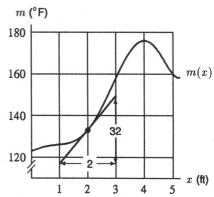

TABLE 13.1: *Temperature $m(x)$ of the rod*

x (ft)	0	1	2	3	4	5
$m(x)$ (°F)	125	128	135	160	175	160

Figure 13.1: Graph of $m(x)$

Example 1   Estimate the derivative $m'(2)$ numerically using Table 13.1, graphically using Figure 13.1, and explain what the answer means in terms of temperature.

Solution   The derivative $m'(2)$ is defined as a limit of difference quotients:

$$m'(2) = \lim_{h \to 0} \frac{m(2+h) - m(2)}{h}.$$

Choosing $h = 1$, we get

$$m'(2) \approx \frac{m(2+1) - m(2)}{1} = \frac{160 - 135}{1} = 25.$$

From the graph in Figure 13.1, we get a more accurate estimate: $m'(2)$ is the slope at $x = 2$, which is about 16, so

$$m'(2) \approx 16.$$

This means that the temperature increases at a rate of 16° F/ft as you go from left to right, past $x = 2$.

Temperature in a Metal Plate: a Two Variable Problem

Now imagine an unevenly heated thin rectangular metal plate, whose temperature (in °F) at the point x feet in from the left and y feet above the bottom is $T(x, y)$. See Figure 13.2 and Table 13.2. How does T vary near the point $(2, 1)$? Since we know how to measure the rate of change of temperature along a rod, we will consider the horizontal line $y = 1$ containing $(2, 1)$, which is like a rod. See Figure 13.3 and Table 13.3. The temperature along the line is the section of T with $y = 1$. In fact,

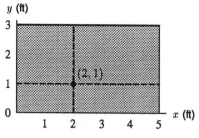

Figure 13.2: A metal plate

TABLE 13.2: *The temperature of the metal plate*

y						
3	85	90	**110**	135	155	180
2	100	110	**120**	145	190	170
1	**125**	**128**	**135**	**160**	**175**	**160**
0	120	135	**155**	160	160	150
	0	1	2	3	4	5

x

it is the same function m we studied in Example 1, so that the rod in that example is actually part of the plate, and $m(x) = T(x, 1)$.

What is the meaning of the derivative $m'(2)$ in this context? It is the rate of change of temperature T *in the x-direction* at the point $(2, 1)$, keeping y fixed. Denote this rate of change by $T_x(2, 1)$, so that

$$T_x(2, 1) = m'(2) = \lim_{h \to 0} \frac{m(2 + h) - m(2)}{h}$$
$$= \lim_{h \to 0} \frac{T(2 + h, 1) - T(2, 1)}{h}.$$

We call $T_x(2, 1)$ the *partial derivative of T with respect to x at the point $(2, 1)$*. The one-variable estimate $m'(2) \approx 16$, which we found from the graph in Figure 13.1, gives $T_x(2, 1) \approx 16$. The fact that $T_x(2, 1)$ is positive means that the temperature of the plate is increasing as you move on the plate past the point $(2, 1)$ in the direction of increasing x (that is, horizontally from left to right in Figure 13.3).

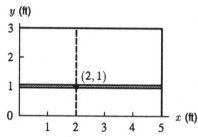

Figure 13.3: The plate and the line $y = 1$

TABLE 13.3: *Temperature along the line $y = 1$*

x (ft)	0	1	2	3	4	5
$m(x)$ (°F)	125	128	135	160	175	160

Example 2   Estimate the rate of change of T in the y-direction at the point $(2, 1)$.

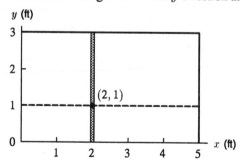

Figure 13.4: The plate and the line $x = 2$

TABLE 13.4: *Temperature along the line $x = 2$*

y (ft)	$n(y)$ (°F)
3	110
2	120
1	135
0	155

Solution   The temperature along the line $x = 2$ is the section of T with $x = 2$, that is, the function $n(y) = T(2, y)$. See Figure 13.4 and Table 13.4. Denote the rate of change of T in the y-direction at $(2, 1)$ by $T_y(2, 1)$. Then

$$T_y(2, 1) = n'(1) = \lim_{h \to 0} \frac{n(1 + h) - n(1)}{h}$$
$$= \lim_{h \to 0} \frac{T(2, 1 + h) - T(2, 1)}{h}.$$

We call $T_y(2, 1)$ the *partial derivative of T with respect to y at the point* $(2, 1)$. Taking $h = 1$ gives

$$T_y(2, 1) \approx \frac{T(2, 1 + 1) - T(2, 1)}{1} = \frac{120 - 135}{1} = -15,$$

and taking $h = -1$ gives

$$T_y(2, 1) \approx \frac{T(2, 1 - 1) - T(2, 1)}{-1} = \frac{155 - 135}{-1} = -20.$$

The average of these two approximations yields the estimate $T_y(2, 1) \approx -17.5$. The fact that $T_y(2, 1)$ is negative means that the temperature decreases as y increases.

Definition of the Partial Derivative

For any function $f(x, y)$ we can use the method of the previous section to calculate the rates of change of $f(x, y)$ with respect to x and y. The idea is to study *separately* the influence of x and y on the value $f(x, y)$, by holding one fixed and letting the other vary. This gives us the *partial derivatives of f at the point* (a, b).

Partial derivatives of f with respect to x and y at (a, b)

We define

$$f_x(a, b) = \quad \begin{array}{c} \text{Rate of change of } f \text{ with respect to } x \\ \text{at the point } (a, b) \end{array} \quad = \lim_{h \to 0} \frac{f(a + h, b) - f(a, b)}{h}$$

$$f_y(a, b) = \quad \begin{array}{c} \text{Rate of change of } f \text{ with respect to } y \\ \text{at the point } (a, b) \end{array} \quad = \lim_{h \to 0} \frac{f(a, b + h) - f(a, b)}{h},$$

whenever the limits exist.

In other words, $f_x(a, b)$ is the ordinary derivative at $x = a$ of the section of f with $y = b$, and $f_y(a, b)$ is the ordinary derivative at $y = b$ of the section of f with $x = a$. Just as with ordinary derivatives, there is an alternative notation:

Alternative notation

For $z = f(x, y)$ we can write

$$f_x(a, b) = \left.\frac{\partial z}{\partial x}\right|_{(a,b)} = \left.\frac{\partial f}{\partial x}\right|_{(a,b)} \qquad \text{and} \qquad f_y(a, b) = \left.\frac{\partial z}{\partial y}\right|_{(a,b)} = \left.\frac{\partial f}{\partial y}\right|_{(a,b)}$$

We use the symbol ∂ to distinguish partial derivatives from ordinary derivatives of one variable functions. In cases where the independent variables have names different from x and y, we adjust the notation accordingly. For example, the partial derivatives of $f(u, v)$ are denoted by f_u and f_v.

Example 3   Let

$$f(x, y) = \frac{x^2}{y + 1}.$$

Estimate $f_x(3, 2)$ and $f_y(3, 2)$ using difference quotients.

Solution   If h is small, then

$$f_x(3, 2) \approx \frac{f(3 + h, 2) - f(3, 2)}{h}.$$

With $h = 0.01$, we find

$$f_x(3, 2) \approx \frac{f(3.01, 2) - f(3, 2)}{0.01}$$

$$= \frac{\frac{3.01^2}{(2+1)} - \frac{3^2}{(2+1)}}{0.01} = 2.00333.$$

With $h = 0.0001$, we get

$$f_x(3, 2) \approx \frac{f(3.0001, 2) - f(3, 2)}{0.0001}$$

$$= \frac{\frac{3.0001^2}{(2+1)} - \frac{3^2}{(2+1)}}{0.0001} = 2.0000333.$$

Since the difference quotient seems to be approaching 2 as h gets smaller, we conclude

$$f_x(3, 2) \approx 2.$$

To estimate $f_y(3, 2)$, we use

$$f_y(3, 2) \approx \frac{f(3, 2 + h) - f(3, 2)}{h}.$$

With $h = 0.01$, we get

$$f_y(3, 2) \approx \frac{f(3, 2.01) - f(3, 2)}{0.01}$$

$$= \frac{\frac{3^2}{(2.01+1)} - \frac{3^2}{(2+1)}}{0.01} = -0.99668.$$

With $h = 0.0001$, we get

$$f_y(3, 2) \approx \frac{f(3, 2.0001) - f(3, 2)}{0.0001}$$

$$= \frac{\frac{3^2}{(2.0001+1)} - \frac{3^2}{(2+1)}}{0.0001} = -0.9999667.$$

Thus, it seems that the difference quotient is approaching -1, so we estimate

$$f_y(3, 2) \approx -1.$$

Visualizing Partial Derivatives on a Graph

The ordinary derivative of a one-variable function is the slope of its graph. Is there a way of visualizing the partial derivative of a two-variable function $f(x, y)$ as a slope? We have already seen that $f_x(a, b)$ is the derivative at $x = a$ of the section of f with $y = b$. The graph of this one-variable function is the curve where the vertical plane $y = b$ cuts through the graph of $f(x, y)$. See Figure 13.5. Thus, $f_x(a, b)$ is the slope of the tangent line to this curve at $x = a$.

Similarly, the graph of the section of f with $x = a$ is the curve where the vertical plane $x = a$ cuts through the graph of f, and the partial derivative $f_y(a, b)$ is the slope of this curve at $y = b$. See Figure 13.6.

Picture standing on the hillside $z = f(x, y)$. Facing eastward (say) in the direction of the x-axis, there is one slope, f_x, whereas facing northward along the y-axis, there is another slope, f_y. In fact we will see shortly that there can be a different slope in every direction.

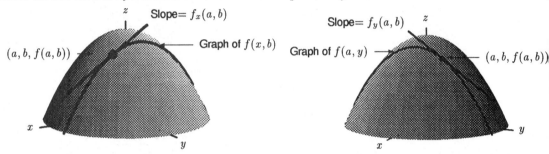

Figure 13.5: The curve $z = f(x, b)$ on the graph of f *Figure 13.6*: The curve $z = f(a, y)$ on the graph of f

---

Example 4   At each point labeled on the graph of the surface $z = f(x, y)$ in Figure 13.7, say whether each partial derivative is positive or negative.

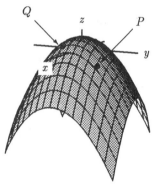

Figure 13.7: Decide the signs of f_x
and f_y at P and Q

Solution   The positive x-axis points out of the page. Imagine heading off in this direction from the point marked P; you would find yourself descending quite steeply. So the partial derivative with respect to x is negative at P, with quite a large absolute value. The same is true for the partial derivative with respect to y at P, since you would also have quite a steep descent in the positive y-direction.

At the point marked Q, heading in the positive x-direction will result in a gentle descent, whereas heading in the positive y-direction will result in a gentle ascent, so the partial derivative f_x is negative but small at Q, and the partial derivative f_y is positive but small.

Estimating Partial Derivatives from a Contour Diagram

If we move parallel to one of the axes on a contour diagram, the partial derivative is the rate of change of the value of the function on the contours. For example, if the values on the contours are increasing, then the partial derivative must be positive.

Example 5   Figure 13.8 shows the contour diagram for the temperature $H(x, t)$ (in °F) in a room as a function of distance x (in feet) from a heater and time t (in minutes) after the heater has been turned on. What are the signs of $H_x(10, 20)$ and $H_t(10, 20)$? Estimate these partial derivatives and explain your answer in practical terms.

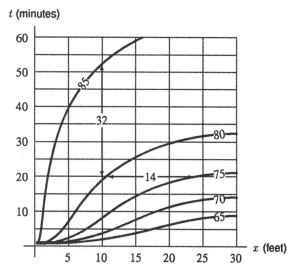

Figure 13.8: Temperature in a heated room

Solution   The point $(10, 20)$ is on the $H = 80$ contour. As x increases you move towards the $H = 75$ contour, so H is decreasing and $H_x(10, 20)$ is negative. This makes sense because as you move further from the heater, the temperature drops. On the other hand, as t increases you move towards the $H = 85$ contour, so H is increasing and $H_t(10, 20)$ is positive. This also makes sense, because it says that as time goes on, the room warms up.

To estimate the partial derivatives, we use a difference quotient. Looking at the contour diagram, we see there is a point on the $H = 75$ contour about 14 units to the right of $(10, 20)$. Hence, H decreases by 5 when x increases by 14, so the rate of change of H with respect to x is about $-5/14 \approx -0.4$. Thus, we find

$$H_x(10, 20) \approx -0.4° \text{F/ft}.$$

This means that the temperature drops about half a degree for each foot you move away from the heater.

To estimate $H_t(10, 20)$, we look again at the contour diagram and notice that the $H = 85$ contour is about 32 units directly above $(10, 20)$. So H increases by 5 when t increases by 32. Hence,

$$H_t(10, 20) \approx \frac{5}{32} = 0.16.$$

This means that on the average the temperature goes up about 1/6 of a degree each minute.

Using Units to Interpret Partial Derivatives

The meaning of a partial derivative can often be explained using units.

Example 6   Suppose that your weight w (in pounds) is a function $f(c, n)$ of the number c of calories you consume daily and the number n of minutes you exercise daily. Interpret the equations

$$\frac{\partial w}{\partial c}(2000, 15) = 0.02 \quad \text{and} \quad \frac{\partial w}{\partial n}(2000, 15) = -0.05$$

in everyday terms using the units for w, c and n.

Solution   The units for $\partial w / \partial c$ are pounds per calorie. The equation

$$\frac{\partial w}{\partial c}(2000, 15) = 0.02$$

means that if you are presently consuming 2000 calories daily and exercising 15 minutes daily, you will weigh 0.02 pounds more for each extra calorie you consume daily, or about 2 pounds for each extra 100 calories. The equation

$$\frac{\partial w}{\partial n}(2000, 15) = -0.25$$

means for the same calorie consumption and number of minutes of exercise, you will weigh 0.25 pounds less for each extra minute you exercise daily, or about 1 pound less for each extra 4 minutes.

Problems for Section 13.1

1. The monthly mortgage payment in dollars, P, for a house is a function of three variables

    $$P = f(A, r, N)$$

    where A is the amount borrowed in dollars, r is the interest rate, and N is the number of years before the mortgage is paid off.

    (a) Suppose $f(92000, 14, 30) = 1090.08$. What does this tell you, in financial terms?

    (b) Suppose $\dfrac{\partial P}{\partial r}(92000, 14, 30) = 72.82$. What is the financial significance of the number 72.82?

    (c) Would you expect $\partial P / \partial A$ to be positive or negative? Why?

    (d) Would you expect $\partial P / \partial N$ to be positive or negative? Why?

2. Suppose you borrow $\$A$ at an interest rate of $r\%$ (per month), and pay it off over t months by making monthly payments of $\$P$. Then $P = g(A, r, t)$. In financial terms, what do the following statements tell you?

    (a) $g(8000, 1, 24) \approx 376.59$ (b) $\dfrac{\partial g}{\partial A}(8000, 1, 24) \approx 0.047$

    (c) $\dfrac{\partial g}{\partial r}(8000, 1, 24) \approx 44.83$

3. Suppose that x is the average price of a new car, and that y is the average price of a gallon of gasoline. Then q_1, the number of new cars bought in a year, depends on both x and y, so $q_1 = f(x, y)$. Similarly, if q_2 is the quantity of gas bought in a year, $q_2 = g(x, y)$.

    (a) What do you expect the signs of $\partial q_1 / \partial x$ and $\partial q_2 / \partial y$ to be? Explain.

    (b) What do you expect the signs of $\partial q_1 / \partial y$ and $\partial q_2 / \partial x$ to be? Explain.

4. A drug is injected into a patient's blood vessel. The function $c = f(x, t)$ represents the concentration of the drug at a distance x in the direction of the blood flow measured from the point of injection and at time t since the injection. What are the units of the following partial derivatives? What are their practical interpretations? What do you expect their signs to be?
   (a) $\partial c / \partial x$.   (b) $\partial c / \partial t$.

5. Suppose $\$P$ is your monthly car payment, and $P = f(P_0, t, r)$ where $\$P_0$ is the amount you borrowed, t is the number of months it takes to pay off the loan, and $r\%$ is the interest rate. What are the units, the practical meanings (in terms of money), and the signs of $\partial P / \partial t$ and $\partial P / \partial r$?

6. Part of the surface $z = f(x, y)$ is shown in Figure 13.9. The points A and B are in the xy-plane.

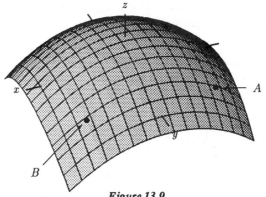

Figure 13.9

   (a) What is the sign of $f_x(A)$?   (b) What is the sign of $f_y(A)$?
   (c) Suppose P is a point in the xy-plane which moves along a straight line from A to B. How does the sign of $f_x(P)$ change? How does the sign of $f_y(P)$ change?

7. Consider the surface $z = f(x, y)$, graphed in Figure 13.10.
   (a) What is the sign of $f_x(0, 5)$?   (b) What is the sign of $f_y(0, 5)$?

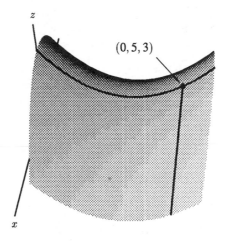

Figure 13.10

8.   Estimate $z_x(1,0)$, $z_x(0,1)$, $z_y(0,1)$ from Figure 13.11.

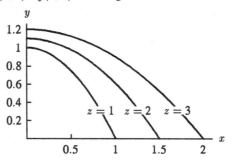

Figure 13.11

For Problems 9 – 11 refer to Table 13.5 giving the wind-chill factor, C, as a function $f(w, T)$ of the wind speed, w, and the temperature, T.

TABLE 13.5: *Wind-chill factor (°F)*

		Temperature (°F)						
		35	30	25	20	15	10	5	0
	5	33	27	21	16	12	7	0	-5
	10	22	16	10	3	-3	-9	-15	-22
Wind (mph)	15	16	9	2	-5	-11	-18	-25	-31
	20	12	4	-3	-10	-17	-24	-31	-39
	25	8	1	-7	-15	-22	-29	-36	-44

9.   Estimate $f_w(10, 25)$. What does your answer mean in practical terms?

10.   Estimate $f_T(5, 20)$. What does your answer mean in practical terms?

11.   From the table you can see that when the temperature is 20° F, the wind-chill factor drops by an average of about 2.6° F with every 1 mph increase in wind speed from 5 mph to 10 mph. Which partial derivative is this telling you about?

For Problems 12 and 13 refer to Table 13.6 giving the heat index, I, as a function $f(h, T)$ of the humidity, h, and the temperature, T.

TABLE 13.6: *Heat index (°F)*

		Temperature (°F)									
		70	75	80	85	90	95	100	105	110	115
	0	64	69	73	78	83	87	91	95	99	103
	10	65	70	75	80	85	90	95	100	105	111
Relative	20	66	72	77	82	87	93	99	105	112	120
humidity	30	67	73	78	84	90	96	104	113	123	135
(%)	40	68	74	79	86	93	101	110	123	137	151
	50	69	75	81	88	96	107	120	135	150	
	60	70	76	82	90	100	114	132	149		

12. Estimate $\partial I/\partial h$ and $\partial I/\partial T$ for typical weather conditions in Tucson in summer ($h = 10$, $T = 100$). What do your answers mean in practical terms for the residents of Tucson?

13. Answer the question in Problem 12 for Boston in summer ($h = 50$, $T = 80$).

14. Use difference quotients with $\Delta x = 0.1$ and $\Delta y = 0.1$ to estimate $f_x(1,3)$ and $f_y(1,3)$ where
$$f(x,y) = e^{-x}\sin y.$$
Then give better estimates by using $\Delta x = 0.01$ and $\Delta y = 0.01$.

15. Figure 13.12 shows a contour diagram for the monthly payment P as a function of the interest rate $r\%$ and the amount L of the loan. Estimate $\partial P/\partial r$ and $\partial P/\partial L$ at the given points. In each case, give the units, and explain the everyday meaning of your answer.

   (a) $r = 8$, $L = 4000$
   (b) $r = 8$, $L = 6000$
   (c) $r = 13$, $L = 7000$

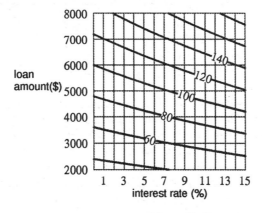

Figure 13.12

16. Figure 13.13 gives a contour diagram for the number n of foxes per square kilometer in southwestern England. Estimate $\partial n/\partial x$ and $\partial n/\partial y$ at the marked points, where x is kilometers east-west and y is kilometers north-south.

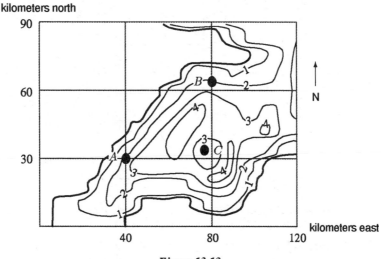

Figure 13.13

17.   Figure 13.14 gives a contour diagram for $f(x, y) = x^2 + y^2$. Use it to estimate $f_x(2, 1)$ and $f_y(2, 1)$. Compare your answers with estimates you get from a table of values for f with $x = 1.9, 2, 2.1$ and $y = 0.9, 1, 1.1$.

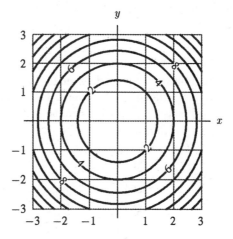

Figure 13.14

18.   An airport can be cleared of fog by heating the air. The amount of heat required to do the job depends on the air temperature and the wetness of the fog. The graph in Figure 13.15 shows the heat $H(t, w)$ required (in calories per cubic meter of fog) as a function of the temperature t (in degrees Celsius) and the water content w (in grams per cubic meter of fog). Note that Figure 13.15 is not a contour diagram, but shows sections of H with w fixed at $0.05, 0.1, 0.2, 0.3, 0.4,$ and 0.5. Use this information to find an approximate value for $H_t(10, 0.1)$. Interpret the partial derivative in practical terms.

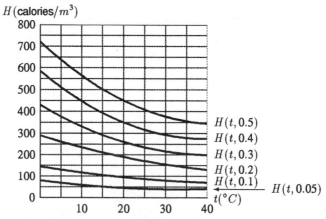

Figure 13.15: Heat needed to dissipate fog as a function of water content and temperature

19.   Make a table of values for $H(t, w)$ from Figure 13.15, and use it to estimate $H_t(t, w)$ for $t = 10, 20,$ and 30 and $w = 0.1, 0.2,$ and 0.3.

20.   Repeat Problem 19 for $H_w(t, w)$ at $t = 10, 20,$ and 30 and $w = 0.1, 0.2,$ and 0.3. What is the practical meaning of these partial derivatives?

21. Suppose that c represents the cardiac output, which is the volume of blood flowing through a person's heart, and that s represents the systemic vascular resistance (SVR), which is the resistance to blood flowing through veins and arteries. Let p be a person's blood pressure. Then $p = f(c, s)$ is a function of c and s.

   (a) What does $\partial p/\partial c$ represent?

   Suppose now that $p = kcs$, where k is a constant.

   (b) Sketch the level curves of p. What do they represent? Label your axes.

   (c) For a person with a weak heart, it is desirable to have the heart pumping against less resistance, while maintaining the same blood pressure. Such a person is given the drug Nitroglycerine to decrease the SVR and the drug Dopamine to increase the cardiac output. Represent this on a graph showing level curves. Put a point A on the graph representing the person's state before drugs are given and a point B for after.

   (d) Right after a heart attack, a patient's cardiac output drops, thereby causing the blood pressure to drop. A common mistake made by medical residents is to get the patient's blood pressure back to normal by using drugs to increase the SVR, rather than by increasing the cardiac output. On a graph of the level curves of p, put a point D representing the patient before the heart attack, a point E representing the patient right after the heart attack, and a third point F representing the patient after the resident has given the drugs to increase the SVR.

13.2 COMPUTING PARTIAL DERIVATIVES ALGEBRAICALLY

Since the partial derivative $f_x(a, b)$ is the ordinary derivative $m'(a)$, where $m(x) = f(x, b)$, and the partial derivative $f_y(a, b)$ is $n'(b)$ where $n(y) = f(a, y)$, you can use all the techniques of differentiation from one-variable calculus to find partial derivatives.

Example 1 Let $f(x, y) = \dfrac{x^2}{y+1}$. Find $f_x(3, 2)$ and $f_y(3, 2)$ algebraically.

Solution These are the same partial derivatives we found numerically in Example 3 on page 113. Now we will use the fact that $f_x(3, 2)$ equals the derivative of $m(x) = f(x, 2)$ at $x = 3$. Since

$$f(x, 2) = \frac{x^2}{(2+1)} = \frac{x^2}{3},$$

we have

$$m'(x) = \frac{2x}{3},$$

and so $f_x(3, 2) = m'(3) = 2$. Notice that the approximations we computed with difference quotients in Example 3 on page 113 were quite good. Similarly, $f_y(3, 2)$ equals the derivative of $f(3, y)$ at $y = 2$. Let

$$n(y) = f(3, y) = \frac{9}{(y+1)}.$$

Then

$$n'(y) = \frac{-9}{(y+1)^2},$$

and so

$$f_y(3, 2) = n'(2) = -1.$$

Again, this agrees with the approximations in Example 3 on page 113.

Example 2   Imagine a vibrating guitar string 1 meter long, and let x be the distance in meters from the left end of the string, as in Figure 13.16. At time t seconds the point x has been displaced $f(x, t)$ meters from its rest position, where

$$f(x, t) = 0.003 \sin(\pi x) \sin(2765 t).$$

(a)   Evaluate $f_x(0.3, 1)$ and explain what it means in practical terms.
(b)   Evaluate $f_t(0.3, 1)$ and explain what it means in practical terms.

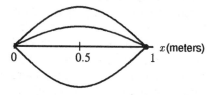

Figure 13.16: A vibrating guitar string

Solution   (a)   Write

$$m(x) = f(x, 1) = 0.003 \sin(\pi x) \sin(2765)$$
$$= 0.0011 \sin(\pi x),$$

so

$$f_x(x, 1) = m'(x) = 0.0011\pi \cos(\pi x).$$

In particular, $f_x(0.3, 1) = m'(0.3) = 0.0011\pi \cos(\pi(0.3)) \approx 0.002$. To see what $f_x(0.3, 1)$ means, think about the function $f(x, 1)$ of which it is the derivative. The graph of $f(x, 1)$ in Figure 13.17 is a snapshot of the string at the time $t = 1$. Thus, the derivative $f_x(0.3, 1)$ is the slope of the string at the point $x = 0.3$ (at the instant when $t = 1$).

(b)   Write

$$n(t) = f(0.3, t) = 0.003 \sin(\pi(0.3)) \sin(2765 t)$$
$$= 0.0024 \sin(2765 t),$$

so

$$f_t(0.3, t) = n'(t) = (0.0024)(2765) \cos(2765 t) = 6.7 \cos(2765 t).$$

Then

$$f_t(0.3, 1) = n'(1) = 6.7 \cos(2765(1)) \approx 6.$$

To see what $f_t(0.3, 1)$ means, think about the function $f(0.3, t)$ of which it is the derivative. The graph of $f(0.3, t)$ is a position-versus-time graph that tracks the movement up and down of a single point on the string, the point where $x = 0.3$. See Figure 13.18. The derivative $f_t(0.3, 1) = 6$ m/sec is thus the velocity of that point on the string at time 1 sec. The fact that $f_t(0.3, 1)$ is positive indicates that the point is moving upward when $t = 1$.

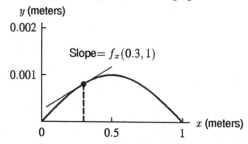

Figure 13.17: The shape of the string at $t = 1$ sec

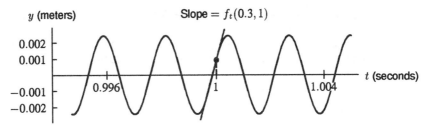

Figure 13.18: Position versus time graph of the point $x = 0.3$ m

The Partial Derivative Functions

So far we have focused on the partial derivatives of a function $z = f(x, y)$ at a fixed point (a, b). By thinking of a and b as variables, $a = x$ and $b = y$, we get the partial derivative functions $f_x(x, y)$ and $f_y(x, y)$. We also have the alternative notation:

$$f_x(x, y) = \frac{\partial z}{\partial x} \quad \text{and} \quad f_y(x, y) = \frac{\partial z}{\partial y}.$$

If a function $f(x, y)$ is given by a formula, then a formula for $f_x(x, y)$ can be found by regarding y as a constant and differentiating f with respect to x, just as you would for a function of the single variable x. Similarly, a formula for $f_y(x, y)$ can be found by regarding x as a constant and differentiating f with respect to y.

Example 3 Compute both partial derivatives for
(a) $f(x, y) = y^2 e^{3x}$ (b) $z = (3xy + 2x)^5$ (c) $g(x, y) = e^{x+3y} \sin(xy)$.

Solution (a) This is the product of a function of x (namely e^{3x}) and a function of y (namely y^2). When we differentiate with respect to x, we can think of the function of y as a constant, and vice versa.

$$f_x(x, y) = y^2 \frac{\partial}{\partial x}\left(e^{3x}\right) = 3y^2 e^{3x},$$

$$f_y(x, y) = e^{3x} \frac{\partial}{\partial y}(y^2) = 2ye^{3x}$$

(b) Here we use the chain rule:

$$\frac{\partial z}{\partial x} = 5(3xy + 2x)^4 \frac{\partial}{\partial x}(3xy + 2x) = 5(3xy + 2x)^4(3y + 2),$$

$$\frac{\partial z}{\partial y} = 5(3xy + 2x)^4 \frac{\partial}{\partial y}(3xy + 2x) = 5(3xy + 2x)^4 3x = 15x(3xy + 2x)^4.$$

(c) Since each function in the product is a function of both x and y, we need to use the product rule for each partial derivative:

$$g_x(x, y) = \frac{\partial}{\partial x}(e^{x+3y}) \sin(xy) + e^{x+3y} \frac{\partial}{\partial x}(\sin(xy))$$

$$= e^{x+3y} \sin(xy) + e^{x+3y} y \cos(xy)$$

$$g_y(x, y) = \frac{\partial}{\partial y}(e^{x+3y}) \sin(xy) + e^{x+3y} \frac{\partial}{\partial y}(\sin(xy))$$

$$= 3e^{x+3y} \sin(xy) + e^{x+3y} x \cos(xy).$$

Partial Derivatives for Functions of More Than Two Variables

For functions of three or more variables, we find partial derivatives by differentiating with respect to one variable, regarding all the other variables as constants.

Example 4 Find all the partial derivatives of $f(x, y, z) = \dfrac{x^2 y^3}{z}$.

Solution To find $f_x(x, y, z)$, we consider y and z as fixed, giving

$$f_x = \frac{2xy^3}{z}.$$

Similarly we have

$$f_y = \frac{3x^2 y^2}{z},$$

$$f_z = -\frac{x^2 y^3}{z^2}.$$

Problems for Section 13.2

1. Let $f(u, v) = u(u^2 + v^2)^{3/2}$.
   (a) Use a difference quotient to approximate $f_u(1, 3)$ with $h = 0.001$.
   (b) Now evaluate $f_u(1, 3)$ exactly. Was the approximation in part (a) reasonable?

2. Let $f(r, s) = r^2 s/(r^2 - s^2)$.
   (a) Use a difference quotient to approximate $f_s(2, 4)$ with $h = 0.001$.
   (b) Now evaluate $f_s(2, 4)$ exactly. Was the approximation in part (a) reasonable?

Find the indicated partial derivatives for Problems 3–34. Assume the variables are restricted to a domain on which the function is defined.

3. $\dfrac{\partial A}{\partial h}$ if $A = \frac{1}{2}(a + b)h$

4. $\dfrac{\partial}{\partial m}\left(\dfrac{1}{2}mv^2\right)$

5. $\dfrac{\partial}{\partial B}\left(\dfrac{1}{u_0}B^2\right)$

6. $\dfrac{\partial}{\partial r}\left(\dfrac{2\pi r}{v}\right)$

7. F_v if $F = \dfrac{mv^2}{r}$

8. u_E if $u = \dfrac{1}{2}\epsilon_0 E^2 + \dfrac{1}{2\mu_0}B^2$

9. $\dfrac{\partial}{\partial v_0}(v_0 + at)$

10. $\dfrac{\partial F}{\partial m_2}$ if $F = \dfrac{Gm_1 m_2}{r^2}$

11. a_v if $a = v^2/r$

12. F_m if $F = mg$

13. $\dfrac{\partial}{\partial T}\left(\dfrac{2\pi r}{T}\right)$

14. $\dfrac{\partial}{\partial t}\left(v_0 t + \dfrac{1}{2}at^2\right)$

15. z_x if $z = x^2y + 2x^5y$

16. $\dfrac{\partial f_0}{\partial L}$ if $f_0 = \dfrac{1}{2\pi\sqrt{LC}}$

17. $\dfrac{\partial y}{\partial t}$ if $y = \sin(ct - 5x)$

18. $\dfrac{\partial}{\partial y}(3x^5y^7 - 32x^4y^3 + 5xy)$

19. z_x if $z = \sin(5x^3y - 3xy^2)$

20. $\dfrac{\partial}{\partial x}(a\sqrt{x})$

21. $\dfrac{\partial}{\partial M}\left(\dfrac{2\pi r^{3/2}}{\sqrt{GM}}\right)$

22. z_x if $z = \dfrac{1}{2x^2ay} + \dfrac{3x^5abc}{y}$

23. g_x if $g(x, y) = \ln(ye^{xy})$

24. $\dfrac{\partial\alpha}{\partial\beta}$ if $\alpha = e^{x\beta-3}/(2y\beta + 5)$

25. $\dfrac{\partial}{\partial T}\left(\ln\dfrac{T+3}{V}\right)$

26. $\dfrac{\partial L}{\partial c}$ if $L = 6x^2y^5c^2e^{c^2-3c-7}$

27. $\dfrac{\partial}{\partial\theta}\left(3t\cos(5\theta - 1) - \tan(7t\theta^2)\right)$

28. $\dfrac{\partial}{\partial x}(xe^{\sqrt{xy}})$

29. z_y if $z = \dfrac{3x^2y^7 - y^2}{15xy - 8}$

30. $\dfrac{\partial}{\partial\lambda}\left(\dfrac{x^2y\lambda - 3\lambda^5}{\sqrt{\lambda^2 - 3\lambda + 5}}\right)$

31. $\dfrac{\partial m}{\partial v}$ if $m = \dfrac{m_0}{\sqrt{1 - v^2/c^2}}$

32. $\dfrac{\partial}{\partial w}(\sqrt{2\pi xyw} - 13x^7y^3v)$

33. z_c if $z = \tan\sqrt{5wc^7 - 2c^4}$

34. $\dfrac{\partial}{\partial w}\left(\dfrac{x^2yw - xy^3w^7}{w - 1}\right)^{-7/2}$

For Problems 35–38 find the partial derivatives specified. Assume the variables are restricted to a domain on which the function is defined.

35. z_x and z_y for $z = x^7 + 2^y + x^y$

36. $\dfrac{\partial H}{\partial t}$ if $H = \ln(e^{t+p} + t) + \sin(p^2t)$

37. $z_x(2, 3)$ if $z = (\cos x) + y$. Estimate your answer to one decimal place using a calculator.

38. $\dfrac{\partial f}{\partial x}\bigg|_{(\pi/3,1)}$ if $f(x, y) = x\ln(y\cos x)$

39. Show that the Cobb-Douglas function

$$Q = bK^\alpha L^{1-\alpha} \quad \text{where} \quad 0 < \alpha < 1$$

satisfies Euler's theorem, which states that:

$$K\dfrac{\partial Q}{\partial K} + L\dfrac{\partial Q}{\partial L} = Q.$$

40. Suppose you know that

$$f_x(x, y) = 4x^3y^2 - 3y^4,$$
$$f_y(x, y) = 2x^4y - 12xy^3.$$

Can you find a function f which has these partial derivatives? If so, are there any others?

13.3 LOCAL LINEARITY AND THE DIFFERENTIAL

In Section 13.1 we learned how to study a function of two variables by changing one variable at a time. In effect, we found a way to use one-variable calculus to get some information about the rate of change of a two-variable function at a point, at least along the two lines through the point where one variable is constant. We now go off those lines, entering a truly multivariable world. Let's see how.

Zooming In to See Local Linearity

Appendix A gives a review of local linearity for one variable. We want to develop a linear approximation for functions of two variables that is analogous to the local linearization for functions of one variable. For a function of one variable, local linearity means that as we zoom in on the graph it looks like a straight line. So it is natural to ask: What happens if we zoom in on the graph of a two-variable function? Figure 13.19 shows three successive views of the graph of such a function, each one a closer view than the previous one. Notice that each view is flatter than the previous one: The closer we zoom in, the more the graph looks like a plane, which is the graph of a linear function.

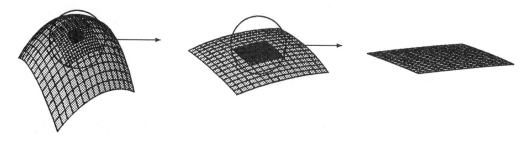

Figure 13.19: Zooming in on a function of two variables until the graph is almost flat

The same effect is seen by zooming in on the contour diagram of the function. Figure 13.20 shows three successive views of the contours near a point. Notice that as you zoom in, the contours look more like equally spaced parallel lines, that is like the contours of a linear function. (As you zoom in, you have to add more contours.)

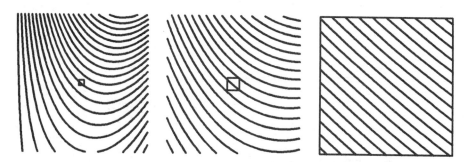

Figure 13.20: Zooming in on a contour diagram until the lines look parallel and equally spaced

This effect can also be seen numerically by zooming in with tables of values. Table 13.7 shows three tables of values for $f(x, y) = x^2 + y^3$ near $x = 2$, $y = 1$, each one a closer view than the previous one. Notice how each table looks more like the table of a linear function.

TABLE 13.7: *Zooming in on values of* $f(x, y) = x^2 + y^3$ *near* $(2, 1)$ *until the table looks linear.*

		y		
		0	1	2
x	1	1	2	9
	2	4	5	12
	3	9	10	17

		y		
		0.9	1.0	1.1
x	1.9	4.34	4.61	4.94
	2.0	4.73	5.00	5.33
	2.1	5.14	5.41	5.74

		y		
		0.99	1.00	1.01
x	1.99	4.93	4.96	4.99
	2.00	4.97	5.00	5.03
	2.01	5.01	5.04	5.07

The Tangent Plane

We have seen that as we zoom in on a surface it looks more and more like a plane. This plane is called the *tangent plane* to the surface at the point. The tangent plane is to functions of two variables what the tangent line is to functions of one variable. Figure 13.21 shows the graph of a function with its tangent plane at a point.

What is the equation of the tangent plane? The x-slope of the graph of f at $(a, b, f(a, b))$ is the partial derivative $f_x(a, b)$, and the y-slope is $f_y(a, b)$. Thus, using the equation for a plane on page 44 of Chapter 11, we have the following equation:

The Tangent Plane to the Surface $z = f(x, y)$ **at the point** (a, b)

$$z = f(a, b) + f_x(a, b)(x - a) + f_y(a, b)(y - b).$$

Here we are thinking of a and b as fixed, so $f(a, b)$, $f_x(a, b)$, and $f_y(a, b)$ are constants; thus, the right side is a linear function of x and y.

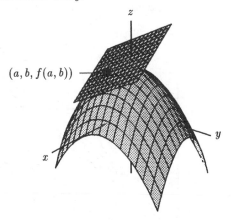

Figure 13.21: The tangent plane

Example 1   Find the equation for the tangent plane to the surface $z = x^2 + y^2$ at the point $(x, y) = (3, 4)$.

Solution   We have $f_x(x, y) = 2x$, so $f_x(3, 4) = 6$, and $f_y(x, y) = 2y$, so $f_y(3, 4) = 8$. Also, $f(3, 4) = 3^2 + 4^2 = 25$. Thus, the equation for the tangent plane at $(3, 4)$ is

$$z = 25 + 6(x - 3) + 8(y - 4) = -25 + 6x + 8y.$$

The Local Linearization of a Function of Two Variables

We now approximate the values of f by z-values from the tangent plane, giving the following result:

The Tangent Plane Approximation

For (x, y) near (a, b),

$$f(x, y) \approx f(a, b) + f_x(a, b)(x - a) + f_y(a, b)(y - b).$$

We are thinking of a and b as being fixed, so the expression on the right side is linear in x and y. This "almost equality" is called the **local linearization** of f near $x = a$, $y = b$.

We also write this as

$$\Delta f \approx f_x(a, b)\Delta x + f_y(a, b)\Delta y,$$

where

$$\text{Change in } f = \Delta f = f(x, y) - f(a, b)$$
$$\text{Change in } x = \Delta x = x - a$$
$$\text{Change in } y = \Delta y = y - b.$$

Since the local linearization contains both variables x and y, it can be used to approximate f at points where neither $x = a$ nor $y = b$. See Figure 13.22.

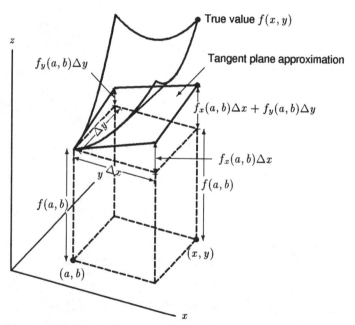

Figure 13.22: Local linearization: Approximation by the tangent plane

Example 2   Consider the heated plate of Section 13.1 on page 110 with temperature $T(x, y)$ at the point (x, y). Use the fact that $T(2, 1) = 135$, $T_x(2, 1) = 16$, and $T_y(2, 1) = -17.5$ to estimate the temperature at the point $(2.04, 0.97)$.

Solution   Local linearization gives us the approximation

$$T(x, y) \approx T(2, 1) + T_x(2, 1)(x - 2) + T_y(2, 1)(y - 1)$$
$$T(x, y) \approx 135 + 16(x - 2) - 17.5(y - 1).$$

Thus,

$$T(2.04, 0.97) \approx 135 + 16(2.04 - 2) - 17.5(0.97 - 1) = 136.17°\text{F}.$$

Example 3   The volume V (in ft^3) of one pound of steam is a function $f(T, p)$ of the temperature T (in °F) and pressure p (in lb/in^2) of the steam. Detailed knowledge of f is critical to the safe design of boilers. Extensive tables, called steam tables, have been published, of which Table 13.8 forms a small portion.

(a) Use the table to give a linear approximation of $V = f(T, p)$ for T near $500°$F and p near 24 lb/in^2.

(b) Estimate the volume of one pound of steam at a temperature of $510°$F and a pressure of 25 lb/in^2.

TABLE 13.8: *Volume (in cubic feet) of one pound of steam at various temperatures and pressures*

		Pressure p (lb/in^2)			
		20	22	24	26
	480	27.85	25.31	23.19	21.39
Temperature	500	28.46	25.86	23.69	21.86
t	520	29.06	26.41	24.20	22.33
(°F)	540	29.66	26.95	24.70	22.79

Solution   (a) We use local linearization around the point $T = 500$, $p = 24$, giving

$$f(T, p) \approx f(500, 24) + f_T(500, 24)(T - 500) + f_p(500, 24)(p - 24).$$

Directly from the table, we read the value $f(500, 24) = 23.69$. Next we approximate $f_T(500, 24)$ by a difference quotient. We look at the $p = 24$ column and compute the average rate of change between $T = 500$ and $T = 520$:

$$f_T(500, 24) \approx \frac{f(520, 24) - f(500, 24)}{520 - 500}$$
$$= \frac{24.20 - 23.69}{20} = 0.0255.$$

Note that $f_T(500, 24)$ is positive, because steam expands when heated.

Next we approximate $f_p(500, 24)$ by looking at the $T = 500$ row and computing the average rate of change between $p = 24$ and $p = 26$:

$$f_p(500, 24) \approx \frac{f(500, 26) - f(500, 24)}{26 - 24}$$
$$= \frac{21.86 - 23.69}{2} = -0.915.$$

Note that $f_p(500, 24)$ is negative, because increasing the pressure decreases the volume.

Using these approximations for the partial derivatives, we obtain the local linearization:

$$V = f(T, p) \approx 23.69 + 0.0255(T - 500) - 0.915(p - 24) \text{ ft}^3 \quad \left(\begin{array}{l} \text{for } T \text{ near } 500\,^\circ\text{F} \\ \text{and } p \text{ near } 24 \text{ lb/in}^2. \end{array}\right)$$

(b)   When $T = 510$ and $p = 25$, we have

$$V \approx 23.69 + 0.0255(510 - 500) - 0.915(25 - 24) = 23.03 \text{ ft}^3$$

Example 4   Find the local linearization of $f(x, y) = x^2 + y^2$ at the point $(3, 4)$. See how close an estimate of $f(2.9, 4.2)$ it gives.

Solution   Let $z = f(x, y) = x^2 + y^2$. In Example 1 on page 127, we found the equation of the tangent plane

$$z = 25 + 6(x - 3) + 8(y - 4).$$

Therefore, for (x, y) near $(3, 4)$, we have

$$f(x, y) \approx 25 + 6(x - 3) + 8(y - 4),$$

so

$$f(2.9, 4.2) \approx 25 + 6(-0.1) + 8(.2) = 26.$$

This compares favorably with the true value $f(2.9, 4.2) = (2.9)^2 + (4.2)^2 = 26.05$. Notice, however, that the local linearization does not give a good approximation at points far away from $(3, 4)$. For example, if $x = 2$, $y = 2$, the local linearization gives

$$f(2, 2) \approx 25 + 6(-1) + 8(-2) = 3,$$

whereas the true value of the function is $f(2, 2) = 2^2 + 2^2 = 8$.

What If There are Three or More Variables?

Local linear approximations for functions of three or more variables follow the same pattern as for functions of two variables. The linear approximation of $f(x, y, z)$ at (a, b, c) is

$$f(x, y, z) \approx f(a, b, c) + f_x(a, b, c)(x - a) + f_y(a, b, c)(y - b) + f_z(a, b, c)(z - c).$$

The Differential

Often we are interested not so much in the value of a function at a point, as in the change in the value of a function between two points. We approximate the change using local linearity in the form

$$\Delta f \approx f_x(a, b)\Delta x + f_y(a, b)\Delta y,$$

where

$$\text{Change in } f = \Delta f = f(x, y) - f(a, b)$$
$$\text{Change in } x = \Delta x = x - a$$
$$\text{Change in } y = \Delta y = y - b.$$

For fixed a and b, the right hand side of this form of local linearity is a linear function of Δx and Δy that can be used to estimate the effect on f of various different changes in x and y. For instance, different models of climatic change predict different small changes in rainfall and temperature from the fixed values representing current climatic conditions. Given corn production as a function of rainfall and temperature, we might want to estimate the change in corn production caused by these various changes. We call the linear function occurring in this form of local linearity the *differential*. To define the differential in general, we introduce new variables dx and dy to represent changes in x and y.

The Differential

For a function $z = f(x, y)$, the **differential**, df (or dz), at a point (a, b) is the linear function of dx and dy given by the formula

$$df = f_x(a, b)\, dx + f_y(a, b)\, dy.$$

The differential at a general point is often written $df = f_x\, dx + f_y\, dy$.

Example 5   Figure 13.23 shows the corn production function $C = f(R, T)$ discussed on page 26. To estimate the effect of climatic change on agriculture, we are interested in the effect of small deviations from the current climate, which is represented by the point $(15, 30)$. Use the differential at $(15, 30)$ to study this effect.

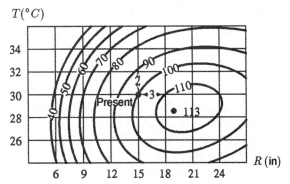

Figure 13.23: Corn production as a function of rainfall and temperature

Solution   The differential is

$$df = f_R(15, 30)\, dR + f_T(15, 30)\, dT.$$

We estimate the partial derivatives from the contour diagram:

$$f_R(15, 30) \approx \frac{f(18, 30) - f(15, 30)}{3} = \frac{110 - 100}{3} = \frac{10}{3} = 3.3$$

$$f_T(15, 30) \approx \frac{f(15, 32) - f(15, 30)}{2} = \frac{90 - 100}{2} = \frac{-10}{2} = -5.$$

Thus, the differential is

$$df = 3.3\, dR - 5\, dT.$$

This can be used to estimate the effect of small climatic changes. For example, if rainfall increases by half an inch and temperature decreases by one degree, then we substitute $dR = 0.5$ and $dT = -1$ into the differential:

$$(3.3)(0.5) - (5)(-1) = 6.65.$$

Local linearity says that this is approximately the corresponding change in f. So corn production will increase by about 6.65%.

Example 6   Compute the differentials of the following functions.

   (a) $f(x, y) = x^2 e^{5y}$ (b) $z = x\sin(xy)$ (c) $f(x, y) = x\cos(2x)$.

Solution (a) Since $f_x(x, y) = 2xe^{5y}$ and $f_y(x, y) = 5x^2 e^{5y}$, we have

$$df = 2xe^{5y}\,dx + 5x^2 e^{5y}\,dy.$$

   (b) Since $\partial z/\partial x = \sin(xy) + xy\cos(xy)$ and $\partial z/\partial y = x^2\cos(xy)$,

$$dz = (\sin(xy) + xy\cos(xy))\,dx + x^2\cos(xy)\,dy.$$

   (c) Since $f_x(x, y) = \cos(2x) - 2x\sin(2x)$ and $f_y(x, y) = 0$,

$$df = (\cos(2x) - 2x\sin(2x))\,dx + 0\,dy = (\cos(2x) - 2x\sin(2x))\,dx.$$

Where Does the Notation for the Differential Come From?

We have defined the differential as a linear function of the variables dx and dy. You may wonder why we chose such strange names for our variables. The reason is historical: people used to think of dx and dy as infinitesimal changes in x and y. The equation

$$df = f_x\,dx + f_y\,dy$$

was regarded as an infinitesimal version of the local linear approximation

$$\Delta f \approx f_x \Delta x + f_y \Delta y.$$

In spite of the problems with defining exactly what 'infinitesimal' means, many mathematicians, scientists, and engineers still think of the differential informally in terms of infinitesimals.

  Figure 13.24 illustrates a way of thinking about differentials that combines the formal definition with this informal point of view. It shows the graph of f along with a blown-up view of what the graph around a fixed point would look like under a microscope. Since f is locally linear at the point, the blown-up view looks like the tangent plane. Under the microscope, you use a coordinate system with its origin at the point and with axes for the coordinates dx, dy, and dz. The graph of the differential df is the tangent plane, which has equation $dz = f_x(a, b)\,dx + f_y(a, b)\,dy$ in the local coordinates.

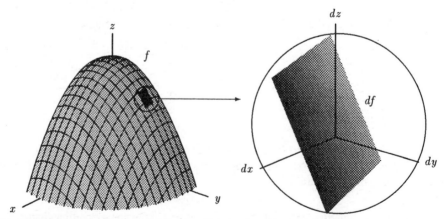

Figure 13.24: The graph of f, with a view through a microscope showing the tangent plane in the local coordinate system

Example 7 The density ρ(in g/cm^3) of carbon dioxide gas CO_2 depends upon its temperature T (in ° C) and pressure P (in atmospheres). The ideal gas model for CO_2 gives what is called the state equation

$$\rho = \frac{0.5363P}{T + 273.15}.$$

(a) Compute the differential $d\rho$.
(b) Use the differential to estimate the change in density that the gas would undergo if its temperature changed from $50°C$ to $45°C$ and its pressure changed from 5 atm to each of the two pressures 5.1 atm and 5.2 atm.

Solution (a) The definition of the differential tells us that for $\rho = f(T, P)$

$$d\rho = f_T(T, P)\, dT + f_P(T, P)dP$$
$$= \frac{-0.5363P}{(T + 273.15)^2}\, dT + \frac{0.5363}{T + 273.15}\, dP.$$

(b) Since we will make both approximations using the differential $d\rho$ at the same point $(T, P) = (50, 5)$, we calculate:

$$d\rho(50, 5) = f_T(50, 5)\, dT + f_P(50, 5)dP$$
$$= \frac{-0.5363(5)}{(50 + 273.15)^2}\, dT + \frac{0.5363}{50 + 273.15}\, dP$$
$$= -2.57 \cdot 10^{-5}\, dT + 1.66 \cdot 10^{-3}\, dP.$$

Notice that the coefficient of dT is negative because increasing the temperature will expand the gas (if the pressure is kept constant) and therefore decrease its density. The coefficient of dP is positive, because increasing the pressure will compress the gas (if the temperature is kept constant) and therefore increase its density.

Now we compute the two estimates asked for:

(i) The change in T is -5 and the change in P is 0.1. So the change in ρ is approximately

$$-2.57 \cdot 10^{-5}(-5) + 1.66 \cdot 10^{-3}(0.1) \approx 0.0003 \text{ gm/cm}^3.$$

(ii) This time the change in P is 0.2, so the change in ρ is approximately

$$-2.57 \cdot 10^{-5}(-5) + 1.66 \cdot 10^{-3}(0.2) \approx 0.0005 \text{ gm/cm}^3.$$

Local Linearity and Differentiability

We use the idea of local linearity to define what it means for a function to be differentiable: A function $f(x, y)$ is said to be *differentiable* at a point (a, b) if f is locally linear at (a, b). We have seen what local linearity means graphically and numerically, but how do we define it analytically?

For a function to be locally linear at a point (a, b), the error in the local linear approximation

$$f(x, y) \approx f(a, b) + f_x(a, b)(x - a) + f_y(a, b)(y - b)$$

must be small, even relative to the distance between (x, y) and (a, b), for all points (x, y) near (a, b). Specifically, we have the following definition.

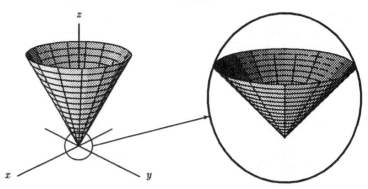

Figure 13.25: The function $f(x, y) = \sqrt{x^2 + y^2}$ is not locally linear at $(0, 0)$

For a function f at a point (a, b), let $E(x, y)$ be the error in the local linear approximation, that is, the absolute value of the difference between the left and right hand sides, and let $d(x, y)$ be the distance between (x, y) and (a, b). Then f is said to be **locally linear**, or **differentiable**, at (a, b) if we can make the ratio $E(x, y)/d(x, y)$ as small as we like by restricting (x, y) to a small enough non-zero distance from (a, b).

Not all functions are locally linear at all points. Consider the function $f(x, y) = \sqrt{x^2 + y^2}$ at the point $(0, 0)$. The graph of this function, as we saw on page 31 of Chapter 11, is a cone. If we zoom in on the graph at $(0, 0)$ as shown in Figure 13.25, the sharp point stays there; the graph never flattens out.

However, all the functions we consider in this book will be differentiable everywhere except perhaps at a few points, such as the cone tip. Strangely enough, it is possible to have a function whose partial derivatives exist at a point P but which is not differentiable at P. However, if the partial derivatives are continuous, then the function is differentiable.

Problems for Section 13.3

1. Find the equation of the tangent plane to $z = e^y + x + x^2 + 6$ at the point $(1, 0, 9)$.

2. Find the equation of the tangent plane to $z = ye^{x/y}$ at the point $(1, 1, e)$.

3. Find the equation of the tangent plane to $z = \frac{1}{2}(x^2 + 4y^2)$ at the point $(2, 1, 4)$.

4. A student was asked to find the equation of the tangent plane to the surface $z = x^3 - y^2$ at the point where $(x, y) = (2, 3)$. His answer was

$$z = 3x^2(x - 2) - 2y(y - 3) - 1.$$

   (a) At a glance, how do you know this is wrong?
   (b) What mistake did the student make?
   (c) Answer the question correctly.

5. Consider the tangent plane to the surface $z = x^2 + 2y^2$ at the point where $(x, y) = (1, 2)$. As x and y decrease to 0, your height on both the surface and the tangent plane also decreases to 0. Do you get to 0 first on the surface or on the plane? How do you know?

6. Verify the local linearity of $f(x,y) = e^{-x} \sin y$ near $x = 1$, $y = 2$ by giving a table of values of f for $x = 0.9$, 1.0 1.1 and $y = 1.9$, 2.0, 2.1. Express values of f with 4 digits after the decimal point. Then give a table of values for $x = 0.99$, 1.00, 1.01 and $y = 1.99$, 2.00. 2.01, showing again 4 digits after the decimal point. Do both tables look nearly linear? Does the second table look more linear than the first? Give the local linearization of f at $(1, 2)$, first using your tables, and second using $f_x = -e^{-x} \sin y$ and $f_y = e^{-x} \cos y$.

7. Give the local linearization for the monthly car-loan payment function at each of the points investigated in Problem 15 on Page 119.

8. Find the local linearization of the function $f(x, y) = x^2 y$ at the point $(3, 1)$.

9. In Example 3 on page 129 we found a linear approximation for $V = f(T, p)$ near $(500, 24)$. Now find a linear approximation near $(480, 20)$.

10. In Example 3 on page 129 we found a linear approximation for $V = f(T, p)$ near $(500, 24)$.
    (a) Test the accuracy of this approximation by comparing its predicted value with the nearby values in the table. Do you notice something? Explain your answer.
    (b) Can you suggest a way of finding a linear approximation for $f(T, p)$ near $(500, 24)$ that does not have the property you noticed in (a)? [Hint: Estimate the partial derivatives in a different way.]

11. The productive capacity P of the United States as a function of capital K and labor L is modeled by the Cobb-Douglas production function

$$P(K, L) = AK^{0.21}L^{0.79},$$

where A is a constant. Suppose that capital increases by 1% but that labor use remains unchanged. Use local linearization to estimate the percentage increase in production.

12. Figure 13.26 shows a schematic diagram for a transistor in common-emitter configuration. The transistor would be wired into an electrical circuit at the three points marked B (base), C (collector), and E (emitter). It's state at any time is determined by the three currents i_b, i_c, and i_e and the two voltages v_b and v_c. All of these can be determined from measurements of i_b and v_c alone, because there are functions f and g (called the *characteristics* of the transistor) such that $i_c = f(i_b, v_c)$ and $v_b = g(i_b, v_c)$, and $i_e = -i_b - i_c$. The units are microamps (μA) for i_b, volts (V) for v_c, and milliamps (mA) for i_c.

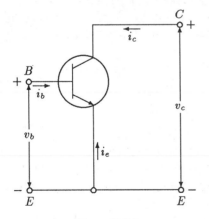

Figure 13.26

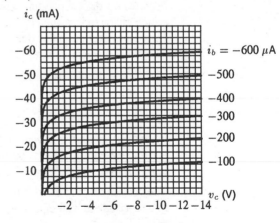

Figure 13.27:

Figure 13.27 shows the graphs of f as a function of v_c with i_b fixed at the values -100, -200, -300, -400, -500, and -600. Find a linear approximation for f that is valid when the transistor is made to operate with i_b near $-300 \, \mu A$ and v_c near $-8V$.

13.   If $z = e^{-x} \cos(y)$ find dz.

Find the differentials of the functions in Problems 14–16.

14. $f(x, y) = \sin(xy)$.

15. $g(u, v) = u^2 + uv$.

16. $h(x, t) = e^{-3t} \sin(x + 5t)$.

Find differentials of the functions in Problems 17– 20 at the given point.

17. $f(x, y) = xe^{-y}$ at $(1, 0)$

18. $g(x, t) = x^2 \sin(2t)$ at $(2, \pi/4)$

19. $P(K, L) = 1.01K^{0.75}L^{0.25}$ at $(1, 100)$

20. $F(m, r) = Gm/r^2$ at $(100, 10)$

21.   For the Cobb-Douglas production function $P = 40L^{0.25}K^{0.75}$, find the differential dP when $L = 2$ and $K = 16$.

22.   One mole of ammonia gas is contained in a vessel which is capable of changing its volume (a piston, for example). The total energy U (in Joules) of the ammonia, is a function of the volume V (in m^3) of the container, and the temperature T (in ° K) of the gas. The differential dU is given by

$$dU = 840 \, dV + 27.32 \, dT.$$

(a)   How does the energy change if the volume is held constant and the temperature is increased slightly?

(b)   How does the energy change if the temperature is held constant and the volume is increased slightly?

(c)   Find the approximate change in energy if the gas is compressed by 100 cm^3 and heated by 2° K.

23.   The area of a rectangle with sides x and y is given by $A = xy$. Find the approximate maximum error ΔA if the measured lengths of the sides are $x = 5$ feet and $y = 6$ feet with maximum errors of 0.01 feet each.

24.   The period of oscillation of a pendulum clock is $2\pi\sqrt{l/g}$, where l, the length of the pendulum, depends on the temperature according to the formula $l = l_0(1 + \alpha t)$ for some value of α which characterizes the clock. The clock is set to the correct period at the temperature t_0. How many seconds a day does the clock lose when the temperature is slightly greater than t_0? Show that this loss is independent of l_0.

25.   In a pendulum experiment, g was determined using the formula

$$g = \frac{4\pi^2 l}{T^2}$$

where the length l is given by $l = s + k^2/s$, $k < s$. If the measurements of k and s are accurate to within 1% find the maximum percentage error in l. If the measurement of T is accurate to 0.5% find the maximum percentage error in the computed value of g.

26.   The coefficient of thermal expansion of a liquid, β, relates the change in its volume V (in m^3) to an increase in its temperature T (in ° C):

$$dV = \beta V \, dT.$$

(a)   Let ρ be the density (in kg/m^3) of 1 kg of water as a function of temperature. Write an expression for $d\rho$ in terms of ρ and dT.

(b) The graph below shows density of water as a function of temperature. Use it to estimate β when $T = 20°C$ and when $T = 80°C$.

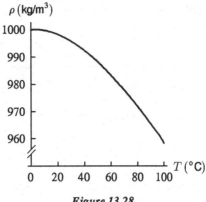

Figure 13.28

27. (a) Write a formula for π using only the perimeter L and the area A of a circle.
    (b) Suppose that L and A are determined empirically. Show that if the percentage errors in the measures values of L and A are λ and μ respectively then the resulting percentage error in π is $2\lambda - \mu$.

13.4 DIRECTIONAL DERIVATIVES

The partial derivatives of a multivariable function f tell you the rate of change of f in the direction of each of the coordinate axes. In this section we will see how to compute the rate of change of f in any direction.

Example 1   Figure 13.29 is a topographical map, showing the elevation $H(x, y)$ of the point (x, y) in feet above sea level. Estimate the slope of the terrain as you walk from A towards B.

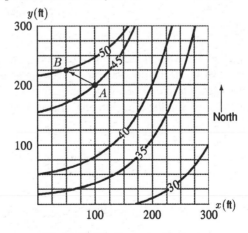

Figure 13.29: A topographical map

Solution   Initially you are on the $H = 45$ contour. By the time you reach B you are on the $H = 50$ contour. The displacement vector from A to B has x component $-50\vec{i}$ and y component $25\vec{j}$, so its length is $\sqrt{(-50)^2 + 25^2} \approx 56$. Thus you have risen 5 feet for a run of 56 feet, so the slope of the terrain in that direction is $5/56 \approx 0.09$.

Definition of the Directional Derivative

Now we use a difference quotient to define the *directional derivative* at any point (a, b) in the direction of a unit vector $\vec{u} = u_1\vec{i} + u_2\vec{j}$. We use a unit vector because this will make the definition easier.

Directional Derivative of f at (a, b) in the direction of $\vec{u}$

If $\vec{u} = u_1\vec{i} + u_2\vec{j}$ is a unit vector, we define

$$f_{\vec{u}}(a, b) = \lim_{h \to 0} \frac{f(a + hu_1, b + hu_2) - f(a, b)}{h},$$

whenever the limit exists.

The numerator in the difference quotient gives the change in f between the point (a, b) and a neighboring point which is displaced by an amount $h\vec{u}$ from (a, b). See Figure 13.30.

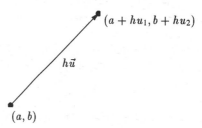

$(a + hu_1, b + hu_2)$

$h\vec{u}$

(a, b)

Figure 13.30: Displacement of $h\vec{u}$ from
the point (a, b)

Since $\|\vec{u}\| = 1$, the displacement $h\vec{u}$ corresponds to moving a distance h (positive or negative) in the direction of $\vec{u}$. Thus, dividing by h gives the average rate of change, and taking the limits gives the instantaneous rate of change in the direction of $\vec{u}$.

Notice that if $\vec{u}$ is in the direction of $\vec{i}$, that is, $u_1 = 1$, $u_2 = 0$, then the directional derivative is the same as f_x, since

$$f_{\vec{i}}(a, b) = \lim_{h \to 0} \frac{f(a + h, b) - f(a, b)}{h} = f_x(a, b).$$

Similarly, if $\vec{u}$ is in the direction $\vec{j}$ then the directional derivative $f_{\vec{j}} = f_y$.

Example 2   For each of the functions f, g, and h in Figure 13.31, decide whether the directional derivative at the indicated point is positive, negative, or zero, in the direction of:
(a)   the vector $\vec{v} = \vec{i} + 2\vec{j}$,   (b)   the vector $\vec{w} = 2\vec{i} + \vec{j}$.

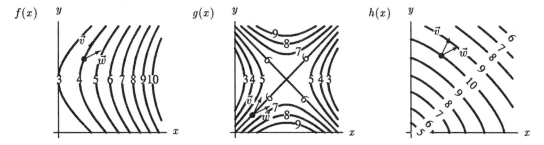

$f(x)$ $g(x)$ $h(x)$

Figure 13.31: Three functions with direction vectors $\vec{v} = \vec{i} + 2\vec{j}$ and $\vec{w} = 2\vec{i} + \vec{j}$ marked.

Solution   On the contour diagram for f, the vector $\vec{v} = \vec{i} + 2\vec{j}$ points in the direction tangent to the contour, so the function is not changing in that direction. Thus, the directional derivative is zero in that direction. The vector $\vec{w} = 2\vec{i} + \vec{j}$ points from the contour marked 4 towards the contour marked 5, so the values of the function are increasing, and thus the directional derivative is positive in that direction.

On the contour diagram for g, the vector $\vec{v} = \vec{i} + 2\vec{j}$ points from the contour marked 6 towards the contour marked 5, so the function is decreasing in that direction. Thus, the directional derivative is negative. On the other hand, the vector $\vec{w} = 2\vec{i} + \vec{j}$ points in the direction of the contour marked 7, and hence the directional derivative is positive.

Finally, on the contour diagram for h, both vectors point from the $h = 10$ contour to the $h = 9$ contour, so both directional derivatives are negative.

Example 3   Use a difference quotient to estimate the directional derivative of $f(x, y) = 2x^2 + y^2$ at the point $(2, 1)$ in the direction of the vector $\vec{u} = (\vec{i} + \vec{j})/\sqrt{2}$.

Solution   We compare values of the function at the point $(2, 1)$ and at the point with coordinates $(2 + \frac{h}{\sqrt{2}}, 1 + \frac{h}{\sqrt{2}})$, that is the change of f between $\vec{r} = 2\vec{i} + \vec{j}$ and $\vec{r} + h\vec{u}$. See Figure 13.32. We want to find the average rate of change between the point $(2, 1)$ and the nearby point $(2 + \frac{h}{\sqrt{2}}, 1 + \frac{h}{\sqrt{2}})$ for a small value of h. We will use $h = 0.01$ and consider the point $(2 + \frac{0.01}{\sqrt{2}}, 1 + \frac{0.01}{\sqrt{2}}) = (2.007, 1.007)$. Since $\vec{u}$ is a unit vector,

$$\text{Average rate of change} \approx \frac{f(2 + \frac{h}{\sqrt{2}}, 1 + \frac{h}{\sqrt{2}}) - f(2, 1)}{h}$$

$$\approx \frac{f(2.007, 1.007) - f(2, 1)}{0.01}$$

$$= \frac{(2(2.007)^2 + (1.007)^2) - (2(2^2) + 1^2)}{0.01} \approx 7.01.$$

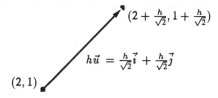

$(2 + \frac{h}{\sqrt{2}}, 1 + \frac{h}{\sqrt{2}})$

$h\vec{u} = \frac{h}{\sqrt{2}}\vec{i} + \frac{h}{\sqrt{2}}\vec{j}$

$(2, 1)$

Figure 13.32: Displacement of $h\vec{u}$ from the point$(2, 1)$

Example 4   Find the directional derivative of $f(x, y) = x^2 + y^2$ in the direction of the vector $\vec{i} + \vec{j}$ at the point $(1, 0)$.

Solution   First we have to find a unit vector in the same direction as the vector $\vec{i} + \vec{j}$. Since this vector has magnitude $\sqrt{2}$, a unit vector is

$$\vec{u} = \frac{1}{\sqrt{2}}(\vec{i} + \vec{j}) = \left(\frac{1}{\sqrt{2}}\vec{i} + \frac{1}{\sqrt{2}}\vec{j} \right)$$

Thus,

$$\begin{aligned}
f_{\vec{u}}(1, 0) &= \lim_{h \to 0} \frac{f(1 + \frac{h}{\sqrt{2}}, \frac{h}{\sqrt{2}}) - f(1, 0)}{h} \\
&= \lim_{h \to 0} \frac{(1 + \frac{h}{\sqrt{2}})^2 + (\frac{h}{\sqrt{2}})^2 - 1}{h} = \lim_{h \to 0} \frac{h^2 + (\frac{2}{\sqrt{2}})h}{h} \\
&= \lim_{h \to 0} (h + \frac{2}{\sqrt{2}}) = \frac{2}{\sqrt{2}} = \sqrt{2}.
\end{aligned}$$

Calculating Directional Derivatives in Cartesian Coordinates

If f is differentiable, we can use local linearity to find a formula for the directional derivative, so that we don't have to compute a limit. The definition says

$$f_{\vec{u}}(a, b) = \lim_{h \to 0} \frac{f(a + hu_1, b + hu_2) - f(a, b)}{h} = \lim_{h \to 0} \frac{\Delta f}{h},$$

where Δf is the change in f between the points (a, b) and $(a + hu_1, b + hu_2)$. Local linearity says

$$\Delta f \simeq f_x(a, b)\Delta x + f_y(a, b)\Delta y,$$

where Δx is the change in x, so

$$\Delta x = (a + hu_1) - a = hu_1$$

and, similarly,

$$\Delta y = (b + hu_2) - b = hu_2.$$

Thus

$$\begin{aligned}
\frac{\Delta f}{h} &\approx \frac{f_x(a, b)\Delta x + f_y(a, b)\Delta y}{h} \\
&= f_x(a, b)\frac{hu_1}{h} + f_y(a, b)\frac{hu_2}{h} \\
&= f_x(a, b)u_1 + f_y(a, b)u_2.
\end{aligned}$$

This approximation gets better and better as $h \to 0$, yielding the following result:

The Directional Derivative in Cartesian Coordinates:

If f is differentiable at (a, b),

$$f_{\vec{u}}(a, b) = f_x(a, b)u_1 + f_y(a, b)u_2,$$

where $\vec{u} = u_1\vec{i} + u_2\vec{j}$ is a unit vector.

Example 5   Use the formula to compute the directional derivative found in Example 4 and check that you get the same answer.

Solution   We have to find $f_{\vec{u}}(1,0)$, where

$$f(x,y) = x^2 + y^2 \quad \text{and} \quad \vec{u} = \frac{1}{\sqrt{2}}\vec{i} + \frac{1}{\sqrt{2}}\vec{j}.$$

The partial derivatives are $f_x(x,y) = 2x$ and $f_y(x,y) = 2y$. So

$$f_{\vec{u}}(1,0) = f_x(1,0)u_1 + f_y(1,0)u_2$$
$$= (2)(\frac{1}{\sqrt{2}}) + (0)(\frac{1}{\sqrt{2}})$$
$$= \frac{2}{\sqrt{2}} = \sqrt{2},$$

which is the answer we got before.

Directional Derivatives for Functions of Three Variables

These work the same way as for functions of two variables.

Example 6   Find the directional derivative of $f(x,y,z) = xy + z$ at the point $(-1,0,1)$ in direction of $\vec{v} = 2\vec{i} + \vec{k}$.

Solution   The magnitude of $\vec{v}$ is $\|\vec{v}\| = \sqrt{2^2 + 1} = \sqrt{5}$, so a unit vector in the same direction is

$$\vec{u} = \frac{\vec{v}}{\|\vec{v}\|} = \frac{2}{\sqrt{5}}\vec{i} + 0\vec{j} + \frac{1}{\sqrt{5}}\vec{k}.$$

The partial derivatives are

$$f_x(x,y,z) = y$$
$$f_y(x,y,z) = x$$
$$f_z(x,y,z) = 1,$$

so,

$$f_{\vec{u}}(-1,0,1) = f_x(-1,0,1)u_1 + f_y(-1,0,1)u_2 + f_z(-1,0,1)u_3$$
$$= (0)(\frac{2}{\sqrt{5}}) + (-1)(0) + (1)(\frac{1}{\sqrt{5}}) = \frac{1}{\sqrt{5}}.$$

Problems for Section 13.4

1. Suppose $f(x,y) = x + \ln y$. Using difference quotients, estimate the following.
   (a) The rate of change of f as you leave the point $(1,4)$ going towards the point $(3,5)$.
   (b) The rate of change of f as you arrive at the point $(3,5)$.

2. Use a difference quotient to estimate the rate of change of $f(x,y) = xe^y$ at the point $(1,1)$ as you move in the direction of the vector $\vec{i} + 2\vec{j}$.

For Problems 3–8 use Figure 13.33, showing level curves of $f(x, y)$, to estimate the directional derivatives.

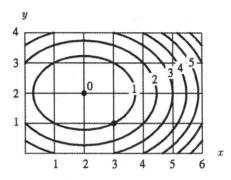

Figure 13.33

3. $f_{\vec{i}}(3, 1)$

4. $f_{\vec{j}}(3, 1)$

5. $f_{\vec{u}}(1, 3)$ where $\vec{u} = (-\vec{i} + \vec{j})/\sqrt{2}$

6. $f_{\vec{u}}(3, 1)$ where $\vec{u} = (\vec{i} - \vec{j})/\sqrt{2}$

7. For what part of the rectangular region shown in Figure 13.33 is $f_{\vec{i}}$ positive?

8. For what part of the rectangular region shown in Figure 13.33 is $f_{\vec{j}}$ negative?

9. The surface $z = g(x, y)$ is shown in Figure 13.34. What is the sign of each of the following directional derivatives?

    (a) $g_{\vec{u}}(2, 5)$ where $\vec{u} = (\vec{i} - \vec{j})/\sqrt{2}$.

    (b) $g_{\vec{u}}(2, 5)$ where $\vec{u} = (\vec{i} + \vec{j})/\sqrt{2}$.

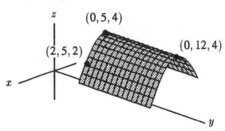

Figure 13.34

10. Using the limit of a difference quotient, compute the rate of change of $f(x, y) = 2x^2 + y^2$ at the point $(2, 1)$ as you move in the direction of the vector $\vec{u} = (\vec{i} + \vec{j})/\sqrt{2}$.

11. True or false? $f_{\vec{u}}(x_0, y_0)$ is a scalar. Explain your answer.

For Problems 12– 14, use the formula on page 140

12. If $f(x, y) = x^2y$ and $\vec{v} = 4\vec{i} - 3\vec{j}$, find the directional derivative at the point $(2, 6)$ in the direction of $\vec{v}$.

13. If $f(x, y, z) = x^2 + 3xy + 2z$, find the directional derivative at the point $(2, 0, -1)$ in the direction of $2\vec{i} + \vec{j} - 2\vec{k}$.

14. Find the directional derivative of $f(x, y, z) = 3x^2y^2 + 2yz$ at the point $(-1, 0, 4)$ in the following directions (a) $\vec{i} - \vec{k}$ (b) $-\vec{i} + 3\vec{j} + 3\vec{k}$.

13.5 THE GRADIENT

The rate of change of a one-variable function $f(x)$ at $x = a$ can be described by a single number $f'(a)$. We have seen that the rate of change of a two-variable function $f(x, y)$ at a point (a, b) cannot be described by a single number, because before you can talk about the rate of change you have to specify a direction. The rate of change in the x-direction is $f_x(a, b)$, the rate of change in the y-direction is $f_y(a, b)$, and, in general, the rate of change in the direction of a unit vector $\vec{u}$ is the directional derivative $f_{\vec{u}}(a, b)$. This suggests that a vector quantity might be better than a scalar quantity for giving complete information about a function's rate of change.

In this section we will define a vector called the *gradient* which describes how the function changes near a point. We use geometric properties to define the gradient vector. Then we show how to compute the components of the gradient vector in Cartesian coordinates.

Geometric Definition of the Gradient Vector

> The **gradient vector** of f at (a, b) is defined to be the vector with the following properties
> * The direction of the gradient vector is the direction in which f is increasing at the greatest rate, if it exists.
> * The magnitude of the gradient vector is the rate of change of f in that direction.
> * If the directional derivative of f at (a, b) is zero in every direction the gradient of f is defined to be the zero vector.
>
> The gradient of f at (a, b) is written grad $f(a, b)$.

Example 1   Figure 13.35 shows the topographical map from Example 1 on page 137. What is the practical significance of the gradient vectors at A and B?

Solution   The gradient vector points in the direction of greatest increase; on a topographical map, that means it points in the direction that will take you directly uphill. The magnitude of the gradient vector measures the steepness of the slope. The gradient vector at A is longer than the gradient vector at B because the contours are closer together at A, so the slope is steeper.

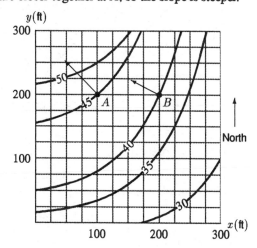

Figure 13.35: A topographical map

Notice that the gradient vector at any point is perpendicular to the contour through that point. This is always the case (if f is differentiable at the point.) To see why, look at Figure 13.36. It shows a close-up of the contours of f around the point (a, b). The contours appear straight, parallel, and equally spaced, because f is locally linear at (a, b). The directional derivative in any direction is approximated by the difference quotient

$$\frac{\text{Change in } f}{\text{Distance traveled}}.$$

Hence, the largest directional derivative is obtained by moving in the direction that takes you to the next contour in the shortest possible distance, which is the direction perpendicular to the contour. Thus we have:

Properties of the Gradient Vector

If f is differentiable at (a, b) then:
- The gradient vector of f at the point (a, b) is perpendicular to the contour of f through (a, b), and points in the direction of increasing f (provided it is not $\vec{0}$).
- The magnitude of the gradient vector, $\| \operatorname{grad} f \|$, is large when the contours are close together and small when they are far apart.

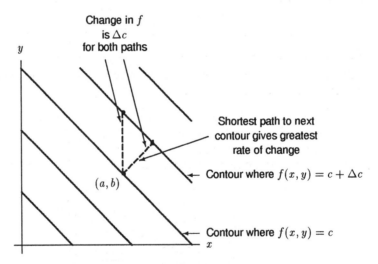

Figure 13.36: The gradient is perpendicular to the contours

Components of the Gradient Vector

According to the definition of grad $f(a, b)$ the directional derivative $f_{\vec{u}}(a, b)$ has its maximum value when $\vec{u}$ is pointing in the direction of grad $f(a, b)$. Furthermore, in that case,

$$f_{\vec{u}}(a, b) = \| \operatorname{grad} f(a, b) \|.$$

On page 140 we saw that if f is differentiable at (a, b) then the directional derivative may be computed by the formula

$$f_{\vec{u}}(a, b) = f_x(a, b)u_1 + f_y(a, b)u_2,$$

where $\vec{u} = u_1\vec{i} + u_2\vec{j}$ is a unit vector. To see when this is at a maximum, we write it as a dot product

$$f_{\vec{u}}(a, b) = (f_x(a, b)\vec{i} + f_y(a, b)\vec{j}) \cdot \vec{u}$$
$$= \|f_x(a, b)\vec{i} + f_y(a, b)\vec{j}\|\|\vec{u}\|\cos\theta$$
$$= \|f_x(a, b)\vec{i} + f_y(a, b)\vec{j}\|\cos\theta,$$

where θ is the angle between $\vec{u}$ and $f_x(a, b)\vec{i} + f_y(a, b)\vec{j}$.

Imagine that (a, b), and hence grad $f(a, b)$, is fixed, but that $\vec{u}$ can rotate. Provided that $f_x(a, b)\vec{i} + f_y(a, b)\vec{j} \neq \vec{0}$, the maximum value of $f_{\vec{u}}(a, b)$ occurs when $\cos\theta = 1$, so $\theta = 0$ and $\vec{u}$ is pointing in the direction of $f_x(a, b)\vec{i} + f_y(a, b)\vec{j}$. Further, the maximum value is

$$f_{\vec{u}}(a, b) = \|f_x(a, b)\vec{i} + f_y(a, b)\vec{j}\|\cos 0 = \|f_x(a, b)\vec{i} + f_y(a, b)\vec{j}\|.$$

Thus the two vectors grad $f(a, b)$ and $(f_x(a, b)\vec{i} + f_y(a, b)\vec{j})$ have the same direction and magnitude. Since two vectors with the same direction and magnitude are the same vector, we have the following result:

Components of the Gradient Vector

If f is differentiable at (a, b), then

$$\text{grad } f(a, b) = f_x(a, b)\vec{i} + f_y(a, b)\vec{j}.$$

If $f_x(a, b)\vec{i} + f_y(a, b)\vec{j} = \vec{0}$, then $f_{\vec{u}}(a, b) = \vec{0}$ for all $\vec{u}$ and so grad $f(a, b) = \vec{0}$. Thus the result is true in that case as well.

Alternative Notation for the Gradient

You can think of $\dfrac{\partial f}{\partial x}\vec{i} + \dfrac{\partial f}{\partial y}\vec{j}$ as the result of applying the vector operator

$$\nabla = \frac{\partial}{\partial x}\vec{i} + \frac{\partial}{\partial y}\vec{j}$$

to the function f. Thus, we get the alternative notation

$$\text{grad } f = \nabla f.$$

Example 2   Find the components of the gradient vector of $f(x, y) = x + e^y$ at the point $(1, 1)$.

Solution   We have

$$\text{grad } f = f_x\vec{i} + f_y\vec{j} = \vec{i} + e^y\vec{j},$$

so at the point $(1, 1)$,

$$\text{grad } f(1, 1) = \vec{i} + e\vec{j}.$$

The Relationship Between the Gradient Vector and Directional Derivatives

In the previous section we saw that

$$f_{\vec{u}}(a, b) = (f_x(a, b)\vec{i} + f_y(a, b)\vec{j}) \cdot \vec{u}$$

and that

$$\text{grad } f(a, b) = f_x(a, b)\vec{i} + f_y(a, b)\vec{j}.$$

Thus we have the following result:

Directional Derivatives and the Gradient Vector

If f is differentiable at (a, b), and if $\vec{u}$ is a unit vector making an angle of θ with the gradient vector:

$$f_{\vec{u}}(a, b) = \text{grad } f(a, b) \cdot \vec{u} = \| \text{grad } f(a, b) \| \cos \theta.$$

In words:
The directional derivative at the point (a, b) in the direction $\vec{u}$ is the dot product of the gradient vector at (a, b) with the vector $\vec{u}$ (provided $\vec{u}$ is a unit vector).

Problem 44 on page 151 shows another way of deriving this formula.

Example 3   Use the gradient to find the rate of change of $f(x, y) = x + e^y$ in the direction of the vector $\vec{i} - \vec{j}$ at the point $(1, 1)$.

Solution   In Example 2 we found
$$\text{grad } f(1, 0) = \vec{i} + e\vec{j}.$$
A unit vector in the direction of $\vec{i} - \vec{j}$ is $\vec{u} = (1/\sqrt{2})\vec{i} - (1/\sqrt{2})\vec{j}$, so

$$f_{\vec{u}}(1, 1) = \text{grad } f(1, 1) \cdot \vec{u} = (\vec{i} + e\vec{j}) \cdot \left(\frac{\vec{i}}{\sqrt{2}} - \frac{\vec{j}}{\sqrt{2}} \right) = \frac{1 - e}{\sqrt{2}} \approx -1.2.$$

The relationship between the directional derivative and the gradient vector tells us a good deal about how a function changes near a point. Imagine as before that grad $f(a, b)$ is fixed, but that the unit vector $\vec{u}$ can rotate. We have already seen that the dot product of grad $f(a, b)$ and $\vec{u}$ is greatest when $\vec{u}$ is in the same direction as grad $f(a, b)$. Then $\theta = 0$ and the maximum rate of change of f at this point is given by

$$f_{\vec{u}}(a, b) = \| \text{grad } f(a, b) \| \cos 0 = \| \text{grad } f(a, b) \|,$$

as we know from the definition of the gradient. What about the minimum rate of change of f, that is, the most negative? This is obtained by taking $\vec{u}$ in the opposite direction to grad f, so that $\theta = \pi$. Thus, the minimum rate of change of f is given by

$$f_{\vec{u}}(a, b) = \| \text{grad } f \| \cos \pi = -\| \text{grad } f(a, b) \|.$$

Finally, whenever $\vec{u}$ is perpendicular to the gradient vector, so $\theta = \pi/2$, the rate of change of f in that direction is zero:

$$f_{\vec{u}}(a, b) = \| \text{grad } f(a, b) \| \cos \frac{\pi}{2} = 0.$$

In summary, we have the following results.

> If f is differentiable at a point and if grad $f \neq \vec{0}$,
> - The maximum rate of change of f at a point is in the direction of the gradient vector at that point, and has magnitude $\|$ grad $f\|$
> - The minimum rate of change of f is in the direction opposite to the gradient, and has value $-\|$ grad $f\|$
> - The rate of change of f in any direction perpendicular to the gradient vector is 0.

The Gradient Vector of a Function of Three Variables

The gradient of a function of three variables is defined the same way as for two variables: It is the vector pointing in the direction of greatest increase of f, whose magnitude is the rate of increase in that direction. In components it is

$$\text{grad } f(a,b,c) = f_x(a,b,c)\vec{i} + f_y(a,b,c)\vec{j} + f_z(a,b,c)\vec{k}.$$

Just as the gradient vector of a function of two variables is perpendicular to the contours, or level curves, so the gradient of a function of three variables is perpendicular to the level surfaces. The reason is the same; the function has the same value all over a level surface, so to change the value as quickly as possible you need to move directly away from the level surface, that is, perpendicular to the surface.

Example 4 Let $f(x,y,z) = x^2 + y^2$ and $g(x,y,z) = -x^2 - y^2 - z^2$. What can you say about the direction of the following vectors?

(a) grad $f(0,1,1)$ (b) grad $f(1,0,1)$ (c) grad $g(0,1,1)$ (d) grad $g(1,0,1)$.

Solution Consider the cylinder $x^2 + y^2 = 1$ in Figure 13.37. This is a level surface of f and contains both the points $(0,1,1)$ and $(1,0,1)$. Since the value of f does not change at all in the z-direction, the gradient vectors are horizontal. They are perpendicular to the cylinder and point outward because the value of f increases as you move out.

Similarly, both points also lie on the same level surface of g, namely the level surface $-x^2 - y^2 - z^2 = -2$, which may also be written $x^2 + y^2 + z^2 = 2$. Part of this level surface is shown in Figure 13.38. Notice that this time the gradient vectors point inward, since the negative signs mean that the function increases (from large negative values to small negative values) as you move inwards.

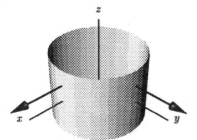

Figure 13.37: The level surface $f(x,y,z) = x^2 + y^2 = 1$ with two gradient vectors

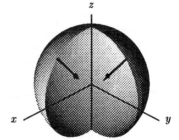

Figure 13.38: The level surface $g(x,y,z) = -x^2 - y^2 - z^2 = -2$ with two gradient vectors

Problems for Section 13.5

1. Suppose that f is differentiable at (a, b). Then there is always a direction in which the rate of change of f at (a, b) is 0. True or false? Explain your answer.

2. If $f(x, y) = x^2 + \ln y$, find (a) grad f, (b) grad f at $(4, 1)$.

In Problems 3–12 compute the gradient of the given function. Assume the variables are restricted to a domain on which the function is defined.

3. $f(m, n) = m^2 + n^2$

4. $f(x, y) = \frac{3}{2}x^5 - \frac{4}{7}y^6$

5. $f(s, t) = \dfrac{1}{\sqrt{s}}(t^2 - 2t + 4)$

6. $f(x, y) = \sin(xy) + \cos(xy)$

7. $f(s, t) = t \cos^2 s$

8. $f(x, y) = y \ln x$

9. $f(\alpha, \beta) = \sqrt{5\alpha^2 + \beta}$

10. $f(\alpha, \beta) = \dfrac{2\alpha + 3\beta}{2\alpha - 3\beta}$

11. $f(x, y) = xye^{\sqrt{x}}$

12. $f(u, v) = \dfrac{u^2 - \sqrt{v}}{u + v}$

In Problems 13–20 find ∇z. Assume the variables are restricted to a domain on which the function is defined.

13. $z = \sin(x/y)$

14. $z = xe^y$

15. $z = (x + y)e^y$

16. $z = \tan^{-1}(x/y)$

17. $z = \sin(x^2 + y^2)$

18. $z = xe^y/(x + y)$

19. $z = \sqrt{x^2 + y^2}$

20. $z = \tan^{-1}(x + y)$

In Problems 21–25 compute the gradient at the specified point.

21. $f(m, n) = 5m^2 + 3n^4$, at $(5, 2)$

22. $f(x, y) = x^2y + 7xy^3$, at $(1, 2)$

23. $f(x, y) = \sqrt{\tan x + y}$, at $(0, 1)$

24. $f(x, y) = \sin(x^2) + \cos y$, at $(\frac{\sqrt{\pi}}{2}, 0)$

25. $f(\alpha, \beta) = \dfrac{5\alpha + 3\beta}{\sqrt{\beta}}$, at $(0, 1)$

26. Let $f(x, y) = (x + y)/(1 + x^2)$ and let $P = (1, -2)$. Find the directional derivative at P in the direction of the vectors
    (a) $\vec{v} = 3\vec{i} - 2\vec{j}$,   (b) $\vec{v} = -\vec{i} + 4\vec{j}$.
    (c) What is the direction of greatest increase at P ?

27. A student was asked to find the directional derivative of $f(x, y) = x^2e^y$ at the point $(1, 0)$ in the direction of $\vec{v} = 4\vec{i} + 3\vec{j}$. His answer was

$$f_{\vec{u}}(1, 0) = \nabla f(1, 0) \cdot \vec{u} = \frac{8}{5}\vec{i} + \frac{3}{5}\vec{j}.$$

    (a) At a glance, how do you know this is wrong?
    (b) What is the correct answer?

28. Find the directional derivative of $f(x, y) = e^x \tan(y) + 2x^2y$ at the point $(0, \pi/4)$ in the following directions (a) $\vec{i} - \vec{j}$ (b) $\vec{i} + \sqrt{3}\vec{j}$.

29. Find the directional derivative of $z = x^2 y$ at the point $(1, 2)$ in the direction making an angle $\theta = 5\pi/4$ with the x-axis. In which direction is the directional derivative the largest?

30. Find the rate of change of $f(x, y) = x^2 + y^2$ at the point $(1, 2)$ in the direction of the vector $\vec{u} = 0.6\vec{i} + 0.8\vec{j}$.

31. The directional derivative of $z = f(x, y)$ at the point $(2, 1)$ in the direction toward the point $(1, 3)$ is $-2/\sqrt{5}$, and the directional derivative in the direction toward the point $(5, 5)$ is 1. Compute $\partial f/\partial x$ and $\partial f/\partial y$ at the point $(2, 1)$.

32. (a) Sketch the surface $z = y^2$ in three dimensions.
    (b) Sketch the level curves on this surface in the xy plane.
    (c) If you are standing on the surface $z = y^2$ at the point $(2, 3, 9)$, in which direction should you move to climb the fastest? (Give your answer as a 2-vector.)

33. Given $z = y - \sin x$,
    (a) Sketch the contours for $z = -1, 0, 1, 2$.
    (b) A bug starts on the surface at the point $(\pi/2, 1, 0)$ and walks on the surface in the direction of the vector $(0, 1, 0)$. Is the bug walking in a valley or on top of a ridge? Explain.
    (c) On the contour for $z = 0$ above, draw the gradients of z at $x = 0$, $x = \pi/2$, and $x = \pi$.

34. Suppose that $F(x, y, z) = x^2 + y^4 + x^2 z^2$ gives the concentration of salt in a fluid at the point (x, y, z), and that you are at the point $(-1, 1, 1)$.
    (a) In which direction should you move if you want the concentration to increase the fastest?
    (b) Suppose you start to move in the direction you found in part (a) at a speed of 4 units/sec. How fast is the concentration changing? Explain your answer.

35. Consider the function $f(x, y)$. If you start at the point $(4, 5)$ and move towards the point $(5, 6)$, the directional derivative is 2. Starting at the point $(4, 5)$ and moving towards the point $(6, 6)$ gives a directional derivative of 3. Find ∇f at the point $(4, 5)$.

36. The temperature at any point in the plane is given by the function

$$T(x, y) = \frac{100}{x^2 + y^2 + 1}.$$

    (a) Where on the plane is it hottest? What is the temperature at that point?
    (b) Find the direction of the greatest increase in temperature at the point $(3, 2)$. What is the magnitude of that greatest increase?
    (c) Find the direction of the greatest decrease in temperature at the point $(3, 2)$.
    (d) Does the vector you found in part (b) point towards the origin?
    (e) Find a direction at the point $(3, 2)$ in which the temperature does not increase or decrease.
    (f) What shape are the level curves of T?

37. Consider S to be the surface represented by the equation $F = 0$, where

$$F(x, y, z) = x^2 - \left(\frac{y}{z^2}\right).$$

    (a) Find all points on S where a normal vector is parallel to the xy-plane.
    (b) Find the tangent plane to S at the points $(0, 0, 1)$ and $(1, 1, 1)$.
    (c) Find the unit vectors $\vec{u}_1$ and $\vec{u}_2$ pointing in the direction of maximum increase of F at the points $(0, 0, 1)$ and $(1, 1, 1)$ respectively.

38. At what point on the surface $z = 1 + x^2 + y^2$ is its tangent plane parallel to the following planes? (a) $z = 5$ (b) $z = 5 + 6x - 10y$.

39. Consider the functions $f(x, y) = x - y^2$ and $g(x, y) = 2x + \ln y$. Sketch three level curves for each function. Show that the level curves of g always intersect the level curves of f at right angles.

40. The sketch in Figure 13.39 shows the level curves of a function $z = f(x, y)$. At the points $(1, 1)$ and $(1, 4)$ on the sketch, draw a vector representing grad f. Explain how you decided the approximate direction and length of each vector.

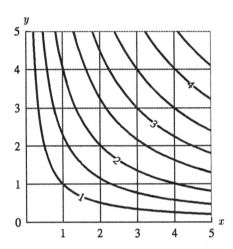

Figure 13.39: Contours of f

41. Figure 13.40 represents the level curves $f(x, y) = c$; the values of f on each curve are marked. In each of the following parts, decide whether the given quantity is positive, negative or zero. Explain your answer.

(a) The value of $\nabla f \cdot \vec{i}$ at P.

(b) The value of $\nabla f \cdot \vec{j}$ at P.

(c) $\partial f / \partial x$ at Q.

(d) $\partial f / \partial y$ at Q.

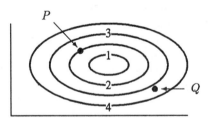

Figure 13.40

42. In Figure 13.40, which is larger: $\|\nabla f\|$ at P or $\|\nabla f\|$ at Q? Explain how you know.

43. In Problem 33 on page 40, we considered the temperature distribution given by the function

$$T(x, y) = 100 - x^2 - y^2$$

and asked in which direction a heat-seeking bug should move from any point in the plane to increase its temperature fastest. Now answer this question using the gradient.

44.   In this problem we will see another way of explaining the formula on page 146 which expresses the directional derivative in terms of the gradient vector. Imagine zooming in on a function $f(x, y)$ at a point (a, b). By local linearity, the contours around (a, b) look like the contours of a linear function. Figure 13.41 shows the contours and the gradient vector at the point (a, b). Suppose you want to find the directional derivative $f_{\vec{u}}(a, b)$ in the direction of a unit vector $\vec{u}$. If you move a small distance h in the direction of $\vec{u}$, the directional derivative is approximated by the difference quotient

$$\frac{\text{Change in } f}{h}.$$

(a)   Use the gradient to to show that

$$\text{Change in } f \approx \| \operatorname{grad} f \|(h \cos \theta).$$

(b)   Use part (a) to justify the formula $f_{\vec{u}}(a, b) = \operatorname{grad} f(a, b) \cdot \vec{u}$.

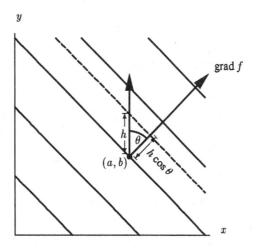

Figure 13.41

13.6 THE CHAIN RULE

How Do We Compose Functions of Many Variables?

The purpose of the chain rule is to enable us to differentiate composite functions, so we start by thinking about what it means to compose functions. To compose functions of one variable we substitute one function into another. A function of more than one variable has more than one place you can substitute a function; for example, if $z = f(x, y)$, then we could substitute two functions, $x = g(t)$ and $y = h(t)$. Then the composite function is $f(g(t), h(t))$. For example, if $f(x, y) = x^2 - y^2$, $g(t) = \cos t$ and $h(t) = \sin t$, then $f(g(t), h(t)) = \cos^2 t - \sin^2 t$. Although it looks complicated, this is now a function of just one variable, t. The next example shows where this sort of substitution arises in practice, and how it leads to the chain rule.

Example 1   Figure 13.42 shows the contour diagram for corn production, C, as a function of temperature, T, and rainfall, R (See Example 5 on page 131). Figure 13.43 and Figure 13.44 show how rainfall and temperature are predicted to vary with time according to a certain model of global warming. Estimate the change in corn production between the year 2000 and the year 2001. Hence, estimate dC/dt when $t = 2000$.

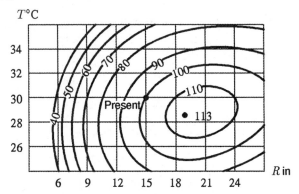

Figure 13.42: Corn production as a function of rainfall and temperature

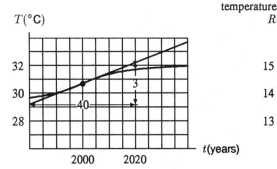

Figure 13.43: Global warming predictions: Temperature as a function of time

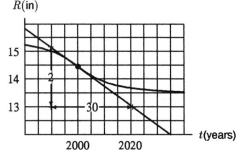

Figure 13.44: Global warming predictions: Rainfall as a function of time

Solution   In Example 5 on page 131 we found that, as long as the temperature and rainfall stay close to their current values of $R = 15$ inches and $T = 30°\text{C}$, a change ΔR in rainfall and a change ΔT in temperature will produce a change in corn production given by

$$\Delta C \approx 3.3\Delta R - 5\Delta T.$$

Now we have a situation where both R and T are functions of time t (in years), and we need to find the effect of a small change in time, Δt, on R and T. Looking at Figure 13.44, we see that the slope of the graph for R versus t is about $-2/30 \approx -0.07$in/year when $t = 2000$. Similarly, the slope of the graph of T is about $3/40 \approx 0.08°C$/year when $t = 2000$. Thus, around the year 2000,

$$\Delta R \approx -0.07\Delta t, \quad \text{and} \quad \Delta T \approx 0.08\Delta t.$$

Substituting these into the equation for ΔC, we get

$$\Delta C \approx (3.3)(-0.07)\Delta t - (5)(0.08)\Delta t \approx -0.6\Delta t.$$

Since at present $C = 100$, corn production will decline by about 0.6 % between the years 2000 and 2001. Now $\Delta C \approx -0.6\Delta t$ tells us that when $t = 2000$,

$$\frac{\Delta C}{\Delta t} \approx -0.6 \quad \text{and therefore that} \quad \frac{dC}{dt} \approx -0.6.$$

Keeping Track of Variables with Tree Diagrams

When you are substituting functions of many variables into other functions of many variables, it is easy to lose track of the relations between the variables. It is helpful to visualize these relations with a diagram. For example, if $z = f(x, y)$, and if $x = g(t)$ and $y = h(t)$, we use the diagram in Figure 13.45. Notice that each line in the diagram is labeled with a derivative relating the variables at its ends.

The diagram helps us keep track of how a change in t propagates through the chain of composed functions, and hence gives us the formula for the chain rule in this situation. There are two paths from t to z, one through x and one through y. For each path, we multiply together the derivatives along the path. Then, to calculate dz/dt, we add up the contributions from both paths. Thus we get

---

If f, g and h are differentiable and if $z = f(x, y)$, $x = g(t)$, $y = h(t)$, then

$$\frac{dz}{dt} = \frac{\partial z}{\partial x}\frac{dx}{dt} + \frac{\partial z}{\partial y}\frac{dy}{dt}.$$

To see why this is true, we substitute the local linearizations

$$\Delta x \approx \frac{dx}{dt}\Delta t$$

and

$$\Delta y \approx \frac{dy}{dt}\Delta t$$

into the local linearization

$$\Delta z \approx \frac{\partial z}{\partial x}\Delta x + \frac{\partial z}{\partial y}\Delta y,$$

yielding

$$\Delta z \approx \frac{\partial z}{\partial x}\frac{dx}{dt}\Delta t + \frac{\partial z}{\partial y}\frac{dy}{dt}\Delta t$$

$$= \left(\frac{\partial z}{\partial x}\frac{dx}{dt} + \frac{\partial z}{\partial y}\frac{dy}{dt}\right)\Delta t.$$

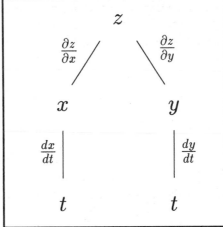

Figure 13.45: Tree diagram for $z = f(x, y)$, $x = f(t)$, $y = g(t)$

Thus

$$\frac{\Delta z}{\Delta t} \approx \frac{\partial z}{\partial x}\frac{dx}{dt} + \frac{\partial z}{\partial y}\frac{dy}{dt}.$$

Taking the limit, we get the chain rule formula above.

Example 2 Suppose that $z = f(x,y) = x \sin y$, where $x = t^2$ and $y = 2t + 1$. Let $z = g(t)$. Compute $g'(t)$ by two different methods.

Solution Since $z = g(t) = f(t^2, 2t + 1) = t^2 \sin(2t + 1)$, it is possible to compute $g'(t)$ directly with one-variable methods:

$$g'(t) = t^2 \frac{d}{dt}(\sin(2t + 1)) + (\frac{d}{dt}(t^2)) \sin(2t + 1)$$
$$= 2t^2 \cos(2t + 1) + 2t \sin(2t + 1).$$

The chain rule provides an alternative route to the same answer. We have

$$\frac{dz}{dt} = \frac{\partial z}{\partial x}\frac{dx}{dt} + \frac{\partial z}{\partial y}\frac{dy}{dt} = (\sin y)(2t) + (x \cos y)(2)$$
$$= 2t \sin(2t + 1) + 2t^2 \cos(2t + 1).$$

The Chain Rule

To find the rate of change of one variable with respect to another in a chain of composed differentiable functions
- Draw a tree diagram expressing the relationship between the variables, and label each line in the diagram with the derivative relating the variables at its ends.
- For each path between the two variables, multiply together all the derivatives along the path.
- Add up the contributions from each path.

This method of finding rates of change works because the tree diagram keeps track of all the ways in which a change in one variable can cause a change in another; the contributions from the diagram are the terms you would get if you made the appropriate substitutions into the local linearizations.

Example 3 Suppose that $z = f(x,y)$, $x = g(u,v)$, and $y = h(u,v)$. Find general formulas for $\frac{\partial z}{\partial u}$ and $\frac{\partial z}{\partial v}$.

Solution Figure 13.46 shows the tree diagram for this situation. Adding the contributions for the two paths from z to u, we get

$$\frac{\partial z}{\partial u} = \frac{\partial z}{\partial x}\frac{\partial x}{\partial u} + \frac{\partial z}{\partial y}\frac{\partial y}{\partial u}.$$

Similarly, looking at the paths from z to v, we get

$$\frac{\partial z}{\partial v} = \frac{\partial z}{\partial x}\frac{\partial x}{\partial v} + \frac{\partial z}{\partial y}\frac{\partial y}{\partial v}.$$

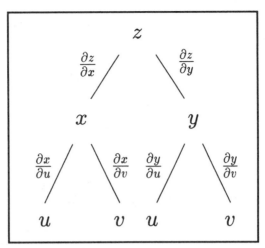

Figure 13.46: Tree diagram for $z = f(x, y)$, $x = g(u, v)$, $y = h(u, v)$

Example 4   Let $w = x^2 e^y$, $x = 4u$, and $y = 3u^2 - 2v$. Compute $\dfrac{\partial w}{\partial u}$ and $\dfrac{\partial w}{\partial v}$ using the chain rule.

Solution   We have

$$\frac{\partial w}{\partial u} = \frac{\partial w}{\partial x}\frac{\partial x}{\partial u} + \frac{\partial w}{\partial y}\frac{\partial y}{\partial u}$$
$$= 2xe^y(4) + x^2 e^y(6u)$$
$$= (8x + 6x^2 u)e^y$$
$$= (32u + 96u^3)e^{3u^2 - 2v},$$

and

$$\frac{\partial w}{\partial v} = \frac{\partial w}{\partial x}\frac{\partial x}{\partial v} + \frac{\partial w}{\partial y}\frac{\partial y}{\partial v}$$
$$= 2xe^y(0) + x^2 e^y(-2)$$
$$= -2x^2 e^y$$
$$= -32u^2 e^{3u^2 - 2v}.$$

Example 5   Suppose a quantity z can be expressed either as a function of x and y, $z = f(x, y)$, or as a function of u and v, $z = g(u, v)$, where the two coordinate systems are related by

$$x = u + v, \quad y = u - v.$$

(a)   Use the chain rule to express $\dfrac{\partial z}{\partial u}$ and $\dfrac{\partial z}{\partial v}$ in terms of $\dfrac{\partial z}{\partial x}$ and $\dfrac{\partial z}{\partial y}$.

(b)   Solve the equations you have just written down for $\dfrac{\partial z}{\partial x}$ and $\dfrac{\partial z}{\partial y}$.

(c)   Show that the expressions you get in part (b) are the same as you would get by expressing u and v in terms of x and y and using the chain rule.

Solution   (a)   Since $\dfrac{\partial x}{\partial u} = 1, \dfrac{\partial x}{\partial v} = 1$, and $\dfrac{\partial y}{\partial u} = 1, \dfrac{\partial y}{\partial v} = -1$, we have

$$\frac{\partial z}{\partial u} = \frac{\partial z}{\partial x}(1) + \frac{\partial z}{\partial y}(1) = \frac{\partial z}{\partial x} + \frac{\partial z}{\partial y}$$

and

$$\frac{\partial z}{\partial v} = \frac{\partial z}{\partial x}(1) + \frac{\partial z}{\partial y}(-1) = \frac{\partial z}{\partial x} - \frac{\partial z}{\partial y}.$$

(b)   Adding together the equations for $\frac{\partial z}{\partial u}$ and $\frac{\partial z}{\partial v}$, we get

$$\frac{\partial z}{\partial u} + \frac{\partial z}{\partial v} = 2\frac{\partial z}{\partial x},$$

so

$$\frac{\partial z}{\partial x} = \frac{1}{2}\frac{\partial z}{\partial u} + \frac{1}{2}\frac{\partial z}{\partial v}.$$

Similarly, subtracting the equations yields

$$\frac{\partial z}{\partial y} = \frac{1}{2}\frac{\partial z}{\partial u} - \frac{1}{2}\frac{\partial z}{\partial v}.$$

(c)   On the other hand, we can solve the equations

$$x = u + v, \quad y = u - v$$

for u and v, which yields

$$u = \frac{1}{2}x + \frac{1}{2}y, \quad v = \frac{1}{2}x - \frac{1}{2}y.$$

Now we can think of z as a function of u and v, and u and v as functions of x and y, and apply the chain rule again. Thus,

$$\frac{\partial z}{\partial x} = \frac{\partial z}{\partial u}\frac{\partial u}{\partial x} + \frac{\partial z}{\partial v}\frac{\partial v}{\partial x} = \frac{1}{2}\frac{\partial z}{\partial u} + \frac{1}{2}\frac{\partial z}{\partial v}$$

and

$$\frac{\partial z}{\partial y} = \frac{\partial z}{\partial u}\frac{\partial u}{\partial y} + \frac{\partial z}{\partial v}\frac{\partial v}{\partial y} = \frac{1}{2}\frac{\partial z}{\partial u} - \frac{1}{2}\frac{\partial z}{\partial v}.$$

These are the same expressions we got in (b).

Problems for Section 13.6

For Problems 1–7, find dz/dt using the chain rule. Assume the variables are restricted to a domain on which the function is defined.

1. $z = xy^2$, $x = e^{-t}$, $y = \sin t$

2. $z = x \sin y + y \sin x$, $x = t^2$, $y = \ln t$

3. $z = \ln(x^2 + y^2)$, $x = 1/t$, $y = \sqrt{t}$

4. $z = \sin(x/y)$, $x = 2t$, $y = 1 - t^2$

5. $z = xe^y$, $x = 2t$, $y = 1 - t^2$

6. $z = (x + y)e^y$, $x = 2t$, $y = 1 - t^2$

7. $z = \tan^{-1}(x/y)$, $x = 2t$, $y = 1 - t^2$

For Problems 8–17, find $\partial z/\partial u$ and $\partial z/\partial v$. Assume the variables are restricted to a domain on which the function is defined.

8. $z = xe^{-y} + ye^{-x}$, $x = u \sin v$,
    $y = v \cos u$

9. $z = \cos(x^2 + y^2)$, $x = u \cos v$,
    $y = u \sin v$

10. $z = xe^y$, $x = \ln u$, $y = v$

11. $z = (x + y)e^y$, $x = \ln u$, $y = v$

12. $z = \tan^{-1}(x/y)$, $x = \ln u$, $y = v$

13. $z = \sin(x/y)$, $x = u^2 + v^2$, $y = u^2 - v^2$

14. $z = xe^y$, $x = u^2 + v^2$, $y = u^2 - v^2$

15. $z = (x+y)e^y$, $x = u^2 + v^2$, $y = u^2 - v^2$

16. $z = \tan^{-1}(x/y)$, $x = u^2 + v^2$, $y = u^2 - v^2$

17. $z = \sin(x/y)$, $x = \ln u$, $y = v$

18. Suppose $f(x, y)$ is a function of x and y and define

$$g(u, v) = f(e^u + \cos v, e^u + \sin v).$$

Find $g_u(0, 0)$ given that

$$
\begin{array}{lllll}
f(0,0) = \pi, & f_x(0,0) = 1, & f_x(1,2) = 3, & f_x(2,1) = 5, & f(2,1) = e, \\
g(0,0) = e, & f_y(0,0) = 2, & f_y(1,2) = 4, & f_y(2,1) = 6, & f(1,2) = \pi^2.
\end{array}
$$

For Problems 19–20, suppose the quantity z can be expressed either as a function of Cartesian coordinates (x, y) or as a function of polar coordinates (r, θ), so that $z = f(x, y) = g(r, \theta)$.

19. (a) Use the chain rule to find $\partial z/\partial r$ and $\partial z/\partial \theta$ in terms of $\partial z/\partial x$ and $\partial z/\partial y$.
    (b) Solve the equations you have just written down for $\partial z/\partial x$ and $\partial z/\partial y$.
    (c) Show that the expressions you get in part (b) are the same as you would get by using the chain rule to find $\partial z/\partial x$ and $\partial z/\partial y$ in terms of $\partial z/\partial r$ and $\partial z/\partial \theta$.

20. Show that
$$\left(\frac{\partial z}{\partial x}\right)^2 + \left(\frac{\partial z}{\partial y}\right)^2 = \left(\frac{\partial z}{\partial r}\right)^2 + \frac{1}{r^2}\left(\frac{\partial z}{\partial \theta}\right)^2.$$

21. Suppose $w = f(x, y, z)$ and that x, y, z are functions of u and v. Draw a tree diagram representing the relation between the variables and use it to derive the chain rule formula for $\frac{\partial w}{\partial u}$ and $\frac{\partial w}{\partial v}$.

22. Let $F(x, y, z)$ be a function and define a function $z = f(x, y)$ implicitly by letting $F(x, y, f(x, y)) = 0$. Use the chain rule to show that

$$\frac{\partial z}{\partial x} = -\frac{\partial F/\partial x}{\partial F/\partial z} \quad \text{and} \quad \frac{\partial z}{\partial y} = -\frac{\partial F/\partial y}{\partial F/\partial z}.$$

13.7 SECOND-ORDER PARTIAL DERIVATIVES

Since the partial derivatives of a function are themselves functions, we can calculate their partial derivatives, called *second-order partial derivatives*. For a function $f(x, y)$, we can differentiate either f_x or f_y with respect to either x or y, so there are four second-order partial derivatives: the two partial derivatives $(f_x)_x$ and $(f_x)_y$ of f_x and the two partial derivatives $(f_y)_x$ and $(f_y)_y$ of f_y. It is convenient to omit the parentheses, writing, for example, f_{xy} instead of $(f_x)_y$. We therefore have f_{xx}, f_{xy}, f_{yx}, and f_{yy}. Similarly, if $z = f(x, y)$, we will usually shorten $\frac{\partial}{\partial x}\left(\frac{\partial z}{\partial x}\right)$ to $\frac{\partial^2 z}{\partial x^2}$ and $\frac{\partial}{\partial y}\left(\frac{\partial z}{\partial x}\right)$ to $\frac{\partial^2 z}{\partial y \partial x}$.

> ### The Second-Order Partial Derivatives of $z = f(x, y)$ are as follows
>
> $$\frac{\partial^2 z}{\partial x^2} = f_{xx} = (f_x)_x \qquad \frac{\partial^2 z}{\partial x \partial y} = f_{yx} = (f_y)_x$$
>
> $$\frac{\partial^2 z}{\partial y \partial x} = f_{xy} = (f_x)_y \qquad \frac{\partial^2 z}{\partial y^2} = f_{yy} = (f_y)_y$$

Example 1 Compute the four second-order partial derivatives of $f(x, y) = y \cos x + 3x^2 e^y$.

Solution From $f_x(x, y) = -y \sin x + 6x e^y$ we get

$$f_{xx}(x, y) = \frac{\partial}{\partial x}(-y \sin x + 6x e^y) = -y \cos x + 6 e^y$$

and

$$f_{xy}(x, y) = \frac{\partial}{\partial y}(-y \sin x + 6x e^y) = -\sin x + 6x e^y.$$

From $f_y(x, y) = \cos x + 3x^2 e^y$ we get

$$f_{yx}(x, y) = \frac{\partial}{\partial x}(\cos x + 3x^2 e^y) = -\sin x + 6x e^y$$

and

$$f_{yy}(x, y) = \frac{\partial}{\partial y}(\cos x + 3x^2 e^y) = 3x^2 e^y.$$

Observe that in this example we have $f_{xy} = f_{yx}$.

What Do the Second-Order Partial Derivatives Tell Us?

Example 2 Let us return to the guitar string of Example 2, page 122. The string is 1 meter long and at time t seconds, the point x meters from one end is displaced $f(x, t)$ meters from its rest position, where

$$f(x, t) = 0.003 \sin(\pi x) \sin(2765 t).$$

Compute the four second-order partial derivatives of f at the point $(x, t) = (0.3, 1)$ and describe their meaning in practical terms.

Solution First we compute $f_x(x, t) = 0.003\pi \cos(\pi x) \sin(2765 t)$, from which we get

$$f_{xx}(x, t) = \frac{\partial}{\partial x}(f_x(x, t)) = -0.003\pi^2 \sin(\pi x) \sin(2765 t),$$
$$f_{xx}(0.3, 1) \approx -0.01;$$

and

$$f_{xt}(x, t) = \frac{\partial}{\partial t}(f_x(x, t)) = (0.003)(2765)\pi \cos(\pi x) \cos(2765 t),$$
$$f_{xt}(0.3, 1) \approx 14.$$

In Example 2 on page 122 we saw that $f_x(0.3, 1)$ is the slope of the guitar string at the point $x = 0.3$ and at the instant $t = 1$. More generally, $f_x(x, t)$ gives the slope of the string at any point and time.

Therefore, $f_{xx}(x, t)$ is the concavity of the string. The fact that $f_{xx}(0.3, 1)$ is negative means the string is concave down at $t = 1$. See Figure 13.47.

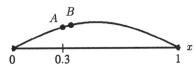

The slope at B is less than the slope at A

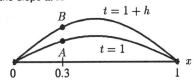

The slope at B is greater than the slope at A

$t = 1 + h$

$t = 1$

Figure 13.47: The concavity of the string at $t = 1$

Figure 13.48: The slope of one point on the string at two different times

On the other hand, $f_{xt}(0.3, 1)$ is the rate of change of the slope of the string with respect to time; $f_{xt}(0.3, 1) = 14$ means that at time $t = 1$ the slope at the point $x = 0.3$ is increasing. See Figure 13.48.

Now we compute

$$f_t(x, t) = (0.003)(2765) \sin(\pi x) \cos(2765t),$$

from which we get

$$f_{tx}(x, t) = \frac{\partial}{\partial x}(f_t(x, t)) = (0.003)(2765)\pi \cos(\pi x) \cos(2765t),$$
$$f_{tx}(0.3, 1) \approx 14;$$

and

$$f_{tt}(x, t) = \frac{\partial}{\partial t}(f_t(x, t)) = -(0.003)(2765)^2 \sin(\pi x) \sin(2765t),$$
$$f_{tt}(0.3, 1) \approx -7200.$$

On page 122 we saw that $f_t(0.3, 1)$ is the velocity of the guitar string at the point $x = 0.3$ and at the instant $t = 1$. More generally, $f_t(x, t)$ gives the velocity of the string at any point and time. Therefore, $f_{tx}(x, t)$ and $f_{tt}(x, t)$ will both be rates of change of velocity. That $f_{tx}(0.3, 1) = 14$ means that at time $t = 1$ the velocities of points just to the right of $x = 0.3$ are greater than the velocity at $x = 0.3$. See Figure 13.49. That $f_{tt}(0.3, 1) = -7200$ means that the velocity of the point $x = 0.3$ is decreasing at time $t = 1$. Thus, $f_{tt}(0.3, 1) = -7200$ m/sec^2 is the acceleration of the point. See Figure 13.50.

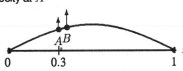

The velocity at B is greater than the velocity at A

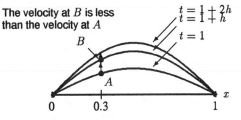

The velocity at B is less than the velocity at A

$t = 1 + 2h$
$t = 1 + h$
$t = 1$

Figure 13.49: The velocity of different points on the string at $t = 1$

Figure 13.50: The velocity of one point on the string at three different times

The Mixed Partials are Equal

Notice that in both Example 1 and Example 2 the mixed partials f_{xy} and f_{yx} or f_{xt} and f_{tx} are equal. This is no accident. We have the following result:

If f_{xy} and f_{yx} are continuous at (a, b), then

$$f_{xy}(a, b) = f_{yx}(a, b)$$

or, in the other notation, that

$$\frac{\partial^2 z}{\partial y \partial x} = \frac{\partial^2 z}{\partial x \partial y}.$$

In words, if we differentiate first with respect to x, getting f_x and then with respect to y, getting f_{xy}, the result will be the same as first differentiating with respect to y, getting f_y, and then differentiating with respect to x, getting f_{yx}. The order in which we take the derivatives does not affect the final outcome.

To see why this is true, use the definitions

$$f_x(a, b) = \lim_{h \to 0} \frac{f(a + h, b) - f(a, b)}{h},$$

and

$$\begin{aligned} f_{xy}(a, b) &= \lim_{k \to 0} \frac{f_x(a, b + k) - f_x(a, b)}{k} \\ &= \lim_{k \to 0} \frac{1}{k} \left(\lim_{h \to 0} \frac{f_x(a + h, b + k) - f_x(a, b + k)}{h} - \lim_{h \to 0} \frac{f_x(a + h, b) - f_x(a, b)}{h} \right) \\ &= \lim_{k \to 0} \lim_{h \to 0} \frac{f(a + h, b + k) - f(a, b + k) - f(a + h, b) + f(a, b)}{hk}. \end{aligned}$$

Similarly,

$$f_{yx}(a, b) = \lim_{h \to 0} \lim_{k \to 0} \frac{f(a + h, b + k) - f(a + h, b) - f(a, b + k) + f(a, b)}{hk}.$$

The numerators in these two expressions are the same (just swap the middle terms), so the only difference between them is the order in which the limits are taken. Swapping limits can be a tricky business, but it can be done in this case if f_{xy} and f_{yx} are continuous. [1] Thus the two mixed partials are equal.

Most of the functions we will encounter not only have f_{xy} and f_{yx} continuous, but all their higher order partial derivatives (e.g. f_{xyx}, or f_{xxxx}) exist and are continuous. Such functions are called *smooth*. Smooth functions have many nice properties, for example, they are differentiable.

Problems for Section 13.7

In Problems 1–8, calculate all four second-order partial derivatives and show that $z_{xy} = z_{yx}$. Assume the variables are restricted to a domain on which the function is defined.

1. $z = \sin(x/y)$
2. $z = xe^y$
3. $z = (x + y)e^y$
4. $z = \tan^{-1}(x/y)$

---

[1] Proving this requires the notions of uniform continuity and uniform convergence, which are treated in advanced calculus books.

5. $z = \sin(x^2 + y^2)$

6. $z = xe^y/(x+y)$

7. $z = \sqrt{x^2 + y^2}, \quad x^2 + y^2 \neq 0$

8. $z = \tan^{-1}(x+y)$

In Problems 9–12, calculate all four second-order partial derivatives and show that $f_{xy} = f_{yx}$. Assume the variables are restricted to a domain on which the function is defined.

9. $f(x, y) = x^y$

10. $f(x, y) = \ln(xy)$

11. $f(x, y) = (x+y)^2$

12. $f(x, y) = (x+y)^3$

13. If $z = f(x) + yg(x)$, what can you say about z_{yy}? Why?

14. If $z_{xy} = 4y$, what can you say about the value of (a) z_{yx}?   (b) z_{xyx}?   (c) z_{xyy}?

In Problems 15–26, use the level curves of the function $z = f(x, y)$ to decide the sign (positive, negative, or zero) of each of the following partial derivatives at the point P. Assume the x- and y-axes are in the usual positions.
(a) $f_x(P)$ (b) $f_y(P)$ (c) $f_{xx}(P)$ (d) $f_{yy}(P)$ (e) $f_{xy}(P)$

15.

16.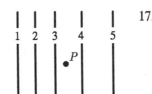

17.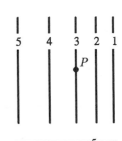

18.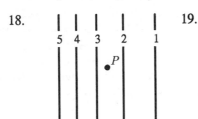

19.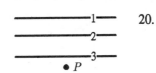

20.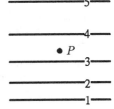

21.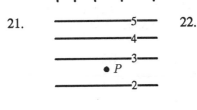

22.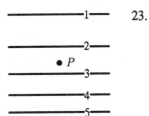

23.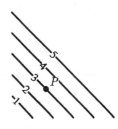

24.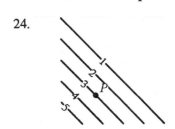

25.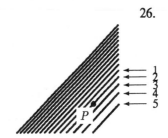

26.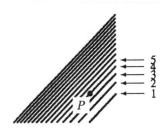

13.8 PARTIAL DIFFERENTIAL EQUATIONS

Heat Flow

Imagine a room heated by a radiator along one wall. What will happen after a window on the opposite wall is opened on a cold day? The temperature in the room will begin to drop, quickly near the window, more slowly near the heater. Eventually the temperature will stabilize, with the room temperature near the window close to the outside temperature, the temperature near the radiator close to what it was before the window was opened, and the temperature in the middle of the room somewhere in between. The temperature $H = u(x, t)$ at any point in the room is a function of the distance x in meters from the heated wall and the time t in minutes since the window was opened. Figure 13.51 shows how the temperature, H, might look as a function of x at various values of t.

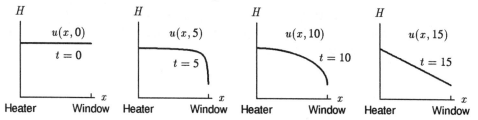

Figure 13.51: Temperature of a room at four different times

The partial derivative $u_t(x, t)$ gives the rate of change of temperature with respect to time at any point and time in the room. The temperature in a region of the room will be increasing if the heat flow into it from the nearby hotter parts of the room is greater than the heat flow out of it to the nearby colder parts and it will be decreasing if the inflow is less than the outflow.

The heat flow between two points will depend on the temperature difference between them: the greater the temperature difference, the greater the rate of flow. A more precise formulation of this statement is Newton's Law of Cooling, which predicts that the rate of heat flow past a point x at time t is proportional to the partial derivative $u_x(x, t)$. This makes sense, since $u_x(x, t)$ measures the rate of change of u with respect to x at a fixed time t; in other words, it tells us how large the temperature difference between neighboring points is.

Example 1   Figure 13.52 gives the graph of temperature $H = u(x, t)$ at a fixed time t. Use Newton's Law of Cooling to determine whether the temperature is increasing or decreasing

(a)   at the point p (b)   at the point q.

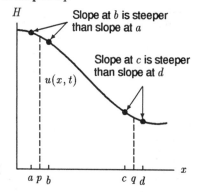

Figure 13.52: The temperature at time t

Solution (a) The question is whether the small region $a < x < b$ that contains p is gaining or losing energy. (See Figure 13.52.) Since energy flows from hotter regions to colder regions, the energy flow is from left to right on the graph. We might say that energy flows downhill on a temperature graph; Newton's Law of Cooling asserts that the steeper the slope, which equals $u_x(x, t)$, the greater the rate of flow. Energy flows into the region $a < x < b$ at the point a, and it flows out at b. The outflow is greater than the inflow because the tangent line at b is steeper than the tangent line at a. The region $a < x < b$ is losing energy and so its temperature is decreasing. Thus, $u_t(p, t) < 0$.

(b) Energy flows into the small region $c < x < d$ at the point c and flows out at d, because the slopes $u_x(c, t)$ and $u_x(d, t)$ are negative. But the inflow is greater than the outflow, by Newton's Law of Cooling, because the tangent line at c is steeper than the tangent line at d. The region $c < x < d$ is gaining energy and so its temperature is increasing. Thus, $u_t(q, t) > 0$.

The Heat Equation

In a problem about heat flow, whether the temperature is increasing or decreasing at some point is determined by the sign of the partial derivative u_t. Example 1 shows that the sign of u_t can be found by studying the shape of the graph of $u(x, t)$ against x for fixed t. The relevant feature of the graph is the way in which the slope $u_x(x, t)$ changes when x varies, and that is determined by the concavity of the graph, which is measured by the second derivative $u_{xx}(x, t)$. In Example 1 the graph of u is concave down at $x = p$, so $u_{xx}(p, t) < 0$, and we found that $u_t(p, t) < 0$. At the point q we have $u_{xx}(q, t) > 0$ and $u_t(q, t) > 0$. In a heat flow problem the two derivatives u_t and u_{xx} are definitely related, for they always have the same sign. In many situations the two derivatives u_t and u_{xx} are actually proportional.

So the function $u(x, t)$ satisfies the following equation:

The One-Dimensional Heat Equation (or Diffusion Equation)

$$u_t(x, t) = A u_{xx}(x, t) \quad \text{where } A \text{ is a positive constant.}$$

The value of the constant of proportionality depends on the situation. The heat equation is an example of a partial differential equation (PDE), that is, an equation involving one or more of the partial derivatives of a function. The unknown quantity in a PDE is the function itself.

Example 2 Which of the following two functions satisfies the heat equation $u_t = u_{xx}$?

(a) $u(x, t) = e^{-4t} \sin(2x)$
(b) $u(x, t) = \sin(x + t)$

Solution (a) $u_t = -4e^{-4t} \sin(2x)$, $u_x = 2e^{-4t} \cos(2x)$, $u_{xx} = -4e^{-4t} \sin(2x)$, so $u_t = u_{xx}$. Thus, $u(x, t) = e^{-4t} \sin(2x)$ is a solution.

(b) $u_t = \cos(x + t)$, $u_x = \cos(x + t)$, $u_{xx} = -\sin(x + t)$, so $u_t \neq u_{xx}$. Thus, $u(x, t) = \sin(x + t)$ is not a solution.

Example 3   A 10 cm metal rod is wrapped with insulation so that heat can flow along the rod but cannot radiate into the air except at the ends. The temperature H (°C) at x cm from one end and at time t seconds is a function $H = u(x, t)$ that satisfies the heat equation $u_t(x, t) = 0.1u_{xx}(x, t)$. At $t = 0$ the temperature at several points is given in Table 13.9.

TABLE 13.9: *Temperature in metal rod at time $t = 0$*

x (cm)	0	2	4	6	8	10
$u(x, 0)$ (°C)	50	52	56	62	70	80

(a)   Is the temperature at the point $x = 6$ increasing or decreasing at $t = 0$?

(b)   Make an estimate of the temperature $H = u(6, 1)$ at the point $x = 6$ at time $t = 1$.

Solution   (a)   The graph of $u(x, 0)$ in Figure 13.53 shows that $u(x, 0)$ is concave up, so $u_{xx}(6, 0) > 0$. Since $u_t(6, 0) = 0.1u_{xx}(6, 0)$ we have $u_t(6, 0) > 0$ also. Since $u_t(6, 0)$ gives the rate of change of temperature at $x = 6$ with respect to time, the fact that it is positive indicates that the temperature at $x = 6$ is increasing. You might have guessed this intuitively from the observation that the temperature 62°C at $x = 6$ is below the average $(56+70)/2 = 63$°C of the temperatures of the neighboring points at $x = 4$ and $x = 8$ (which is because $u_{xx}(6, 0) > 0$). The average effect of the neighboring points is to heat up the point at $x = 6$.

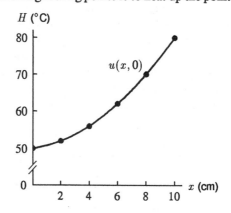

Figure 13.53: Temperature, H, at time $t = 0$

(b)   To estimate the temperature at $t = 1$ from the temperature at $t = 0$ we must first estimate the rate of change of temperature with respect to time, $u_t(6, 0)$. Since $u_t(6, 0) = 0.1u_{xx}(6, 0)$, we will estimate $u_{xx}(6, 0)$. Since $u_{xx}(6, 0)$ is approximated by a difference quotient of $u_x(x, 0)$, we first estimate u_x at two points near $x = 6$.

$$u_x(5, 0) \approx \frac{u(6, 0) - u(4, 0)}{6 - 4} = \frac{62 - 56}{2} = 3, \quad \text{and} \quad u_x(7, 0) \approx \frac{u(8, 0) - u(6, 0)}{8 - 6} = 4.$$

Therefore,

$$u_{xx}(6, 0) \approx \frac{u_x(7, 0) - u_x(5, 0)}{7 - 5} \approx \frac{4 - 3}{2} = 0.5.$$

Thus,

$$u_t(6, 0) \approx 0.1u_{xx}(6, 0) \approx (0.1)(0.5) = 0.05 \text{ deg/sec}.$$

Finally we can make the local linear approximation for the temperature at $t = 1$ that we want:

$$u(6, 1) \approx u(6, 0) + u_t(6, 0)(1) \approx 62 + (0.05)(1) = 62.05°C.$$

Warning

Getting accurate numerical approximations of solutions to PDEs is generally quite difficult. Example 3 was given to show that quantitative information can be extracted from a PDE, not as a practical way to get accurate approximations to solutions.

Boundary Conditions

The heat equation $u_t = u_{xx}$ has many solutions. More information than this equation alone is required in order to specify a single solution, information analogous to the initial conditions we are familiar with from the study of ordinary differential equations. In Example 1, for example, we would need to know the temperature in the room at the time the window was opened, the outside temperature at all times, and the temperature very near the heater (which tells what the heater is doing). This information is collectively referred to as the *boundary conditions*. The problems at the end of this section contain examples of how the boundary conditions can be used.

A Traveling Wave

Think about a bottle on a quiet sea that floats up then down as a wave rolls through. The vertical velocity of the bottle will depend on the shape and horizontal velocity of the wave.

Example 4   Suppose that the height y of the sea (and hence of a floating bottle) above normal at time t seconds and distance x meters from a reference point is given by the function $y = u(x, t)$, which is graphed at $t = 0$ in Figure 13.54. Suppose that the wave is moving in the direction of increasing x.
   (a)   Determine whether $u_x(x, 0)$ and $u_t(x, 0)$ are positive or negative at the following points:
      (i) $x = p$,   (ii) $x = q$,   (iii) $x = r$.
   (b)   Would $u_t(p, 0)$ be greater or smaller if the wave were traveling faster?

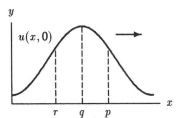

Figure 13.54: The height of the wave
at $t = 0$; wave moving to the right

Solution   (a)   (i)   The partial derivative $u_x(p, 0)$ gives the slope of the tangent line to the graph of $u(x, 0)$ at $x = p$. From Figure 13.54 it is clear that the slope is negative, so $u_x(p, 0) < 0$. On the other hand, $u_t(p, 0)$ equals the vertical velocity of a bottle floating at the point $x = p$ at the time $t = 0$. Since the bottle is rising as the wave rolls by, that velocity is positive. Thus, $u_t(p, 0) > 0$.
      (ii)   We have $u_x(q, 0) = 0$, because the tangent to the wave at $x = q$ and $t = 0$ is horizontal, slope 0. A bottle at $x = q$ and $t = 0$ would be exactly at the crest of the wave, momentarily stopped as it reverses direction from rising to falling, so its velocity would be zero. Thus, $u_t(q, 0) = 0$.
      (iii)   At $t = 0$ the point r is on the back side of the wave, so $u_x(r, 0) > 0$. A bottle at $x = r$ would be falling at $t = 0$, so $u_t(r, 0) < 0$.
   (b)   If the wave were moving faster, then a floating bottle would rise and fall much faster, so the positive velocity $u_t(p, 0)$ would be greater than for the original, slower wave.

The Traveling Wave Equation

Observe that in Example 4 the derivatives $u_x(x, 0)$ and $u_t(x, 0)$ are of opposite sign or are both zero together, for all x. In this example there is no x for which $u_x(x, 0)$ and $u_t(x, 0)$ are both positive or both negative, suggesting that the two derivatives u_x and u_t are related.

To investigate the relationship, let us suppose that a wave is moving to the right with velocity c, and that its positions at two nearby times t and $t + \Delta t$ are as shown in Figure 13.55. If Δt is small enough, $u_x(p, t)$, the slope of the graph at B, is well approximated by the slope of the secant line between the points A and B. Note that during time interval Δt the wave moves horizontally a distance of $c\,\Delta t$. Thus, we have

$$
\begin{aligned}
u_x(p, t) &= \text{Slope of tangent at } B \\
&\approx \text{Slope of secant between } A \text{ and } B \\
&= \frac{u(p, t) - u(p, t + \Delta t)}{c\Delta t} \\
&= -\frac{1}{c}\frac{u(p, t + \Delta t) - u(p, t)}{\Delta t} \\
&\approx -\frac{1}{c}u_t(p, t)
\end{aligned}
$$

If $u(x, t)$ represents a wave moving in the positive x-direction with speed c then it satisfies the following PDE:

The Traveling Wave Equation

$$u_t(x, t) = -cu_x(x, t) \quad \text{where } c \text{ is a positive constant.}$$

Analytic Representation of Traveling Waves

Suppose that a traveling wave is moving in the positive x-direction with speed c and that the wave has a known shape $y = f(x)$ at time $t = 0$, which means that $u(x, 0) = f(x)$. By time t the wave will have moved a distance ct (distance = rate × time) to the right, so at time t it will have shape $y = f(x - ct)$. In other words, $u(x, t) = f(x - ct)$.

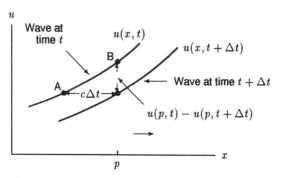

Figure 13.55: A traveling wave moving to the right at
speed c

Example 5   (a)   Write an equation for $u(x, t)$ that describes a wave whose shape at time $t = 0$ is $y = \sin x$ and that is moving in the positive direction with speed 0.5.
(b)   Show that the function $u(x, t)$ found in part (a) satisfies the traveling wave equation.
(c)   Sketch the graphs of $u(x, t)$ against x for $t = 0, 1, 2$.

Solution   (a) $u(x, t) = \sin(x - ct) = \sin(x - 0.5t)$.
(b)   Since $u_t(x, t) = -0.5 \cos(x - 0.5t)$ and $u_x(x, t) = \cos(x - 0.5t)$, the function $u(x, t)$ satisfies the traveling wave equation with $c = 0.5$:

$$u_t(x, t) = -0.5 u_x(x, t).$$

(c)   The graphs are in Figure 13.56. Notice that the forward motion of the wave is clearly visible.

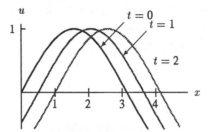

Figure 13.56: Graph of
$u(x, t) = \sin(x - 0.5t)$ at three times

A String Under Tension

What is the mathematics of a vibrating string under tension, such as a plucked guitar string? Let $y = u(x, t)$ be the displacement at time t of the point on the string x units from one end, where displacement is measured from the equilibrium position on the straight string. Then it can be shown that u satisfies the following equation.

The One-Dimensional Wave Equation:

$$u_{tt} = c^2 u_{xx} \quad \text{where } c \text{ is a positive constant.}$$

Let's see what this equation says. The function $u(x, t)$ describes the motion of a mass (the string) under the influence of a force (tension). Thus, Newton's Second Law of Motion must apply, so we expect an equation of the type $F = ma$, where F is force, m is mass, and a is acceleration. The term $u_{tt}(x, t)$ in the wave equation is the acceleration of the point x at time t. The term $u_{xx}(x, t)$ is closely related to the force, for it measures the concavity of the string at the point x at time t. The greater the concavity, the stronger the force back towards equilibrium, as anyone knows who has shot an arrow from a bow.

Problems for Section 13.8 ━━━━━━━━━━━━━━━━━━

1.   (a)   Find an estimate for $u(4, 1)$ and $u(8, 1)$ where $u(x, t)$ is as in Example 3.
     (b)   Use your answers to part (a) and the estimate for $u(6, 1)$ worked out in Example 3 to make an estimate for $u(6, 2)$.

2.   Sketch the graph of $u(x,t) = 1 - (x - 2t)^2$ for $t = 0, 1, 2$. Do you see a traveling wave? What is its speed?

3.   We will model the spread of an epidemic through a region. Suppose $I(x, y, t)$ represents the density of sick people per unit area at the point (x, y) in the plane at time t. Suppose I satisfies the diffusion equation:

$$\frac{\partial I}{\partial t} = D\left(\frac{\partial^2 I}{\partial x^2} + \frac{\partial^2 I}{\partial y^2}\right)$$

where D is a constant. Suppose we know that for some particular epidemic

$$I = e^{ax+by+ct}.$$

What can you say about the relationship between a, b, and c?

4.   For what values of the constants a and b will the function $u = (x + y)e^{ax+by}$ satisfy the partial differential equation

$$\frac{\partial^2 u}{\partial x \partial y} - \frac{\partial u}{\partial x} - \frac{\partial u}{\partial y} + u = 0?$$

Show that the functions in Problems 5–8 satisfy Laplace's equation: $F_{xx} + F_{yy} = 0$.

5. $F(x, y) = e^{-x} \sin y$.

6. $F(x, y) = \arctan \frac{y}{x}$.

7. $F(x, y) = e^x \sin(y) + e^y \sin(x)$.

8. $F(x, y) = e^{ax+by}$, if $a^2 + b^2 = 0$.

9.   The function $f(x, t) = 0.003 \sin(\pi x) \sin(2765t)$ was used in Example 2 of Section 13.7 to describe a vibrating guitar string. Show that it is a solution of the wave equation. $f_{tt} = c^2 f_{xx}$ for some c.

10.   Suppose that f is any differentiable function of one variable. Define V, a function of two variables, by

$$V(x, t) = f(x + ct).$$

Show that V satisfies the partial differential equation

$$\frac{\partial V}{\partial t} = c\frac{\partial V}{\partial x}.$$

11.   Show that $u = x^2 y - \frac{1}{2}xy^2 + 2\sin x + 3y^4 - 5$ satisfies the second-order partial differential equation

$$\frac{\partial^2 u}{\partial x \partial y} = 2x - y.$$

12.   If $V = f(u, v, w)$, and if $u^3 = x^3 + y^3 + z^3$, $v^2 = x^2 + y^2 + z^2$, $w = x + y + z$, show that

$$xV_x + yV_y + zV_z = uV_u + vV_v + wV_w.$$

13.   Show that the partial differential equation

$$z_{xx} + 2kz_{xy} + k^2 z_{yy} = 0, \qquad k \neq 1,$$

can be written as $z_{vv} = 0$ by setting $u = y + \alpha x$, $v = y - x$ and choosing a suitable value for α.

14. Consider the wave equation:

$$\frac{\partial^2 y}{\partial t^2} = a^2 \frac{\partial^2 y}{\partial x^2}$$

where a is a constant, and y is the displacement of the wave at any point x and at any time t (so y is a function of x and t). Write the boundary conditions for a vibrating string of length L for which:

(a) The ends (at $x = 0$ and $x = L$) are fixed.
(b) The initial shape is given by $f(x)$.
(c) The initial velocity distribution is given by $g(x)$.

15. Figure 13.57 is the graph of the temperature versus position at one instant in a metal rod, where temperature $T = u(x, t)$ satisfies the heat equation $u_t = u_{xx}$. Determine for which x the temperature is increasing, for which it is decreasing, and use this information to sketch a graph of temperature at a slightly later time.

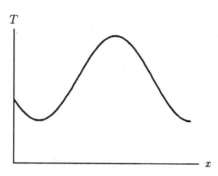

Figure 13.57: Temperature at time t

16. The gravitational potential, V, at any point (x, y, z) outside a spherically symmetric mass m located at the point (x_0, y_0, z_0), is defined as $-Gm/r$, where r is the distance from (x, y, z) to (x_0, y_0, z_0) and G is a constant. Show that V satisfies Laplace's equation:

$$\frac{\partial^2 V}{\partial x^2} + \frac{\partial^2 V}{\partial y^2} + \frac{\partial^2 V}{\partial z^2} = 0,$$

for all points outside the mass.

17. Assuming the solution to the wave equation

$$\frac{\partial^2 y}{\partial t^2} = 4 \frac{\partial^2 y}{\partial x^2}$$

is of the form $y = F(x + 2t) + G(x - 2t)$, find a solution satisfying the boundary conditions:

$$y(0, t) = y(5, t) = 0, \quad y(x, 0) = 0, \quad \left.\frac{\partial y}{\partial t}\right|_{t=0} = 5 \sin \pi x, \quad 0 < x < 5, \quad t > 0.$$

18. Find the relationship between a and b that must hold for $u(x, t) = e^{at} \sin(bx)$ to satisfy the heat equation $u_t = u_{xx}$.

19. (a) Suppose you wish to study heat conduction in a 1 meter metal rod, $0 \le x \le 1$, wrapped in insulation, whose ends are maintained at $0°$ C at all times (for instance because they are stuck into ice baths). The conditions at the ends of the rod represent a boundary condition on the possible functions $u(x, t)$ that could describe the temperature in the rod. The boundary condition must hold in addition to the PDE $u_t = u_{xx}$. State the boundary condition as a pair of equations.

    (b) Determine all possible values of a and b such that $u(x, t) = e^{at} \sin(bx)$ satisfies both the PDE and the boundary conditions of part (a).

20. (a) Show that $u(x, t) = ax + b$ satisfies the heat equation.

    (b) The solution in part (a) is called a steady-state solution of the heat equation because it represents a temperature that does not depend on time. Find and graph the steady state solution such that $u(0, t) = 20$ and $u(10, t) = 50$ for all t.

21. (a) Verify that the function

$$u(x, t) = \frac{1}{2\sqrt{\pi t}} e^{-x^2/(4t)}$$

satisfies the heat equation $u_t = u_{xx}$ for $t > 0$ and all x.

    (b) Sketch the graphs of $u(x, t)$ against x for $t = 0.01, 0.1, 1, 10$. These graphs represent heat conduction in an infinitely long insulated rod that at $t = 0$ is $0°$C everywhere except at the origin $x = 0$, and that is infinitely hot at $t = 0$ at the origin.

22. The temperature T of a metal plate can be described by a function $T = u(x, y, t)$ of three variables, the two space variables x and y and the time variable t. In many situations this function will satisfy the two-dimensional heat equation:

$$u_t = A(u_{xx} + u_{yy}) \quad \text{where } A \text{ is a positive constant.}$$

Find conditions on a, b, and c such that

$$u(x, y, t) = e^{-at} \sin(bx) \sin(cy)$$

satisfies this equation.

23. (a) Show that if a function $u(x, t)$ represents a wave that is traveling in the negative x-direction with speed c, then $u_t(x, t) = cu_x(x, t)$.

    (b) Show that a wave traveling in the negative x-direction with speed c can be represented by the formula $u(x, t) = f(x + ct)$, where $f(x) = u(x, 0)$.

24. By focusing on the difference between the traveling wave equation $u_t = -cu_x$ and the modified equation $u_t = -uu_x$, explain why the modified equation might be used to describe a breaking wave.

25. (a) Verify that $u(x, t) = \sin(ax) \sin(at)$ satisfies the wave equation $u_{tt} = u_{xx}$. This solution represents a vibration with period $2\pi/a$, since $u(x, t + 2\pi/a) = u(x, t)$.

    (b) Suppose that you wish to study the vibrations of a 1-meter string with fixed ends (such as a guitar string), so that $u(0, t) = 0$ and $u(1, t) = 0$ for all t, and such that $u_{tt} = u_{xx}$. The condition on the ends is a boundary condition. Find all $a > 0$ such that $u(x, t) = \sin(ax) \sin(at)$ satisfies both the PDE and the boundary condition. (It has been known to musicians since at least the time of Pythagoras that stretched strings can be made to vibrate at only special frequencies.)

26. You can generate a traveling wave on a string under tension by giving one end a snap. This suggests that there should be traveling wave solutions of the wave equation.

    (a) Show that if f is an arbitrary function, then $u(x, t) = f(x - ct)$ is a solution of the wave equation $u_{tt} = c^2 u_{xx}$.

(b)   Show that if g is an arbitrary function, then $u(x,t) = g(x+ct)$ is a solution of the wave equation $u_{tt} = c^2 u_{xx}$.

(c)   Show that if f and g are arbitrary functions, then $u(x,t) = f(x-ct) + g(x+ct)$ is also a solution of the wave equation. In fact, all solutions of the wave equation can be written in this form, as the sum of a forward traveling wave and a backward traveling wave.

27.   (a)   Show that $u(x,t) = \sin t \sin x$ is a solution of the wave equation $u_{tt} = u_{xx}$.

(b)   Use the identities

$$\cos(x+t) = \cos x \cos t - \sin x \sin t$$
$$\cos(x-t) = \cos x \cos t + \sin x \sin t$$

to express the function $u(x,t)$ of part (a) as the sum of a forward traveling wave and a backward traveling wave.

28.   The vibration of a two dimensional object under tension, such as a drum head, is described by a function $u(x,y,t)$ of three variables, two space variables x and y and one time variable t. Such a function often satisfies the two dimensional wave equation: $u_{tt} = c^2(u_{xx} + u_{yy})$. Find conditions on a, b, and e such that $u(x,y,t) = \sin(ax)\sin(by)\sin(et)$ satisfies this equation.

29.   Let u satisfy Burger's equation $u_t = u_{xx} + u_x^2$ and set $\omega = e^u$. Show that ω satisfies the heat equation $\omega_t = \omega_{xx}$. This change of variables is known as the Hopf-Cole transformation.

13.9   NOTES ON QUADRATIC APPROXIMATIONS

Just as a function of one variable can usually be better approximated by a quadratic function than by a linear function, so can a function of several variables. In Section 13.3, we saw how to approximate $f(x,y)$ by a linear function (its local linearization). In this section, we see how to improve this approximation of $f(x,y)$ using a quadratic function.

Quadratic Approximations Near (0,0)

The local linearization of a function is the best linear approximation to the function at a point. For a function of a single variable, local linearity tells us that

$$f(x) \approx f(a) + f'(a)(x-a) \quad \text{for } x \text{ near } a.$$

We can improve on this by taking a second-order Taylor polynomial

$$f(x) \approx f(a) + f'(a)(x-a) + \frac{f''(a)}{2}(x-a)^2.$$

The local linearization of a function of two variables near (a,b) is

$$f(x,y) \approx f(a,b) + f_x(a,b)(x-a) + f_y(a,b)(y-b),$$

or, if $(a,b) = (0,0)$,

$$f(x,y) \approx f(0,0) + f_x(0,0)x + f_y(0,0)y.$$

To get a better approximation to a function $f(x,y)$ we use a quadratic polynomial $P(x,y)$

$$f(x,y) \approx P(x,y) = A + Bx + Cy + Dx^2 + Exy + Fy^2.$$

What quadratic polynomial should we choose? We would certainly want the approximation to be exact at $(0,0)$, which means that

$$f(0,0) = P(0,0) = A.$$

If we arrange for the first partials of f and P to agree at $(0,0)$, that is

$$f_x(0,0) = P_x(0,0) = B \quad \text{and} \quad f_y(0,0) = P_y(0,0) = C,$$

then f and P will stay close near $(0,0)$ because they will have the same local linearizations at $(0,0)$. The natural next step is to insist that the second partials of f and P be equal at $(0,0)$. In other words, we ask that $P_{xx}(0,0) = f_{xx}(0,0)$, $P_{xy}(0,0) = f_{xy}(0,0)$, and $P_{yy}(0,0) = f_{yy}(0,0)$. Then

$$f_{xx}(0,0) = P_{xx}(0,0) = 2D \quad \text{and} \quad f_{xy}(0,0) = P_{xy}(0,0) = E \quad \text{and} \quad f_{yy}(0,0) = P_{yy}(0,0) = 2F$$

Hence, we can find the coefficients $A, B, \ldots, F$, and thus the best quadratic approximation for f near $(0,0)$ is given by the following polynomial:

Taylor Polynomial of Degree 2 Approximating $f(x, y)$ for (x, y) near (0,0)

$$f(x,y) \approx P(x,y)$$
$$= f(0,0) + f_x(0,0)x + f_y(0,0)y + \frac{f_{xx}(0,0)}{2}x^2 + f_{xy}(0,0)xy + \frac{f_{yy}(0,0)}{2}y^2.$$

Example 1 Let $f(x, y) = \sqrt{x + 2y + 1}$.

(i) Compute the local linearization of f at $(0,0)$.

(ii) Compute the Taylor polynomial of degree 2 for f at $(0,0)$.

(iii) Compare the linear and quadratic approximations in part (a) and part (b) with the true values for $f(x, y)$ at the points $(0.1, 0.1)$, $(-0.1, 0.1)$, $(0.1, -0.1)$, $(-0.1, -0.1)$.

Solution Let us first collect the computations that we will need.

$$
\begin{aligned}
f(x, y) &= (x + 2y + 1)^{1/2} & f(0,0) &= 1 \\
f_x(x, y) &= \tfrac{1}{2}(x + 2y + 1)^{-1/2} & f_x(0,0) &= 1/2 \\
f_y(x, y) &= (x + 2y + 1)^{-1/2} & f_y(0,0) &= 1 \\
f_{xx}(x, y) &= -\tfrac{1}{4}(x + 2y + 1)^{-3/2} & f_{xx}(0,0) &= -1/4 \\
f_{xy}(x, y) &= -\tfrac{1}{2}(x + 2y + 1)^{-3/2} & f_{xy}(0,0) &= -1/2 \\
f_{yy}(x, y) &= -(x + 2y + 1)^{-3/2} & f_{yy}(0,0) &= -1.
\end{aligned}
$$

(i) The local linearization $L(x, y)$ of f at $(0,0)$ is given by

$$f(x,y) \approx L(x,y) = f(0,0) + f_x(0,0)x + f_y(0,0)y = 1 + \frac{1}{2}x + y.$$

(ii) The second-degree Taylor polynomial, $P(x, y)$, for f at $(0, 0)$ is given by

$$f(x, y) \approx P(x, y)$$
$$= f(0, 0) + f_x(0, 0)x + f_y(0, 0)y + \frac{f_{xx}(0, 0)}{2}x^2 + f_{xy}(0, 0)xy + \frac{f_{yy}(0, 0)}{2}y^2$$
$$= 1 + \frac{1}{2}x + y - \frac{1}{8}x^2 - \frac{1}{2}xy - \frac{1}{2}y^2.$$

Notice that the local linearization of f is the same as the linear part of the Taylor polynomial of degree 2 for f. The extra terms in the Taylor polynomial of degree 2 can be thought of as "correction terms" to the linear approximation.

(iii) Table 13.10 records the values of $f(x, y)$, $L(x, y)$, and $P(x, y)$. Observe that the quadratic approximations $P(x, y)$ are closer to the true values $f(x, y)$ than are the linear approximations $L(x, y)$. Of course both approximations are exact at $(0, 0)$.

TABLE 13.10: *Linear and quadratic approximations to f near* $(0, 0)$

Point	Linear	Quadratic	True
(x, y)	$L(x, y)$	$P(x, y)$	$f(x, y)$
$(0, 0)$	1	1	1
$(0.1, 0.1)$	1.15	1.13875	1.140175
$(-0.1, 0.1)$	1.05	1.04875	1.048809
$(0.1, -0.1)$	0.95	0.94875	0.948683
$(-0.1, -0.1)$	0.85	0.83875	0.836660

Quadratic Approximations near (a, b)

Recall that the local linearization for a function $f(x, y)$ at a point (a, b) is

$$f(x, y) \approx f(a, b) + f_x(a, b)(x - a) + f_y(a, b)(y - b).$$

This suggests that a quadratic polynomial approximation $P(x, y)$ for $f(x, y)$ near a point (a, b) should be written in terms of $(x - a)$ and $(y - b)$ instead of x and y. If we require that $P(a, b) = f(a, b)$ and that the first- and second-order partial derivatives of P and f at (a, b) be equal, then after some algebra, we get the following polynomial:

Taylor Polynomial of Degree 2 Approximating $f(x, y)$ for (x, y) near (a, b)

$$f(x, y) \approx P(x, y)$$
$$= f(a, b) + f_x(a, b)(x - a) + f_y(a, b)(y - b)$$
$$+ \frac{f_{xx}(a, b)}{2}(x - a)^2 + f_{xy}(a, b)(x - a)(y - b) + \frac{f_{yy}(a, b)}{2}(y - b)^2.$$

These coefficients are derived in the same way as for $(a, b) = (0, 0)$.

Example 2   Find the Taylor polynomial of degree 2 at the point $(1, 2)$ for the function

$$f(x, y) = \frac{1}{xy}.$$

Solution   First we calculate all the necessary partial derivatives and their values at the point $(1, 2)$. See Table 13.11.

TABLE 13.11: *Partial derivatives of* $f(x, y) = 1/(xy)$

Derivative	Formula	Value at $(1, 2)$
$f(x, y)$	$1/xy$	$1/2$
$f_x(x, y)$	$-1/x^2y$	$-1/2$
$f_y(x, y)$	$-1/xy^2$	$-1/4$
$f_{xx}(x, y)$	$2/x^3y$	1
$f_{xy}(x, y)$	$1/x^2y^2$	$1/4$
$f_{yy}(x, y)$	$2/xy^3$	$1/4$

So that the Taylor polynomial at (1,2) is

$$\frac{1}{xy} \approx P(x, y)$$

$$= \frac{1}{2} - \frac{1}{2}(x - 1) - \frac{1}{4}(y - 2) + \frac{1}{2}(1)(x - 1)^2 + \frac{1}{4}(x - 1)(y - 2) + \left(\frac{1}{2}\right)\left(\frac{1}{4}\right)(y - 2)^2$$

$$= \frac{1}{2} - \frac{x - 1}{2} - \frac{y - 2}{4} + \frac{(x - 1)^2}{2} + \frac{(x - 1)(y - 2)}{4} + \frac{(y - 2)^2}{8}.$$

Problems for Section 13.9

Find the quadratic Taylor polynomials about $(0, 0)$ for the functions in Problems 1–3.

1. $e^{-2x^2 - y^2}$ 2. $\sin 2x + \cos y$ 3. $\ln(1 + x^2 - y)$

For each of the functions in Problems 4–11, find a quadratic approximation valid near $(1, 1)$. Evaluate the approximation at $(1.1, 1.1)$ and compare it with the exact value of the function.

4. $z = \sin(x/y)$ 5. $z = xe^y$ 6. $z = (x + y)e^y$

7. $z = \sin(x^2 + y^2)$ 8. $z = \sqrt{x^2 + y^2}$ 9. $z = \arctan(x + y)$

10. $z = \arctan(x/y)$ 11. $z = \dfrac{xe^y}{x + y}$

REVIEW PROBLEMS FOR CHAPTER THIRTEEN

For Problems 1–4, find the indicated partial derivatives. Assume the variables are restricted to a domain on which the function is defined.

1. $\dfrac{\partial z}{\partial x}$ and $\dfrac{\partial z}{\partial y}$ if $z = (x^2 + x - y)^7$

2. $\dfrac{\partial F}{\partial L}$ if $F(L, K) = 3\sqrt{LK}$

3. $\dfrac{\partial f}{\partial p}$ and $\dfrac{\partial f}{\partial q}$ if $f(p, q) = e^{p/q}$

4. $\dfrac{\partial f}{\partial x}$ if $f(x, y) = (\ln y)e^{xy}$

Find both partial derivatives for the functions in Problems 5–8. Assume the variables are restricted to a domain on which the function is defined.

5. $z = x^4 - x^7 y^3 + 5xy^2$

6. $z = \tan(\theta)/r$

7. $w = s\ln(s + t)$

8. $w = \arctan(ue^{-v})$

9. Let $f(w, z) = e^{w \ln z}$.
   (a) Use difference quotients with $h = 0.01$ to approximate $f_w(2, 2)$ and $f_z(2, 2)$.
   (b) Now evaluate $f_w(2, 2)$ and $f_z(2, 2)$ exactly.

10. Find the rate of change of $f(x, y) = xe^y$ at the point $(1, 1)$ in the direction of the vector $\vec{i} + 2\vec{j}$.

11. Find the directional derivative of $z = x^2 - y^2$ at the point $(3, -1)$ in the direction making an angle $\theta = \frac{\pi}{4}$ with the x-axis. In which direction is the directional derivative the largest?

12. Figure 13.58 shows a contour diagram for the temperature T along a wall in a heated room as a function of distance x along the wall and time t. Estimate $\partial T/\partial x$ and $\partial T/\partial t$ at the given points. Give units for your answers and say what the answers mean.
   (a) $x = 15, t = 20$
   (b) $x = 5, t = 12$

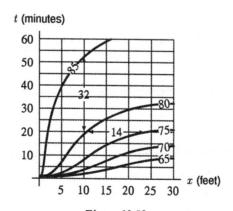

Figure 13.58

13. Find the quadratic Taylor polynomial about $(0, 0)$ for $f(x, y) = \cos(x + 2y)\sin(x - y)$.

Are the statements in Problems 14–18 true or false? Explain your answer.

14. $f_{\vec{u}}(a, b) = \|\nabla f(a, b)\|$

15. grad $f \cdot \vec{u} = 0$ where $\vec{u}$ is tangent to the level curves of f at the point where grad f is computed.

16. There is a point where $\|\operatorname{grad} f\| = 0$ and where there is a directional derivative which is not zero.

17. There is a function with $\|\operatorname{grad} f\| = 4$ and $f_{\vec{i}} = 5$ at some point.

18. There is a function with $\|\operatorname{grad} f\| = 4$ and $f_{\vec{j}} = -3$ at some point.

19. Figure 13.59 shows a contour diagram for the vibrating string function:

$$f(x, t) = \cos t \sin x, \quad 0 \le x \le \pi.$$

(a) Is $f_t(\pi/2, \pi/2)$ positive or negative? How about $f_t(\pi/2, \pi)$? What does the sign of $f_t(\pi/2, b)$ tell you about the motion of the point on the string at $x = \pi/2$ when $t = b$.

(b) Find all t for which f_t is positive, for $0 \le t \le 5\pi/2$.

(c) Find all x and t such that f_x is positive.

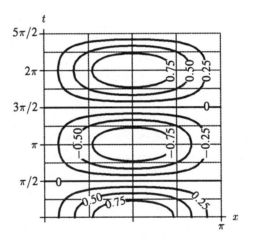

Figure 13.59

20. The quantity Q (in pounds) of beef that a certain community buys during a week is a function $Q = f(b, c)$ of the prices of beef, b, and chicken, c, during the week. Is $\partial Q/\partial b$ positive or negative? What about $\partial Q/\partial c$?

21. Suppose the cost of producing one unit of a certain product is given by

$$c = a + bx + ky$$

where x is the amount of labor used (in man hours) and y is the amount of raw material used (by weight) and a and b are constants. What does $\partial c/\partial x = b$ mean? What is the practical interpretation of b?

22. People commuting to a city can choose to go either by bus or by train. The number of people who choose either method depends in part upon the price of each. Let $f(P_1, P_2)$ be the number of people who take the bus when P_1 is the price of a bus ride and P_2 is the price of a train ride. What can you say about the signs of $\partial f/\partial P_1$ and $\partial f/\partial P_2$? Explain the reasons for your answers.

23. A soft drink company is interested in seeing how the demand for its products is affected by prices. The company believes that the quantity, q, of its soft drinks sold depends on p_1, the average price of the company's soft drinks, p_2, the average price of competing soft drinks, and p_3, the average amount of money spent by the company on advertising. Suppose also that

$$q = C - 8 \cdot 10^6 p_1 + 4 \cdot 10^6 p_2 + p_3.$$

   (a) What does the constant C represent?
   (b) Find the marginal demand for soft drinks with respect to changes in p_1, p_2, and p_3. Explain why the signs and relative magnitudes of your answers are reasonable. [Hint: The marginal demand is the rate of change of quantity demanded with price.]

24. Imagine a box-shaped building with a flat roof. Suppose it has length x feet, width y feet, height z feet and is oriented as shown in Figure 13.60.

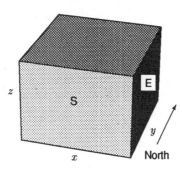

Figure 13.60: A box-shaped building

   Each of the walls, the roof and the floor loses heat at a different rate because they are made of different materials and have different exposures. (For example, the southern exposure loses less heat because it receives more of the sun's rays.) The heat loss (measured in BTUs per square foot per hour, where a BTU is a unit of heat) is given in Table 13.12. Let $H(x, y, z)$ represent the total heat loss per hour from the house.

TABLE 13.12

Exposure	N	S	E	W	Floor	Roof
Heat loss per ft^2 per hour (BTU)	8	3	6	7	1	12

   (a) Find a formula for H in terms of x, y, z.
   (b) Find $\partial H/\partial x$.
   (c) Interpret $\partial H/\partial x$ in terms of heat loss. Explain the presence of each term in $\partial H/\partial x$ as a contribution from some parts of the house. [Hint: Think geometrically about how the house changes as x varies.]

25. Suppose that x is the price of one brand of gasoline and y is the price of a competing brand. Then q_1, the quantity of the first brand sold in a fixed time period, depends on both x and y, so $q_1 = f(x, y)$. Similarly, if q_2 is the quantity of the second brand sold in the same time, $q_2 = g(x, y)$.

   (a) What do you expect the signs of $\partial q_1/\partial x$ and $\partial q_2/\partial y$ to be? Explain.
   (b) What do you expect the signs of $\partial q_1/\partial y$ and $\partial q_2/\partial x$ to be? Explain.

26. In the 1940s the quantity, q, of beer sold each year in Britain was found to depend on I (the aggregate personal income, adjusted for taxes and inflation), p_1 (the average price of beer) and p_2 (the average price of all other goods and services). Would you expect $\partial q/\partial I, \partial q/\partial p_1, \partial q/\partial p_2$ to be positive or negative? Give reasons for your answers.

27. The acceleration g due to gravity, at a distance r from the center of a planet of mass m, is given by

$$g = \frac{Gm}{r^2}$$

where G is the universal gravitational constant.

   (a) Find $\partial g/\partial m$ and $\partial g/\partial r$.
   (b) Interpret each of the partial derivatives you found in part (a) as the slope of a graph in the plane and sketch the graph.

28. Suppose that the function $P = f(K, L)$ expresses the production of a firm as a function of the capital invested, K, and its labor costs, L.

   (a) Suppose

$$f(K, L) = 60K^{1/3}L^{2/3}.$$

   Find the relationship between K and L if the marginal productivity of capital (that is, the rate of change of production with capital) equals the marginal productivity of labor cost (that is, the rate of change of production with labor cost). Put your answer in the simplest form possible.

   (b) Now suppose

$$f(K, L) = cK^a L^b$$

   $(a, b, c$ positive constants). What must be true of a, b and c if the relationship between K and L is to be the same as you found in part (a)?

29. In analyzing a factory and deciding whether or not to hire more workers, it is useful to know under what circumstances productivity increases. Suppose $P = f(x_1, x_2, x_3)$ is the total quantity produced as a function of x_1, the number of workers, and any other variables x_2, x_3. We define the average productivity of a worker as P/x_1. Show that the average productivity increases as x_1 increases when marginal production, $\partial P/\partial x_1$, is greater than the average productivity, P/x_1.

30. Figure 13.61 shows the level curves of a function $f(x, y)$. Give the approximate value of $f_{\vec{u}}(3, 1)$ with $\vec{u} = (-2\vec{i} + \vec{j})/\sqrt{5}$. Explain your answer.

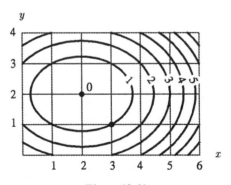

Figure 13.61

31. The level curves representing the height of a surface above the xy-plane are shown in Figure 13.62. Sketch the paths that would be followed by freely rolling marbles starting at P and at Q.

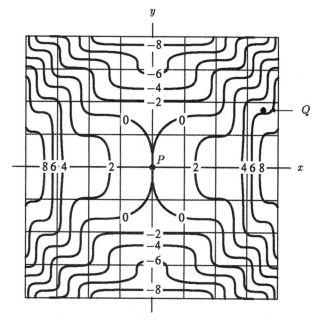

Figure 13.62

32. Show that if F is any differentiable function of one variable, then $V(x, y) = xF(2x + y)$ satisfies

$$x\frac{\partial V}{\partial x} - 2x\frac{\partial V}{\partial y} = V.$$

33. Find a particular solution to the differential equation in Problem 32 satisfying

$$V(1, y) = y^2.$$

34. Using Figure 13.63, develop a linear function that gives you approximately the monthly payment, m, on a 5-year car loan if you borrow P dollars at r percent interest. What is the practical significance of the constants in your formula?

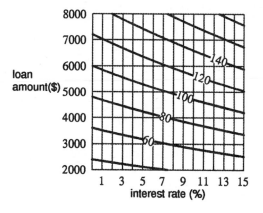

Figure 13.63

35. Suppose that the values of the function $f(x, y)$ near the point $x = 2$, $y = 3$ are given in Table 13.13. Estimate the following.

    (a) $\dfrac{\partial f}{\partial x}\bigg|_{(2,3)}$ and $\dfrac{\partial f}{\partial y}\bigg|_{(2,3)}$.

    (b) The rate of change of f at $(2, 3)$ in the direction of the vector $\vec{i} + 3\vec{j}$.

    (c) The maximum possible rate of change of f as you move away from the point $(2, 3)$. In which direction should you move to obtain this rate of change?

    (d) Write an equation for the level curve through the point $(2, 3)$.

    (e) Find a vector tangent to the level curve of f through the point $(2, 3)$.

    (f) Find the differential of f at the point $(2, 3)$. If $dx = 0.03$, $dy = 0.04$, find df. What does df represent?

TABLE 13.13

		\(x\)	
		2.00	2.01
\(y\)	3.00	7.56	7.42
	3.02	7.61	7.47

36. Find the equation of the tangent plane to $z = \sqrt{17 - x^2 - y^2}$ at the point $(3, 2, 2)$.

37. Find the equation of the tangent plane to $z = 8/(xy)$ at the point $(1, 2, 4)$.

38. Show that a normal vector to the surface $z = F(x, y)$ at a point (x_0, y_0, z_0) is given by

$$(F_x(x_0, y_0))\vec{i} + (F_y(x_0, y_0))\vec{j} - \vec{k}$$

39. Find an equation of the tangent plane and of a normal vector to the surface $x = y^3 z^7$ at the point $(1, -1, -1)$.

40. Find the point(s) on $x^2 + y^2 + z^2 = 8$ where the tangent plane is parallel to the plane $x - y + 3z = 0$.

41. Show that the tangent plane to

$$z = \frac{x^2}{a^2} + \frac{y^2}{b^2}$$

at the point (x_1, y_1, z_1) is given by

$$z + z_1 = \frac{2xx_1}{a^2} + \frac{2yy_1}{b^2}.$$

42. Two surfaces are said to be tangential at a point P if they have the same tangent plane. Show that the surfaces $z = \sqrt{2x^2 + 2y^2 - 25}$ and $z = \frac{1}{5}(x^2 + y^2)$ are tangential at the point $(4, 3, 5)$.

43. Two surfaces are said to be orthogonal to each other at a point P if the normals to their tangent planes are perpendicular at P. Show that $z = \frac{1}{2}(x^2 + y^2 - 1)$ and $z = \frac{1}{2}(1 - x^2 - y^2)$ are orthogonal at all points of intersection.

44. The radius of a right circular cone is 4 inches and its altitude is 9 inches. If the maximum error in the radius is 0.01 inches and if the altitude is 0.02 inches, estimate the error in the computed volume.

45. The area of a triangle can be calculated from the formula $S = \frac{1}{2}ab \sin C$. Show that if an error of 10′ (or $\pi/1080$ radians) is made in measuring C then the error in S is approximately $\pi S/(1080 \tan C)$.

46. Find the differential of $f(x, y) = \sqrt{x^2 + y^3}$ at the point $(1, 2)$. Use it to estimate $f(1.04, 1.98)$.

47. The gas equation for one mole of oxygen relates its pressure P (in atmospheres), its temperature T (in °K), and its volume V (in cubic decimeters):

$$T = 16.574\frac{1}{V} - 0.52754\frac{1}{V^2} + 0.3879P + 12.187VP.$$

   (a) Find the temperature T and differential dT if the volume of gas is 25 dm³ and the pressure is 1 atmosphere.
   (b) Use your answer to part (a) to estimate how much the volume of oxygen would have to change if the pressure increased by 0.1 atmosphere and the temperature remained constant.

48. Each diagram (I) – (IV) in Figure 13.64 represents the level curves of a function $f(x, y)$. For each function f, consider the point above P on the surface $z = f(x, y)$ and choose from the lists which follow:

   (a) A vector which could be the normal to the surface at that point;
   (b) An equation which could be the equation of the tangent plane to the surface at that point.

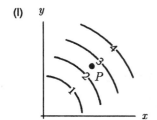 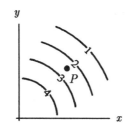 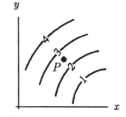 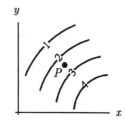

Figure 13.64

Vectors

(E) $2\vec{i} + 2\vec{j} - 2\vec{k}$
(F) $2\vec{i} + 2\vec{j} + 2\vec{k}$
(G) $2\vec{i} - 2\vec{j} + 2\vec{k}$
(H) $-2\vec{i} + 2\vec{j} + 2\vec{k}$

Equations

(J) $x + y + z = 4$
(K) $2x - 2y - 2z = 2$
(L) $-3x - 3y + 3z = 6$
(M) $-\dfrac{x}{2} + \dfrac{y}{2} - \dfrac{z}{2} = -7$

CHAPTER FOURTEEN

OPTIMIZATION

In one-variable calculus we saw how to find the maximum and minimum values of a function of one variable. In real life, there are often several variables in an optimization problem. For example, you may have $10,000 to invest in new equipment and advertising for your business. What combination of equipment and advertising will yield the greatest extra income? Or, what combination of drugs will lower a patient's temperature the most? In this chapter we consider optimization problems, where the variables are completely free to vary (unconstrained optimization) and where there is a constraint on the variables (for example, a budget constraint).

14.1   LOCAL AND GLOBAL EXTREMA

Functions of several variables, like functions of one variable, can have *local* and *global* extrema. (That is, local and global maxima and minima.) A review of the one-variable case is given in Appendix B. A function has a local extremum at a point where it takes on the largest or smallest values in a small region around the point. Global extrema are the largest or smallest values anywhere on the domain. For example, see Figure 14.1.

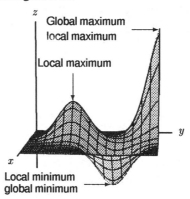

Figure 14.1: Local and global extrema
for a function of two variables on
$0 \leq x \leq a, 0 \leq y \leq b$

More precisely, assuming (x, y) is a point at which f is defined, we say:

- f has a **local maximum** at the point (x_0, y_0) if $f(x, y) \leq f(x_0, y_0)$ for all (x, y) near (x_0, y_0).
- f has a **local minimum** at (x_0, y_0) if $f(x, y) \geq f(x_0, y_0)$ for all (x, y) near (x_0, y_0).
- f has a **global maximum** at (x_0, y_0) if $f(x, y) \leq f(x_0, y_0)$ for all (x, y).
- f has a **global minimum** at (x_0, y_0) if $f(x, y) \geq f(x_0, y_0)$ for all (x, y).

Thus, a global extremum is also a local extremum, although the converse is not necessarily true. Notice that the definitions of local and global extrema extend to functions of three or more variables.

How Do We Detect a Local Maximum or Minimum?

In one-variable calculus, the local extrema of a function occur at critical points, namely points where the derivative is zero or undefined. How does this generalize to the case of functions of two or more variables? Recall that if the gradient vector of a function is defined and nonzero, then it points in a direction in which the function is increasing. Suppose that a function $f(x, y)$ has a local maximum at a point (x_0, y_0) which is not on the boundary of the domain. If the vector grad $f(x_0, y_0)$ were defined and nonzero, then we could increase f by moving in the direction of the vector grad $f(x_0, y_0)$. Since f has a local maximum at (x_0, y_0), there is no direction in which f is increasing, so we must have

$$\text{grad } f(x_0, y_0) = \vec{0}.$$

Similarly, suppose $f(x, y)$ has a local minimum at the point (x_0, y_0). If the gradient grad $f(x_0, y_0)$ were defined and nonzero, then we could decrease f by moving in the direction of $-$ grad $f(x_0, y_0)$, and so we must again have grad $f(x_0, y_0) = \vec{0}$. Therefore, we arrive at the following conclusion:

> If a function f has a local maximum or minimum at a point (x_0, y_0), not on the boundary of the domain, then grad $f(x_0, y_0)$ is either zero or undefined.

Points where the gradient is either zero or undefined are called *critical points* of the function. For a function of two variables, we can also see that the gradient vector must be zero at a local maximum by looking at its contour diagram, Figure 14.2, and at Figure 14.3, which shows a plot of its gradient vectors. Around the maximum the vectors are all pointing inward, perpendicularly to the contours. At the maximum there is nowhere to point, so the gradient vector must be zero. We have similar contour and gradient vector plots near a local minimum.

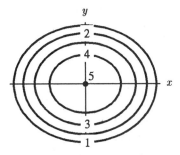

Figure 14.2: Contour diagram around a local maximum of a function

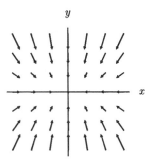

Figure 14.3: Gradients pointing toward the local maximum of the function in Figure 14.2

Again, notice that the rule for finding the critical points of a function is easy to extend to functions of three or more variables.

How Do You Find Critical Points?

To find critical points we set grad f equal to zero. Since grad $f = f_x\vec{i} + f_y\vec{j} + f_z\vec{k}$, we set all the partial derivatives equal to zero. We must also look for the points where one or more of them is undefined.

Example 1   Find and analyze the critical points of $f(x, y) = x^2 - 2x + y^2 - 4y + 5$.

Solution   Since
$$\text{grad } f = (2x - 2)\vec{i} + (2y - 4)\vec{j},$$
grad $f = \vec{0}$ when $2x - 2 = 0$, $2y - 4 = 0$, that is, when $(x, y) = (1, 2)$. Hence, f has only one critical point, namely $(1, 2)$. What is the behavior of f near $(1, 2)$? Look at the values of the function in Table 14.1.

TABLE 14.1: *Values of $f(x, y)$ near the point $(1, 2)$*

		x			
		0.8	0.9	1.0	1.1	1.2
	1.8	0.08	0.05	0.04	0.05	0.08
	1.9	0.05	0.02	0.01	0.02	0.05
y	2.0	0.04	0.01	0.00	0.01	0.04
	2.1	0.05	0.02	0.01	0.02	0.05
	2.2	0.08	0.05	0.04	0.05	0.08

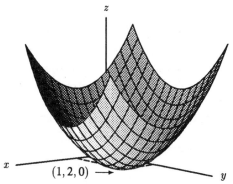

Figure 14.4: The graph of
$f(x, y) = x^2 - 2x + y^2 - 4y + 5$ with a local
minimum at the point $(1, 2)$

The table suggests that the function has a minimum value of 0 at $(1, 2)$. We verify this by completing the square:

$$f(x, y) = x^2 - 2x + y^2 - 4y + 5 = (x - 1)^2 + (y - 2)^2.$$

Figure 14.4 shows that the graph of f is a parabolic bowl with vertex at the point $(1, 2, 0)$. It is the same shape as the graph of $z = x^2 + y^2$ (shown in Figure 11.17 on page 16), except that the vertex has been shifted to $(1, 2)$. So the local minimum of f (as well as a global minimum) must be at the point $(1, 2)$.

Example 2   Find any critical points of $f(x, y) = -\sqrt{x^2 + y^2}$.

Solution   We look for points where grad $f = \vec{0}$ or is undefined. The partial derivatives are given by

$$\frac{\partial f}{\partial x} = -\frac{x}{\sqrt{x^2 + y^2}},$$
$$\frac{\partial f}{\partial y} = -\frac{y}{\sqrt{x^2 + y^2}}.$$

These are *never* both zero; but they are both undefined at $(x, y) = (0, 0)$. Thus, $(0, 0)$ is a critical point and a possible extreme point. The graph of f (see Figure 14.5) is an upward pointing cone, with vertex at $(0, 0)$. So $(0, 0)$ is a local and global maximum.

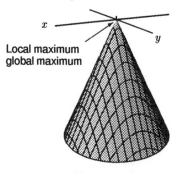

Local maximum
global maximum

Figure 14.5: Graph of
$f(x, y) = -\sqrt{x^2 + y^2}$

Example 3   Find the local extrema of the function $f(x, y) = 8y^3 + 12x^2 - 24xy$.

Solution   We begin by looking for critical points:

$$\frac{\partial f}{\partial x} = 24x - 24y,$$

$$\frac{\partial f}{\partial y} = 24y^2 - 24x.$$

Setting these expressions equal to zero gives the system of equations

$$x = y, \qquad x = y^2,$$

which has two solutions, $(0, 0)$ and $(1, 1)$. Are these maxima, minima or neither? Let's look at the contours near the points: Figure 14.6 shows the contour diagram of this function. Notice that $f(1, 1) = -4$ and that there is no other -4 contour. The contours near $P = (1, 1)$ appear oval in shape and show that f is increasing in value no matter in which direction you move away from P. This suggests that f has a local minimum at the point $(1, 1)$.

The level curves near $Q = (0, 0)$ show a very different behavior. While $f(0, 0) = 0$, we see that f takes on both positive and negative values at nearby points. Thus, the point $(0, 0)$ is a critical point which is neither a local maximum or minimum.

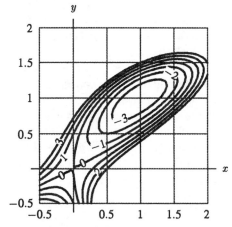

Figure 14.6: Contour diagram of
$f(x, y) = 8y^3 + 12x^2 - 24xy$

Saddle Points

In one-variable calculus, critical points can be local maxima, minima, or inflection points of the function. The previous example shows that the situation is similar in multivariable calculus: critical points can occur at local maxima or minima, or at points which are neither — the value of the function is larger in some directions and smaller in others.

> A **saddle point** is a critical point (x_0, y_0) for which, within any distance, no matter how small, there are points, (x_1, y_1) and (x_2, y_2), with
>
> $$f(x_1, y_1) > f(x_0, y_0) \quad \text{and} \quad f(x_2, y_2) < f(x_0, y_0).$$

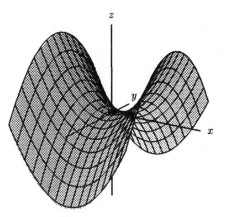

Figure 14.7: Graph of $g(x, y) = x^2 - y^2$,
showing a saddle point at the origin

Thus, we see from Figure 14.6 that the function $f(x, y) = 8y^3 + 12x^2 - 24xy$ in Example 3 has a saddle point at the origin. For another example, look at the function $g(x, y) = x^2 - y^2$, whose graph is in Figure 14.7. This function has a critical point and value zero at $(0, 0)$. Moving along the y-axis in either direction from $(0, 0)$ gives values for g that are negative, while moving along the x-axis in either direction gives values of g that are positive. Thus, the origin is a saddle point. Notice that the graph of g looks like a saddle there.

Let's look at the level curves of g near the saddle point $(0, 0)$. See Figure 14.8. They are hyperbolas; the ones intersecting the x-axis correspond to positive values of g, and the ones intersecting the y-axis correspond to negative values of g. You can see that near $(0, 0)$ there are always values of g which are above and below the value of g at $(0, 0)$.

Contrast this with the appearance of level curves near a local maximum or minimum, for example, near $(0, 0)$ of $h(x, y) = x^2 + y^2$. See Figure 14.9. These are circles which surround the point, so that near $(0, 0)$ the function is increasing or decreasing (in this case, increasing) no matter in which direction you look.

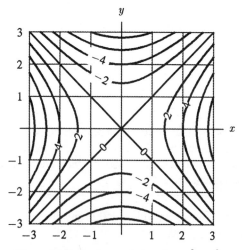

Figure 14.8: Contours of $g(x, y) = x^2 - y^2$,
showing a saddle point at the origin

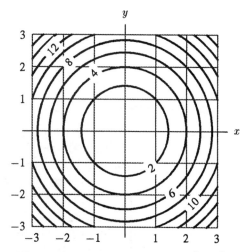

Figure 14.9: Contours of $h(x, y) = x^2 + y^2$,
showing a local minimum at the origin

Is a Critical Point a Local Maximum, Local Minimum, or Saddle Point?

We can see whether a critical point is a maximum, minimum, or saddle point by looking at the contour diagram. Now we develop an analytic method for making the distinction if the critical point is one at which the gradient is zero. We shall describe a generalization of the second derivative test of one-variable calculus. (See Appendix B.)

Quadratic Functions of the form $f(x, y) = ax^2 + bxy + cy^2$

Let's begin by looking at what can happen at critical points of quadratic functions of the form $f(x, y) = ax^2 + bxy + cy^2$, where a, b and c are constants. You may think that these are very special functions, but we can use the Taylor formula to approximate most functions by quadratic functions. The approximation is usually good enough that the function and its quadratic Taylor polynomial have the same classification of the critical point into maximum, minimum, saddle point, or none of these. So the key point is to understand quadratic functions.

Example 4 Find and analyze the local extrema of the function $f(x, y) = x^2 + xy + y^2$.

Solution The partial derivatives are

$$\frac{\partial f}{\partial x} = 2x + y,$$
$$\frac{\partial f}{\partial y} = x + 2y.$$

The only critical point is $(0, 0)$, and the value of the function there is $f(0, 0) = 0$. So, if f is always positive or zero near $(0, 0)$, then $(0, 0)$ is a local minimum, and if f is always negative or zero near $(0, 0)$, it is a local maximum, and if f takes both positive and negative values it is a saddle point. One way to find out is to look at the graph in Figure 14.10, which shows that $(0, 0)$ is a local minimum.

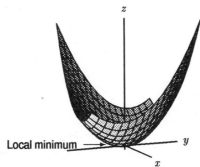

Figure 14.10: Graph of
$f(x, y) = x^2 + xy + y^2 = (x + \frac{1}{2}y)^2 + \frac{3}{4}y^2$
showing local minimum at the origin

How could we have known the graph would look like this without drawing it? There is an algebraic way to determine if a quadratic function is always negative, always positive, or neither, and that is to complete the square. If we write

$$f(x, y) = x^2 + xy + y^2 = \left(x + \frac{1}{2}y\right)^2 + \frac{3}{4}y^2,$$

then we can see that $f(x, y)$ is a sum of two squares, so it must always be greater than or equal to zero. Thus, the critical point must be a minimum.

The Shape of the Graph of $f(x, y) = ax^2 + bxy + cy^2$

In general, to analyze the graph of a quadratic function, we complete the square. Assuming $a \neq 0$, we write

$$
\begin{aligned}
ax^2 + bxy + cy^2 &= a\left[x^2 + \frac{b}{a}xy + \frac{c}{a}y^2\right] \\
&= a\left[\left(x + \frac{b}{2a}y\right)^2 + \left(\frac{c}{a} - \frac{b^2}{4a^2}\right)y^2\right] \\
&= a\left[\left(x + \frac{b}{2a}y\right)^2 + \left(\frac{4ac - b^2}{4a^2}\right)y^2\right].
\end{aligned}
$$

The basic shape of the graph of f depends on whether the coefficient of y^2 is positive, negative, or zero. The sign of $4ac - b^2$ determines the sign of this coefficient. The quantity $D = 4ac - b^2$ is called the *discriminant*.

- If $D = 4ac - b^2 > 0$, then the coefficients of both squares inside the brackets are positive, so the function has a maximum or a minimum. It has a minimum if a is positive, since then it will be a right-side-up paraboloid (like $x^2 + y^2$), and it has a maximum if a is negative, since it will then be an upside-down paraboloid (like $-x^2 - y^2$).

- If $D = 4ac - b^2 < 0$ then the coefficients of the two squares have different signs; this means the function goes up in some directions and goes down in others (like $x^2 - y^2$), and hence it has a saddle point at $(0, 0)$.

- If $D = 4ac - b^2 = 0$, then the quadratic function is $a(x + by/2a)^2$, whose graph is a parabolic cylinder.

More generally, if f is a quadratic function of the form $f(x, y) = a(x - x_0)^2 + b(x - x_0)(y - y_0) + c(y - y_0)^2 + d$, then its graph will have exactly the same shape as the graph of $z = ax^2 + bxy + cy^2$, except that it will be centered at (x_0, y_0) rather than $(0, 0)$ and shifted vertically by d units. Therefore, the discriminant test gives the same results for the behavior of f near (x_0, y_0). In particular, provided $D \neq 0$, the signs of D and a tell us that the shape of the graph of $z = f(x, y)$ near a critical point is either *concave up*, *concave down*, or *saddle-shaped*. See Figures 14.11–14.13. [1]

Figure 14.11: Concave down: $D > 0$ and $a < 0$

Figure 14.12: Concave up: $D > 0$ and $a > 0$

Figure 14.13: Saddle-shaped: $D < 0$

Classifying the Critical Points of a Function

The reason we spent so much time studying functions of the form $ax^2 + bxy + cy^2$ is that, near a critical point, most functions of two variables behave like such quadratic functions. This means that we can apply the discriminant analysis to any function with continuous second order partial derivatives to classify its critical points as local maxima, minima, or saddle points.

---

[1] We assumed that $a \neq 0$; it turns out that these rules also cover the case where $a = 0$.

To see why this is so, suppose that a function f has a critical point at the origin with grad $f(0,0) = \vec{0}$. We recall from page 173 that f is approximated by its quadratic Taylor polynomial near $(0,0)$:

$$f(x,y) \approx f(0,0) + f_x(0,0)x + f_y(0,0)y$$
$$+ \frac{1}{2}f_{xx}(0,0)x^2 + f_{xy}(0,0)xy + \frac{1}{2}f_{yy}(0,0)y^2.$$

At critical points, where $f_x(0,0) = f_y(0,0) = 0$, the expansion simplifies to

$$f(x,y) \approx f(0,0) + \frac{1}{2}f_{xx}(0,0)x^2 + f_{xy}(0,0)xy + \frac{1}{2}f_{yy}(0,0)y^2,$$

or

$$f(x,y) - f(0,0) \approx \frac{1}{2}f_{xx}(0,0)x^2 + f_{xy}(0,0)xy + \frac{1}{2}f_{yy}(0,0)y^2.$$

In other words, this says that $f(x,y)$ differs from $f(0,0)$ by an amount that behaves like a quadratic function. In this case, the discriminant is

$$D = 4\left(\frac{1}{2}f_{xx}\right)\left(\frac{1}{2}f_{yy}\right) - f_{xy}{}^2,$$

which simplifies to

$$D = f_{xx}f_{yy} - f_{xy}{}^2,$$

where the partial derivatives are evaluated at $(0,0)$. Similarly, if f has a critical point at (x_0, y_0), where grad $f(x_0, y_0) = 0$, then we can use the second-order Taylor approximation to compare the behavior of f near (x_0, y_0) with the quadratic function $a(x-x_0)^2 + b(x-x_0)(y-y_0) + c(y-y_0)^2 + d$. We get the following result:

Second Derivative Test for Functions of Two Variables

Suppose (x_0, y_0) is a point where grad $f(x_0, y_0)$ is zero. Let

$$D = f_{xx}(x_0, y_0)f_{yy}(x_0, y_0) - f_{xy}(x_0, y_0)^2.$$

- If $D > 0$ and $f_{xx}(x_0, y_0) > 0$, then (x_0, y_0) is a local minimum.
- If $D > 0$ and $f_{xx}(x_0, y_0) < 0$, then (x_0, y_0) is a local maximum.
- If $D < 0$, then (x_0, y_0) is a saddle point.
- If $D = 0$, anything can happen.

Example 5   Find the local maxima, minima, and saddle points of the function

$$f(x,y) = \frac{x^2}{2} + 3y^3 + 9y^2 - 3xy + 9y - 9x.$$

Solution   The partial derivatives are $f_x = x - 3y - 9$ and $f_y = 9y^2 + 18y - 3x + 9$. Setting $f_x = 0$ and $f_y = 0$ gives

$$9y^2 + 18y + 9 - 3x = 0,$$
$$x - 3y - 9 = 0.$$

Eliminating x gives

$$9y^2 + 9y - 18 = 0,$$

which has solutions $y = -2$ and $y = 1$.

Therefore, the critical points are $(3, -2)$ and $(12, 1)$. The discriminant is

$$D(x, y) = f_{xx}f_{yy} - f_{xy}^2 = (1)(18y + 18) - (-3)^2$$
$$= 18y + 9.$$

Since $D(3, -2) = -36 + 9 < 0$, therefore $(3, -2)$ is a saddle point. Since $D(12, 1) = 18 + 9 > 0$ and $f_{xx}(12, 1) = 1 > 0$, therefore $(12, 1)$ is a local minimum.

The second derivative test does not give any information in the case $D = 0$. However, as the following example illustrates, we can still classify the critical points by looking at the graph of the function, just as we do in cases where grad f is not defined and the second derivative test is therefore not applicable.

Example 6   Classify the critical point $(0, 0)$ of the functions $f(x, y) = x^4 + y^4$, $g(x, y) = -x^4 - y^4$, and $h(x, y) = x^4 - y^4$.

Solution   The functions f, g, and h each have zero gradient at $(0, 0)$ and so the origin is a critical point for each of these functions. However, they also have zero second-order partial derivatives at $(0, 0)$. Thus, $D = 0$ for each function. Near the origin, the graphs of f, g and h look like the surfaces in Figures 14.11–14.13, and so we see that f has a minimum at $(0, 0)$, that g has a maximum at $(0, 0)$, and that h has a saddle point at $(0, 0)$.

How Do We Find Global Maxima and Minima?

Not all functions have a global maximum or minimum: it depends on the function and the domain. If there is a global maximum and minimum, it may be at one of the critical points in the domain, or on the boundary. We will see how to investigate points on the boundary in Section 14.3.

If the domain of the function is not stated explicitly, it is understood to be the whole xy-plane. The following examples show that there may or may not be a global maximum or minimum in this case.

Example 7   Investigate the global maxima and minima of $f(x, y) = x^2 - 2x + y^2 - 4y + 5$ and of $g(x, y) = x^2 - y^2$.

Solution   The graph of f in Figure 14.4 on page 186 shows that f has global minimum at the point $(1, 2)$ and no global maximum (because the value of f increases without bound as $x \to \infty$, $y \to \infty$).

The graph of g in Figure 14.7 on page 188 shows that g has no global maximum because $g(x, y) \to \infty$ as $x \to \infty$ if y is constant. Similarly, g has no global minimum because $g(x, y) \to -\infty$ as $y \to \infty$ if x is constant.

Example 8   Does the function f in Example 5 on page 191 have global maxima and minima? If so, what are they?

Solution   Suppose x is fixed. Then for large values of y the sign of f is determined by the highest power of y, namely y^3. Thus,

$$f(x, y) \to \infty \quad \text{as} \quad y \to \infty$$
$$f(x, y) \to -\infty \quad \text{as} \quad y \to -\infty.$$

So f does not have a global maximum or minimum.

How Do We Know Whether a Function Has a Global Extremum?

From one-variable calculus we know that a continuous function, $h(x)$, on a closed interval $a \leq x \leq b$ achieves its maximum and minimum values on that interval. However, if h is continuous on a non-closed interval, such as $a \leq x < b$ or $a < x < b$, or on an interval which is not bounded, such as $a < x < \infty$, then h need not have a maximum or minimum value.

What is the situation for functions of two variables? As it turns out, a similar result is true for functions defined on regions which are closed and bounded, analogous to the closed and bounded interval $a \leq x \leq b$.

In everyday language we say

- a **closed** region is one which contains its boundary;
- a **bounded** region is one which does not stretch to infinity in any direction.

More precise definitions are as follows. Suppose R is a region in two-space. A point (x_0, y_0) is a *boundary point* of R if, for every $r > 0$, the disk $(x - x_0)^2 + (y - y_0)^2 < r^2$ with center (x_0, y_0) and radius r contains both points which are in R and points which are not in R. See Figure 14.14. A point (x_0, y_0) can be a boundary point of the region R without actually belonging to R. A point (x_0, y_0) in R is an *interior point* if it is not a boundary point; thus, for small enough $r > 0$, the disk of radius r centered at (x_0, y_0) lies entirely in the region R. See Figure 14.15. The collection of all the boundary points is the *boundary of R* and the collection of all the interior points is the *interior of R*. The region R is *closed* if it contains its boundary, while it is *open* if every point in R is an interior point.

A region R in 2-space is *bounded* if the distance between every point (x, y) in R and the origin is less than or equal to some number A. The notions of open, closed, and bounded regions in 3-space can be defined in exactly the same manner. The following example illustrates the concepts of open, closed, and bounded regions in 2-space.

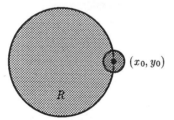

Figure 14.14: Boundary point (x_0, y_0) of R

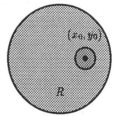

Figure 14.15: Interior point (x_0, y_0) of R

Example 9

a) The square $-1 \leq x \leq 1, -1 \leq y \leq 1$ is closed and bounded.

b) The first quadrant $x \geq 0, y \geq 0$ is closed but is not bounded.

c) The disc $x^2 + y^2 < 1$ is open and bounded, but is not closed.

d) The half-plane $y > 0$ is open, but is neither closed nor bounded.

e) Two-space is both open and closed, but is not bounded.

Our interest in closed and bounded regions stems from the following important result:

If f is a continuous function on a closed and bounded region R, then f has a global maximum at some point (x_0, y_0) in R and a global minimum at some point (x_1, y_1) in R.

The result is also true for functions of three or more variables. Although the result is intuitively reasonable, its proof is difficult and is only discussed in more advanced texts. [2]

If f is not continuous or the region R is not closed and bounded, there is no guarantee that f will achieve a global maximum or global minimum value on R. In Example 8, the function f is continous but does not achieve a global maximum or minimum value in 2-space, a region which is closed but not bounded. The following example illustrates what can go wrong when the region is bounded but not closed.

Example 10   Does the function

$$f(x, y) = \frac{1}{x^2 + y^2}$$

have a global maximum or minimum on the region R given by $0 < x^2 + y^2 \leq 1$?

Solution   The region R is bounded, but it is not closed since it does not contain the boundary point $(0, 0)$. We see from the graph of $z = f(x, y)$ in Figure 14.16 that f has a global minimum on the circle $x^2 + y^2 = 1$, but $f(x, y) \to \infty$ as $(x, y) \to (0, 0)$ and so f has no global maximum.

Figure 14.16: Graph showing $f(x, y) = \frac{1}{x^2+y^2}$ has no global max or min on $0 < x^2 + y^2 \leq 1$

In Section 14.3 we will consider the problem of finding the global maximum or minimum values of a continuous function $f(x, y)$ when the points (x, y) are constrained to lie in a closed and bounded region R. While it may be difficult to find the global maxima or minima in practice, we at least have a guarantee that they exist in this case.

Problems for Section 14.1

1. By looking at the weather map in Figure 11.1 on page 2, find the maximum and minimum daily temperatures in the states of Mississippi, Alabama, Pennsylvania, New York, California, Arizona, and Massachusetts.

2. By looking at the table of UV exposure as a function of latitude and year in Table 25 on page 62, find the latitude and year that will have the worst UV exposure between 1970 and the year 2010.

---

[2]See W. Rudin, *Principles of Mathematical Analysis*, 2nd ed., p. 89, (New York: McGraw-Hill, 1976).

In Problems 3–5, find the global maximum and minimum of the given function over the square $-1 \leq x \leq 1$, $-1 \leq y \leq 1$, and say whether it occurs on the boundary of the square. (Hint: Consider the graph of the function.)

3. $z = x^2 + y^2$ 4. $z = x^2 - y^2$ 5. $z = -x^2 - y^2$

6. Suppose $f(x, y) = A - (x^2 + Bx + y^2 + Cy)$. What values of A, B, and C give $f(x, y)$ a maximum value of 15 at the point $(-2, 1)$?

For Problems 7–9, use Figure 14.17, which shows level curves of some function $f(x, y)$.

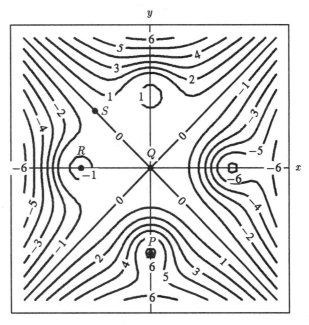

Figure 14.17

7. Decide whether each point is a local maximum, local minimum, saddle point, or none of these:
    (a) P (b) Q (c) R
    (d) S

8. Sketch ∇f at a few points around each of P, Q, and R. Your sketch should make clear the direction of ∇f; don't worry about the magnitude.

9. Put arrows showing the direction of ∇f at the points where $\|\nabla f\|$ is largest.

10. (a) Find all the critical points of the function

$$f(x, y) = 8xy - \frac{1}{4}(x + y)^4.$$

   (b) Are these local maxima, local minima or saddle points? Justify your answer.

11. Find the values of x and y which maximize

$$P(x, y) = 400 - 3x^2 - 4x + 2xy - 5y^2 + 48y.$$

For Problems 12–14, find the local maxima, minima and saddle points of the given function.

12. $f(x, y) = \frac{x^2}{2} + 3y^3 + 9y^2 - 3xy + 9y - 9x$.   13. $f(x, y) = (x + y)(xy + 1)$.

14. $f(x, y) = \sin x + \sin y + \sin(x + y), \quad 0 < x < \pi, \quad 0 < y < \pi.$

For Problems 15 and 16, find the local maxima, local minima, and saddle points of the functions given. Decide if the local maxima or minima are global maxima or minima. Explain.

15. $f(x, y) = x^2 + y^3 - 3xy$

16. $f(x, y) = xy + \ln x + y^2 - 10 \qquad (x > 0)$

17.   Find and classify the critical points of the function

$$E = 1 - \cos x + \frac{y^2}{2}.$$

18.   (a)   Find all the critical points of the function

$$f(x, y) = \sin x \sin y.$$

(b)   Are these local maxima, local minima or saddle points? Justify your answer.

Each function in Problems 19–21 has a critical point at $(0, 0)$. What sort of critical point is it?

19. $f(x, y) = x^6 + y^6$

20. $g(x, y) = x^4 + y^3$

21. $h(x, y) = \cos x \cos y$

22.   (a)   Find all critical points of

$$f(x, y) = e^x(1 - \cos y).$$

(b)   Are these critical points local maxima, local minima, or saddle points?

23.   Suppose $f_x = f_y = 0$ at $(1, 3)$ and $f_{xx} > 0, f_{yy} > 0, f_{xy} = 0$.

(a)   What can you conclude about the behavior of the function near the point $(1, 3)$?

(b)   Sketch a possible contour diagram.

24.   Suppose that for some function $f(x, y)$ at the point (a, b), we have $f_x = f_y = 0, f_{xx} > 0, f_{yy} = 0, f_{xy} > 0$.

(a)   What can you conclude about the shape of the graph of f near the point (a, b)?

(b)   Sketch a possible contour diagram.

14.2 UNCONSTRAINED OPTIMIZATION

In this section we give some examples of the theory of optimization applied to practical problems.

Economic Example: Maximizing Profit

In planning production, a manufacturing company is concerned with how much of a particular item to produce and the price at which the item can be sold. In general, the higher the price, the less that can be sold. To determine how much to produce, the company often chooses the combination of price and quantity that maximizes the profit.

To calculate the maximum we use the fact that

$$\text{Profit} = \text{Revenue} - \text{Cost},$$

and, provided the price is constant,

$$\text{Revenue} = \text{Price} \times \text{Quantity} = pq.$$

In addition, we need to know how the cost and price depend on quantity.

Example 1 A manufacturing company produces two items which are sold in two separate markets. The company's economists analyze the two markets and determine that the quantities demanded by consumers, q_i, and the prices, p_i (in dollars), of each item are related by

$$p_1 = 600 - 0.3q_1 \quad \text{and} \quad p_2 = 500 - 0.2q_2.$$

Thus, if the price for either item increases, the demand for it decreases. The production costs for the two items are related, so that if the company increases production of one item, it has to decrease production of the other. The company's total production cost is given by

$$C = 16 + 1.2q_1 + 1.5q_2 + 0.2q_1q_2.$$

If the company wants to maximize its total profits, how much of each product should it sell? What will be the maximum profit? [3]

Solution The total revenue R is the sum of the revenues, p_1q_1 and p_2q_2, from each market. Substituting for p_1 and p_2, we get

$$
\begin{aligned}
R &= p_1q_1 + p_2q_2 \\
&= (600 - 0.3q_1)q_1 + (500 - 0.2q_2)q_2 \\
&= 600q_1 - 0.3q_1^2 + 500q_2 - 0.2q_2^2.
\end{aligned}
$$

Thus the total profit P is given by

$$
\begin{aligned}
P &= R - C \\
&= 600q_1 - 0.3q_1^2 + 500q_2 - 0.2q_2^2 - (16 + 1.2q_1 + 1.5q_2 + 0.2q_1q_2) \\
&= -16 + 598.8q_1 - 0.3q_1^2 + 498.5q_2 - 0.2q_2^2 - 0.2q_1q_2.
\end{aligned}
$$

To maximize P, we compute partial derivatives:

$$\frac{\partial P}{\partial q_1} = 598.8 - 0.6q_1 - 0.2q_2,$$

$$\frac{\partial P}{\partial q_2} = 498.5 - 0.4q_2 - 0.2q_1.$$

Since grad P is defined everywhere, the only critical points of P are those where grad $P = 0$. Thus, we solve the equations

$$598.8 - 0.6q_1 - 0.2q_2 = 0,$$
$$498.5 - 0.4q_2 - 0.2q_1 = 0,$$

for q_1, q_2, and find that

$$q_1 = 699.1 \quad \text{and} \quad q_2 = 896.7.$$

To see whether or not we have found a maximum point, we compute the second-order partial derivatives:

$$\frac{\partial^2 P}{\partial q_1^2} = -0.6, \quad \frac{\partial^2 P}{\partial q_2^2} = -0.4, \quad \frac{\partial^2 P}{\partial q_1 \partial q_2} = -0.2.$$

Therefore,

$$D = \frac{\partial^2 P}{\partial q_1^2} \frac{\partial^2 P}{\partial q_2^2} - \left(\frac{\partial^2 P}{\partial q_1 \partial q_2}\right)^2 = (-0.6)(-0.4) - (-0.2)^2 = 0.2,$$

and so the second derivative test implies that we have found a local maximum point. The graph of P is an upside down paraboloid and so $(699.1, 896.7)$ is in fact a global maximum point. The maximum profit is therefore $P(699.1, 896.7) \approx \$433,000.$

[3] Adapted from M. Rosser, *Basic Mathematics for Economists*, p. 316, (New York: Routledge, 1993).

Fitting a Line to Data

Another useful application of optimization is to the problem of fitting a line to data. Suppose we have some experimental data, and want to find the "best fitting" line.

The *least squares* method measures the distance from a line to the data points by adding together all of the squares of the vertical distances from each point to the line. The line with a minimal sum of square distances is called the *least squares line*, or sometimes the *regression* line. (The reason we use the squares of the distances is to prevent the contributions from points above the line and below the line from canceling.) If the data is nearly linear, the least squares line will be a good fit; otherwise it may not be. See Figure 14.18.

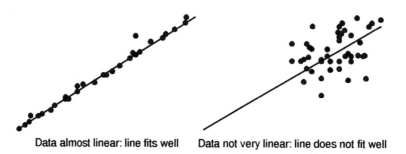

Data almost linear: line fits well   Data not very linear: line does not fit well

Figure 14.18: Fitting lines to data points

Example 2   Find a least squares line for the following data points: $(1, 1), (2, 1)$, and $(3, 3)$.

Solution   A line may be specified by an equation $y = b + mx$. If we can find b and m then we'll have found the line. So, for this problem, b and m are the two variables. What is the function that we want to minimize? It is the sum of the three squared vertical distances from the points to the line (see Figure 14.19).

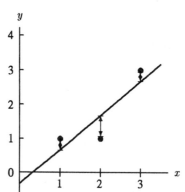

Figure 14.19: Least squares line
minimizes the sum of the squares of
these vertical distances

The y-coordinate of a point on the line has values $b + m$, $b + 2m$ and $b + 3m$ at the values of x given by 1, 2 and 3, respectively. Then the vertical distance of each data point to the line is $1 - (b + m)$, $1 - (b + 2m)$, and $3 - (b + 3m)$, so the sum of squares is

$$f(m, b) = (1 - (b + m))^2 + (1 - (b + 2m))^2 + (3 - (b + 3m))^2.$$

To minimize f we look for critical points. First we differentiate f with respect to m:

$$\frac{\partial f}{\partial m} = 2(1 - (b + m))(-1) + 2(1 - (b + 2m))(-2) + 2(3 - (b + 3m))(-3)$$
$$= -2 + 2b + 2m - 4 + 4b + 8m - 18 + 6b + 18m$$
$$= -24 + 12b + 28m.$$

Now we differentiate with respect to b:

$$\frac{\partial f}{\partial b} = -2(1 - (b + m)) - 2(1 - (b + 2m)) - 2(3 - (b + 3m))$$
$$= -2 + 2b + 2m - 2 + 2b + 4m - 6 + 2b + 6m$$
$$= -10 + 6b + 12m.$$

Setting both partial derivatives equal to zero yields a system of two linear equations in two unknowns:

$$-24 + 12b + 28m = 0,$$
$$-10 + 6b + 12m = 0.$$

The solution to this pair of equations is $m = 1$ and $b = -1/3$. Since

$$D = f_{mm} f_{bb} - (f_{mb})^2 = (28)(6) - 12^2 = 24 \quad \text{and} \quad f_{mm} = 28 > 0$$

the critical point we have found is a local minimum. The graph of $f(m, b)$ is a parabolic bowl, so the local minimum is the global minimum. Thus the regression line is

$$y = x - \frac{1}{3}.$$

Does this make sense? Without the point (2,1) the line that comes as close as possible to the remaining points (1,1) and (3,3) is a line that passes through both of them: it would have slope 1 and y-intercept 0. Introducing the point (2,1) moves the y intercept down (from 0 to $-1/3$).

The general formulas for finding the slope and y-intercept of a least squares line are derived in Problem 12 at the end of this section. Many calculators have these formulas built in, so that all you need to do is enter the data points and out pops the values of m and b. There are other measures of interest that are typically computed at the same time. For example, there is the *correlation coefficient*, which is a measure of how close the data points actually come to fitting the least squares line.

Gradient Search for Finding Local Extrema

So far we have searched for values that maximize or minimize a function $f(x, y)$ by first finding the critical points of f. Finding the critical points amounts to solving the equation grad $f = 0$, which is really a pair of simultaneous equations for x and y:

$$\frac{\partial f}{\partial x}(x, y) = 0 \quad \text{and} \quad \frac{\partial f}{\partial y}(x, y) = 0.$$

However, solving such equations can be a very difficult problem in its own right. In many cases even equations in a single-variable must be solved numerically, say by Newton's method, rather than explicitly. Thus, we often cannot find explicit values for critical points of a function of several variables. In practice, most optimization problems are not solved by solving the equations grad $f = 0$ for the critical point. Instead numerical methods such as the *gradient search* are used.

The gradient search method of optimization can be explained by analogy with a mountain climber who wishes to maximize his elevation by getting to the top of the highest mountain. All he has to do is keep going up, and eventually he will get to the top of some mountain. If he started near enough to the highest mountain, that is probably the mountain he will conquer. If not, there is a chance that he will go up a lower mountain, winding up at a local rather than a global maximum, but if he is lucky, he will still be pretty high.

The gradient search method is illustrated in the next example. It is a minimization problem, so you should imagine a hiker who seeks the lowest valley by always going down.

Example 3   Twenty cubic meters of gravel are to be delivered to a landfill by a trucker. She plans to purchase an open-top box in which to transport the gravel in numerous trips. She figures that the cost to her will be the cost of the box plus $2 per trip. The box must have height 0.5 m, but she can choose the length and width. The cost of the box will be $20/m^2 for the ends and $10/m^2 for the bottom and sides. Notice the tradeoff she faces: A smaller box is cheaper to buy but will require more trips. What size box should she buy to minimize her over-all costs? [4]

Solution   We first get an algebraic expression for the trucker's cost. Let the length and width of the box be x meters and y meters, so that the box measures x m $\times$ y m $\times$ 0.5 m as in Figure 14.20.

The volume of the box is $0.5xy$ m^3, so delivery of 20 m^3 of gravel will require $20/(0.5xy)$ trips. The trucker's cost are itemized in Table 14.2. Our problem is to choose x and y to minimize the total cost function

$$f(x, y) = \frac{80}{xy} + 10x + 10xy + 20y.$$

We pick a starting point (x_0, y_0) that may not minimize f but which we hope is not too far from a minimum point. In this example we start with $(x_0, y_0) = (5, 5)$, which is definitely not a critical point of f because

$$\text{grad } f(5, 5) = 59.4\vec{i} + 69.4\vec{j} \neq 0\vec{i} + 0\vec{j}.$$

We plan to move from (x_0, y_0) to a new point (x_1, y_1) in such a way that f decreases, that is $f(x_1, y_1) < f(x_0, y_0)$. Motion in the direction of grad $f(x_0, y_0)$ increases f as rapidly as possible, so we move in the opposite direction, namely $-$ grad $f(x_0, y_0)$. We continue to move in this direction until the f values begin to increase again. If we move parallel to $-$ grad $f(x_0, y_0)$, our displacement from the original point is of the form $-t$ grad $f(x_0, y_0)$ where t is a scalar. Thus, since grad $f(x_0, y_0) = 59.4\vec{i} + 69.4\vec{j}$, the coordinates of our final point are

$$(x_0 - 59.4t, y_0 - 69.4t).$$

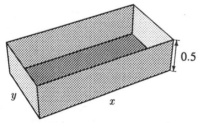

Figure 14.20: A box for transporting gravel

TABLE 14.2: *Trucker's itemized cost*

20/(0.5xy) at $2/trip	$80/(xy)$
2 ends at $20/m^2 $\times$ 0.5 y m^2	$20y$
2 sides at $10/m^2 $\times$ 0.5 x m^2	$10x$
1 bottom at $10/m^2 $\times$ xy m^2	$10\ xy$
Total cost	$f(x, y)$

[4] Adapted from Claude McMillan, Jr., *Mathematical Programming*, 2nd ed., p. 156-157, (New York: Wiley, 1978).

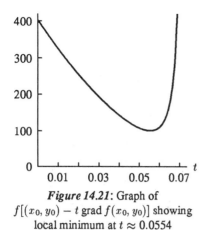

Figure 14.21: Graph of
$f[(x_0, y_0) - t \operatorname{grad} f(x_0, y_0)]$ showing
local minimum at $t \approx 0.0554$

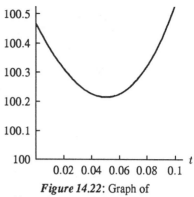

Figure 14.22: Graph of
$f((x_1, y_1) - t \operatorname{grad} f(x_1, y_1))$ showing
local minimum at $t \approx 0.050$

We want to find the minimum value of the function f as t increases. Figure 14.21 gives the graph of

$$f[(x_0, y_0) - t \operatorname{grad} f(x_0, y_0)] = f(5 - 59.4t, 5 - 69.4t)$$

for positive t. The local minimum is at $t \approx 0.0554$. So we take a new point given by

$$(x_1, y_1) = (5 - (59.4)(0.0554), 5 - (69.4)(0.0554)) = (1.71, 1.16).$$

Notice that the cost at the initial point is $f(5, 5) = 403.2$ and the cost at the new point is $f(1.71, 1.16) = 100.47$, so we have decreased the cost function f a lot.

We can further decrease f by moving away from $(1.71, 1.16)$ in the direction opposite to $\operatorname{grad} f(1.71, 1.16) = -1.99\vec{i} + 2.33\vec{j}$. Figure 14.22 gives the graph of

$$f[(x_1, y_1) - t \operatorname{grad} f(x_1, y_1)] = f(1.71 + 1.99t, 1.16 - 2.33t)$$

which achieves a local minimum at $t \approx 0.050$. So we take

$$(x_2, y_2) = (1.71 + (1.99)(0.050), 1.16 - (2.33)(0.050))$$
$$= (1.81, 1.04).$$

Notice that $f(1.81, 1.04) \approx 100.22$ whereas $f(1.71, 1.16) \approx 100.47$. The move from (x_1, y_1) to (x_2, y_2) decreased the cost f by a rather small amount, only $0.25, so while we have not actually achieved a minimum, we may feel that for practical purposes we are close enough. The trucker will round off, buying a box of dimension 1.8m × 1m × 0.5m.

Problems for Section 14.2

1. A company sells two products which are partial subsitutes for each other such that if the price of one product increases then the demand for the other product rises. The market demand quantities q_1, q_2 in terms of the prices p_1, p_2 are given by

$$q_1 = 517 - 3.5p_1 + 0.8p_2 \quad \text{and} \quad q_2 = 770 - 4.4p_2 + 1.4p_1.$$

What prices should the company charge in each market in order to maximize the total sales revenue? [5]

---

[5] Adapted from M. Rosser, *Basic Mathematics for Economists*, Routledge, New York, 1993, p. 318.

2. A company operates two plants which manufacture the same item and whose total cost functions are
$$C_1 = 8.5 + 0.03q_1^2 \quad \text{and} \quad C_2 = 5.2 + 0.04q_2^2,$$
where q_1, q_2 are the quantities produced by each plant. The total market demand quantity $q = q_1 + q_2$ is related to the price p by
$$p = 60 - 0.04q.$$
How much should each plant produce in order to maximize the company's profit? [6]

3. Assume that two products are manufactured in quantities q_1 and q_2 and sold at prices of p_1 and p_2 respectively, and that the cost of producing them is given by
$$C = 2q_1^2 + 2q_2^2 + 10.$$
    (a) Find the maximum profit that can be made, assuming the prices are fixed.
    (b) Find the rate of change of that maximum profit as p_1 increases.

4. A cruise missile has a remote guidance device which is sensitive to both temperature and humidity. Army engineers have worked out a formula to show the range at which the missile can be controlled:
$$\text{Max range in miles} = 12,000 - t^2 - 2ht - 2h^2 + 200t + 260h,$$
where t is the temperature in $°F$ and h is percent humidity. What are the optimal atmospheric conditions for controlling the missile?

5. A mountain climber reaches the peak of a mountain late in the day. After taking photographs of the view from the top, the weather deteriorates, and she has to descend the mountain as quickly as possible to seek shelter at a lower altitude. The altitude of the mountain is given approximately by
$$h(x, y) = 10000 - \frac{1}{24000}(5x^2 + 4xy + 2y^2) \quad \text{feet},$$
where x, y are horizontal coordinates on the earth (in feet), with the mountain summit located above the origin. In fifteen minutes, the climber can reach any point (x, y) on a circle of radius 2000 feet. In which direction should she travel in order to descend as far as possible?

6. A company manufactures a product which requires capital and labor to produce. The quantity, Q, of products manufactured is given by the Cobb-Douglas production function
$$Q = AK^a L^b,$$
where K is the quantity of capital and L is the quantity of labor used and A, a, and b are positive constants with $0 < a < 1$ and $0 < b < 1$. Suppose one unit of capital costs $\$k$ and one unit of labor costs $\$\ell$. The price of the good is fixed at $\$p$.
    (a) Assume that $a + b < 1$. How much capital and labor should the company use to maximize its profit?
    (b) Is there a maximum profit in the case $a + b = 1$? What about $a + b \geq 1$? Explain.
    [Note: See Section 11.4 for a discussion of the Cobb-Douglas production function. The three cases considered above, namely $a + b < 1$, $a + b = 1$, and $a + b > 1$ are, respectively, the cases of *decreasing returns to scale*, *constant returns to scale*, and *increasing returns to scale*.]

---

[6]Adapted from M. Rosser, *Basic Mathematics for Economists*, Routledge, New York, 1993, p. 318.

7. It is a familiar fact that some items are sold at different prices to different groups of people. For example, there are sometimes discounts for senior citizens or for children. The reason is that these groups may be more sensitive to price and so a small discount will have greater impact on their purchasing decisions. The seller faces an optimization problem: How large a discount to offer in order to maximize profits?

   A theater believes that it will sell q_c child tickets and q_a adult tickets at prices p_c and p_a, according to the following demand functions:

$$q_c = rp_c^{-4},$$
$$q_a = sp_a^{-2},$$

   and that its operating costs are proportional to the total number of tickets sold. What should be the relative price of children's and adults' tickets?

8. Show that the function $f(x, y)$ in Example 3 achieves a minimum at the point $(2, 1)$.

9. Design a rectangular milk carton box of width w, length l, and height h which holds 512 cm³ of milk. The sides of the box cost 1 cent/cm² and the bottoms cost 2 cent/cm². Find the dimensions of the box that minimize the total cost of materials used.

10. An international airline has a regulation that each passenger can carry, at no charge, up to two suitcases each having the sum of its width, length and height less than or equal to 54 inches. Find the dimensions of the suitcase of maximum volume that the passenger may carry under this airline regulation.

11. An irrigation canal has a trapezoidal cross-section of area 50 square feet, as in Figure 14.23.

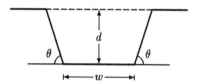

Figure 14.23

   The average flow velocity in the canal is inversely proportional to the wetted perimeter, p, of the canal, that is, to the perimeter of the trapezoid in Figure 14.23, excluding the top. Thus, to maximize the flow we must minimize p. Find the depth d, base width w, and angle θ that give the maximum flow rate.[7]

12. This exercise will lead you to a derivation of general formulas for the slope and y-intercept of a least squares line. Assume that you have n data points $(x_1, y_1), (x_2, y_2), \cdots, (x_n, y_n)$. Let the equation of the sought-after least squares line be $y = b + mx$.

    (a) For each data point (x_i, y_i), show that the corresponding point directly above or below it on the least squares line has y-coordinate $b + mx_i$.

    (b) For each data point (x_i, y_i), show that the square of the vertical distance from it to the point found in part (a) is $(y_i - (b + mx_i))^2$.

    (c) Now form the function $f(b, m)$ which is the sum of all of the n squared distances found in part (b). That is,

$$f(b, m) = \sum_{i=1}^{n} (y_i - (b + mx_i))^2$$

---

[7]Adapted from Robert M. Stark and Robert L. Nichols, *Mathematical Foundations of Design: Civil Engineering Systems*, (New York: McGraw-Hill, 1972).

Show that the partial derivatives $\dfrac{\partial f}{\partial b}$ and $\dfrac{\partial f}{\partial m}$ are given by

$$\frac{\partial f}{\partial b} = -2\sum_{i=1}^{n}(y_i - (b + mx_i))$$

and

$$\frac{\partial f}{\partial m} = -2\sum_{i=1}^{n}(y_i - (b + mx_i)) \cdot x_i.$$

(d)   Now set the partial derivatives equal to zero and solve for m and b. This is easier than it looks: you can simplify the appearance of the equations by temporarily substituting other symbols for the sums. For example, write SY for $\sum y_i$, SX for $\sum x_i$, SXY for $\sum y_i x_i$ and SXX for $\sum x_i^2$. Remember, the x_i and y_i are all constants. You should get a pair of simultaneous linear equations in m and b; solving for m and b will give you formulas in terms of SX, SY, SXY, and SXX.

(e)   Apply these formulas to the data points $(1, 1), (2, 1), (3, 3)$ to verify that you get the same result as in Example 2.

Sometimes you do not expect your data to be linear, but you can transform it in such a way that it looks more linear. For example, suppose you expect that your data points (x, y) fit an exponential equation, say

$$y = Ce^{ax},$$

where a and C are constants. Taking the natural log of both sides, we get

$$\ln y = ax + \ln C.$$

Thus $\ln y$ is a linear function of x. To find a and C, you can use least squares for the plot of $\ln y$ against x.

13.   The population of the United States was about 180 million in 1960, grew to 206 million in 1970, and 226 million in 1980. Assuming that the population was growing at an exponential rate, use the method of least squares to estimate the population in 1990.

14.   Compute the regression line for the points $(-1, 2), (0, -1), (1, 1)$ using least squares.

15.   The following data indicates the increase in the cost of a first class stamp in the United States over the last 70 years.

TABLE 14.3: *Cost of a first class stamp*

Year	1920	1932	1958	1963	1968	1971	1974
Postage	0.02	0.03	0.04	0.05	0.06	0.08	0.10
Year	1975	1978	1981	1985	1988	1991	1995
Postage	0.13	0.15	0.20	0.22	0.25	0.29	0.32

(a)   Find the line of best fit through the data. Using this line, predict the cost of a postage stamp in the year 2000.

(b)   Plot the data. Does it look linear?

(c)   Plot the year against the natural logarithm of the price: Does this look linear? If it is linear, what does that tell you about the price of a stamp as a function of time? Find the line of best fit through this data, and use your answer to again predict the cost of a postage stamp in the year 2000.

16. A biological rule of thumb states that when the area A of an island increases tenfold, the number of species, N, living on it doubles. Consider the following table on some of the islands in the West Indies showing the area of each island and the number of species of reptiles and amphibians living on it.

TABLE 14.4: *Number of species on various islands*

Island	Area	Number
Redonda	1	5
Saba	8	9
Montserrat	75	15
Puerto Rico	3460	75
Jamaica	4240	70
Hispaniola (Haiti and Dominican Rep.)	29520	130
Cuba	44420	125

(a) What sort of function of A is N, according to this rule of thumb?
(b) What sort of function of $\ln A$ is $\ln N$, according to this rule of thumb?
(c) Tabulate $\ln N$ against $\ln A$ and find the line of best fit. Does your answer agree with the biological rule?

17. We wish to find the minimum value of

$$f(x, y) = (x + 1)^4 + (y - 1)^4 + \frac{1}{x^2 y^2 + 1}.$$

(a) Use a computer to plot the contour diagram for f.
(b) Minimize f using the gradient search method.

18. The government wants to build a pipe that will pump water up from a dam to a reservoir, as in Figure 14.24.

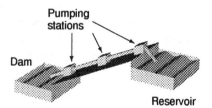

Pumping stations

Dam

Reservoir

Figure 14.24

The cost, C, (in millions of dollars) will depend on the diameter, d, of the pipe (in inches) and the number, n, of pumping stations, according to the following formula[8]:

$$C = 0.15n + 3\left(\frac{d}{50}\right)^{-4.87} + \left(\frac{d}{50}\right)^{1.8} + 3\left(\frac{d}{50}\right)^{1.8} n^{-1}.$$

Using the gradient search method, find the optimal number of pumping stations and pipe diameter.

---

[8]From Douglass J. Wilde, *Globally Optimal Design*, (New York: John Wiley & Sons, 1978).

14.3 CONSTRAINED OPTIMIZATION

Many, perhaps most, real optimization problems are constrained by external circumstances. For example, a city wanting to build a public transportation system has only a limited number of tax dollars it can spend on the project. Any nation trying to maintain its balance of trade cannot spend more on imports than it earns on exports. In this section, we will see how to find an optimum value under such constraints.

A Graphical Approach to Maximizing Production Subject to a Budget Constraint

Let's consider the example of trying to maximize the production of a firm under a budget constraint. Suppose production, f, is a function of two variables, x and y, which could be quantities of two raw materials, or labor and capital, or the number of two different types of workers (doctors and nurses, for example). We suppose that

$$f(x, y) = x^{2/3} y^{1/3}.$$

Suppose that x and y are purchased at prices of p_1 and p_2 per unit. What is the maximum value of the production f that can be obtained with a budget of c dollars?

If we want to maximize f without regard to the budget, we simply increase x and y as far as we can. However, the budget will prevent us from increasing x and y beyond a certain point. Exactly how does the budget constrain us? With prices of p_1 and p_2, the amount spent on x is $p_1 x$ and the amount spent on y is $p_2 y$, so we must have

$$g(x, y) = p_1 x + p_2 y \le c,$$

where $g(x, y)$ is the total cost of the inputs x and y. Let's look at the case when $p_1 = p_2 = 100$ and $c = 378$. Then

$$100x + 100y \le 378$$

so

$$x + y \le 3.78.$$

Graphically, the budget constraint is represented by the line in Figure 14.25. Any point on or below the line represents a pair of values of x and y that we can afford. A point on the line completely exhausts the budget, a point below the line represents values of x and y which can be bought without using up the budget. Any point above the line represents a pair of values that breaks the budget — we cannot afford such a combination.

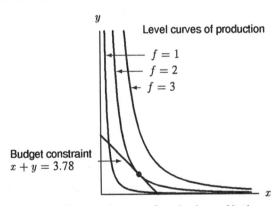

Figure 14.25: Level curves of production and budget constraint

Figure 14.25 also shows some contours of f. Since we want to maximize f, we want to find the point which lies on the level curve with the largest possible f value *and* which lies within the budget. The point we are looking for must lie on the budget line because we should clearly spend all the available money if we want to maximize f. *Unless* we are at the point where the budget constraint is tangent to the contour $f = 2$, we can always increase f by moving along the line representing the budget constraint in Figure 14.25. For example, if we are to the left of the point of tangency, moving right will increase f; if we are to the right of the point of tangency, moving left will increase f. Thus, the maximum value of f on the budget constraint occurs at the point where the budget constraint is tangent to the contour $f = 2$.

In theory, we could find the values of x and y giving this maximum by reading them off the graph. In practice, however, we often use the method of Lagrange multipliers described in this section.

An Analytical Solution: Lagrange Multipliers

We saw from the graphical approach to the constrained optimization problem that the maximum production is achieved at the point where the budget constraint is tangent to a level curve of the production function. The method of Lagrange multipliers uses this fact in algebraic form. Figure 14.26 shows that at the optimum point, the gradient of f and the normal to the budget line $g(x, y) = 3.78$, given by grad g, are parallel, so

$$\text{grad } f = \lambda \text{ grad } g$$

for some scalar λ, called the *Lagrange multiplier*. Since

$$\text{grad } f = \left(\frac{2}{3}x^{-1/3}y^{1/3}\right)\vec{i} + \left(\frac{1}{3}x^{2/3}y^{-2/3}\right)\vec{j} \quad \text{and} \quad \text{grad } g = \vec{i} + \vec{j},$$

we have, after equating the components,

$$\frac{2}{3}x^{-1/3}y^{1/3} = \lambda \quad \text{and} \quad \frac{1}{3}x^{2/3}y^{-2/3} = \lambda.$$

Eliminating λ gives

$$\frac{2}{3}x^{-1/3}y^{1/3} = \frac{1}{3}x^{2/3}y^{-2/3}, \quad \text{so} \quad 2y = x.$$

Since we must also satisfy the constraint $x + y = 3.78$, we have $x = 2.52$ and $y = 1.26$. For these values,

$$f(2.52, 1.26) = (2.52)^{2/3}(1.26)^{1/3} = 2.$$

Thus, as before, we see that the maximum value of f is 2; we also learn that this maximum occurs at $x = 2.52$ and $y = 1.26$.

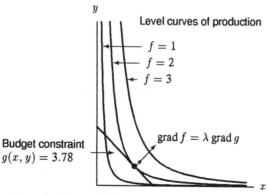

Figure 14.26: grad f is parallel to grad g at maximum production

Lagrange Multipliers in General

Suppose we wish to optimize an *objective function* $f(x, y)$ subject to a *constraint* $g(x, y) = c$. Then, assuming (x, y) is a point at which f is defined, we say

---

- The function f has a **local maximum** at the point (x_0, y_0), **subject to the constraint** $g(x, y) = c$, if $f(x, y) \leq f(x_0, y_0)$ for all (x, y) near (x_0, y_0) such that $g(x, y) = c$.
- The function f has a **local minimum** at (x_0, y_0), **subject to the constraint** $g(x, y) = c$, if $f(x, y) \geq f(x_0, y_0)$ for all (x, y) near (x_0, y_0) such that $g(x, y) = c$.
- The function f has a **global maximum** at the point (x_0, y_0), **subject to the constraint** $g(x, y) = c$, if $f(x, y) \leq f(x_0, y_0)$ for all (x, y) such that $g(x, y) = c$.
- The function f has a **global minimum** at (x_0, y_0), **subject to the constraint** $g(x, y) = c$, if $f(x, y) \geq f(x_0, y_0)$ for all (x, y) such that $g(x, y) = c$.

So, in order to solve the constrained optimization problem, we need to know how to find the local extrema of f subject to the constraint $g = c$. These constrained local extrema are not necessarily related to the unconstrained local extrema.

If we could first solve the equation $g(x, y) = c$ for y as a function of x, say $y = h(x)$, then we could simply substitute for y and use one-variable optimization. However, even in the two-variable constrained optimization problem, the equation $g(x, y) = c$ may be difficult or impossible to solve explicitly. The method of Lagrange multipliers avoids this difficulty.

When we considered the unconstrained optimization problem, we showed that the optimum value of f occurred at one of its critical points—namely, the points where grad f is zero (or is undefined). What is the right notion of a *critical point* for the constrained optimization problem? Suppose that f has a local maximum on the curve C at a point (x_0, y_0), where C is the level curve defined by the constraint $g(x, y) = c$. (The argument in the case of a local minimum is the same.) Let $\vec{u}$ be a unit vector which is tangent to C at (x_0, y_0). See Figure 14.27. If the directional derivative $f_{\vec{u}}(x_0, y_0)$ were positive, then by moving along the curve C in the direction of $\vec{u}$, we could increase f. Similarly, if $f_{\vec{u}}(x_0, y_0)$ were negative, we could increase f by moving along C in the direction $-\vec{u}$. Thus, if $f(x, y)$ has a local maximum on C at (x_0, y_0), we must have

$$f_{\vec{u}}(x_0, y_0) = 0.$$

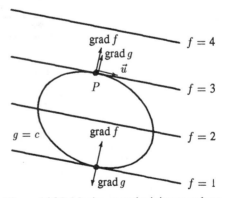

Figure 14.27: Maximum and minimum values of $f(x, y)$ on $g(x, y)$ at points where grad f is parallel to grad g

From Section 13.5 we know that $f_{\vec{u}}(x_0, y_0) = \text{grad } f(x_0, y_0) \cdot \vec{u}$, and so

$$\text{grad } f(x_0, y_0) \cdot \vec{u} = 0.$$

But grad g is perpendicular to the curve C, since C is a level curve of the function g, and so

$$\text{grad } g(x_0, y_0) \cdot \vec{u} = 0.$$

Since the vectors grad $f(x_0, y_0)$ and grad $g(x_0, y_0)$ are both perpendicular to the vector $\vec{u}$, they must be parallel. Thus, if (x_0, y_0) is a local maximum or minimum, subject to the constraint $g(x, y) = c$, and provided grad $g(x_0, y_0)$ is non-zero, there is a scalar λ such that

$$\text{grad } f(x_0, y_0) = \lambda \text{ grad } g(x_0, y_0).$$

In summary, we have: *If $f(x, y)$ has a local maximum or minimum subject to the constraint $g(x, y) = c$ at a point (x_0, y_0) and grad $g(x_0, y_0)$ is non-zero, then there is a scalar λ such that*

$$\text{grad } f(x_0, y_0) = \lambda \text{ grad } g(x_0, y_0).$$

To optimize $f(x, y)$ subject to the constraint $g(x, y) = c$, we solve the following system of three scalar equations

$$f_x(x, y) = \lambda g_x(x, y),$$
$$f_y(x, y) = \lambda g_y(x, y),$$
$$g(x, y) = c,$$

for the three unknowns x, y, λ.

If f has a constrained global maximum or minimum, then it must occur at one of the solutions (x_0, y_0) to this system. (The converse is not necessarily true: solutions to the above system do not necessarily give local maxima or minima of f subject to the constraint.)

Example 1   Find the maximum and minimum values of $x + y$ on the circle $x^2 + y^2 = 4$.

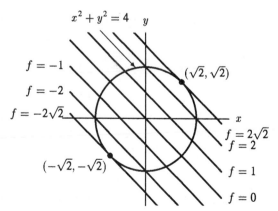

Figure 14.28: Maximum and minimum values of $x + y$
on the circle $x^2 + y^2 = 4$

Solution   The objective function is

$$f(x, y) = x + y,$$

and the constraint is

$$g(x, y) = x^2 + y^2 = 4.$$

Since grad $f = f_x \vec{i} + f_y \vec{j} = \vec{i} + \vec{j}$ and grad $g = g_x \vec{i} + g_y \vec{j} = 2x\vec{i} + 2y\vec{j}$, then grad $f = \lambda$ grad g gives:

$$1 = 2\lambda x,$$
$$1 = 2\lambda y,$$

so

$$x = y.$$

We also know that

$$x^2 + y^2 = 4,$$

giving $x = y = \sqrt{2}$ or $x = y = -\sqrt{2}$.

Since $f(x, y) = x + y$, we see that the maximum value of $x + y$ is $f(\sqrt{2}, \sqrt{2}) = 2\sqrt{2}$, and occurs when $x = y = \sqrt{2}$; the minimum value of $x + y$ is $f(-\sqrt{2}, -\sqrt{2}) = -2\sqrt{2}$, and occurs when $x = y = -\sqrt{2}$. See Figure 14.28.

How to Distinguish Maxima from Minima

In Section 14.1 we developed a second derivative test to classify the critical points of unconstrained optimization problems. There is also a second derivative test for classifying the critical points of constrained optimization problems, but it is rather complicated. [9] However, as you can see from the examples, a graph of the constraint and some level curves can usually make it clear which points are maxima, which points are minima, and which are neither.

Optimization with Inequality Constraints

The production problem that we looked at first was to maximize

$$f(x, y)$$

subject to a budget constraint

$$g(x, y) = p_1 x + p_2 y \le c.$$

This budget constraint is an example of an inequality constraint, because it contains an inequality sign rather than an equals sign. The fact that there is an inequality means that the constraint restricts (x, y) to a region R of the plane rather than a curve in the plane. In principle, we should first check to see whether or not $f(x, y)$ has any critical points in the open region defined by

$$p_1 x + p_2 y < c,$$

the interior of the region R. However, in the case of the budget constraint, it is easy to see that the global maximum of f must occur when the budget was exhausted, and so we can restrict our attention to looking for the maximum value of f on the line

$$p_1 x + p_2 y = c.$$

This line is the boundary of the region R, representing the case where the budget is exhausted.

---

[9]For a discussion of this test, see J. E. Marsden and A. J. Tromba, *Vector Calculus*, 2nd ed., pp. 224–230, (San Francisco: W.H. Freeman, 1981).

General Strategy for Optimizing $f(x, y)$ Subject to the Constraint that (x, y) Lie in a Region, R

If the region R is *open*, find the critical points of f, and then restrict attention to those which lie in R.

If the region R is *closed*, we solve the constrained optimization problem by following these steps:

- Find all points in the interior of R where grad f is zero or undefined.

- Find the local extrema of f restricted to the boundary curve C of the region R.

- Evaluate f at the points found in the previous two steps. Compare the values. The global maximum of f is the largest local maximum and global minimum of f is the smallest local minimum.

Thus, if the region R is defined by an inequality $g(x, y) \leq c$, its interior is the region defined by $g(x, y) < c$ and its boundary curve C is the level curve $g(x, y) = c$ (provided g is continuous). So we first look for the critical points of f in the open region $g(x, y) < c$, using the methods of unconstrained optimization, and then use the method Lagrange multipliers to find the local extrema of f subject to the equality constraint $g(x, y) = c$.

From Section 14.1 we know that if R is both closed and bounded and f is continuous on R, then f attains its global maximum and minimum values on R and so a solution to our optimization problem is guaranteed to exist in this case.

The Meaning of the Lagrange Multiplier

The Meaning of λ in the Production Example

In our previous examples, we never found (or needed) the value of λ. However, λ does have a practical interpretation.

Let's look back at the production problem where we wanted to maximize

$$f(x, y) = x^{2/3}y^{1/3}$$

subject to the constraint

$$g(x, y) = x + y = 3.78.$$

We solve the equations

$$\frac{2}{3}x^{-1/3}y^{1/3} = \lambda,$$

$$\frac{1}{3}x^{2/3}y^{-2/3} = \lambda,$$

$$x + y = 3.78,$$

to get $x = 2.52, y = 1.26$. Continuing to find λ gives us

$$\lambda = 0.53.$$

Suppose now we did another, apparently unrelated calculation. Suppose our budget is increased slightly, from 3.78 to 4.78, giving a new budget constraint of $x + y = 4.78$. Then the corresponding solution is at $x = 3.19$ and $y = 1.59$ and the new maximum value (instead of $P = 2$) is

$$P = (3.19)^{2/3}(1.59)^{1/3} \approx 2.53.$$

Notice that the amount by which f has increased is 0.53, the value of λ. Thus, in this example, the value of λ represents the extra production achieved by increasing the budget by one — in other words, the extra 'bang' you get for an extra 'buck' of budget.

In summary:
- The value of λ is approximately the increase in the optimum value of f when the budget is increased by 1 unit.

More precisely:
- The value of λ represents the rate of change of the optimum value of f as the budget increases.

The Meaning of λ in General

To interpret λ in general, we look at how the optimum value of the objective function f changes as the value c of the constraint function g is varied. The optimum point (x_0, y_0) will, in general, depend on the constraint value c. So, provided x_0 and y_0 are differentiable functions of c, we can use the chain rule to differentiate the optimum value $f(x_0(c), y_0(c))$ with respect to c:

$$\frac{df}{dc} = \frac{\partial f}{\partial x}\frac{dx_0}{dc} + \frac{\partial f}{\partial y}\frac{dy_0}{dc}.$$

At the optimum point (x_0, y_0), we have $f_x = \lambda g_x$ and $f_y = \lambda g_y$, and therefore

$$\frac{df}{dc} = \lambda \left(\frac{\partial g}{\partial x}\frac{dx_0}{dc} + \frac{\partial g}{\partial y}\frac{dy_0}{dc} \right) = \lambda\frac{dg}{dc}.$$

But, as $g(x_0(c), y_0(c)) = c$, we see that $dg/dc = 1$, and so $df/dc = \lambda$. Thus, we have the following interpretation of the Lagrange multiplier λ:

> The value of λ is the rate of change of the optimum value of f as c increases (where $g(x, y) = c$). If the optimum value of f is written as $f(x_0(c), y_0(c))$, then we have
> $$\frac{d}{dc}f(x_0(c), y_0(c)) = \lambda.$$

We can estimate the change, Δf, in the optimum value of f when the constraint value c undergoes a small change from c to $c + \Delta c$ by using local linearity:

$$\Delta f \approx \frac{df}{dc}\Delta c = \lambda\Delta c.$$

Example 2   Suppose the quantity of goods produced according to the function $P(x, y) = x^{2/3}y^{1/3}$ is maximized subject to the constraint $x + y \leq 3.78$. What price must the product sell for if it is to be worth an increased budget for its production?

Solution   On page 211 we found that $\lambda = 0.53$, and therefore increasing the budget by \$1 increases production by about 0.53 unit. In order to make the increase in budget profitable, the extra goods produced must sell for more than \$1. Thus, the price per unit must be at least $1/0.53 = \$1.89$.

More General Constrained Optimization Problems

In many applications we encounter optimization problems where the objective function f is a function of three or more variables and where there may be two or more constraint functions. The strategy for solving such problems, although the algebra may be harder, is the same as the one we used to optimize a function $f(x, y)$ subject to a single constraint $g(x, y) = c$.

For example, suppose the function $f(x, y, z)$ has a local maximum or minimum at the point (x_0, y_0, z_0), subject to the constraints $g_1(x, y, z) = c_1$ and $g_2(x, y, z) = c_2$. Suppose that the two vectors grad $g_1(x_0, y_0, z_0)$ and grad $g_2(x_0, y_0, z_0)$ are not parallel and so define a two-dimensional plane through (x_0, y_0, z_0). We can now use an argument similar to the one we used in the case of a single constraint to show that the vector grad $f(x_0, y_0, z_0)$ lies in the plane defined by grad $g_1(x_0, y_0, z_0)$ and grad $g_2(x_0, y_0, z_0)$. This implies that there are scalars λ_1, λ_2 such that

$$\text{grad } f(x_0, y_0, z_0) = \lambda_1 \text{ grad } g_1(x_0, y_0, z_0) + \lambda_2 \text{ grad } g_2(x_0, y_0, z_0).$$

Therefore, in order to optimize f subject to the constraints $g_1 = c_1$ and $g_2 = c_2$, we solve the system of equations,

$$\text{grad } f(x, y, z) = \lambda_1 \text{ grad } g_1(x, y, z) + \lambda_2 \text{ grad } g_2(x, y, z),$$
$$g_1(x, y, z) = c_1,$$
$$g_2(x, y, z) = c_2,$$

for the five unknowns $x, y, z, \lambda_1, \lambda_2$. We ask you to provide a geometric justification and interpretation of the method of Lagrange multipliers for this example in the problems at the end of this section. [10]

Example 3   The plane $x + y + z = 1$ cuts the cylinder $x^2 + y^2 = 2$ in a curve C. Find the points on C of minimum and maximum height above the xy-plane.

Solution   Since z is the distance of a point above the xy-plane, we want to maximize the objective function $f(x, y, z) = z$ subject to the constraints

$$g_1(x, y, z) = x^2 + y^2 = 2 \quad \text{and} \quad g_2(x, y, z) = x + y + z = 1.$$

Thus, we look for solutions to the equation grad $f = \lambda_1$ grad $g_1 + \lambda_2 g_2$, subject to $g_1(x, y, z) = 2$ and $g_2(x, y, z) = 1$:

$$0 = 2\lambda_1 x + \lambda_2,$$
$$0 = 2\lambda_1 y + \lambda_2,$$
$$1 = \lambda_2,$$
$$x^2 + y^2 = 2,$$
$$x + y + z = 1.$$

From the first two equations, we get $x = y$. Using the third as well gives $x = y = -1/2\lambda_1$. If we now substitute these values for x and y in the fourth, we get

$$\lambda_1 = \pm 1/2.$$

This gives $x = y = \pm 1$ and the last gives $z = -1$ and $z = 3$. Therefore, $P_1 = (-1, -1, 3)$ is the point on C of maximum height above the xy-plane and $P_2 = (1, 1, -1)$ is the point of minimum height.

---

[10] A rigorous justification of the method of Lagrange multipliers requires the *implicit function theorem* and is explained in more advanced undergraduate textbooks on mathematical analysis, such as J. E. Marsden and M. H. Hoffman, *Elementary Classical Analysis*, 2nd ed., (New York: W. H. Freeman, 1993).

The Lagrangian Function

It is often convenient to write constrained optimization problems in terms of a *Lagrangian function* $\mathcal{L}$. For example, suppose we wish to optimize the function $f(x, y)$ subject to the constraint $g(x, y) = c$. The Lagrangian function for this problem is

$$\mathcal{L}(x, y, \lambda) = f(x, y) - \lambda(g(x, y) - c).$$

If we compute the partial derivatives of $\mathcal{L}$,

$$\frac{\partial \mathcal{L}}{\partial x} = \frac{\partial f}{\partial x} - \lambda \frac{\partial g}{\partial x},$$
$$\frac{\partial \mathcal{L}}{\partial y} = \frac{\partial f}{\partial y} - \lambda \frac{\partial g}{\partial y},$$
$$\frac{\partial \mathcal{L}}{\partial \lambda} = -(g(x, y) - c),$$

we notice that if (x_0, y_0) is a critical point of $f(x, y)$ subject to the constraint $g(x, y) = c$ and λ_0 is the corresponding Lagrange multiplier, then at the point (x_0, y_0, λ_0) we have

$$\operatorname{grad} \mathcal{L} = \left(\frac{\partial \mathcal{L}}{\partial x}, \frac{\partial \mathcal{L}}{\partial y}, \frac{\partial \mathcal{L}}{\partial \lambda} \right) = 0.$$

In other words, (x_0, y_0) is a critical point for the problem of optimizing $f(x, y)$ subject to the constraint $g(x, y) = c$ if and only if (x_0, y_0, λ_0) is a critical point for the unconstrained problem of optimizing the Lagrangian function $\mathcal{L}(x, y, \lambda)$.

We can also use Lagrangian functions in more complicated optimization problems. For example, if we wish to optimize the function $f(x, y, z)$ subject to the constraints $g_1(x, y, z) = c_1$ and $g_2(x, y, z) = c_2$, the Lagrangian function is

$$\mathcal{L}(x, y, z, \lambda_1, \lambda_2) = f(x, y, z) - \lambda_1(g_1(x, y, z) - c_1) - \lambda_2(g_2(x, y, z) - c_2).$$

In a similar manner, we can write down Lagrangian functions for more general constrained optimization problems with more than three variables or more than two constraints.

Example 4   A company uses three inputs x, y, and z to manufacture a product with production function given by

$$f(x, y, z) = 50x^{2/5}y^{1/5}z^{1/5}.$$

The total budget is \$24,000 and the company can acquire x, y, and z at \$80, \$12, and \$10, respectively, per unit. What combination of inputs will maximize output? [11]

Solution   We need to maximize the objective function

$$f(x, y, z) = 50x^{2/5}y^{1/5}z^{1/5},$$

subject to the constraint

$$g(x, y, z) = 80x + 12y + 10z = 24{,}000.$$

---

[11] Adapted from M. Rosser, *Basic Mathematics for Economists*, p. 363, (New York: Routledge, 1993).

Therefore, the Lagrangian function is

$$\mathcal{L}(x, y, z) = 50x^{2/5}y^{1/5}z^{1/5} - \lambda(80x + 12y + 10z - 24{,}000),$$

and so we look for solutions to the system of equations grad $\mathcal{L} = 0$:

$$\frac{\partial \mathcal{L}}{\partial x} = 20x^{-3/5}y^{1/5}z^{1/5} - 80\lambda = 0,$$

$$\frac{\partial \mathcal{L}}{\partial y} = 10x^{2/5}y^{-4/5}z^{1/5} - 12\lambda = 0,$$

$$\frac{\partial \mathcal{L}}{\partial z} = 10x^{2/5}y^{1/5}z^{-4/5} - 10\lambda = 0,$$

$$\frac{\partial \mathcal{L}}{\partial \lambda} = -(80x + 12y + 10z - 24{,}000) = 0.$$

We simplify this system slightly to give:

$$\lambda = 0.25x^{-3/5}y^{1/5}z^{1/5},$$

$$\lambda = \frac{5}{6}x^{2/5}y^{-4/5}z^{1/5},$$

$$\lambda = x^{2/5}y^{1/5}z^{-4/5},$$

$$80x + 12y + 10z = 24{,}000.$$

We eliminate z from the first two equations, to obtain $x = 0.3y$, and eliminate x from the second and third equations, to give $z = 1.2y$. Substituting for x and z into $80x + 12y + 10z = 24{,}000$ gives

$$80(0.3y) + 12y + 10(1.2y) = 24{,}000,$$

so that $y = 500$. Thus we get $x = 150$ and $z = 600$, and the corresponding value of f is $f(150, 500, 600) = 4{,}622$ units.

Problems for Section 14.3

In Problems 1–7, use Lagrange multipliers to find the maximum and minimum values of $f(x, y)$ subject to the given constraints.

1. $f(x, y) = x + y$, $x^2 + y^2 = 1$

2. $f(x, y) = 3x - 2y$, $x^2 + 2y^2 = 44$

3. $f(x, y) = x^2 + y$, $x^2 - y^2 = 1$

4. $f(x, y) = xy$, $4x^2 + y^2 = 8$

5. $f(x, y) = x^2 + y^2$, $x^4 + y^4 = 2$

6. $f(x, y, z) = x + 3y + 5z$, $x^2 + y^2 + z^2 = 1$

7. $f(x, y, z) = 2x + y + 4z$, $x^2 + y + z^2 = 16$

8. Let $f(x, y)$ be a linear function, so that $f(x, y) = ax + by + c$ where a, b and c are constants, and let R be a region in the xy-plane.

   (a) If R is any disk, show that the maximum and minimum values of f on R occur on the boundary of the disk.

   (b) If R is any rectangle, show that the maximum and minimum values of f on R occur at the corners of the rectangle. They may occur at other points of the rectangle as well.

   (c) Explain, with the aid of a graph of the plane $z = f(x, y)$, why you would have expected the answers you obtained in parts (a) and (b).

9. A company manufactures a product using inputs x, y, and z according to the production function

$$Q(x, y, z) = 20x^{1/2}y^{1/4}z^{2/5}.$$

The input prices per unit are \$20 for x, and \$10 for y, and \$5 for z. What inputs should the company use if it wishes to manufacture 1,200 products at minimum cost?[12]

10. An industry manufactures a product from two raw materials. The quantity produced, Q, can be given by the Cobb-Douglas function:

$$Q = cx^a y^b,$$

where x and y are quantities of each of the two raw materials used and a, b, and c are positive constants. Suppose the first raw material costs \$$P_1$ per unit and the second costs \$$P_2$ per unit. Find the maximum production possible if no more than \$$K$ can be spent on raw materials.

11. Each person tries to balance his or her time between leisure and work. The tradeoff is, as we all know, that as you work less your income falls. Therefore each person has indifference curves which connect the number of hours of leisure, l, and income, s. If, for example, you are indifferent between 0 hours of leisure and an income of \$450 a week on the one hand, and 10 hours of leisure and an income of \$300 a week on the other hand, then the points $l = 0$, $s = 450$, and $l = 10$, $s = 300$ both lie in the same indifference curve. Table 11 below gives information on three indifference curves, I, II, and III.

Weekly Income			Weekly Leisure Hours		
I	II	III	I	II	III
450	500	550	0	20	40
300	350	400	10	30	50
200	250	300	20	40	60
150	200	250	30	50	70
100	150	200	50	70	90

(a) Sketch the three indifference curves on squared paper.
(b) Now suppose you have only 100 hours a week available for work and leisure combined, and that you earn \$4/hour. Write an equation in terms of l and s which represents this constraint.
(c) On the same squared paper, sketch a graph of this constraint.
(d) Estimate from the graph what combination of leisure hours and income you would choose under these circumstances. Give the corresponding number of hours per week you would work. Make very clear on your drawing, or explain in words, how you made this estimate.

12. Figure 14.29 shows ∇f for a function $f(x, y)$ and two curves $g(x, y) = 1$ and $g(x, y) = 2$. Notice that $g = 1$ is the inside curve and $g = 2$ is the outside curve. Mark the following points on a copy of the figure.
(a) The point(s) A where f has a local maximum.
(b) The point(s) B where f has a saddle point.
(c) The point C where f has a maximum on $g = 1$.
(d) The point D where f has a minimum on $g = 1$.
(e) If you had used Lagrange multipliers to find C, what would the sign of λ have been? Why?

---

[12]Adapted from M. Rosser, *Basic Mathematics for Economists*, Routledge, New York, 1993, p. 363.

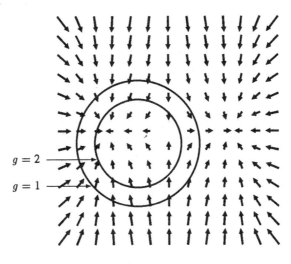

Figure 14.29

13. Design a closed cylindrical container which holds 100 cm³ and has the minimal possible surface area. What should its dimensions be?

14. The quantity, Q, of a product manufactured by a company is given by

$$Q = aK^{0.6}L^{0.4},$$

where a is a positive constant, K is the quantity of capital and L is the quantity of labor used. Capital costs are \$20 per unit, labor costs are \$10 per unit, and the company wants to keep costs for capital and labor combined to be no higher than \$150. Suppose you are asked to consult for the company, and learn that 5 units each of capital and labor are being used.

    (a) What do you advise? Should the plant use more or less labor? More or less capital? If so, by how much?

    (b) Write a one sentence summary that could be used to "sell" your advice to the board of directors.

15. Suppose that the quantity, Q, manufactured of a certain product depends on the number of units of labor, L, and of capital, K, according to the function

$$Q = 900L^{1/2}K^{2/3}.$$

Suppose also that labor costs \$100 per unit and that capital costs \$200 per unit. What combination of labor and capital should be used to produce 36,000 units of the goods at minimum cost? What is that minimum cost?

16. Suppose the quantity, q, of a product manufactured depends on the number of workers, W, and the amount of capital invested, K, and is represented by the Cobb-Douglas function

$$q = 6W^{3/4}K^{1/4}.$$

In addition, labor costs are \$10 per worker and capital costs are \$20 per unit, and the budget is \$3000.

    (a) What are the optimum number of workers and the optimum number of units of capital?

(b) Check that at the optimum values of W and K, the ratio of the marginal productivity of labor to the marginal productivity of capital is the same as the ratio of the cost of a unit of labor to the cost of a unit of capital.

(c) Recompute the optimum values of W and K when the budget is increased by one dollar. Check that increasing the budget by $1 allows the production of λ extra units of the good, where λ is the Lagrange multiplier.

17. A company manufactures x units of one item and y units of another. The total cost in dollars, C, of producing these two items is approximated by the function

$$C = 5x^2 + 2xy + 3y^2 + 800.$$

(a) If the production quota for the total number of items (both types combined) is 39, find the minimum production cost.

(b) Estimate the additional production cost or savings if the production quota is raised to 40 or lowered to 38.

18. The director of a neighborhood health clinic has an annual budget of $600,000. He wants to allocate his budget so as to maximize the number of patient visits, V, which is given as a function of the number of doctors, D, and the number of nurses, N, by:

$$V = 1000D^{0.6}N^{0.3}.$$

Doctors receive a salary of $40,000, while nurses get $10,000.

(a) Set up the director's constrained optimization problem.

(b) Describe, in words, the conditions which must be satisfied by $\partial V / \partial D$ and $\partial V / \partial N$ for V to have an optimum value.

(c) Solve the problem formulated in part (a).

(d) Find the value of the Lagrange multiplier and interpret its meaning in this problem.

(e) What is the marginal cost of a patient visit at the optimum point? Will that marginal cost rise or fall with the number of visits? Why?

19. Consider a firm which manufactures a commodity at two different factories. The total cost of manufacturing depends on the quantities, q_1 and q_2, supplied by each factory, and is expressed by the *joint cost function*, $C = f(q_1, q_2)$. Suppose the joint cost function is approximated by

$$f(q_1, q_2) = 2q_1^2 + q_1q_2 + q_2^2 + 500$$

and that the company's objective is to produce 200 units, at the same time minimizing production costs. How many units should be supplied by each factory?

20. Minimize
$$f(x, y, z) = \sqrt{(x - a)^2 + (y - b)^2 + (z - c)^2},$$
subject to the constraint $Ax + By + Cz + D = 0$. What is the geometric meaning of your solution?

21. The energy required to compress a gas from pressure p_1 to pressure p_{N+1} in N stages is proportional to

$$E = \left(\frac{p_2}{p_1}\right)^2 + \left(\frac{p_3}{p_2}\right)^2 + \cdots + \left(\frac{p_{N+1}}{p_N}\right)^2 - N.$$

Show how to choose the intermediate pressures $p_2, \ldots, p_N$ so as to minimize the energy requirement. [13]

---

[13] Adapted from Rutherford, Aris, *Discrete Dynamic Programming*, p. 35. (New York: Blaisdell, 1964).

REVIEW PROBLEMS FOR CHAPTER FOURTEEN

1. Consider $f(x, y) = x + y + \dfrac{1}{x} + \dfrac{4}{y}$.

   (a) Find and classify the local maxima, minima, and saddle points.
   (b) What are the global maxima and minima? Explain.

2. Suppose $f_x = f_y = 0$ at $(1, 3)$ and $f_{xx} < 0$, $f_{yy} < 0$, $f_{xy} = 0$. Draw a possible contour diagram.

3. The quantity of a product demanded by consumers is affected by its price; the *demand function* gives the dependence of the quantity demanded based on price. In some cases the quantity of one product demanded depends on the price of other products. For example, the demand for tea may be affected by the price of coffee; the demand for cars may be affected by the price of gas. Suppose the quantities demanded, q_1 and q_2, of two products depend on their prices p_1 and p_2 as follows

$$q_1 = 150 - 2p_1 - p_2$$

$$q_2 = 200 - p_1 - 3p_2$$

   (a) What does the fact that q_1 is a function of p_1 and p_2 (instead of p_1 alone) tell you?
   (b) What does the fact that the coefficients of p_1 and p_2 are negative tell you? Give an example of two products that might be related this way.
   (c) Suppose one manufacturer sells both of these products. How should the manufacturer set prices to earn the maximum possible revenue? What is that maximum possible revenue?

4. Find the least squares line for the data points $(0, 4)$, $(1, 3)$, $(2, 1)$.

5. Find the minimum and maximum of the function $z = 4x^2 - xy + 4y^2$ over the closed disc $x^2 + y^2 \leq 2$.

6. An international organization must decide how to spend the $2000 they have been allotted for famine relief in a remote area. They expect to divide the money between buying rice at $5/sack and beans at $10/sack. The number, P, of people who would be fed if they buy x sacks of rice and y sacks of beans is given by

$$P = x + 2y + \frac{x^2 y^2}{2 \cdot 10^8}.$$

   What is the maximum number of people that can be fed, and how should the organization allocate its money?

7. The Cobb-Douglas equation models the total quantity, q, of a commodity produced as a function of the number of workers, W, and the amount of capital invested, K, by the production function

$$q = cW^{1-a}K^a$$

   where a and c are positive constants. Assume labor costs are $\$p_1$ per worker, capital costs are $\$p_2$ per unit, and there is a fixed budget of $\$b$. Show that when W and K are at their optimal levels, the ratio of marginal productivity of labor to marginal productivity of capital equals the ratio of the cost of one unit of labor to one unit of capital.

8. What are the maximum and minimum values of $f(x, y) = -3x^2 - 2y^2 + 20xy$ on the line $x + y = 100$?

9. A family wants to move to a house at a point that is better situated with respect to the children's school and both the parents' places of work. See Figure 14.30.

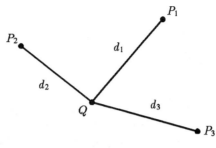

Figure 14.30

Currently they live at Q, the school is at P_1, the mother's work is at P_2, and the father's work is at P_3. They want to minimize $d = d_1 + d_2 + d_3$, where

$$d_1 = \text{distance to school,}$$
$$d_2 = \text{distance to mother's work,}$$
$$d_3 = \text{distance to father's work.}$$

(a) Show that grad d_i is a unit vector pointing directly away from P_i, for $i = 1, 2, 3$. [Hint: Draw contours of d_i, paying careful attention to the spacing of the contours.]

(b) Use your answer to part (a) to draw grad d at the point Q in Figure 14.30. In what direction should the family move to decrease d?

(c) Find as best you can the point on the diagram where grad $d = 0$. What geometric condition characterizes this location?

CHAPTER FIFTEEN

INTEGRATING FUNCTIONS OF MANY VARIABLES

To find the total population of a region, given the population density as a function of position, we need a definite integral. The definite integral represents the limit of a sum, where each term in the sum represents the contribution to the total population from a small subregion. In one-variable calculus we considered problems where the population density only depended on one variable; for example, distance along a turnpike, or distance from the center of the city. If the density depends on more than one variable, we need a multivariable integral to calculate it. In this chapter we will develop the multivariable integral and its interpretations.

15.1 THE DEFINITE INTEGRAL OF A FUNCTION OF TWO VARIABLES

In single-variable calculus, to calculate the total population from a variable population density requires a definite integral. Thus, to investigate integrals of functions of two variables we will start by considering a population density in the plane and how to calculate the total population from that.

Population Density of Foxes in England

England has been largely free of the disease rabies, which is spread by animals. In predicting the spread of an epidemic that might be introduced from Europe at one of the southern ports, it is useful to know the population of animals such as foxes that would be responsible for spreading the disease. The contour diagram in Figure 15.1 shows the population density $D = f(x, y)$ of foxes in the southwestern corner of England, where x and y are in kilometers from the southwest corner of the map, and D is in foxes per square kilometer.[1] It was obtained from estimates of population densities in various parts of England. The bold contour is the coastline (approximately), and may be thought of as the $D = 0$ contour; obviously the density is zero outside it.

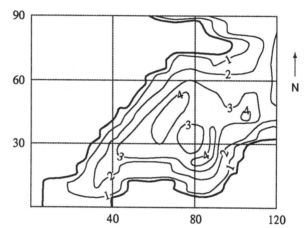

Figure 15.1: Population density of foxes in southwestern
England

Example 1   Estimate the total fox population in the part of England represented by the map in Figure 15.1.

Solution   We want to find upper and lower bounds for the population. We will subdivide the map into the nine rectangles shown in Figure 15.1 and estimate the population in each rectangle. Then we find an upper bound for the population density in each rectangle, multiply it by the area of the rectangle to get an upper bound for the population in that rectangle, then add up all these upper bounds to get an upper bound for the total population. The top left rectangle is in the ocean, so the population is 0 everywhere in that rectangle. The next rectangle to the right just touches the $D = 3$ contour, so the density in that region is at most 3 foxes per square kilometer. For the rectangle immediately below the top left one, we notice that its bottom right corner is halfway between the $D = 2$ and $D = 3$ contours, so we estimate that the maximum population density in that rectangle is 2.5 foxes per square kilometer. In the middle rectangle we see the density goes above 4 inside the $D = 4$

[1] Adapted from J. D. Murray, *Mathematical Biology*, Springer-Verlag, 1989

contour, but not above 5, since there is no $D = 5$ contour. Hence, we choose a maximum of 5 for that rectangle. Continuing in this way, we find upper bounds as tabulated below.

TABLE 15.1: *Upper bounds for the population densities*

0	3	3
2.5	5	5
2.5	5	5

Each region has an area of $30 \times 40 = 1200 \text{ km}^2$, so we obtain an upper bound on the total fox population of $(0 + 3 + 3 + 2.5 + 5 + 5 + 2.5 + 5 + 5)(1200) = 37{,}200$ foxes. What about a lower bound? Since each region contains part of the sea, the minimum density on each region is zero, so our lower bound is 0. So all we know is that the fox population is between 0 and 37,200; taking the average, we estimate it to be 18,600.

Notice that there is a big difference between our lower and upper bounds in the previous example. To do better, we need a finer subdivision.

Example 2   Using the finer grid of 36 subdivisions in Figure 15.2, improve the estimate made in Example 1.

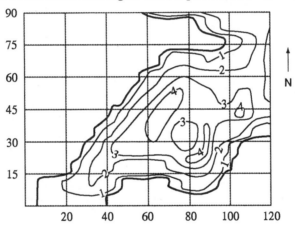

Figure 15.2: A finer grid for the fox population density

Solution   Proceeding in the same way as before, we obtain the following upper and lower estimates.
This time the area of each subdivision is $15 \times 20 = 300 \text{ km}^2$, so the total lower estimate is

$$(0.25 + 0.25 + 1 + 2.25 + 2.25 + 2 + 2 + 1 + 1)(300) = 3600 \text{ foxes,}$$

TABLE 15.2: *Lower estimates of the fox population using the finer grid*

0	0	0	0	0	0.25
0	0	0	0	0	0.25
0	0	0	1	2.25	2.25
0	0	0	2	2	0
0	0	1	1	0	0
0	0	0	0	0	0

TABLE 15.3: *Upper estimates of the fox population using the finer grid*

0	0	0	1.5	3	3
0	0	1	3	3	3
0	0	3	5	3.75	4.5
0	2.25	4.25	5	5	5
0	2.5	4	4.25	5	1
1.25	2.25	2	2	2.25	0

and the total upper estimate is

$$
\begin{aligned}
(1.5 + 3 + 3 + 1 + 3 + 3 + 3 + 3 + 5 + 3.75 + 4.5 + 2.25 + 4.25 + 5 \\
+ 5 + 5 + 2.5 + 4 + 4.25 + 5 + 1 + 1.25 + 2.25 + 2 + 2 + 2.25)(300) = 24{,}525 \ \text{foxes.}
\end{aligned}
$$

Even with 36 subdivisions, there is still a wide disparity between the upper and lower estimates, but we have narrowed the gap; by taking finer and finer subdivisions we could make the upper and lower estimates closer and closer. The average of our two estimates is about 14,000, so we will take that as our estimate for now.

Definition of the Definite Integral

We now give a working definition of a definite integral for a function f of two variables defined on a rectangular region. A review of the one-variable definite integral is given in Appendix D on page 459. Given a function $f(x, y)$ defined on a rectangular region $a \leq x \leq b$ and $c \leq y \leq d$, we construct a Riemann sum by subdividing the region into smaller rectangles. We can do this by subdividing each of the intervals $a \leq x \leq b$ and $c \leq y \leq d$ into n and m equal subintervals, which yields a total of nm subrectangles. See Figure 15.3.

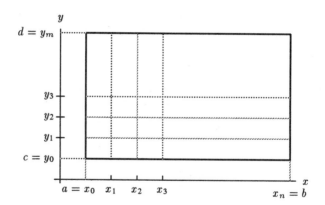

Figure 15.3: Subdivision of a rectangle

To compute the Riemann sum, we multiply the area of each rectangle by the value of the function at a point in the rectangle, and then add all the resulting numbers together. We will call the area of each subrectangle ΔA. Thus, $\Delta A = \Delta x \Delta y$, where $\Delta x = (b - a)/n$ is the width of each subdivision along the x-axis, and $\Delta y = (d - c)/m$ is the width of each subdivision along the y-axis. In the fox density example, we constructed upper and lower sums by taking the maximum or minimum value of the function in each subrectangle. Since this value may be hard to find, it is sometimes more convenient to define a Riemann sum by choosing a definite point at which to evaluate the Riemann sum in each subrectangle; for example, the bottom left hand corner. Then the Riemann sum is

$$
\sum_{i,j} f(x_i, y_j) \, \Delta x \, \Delta y.
$$

Taking the limit as Δx and Δy tend to 0 we get

The **definite integral** of f on a rectangle R is defined to be

$$\int_R f \, dA = \lim_{\Delta x, \Delta y \to 0} \sum_{i,j} f(x_i, y_j) \Delta x \Delta y.$$

Such an integral is called a **two-variable integral**.

Sometimes we use the notation

$$\int_R f \, dA = \int_R f(x, y) \, dx \, dy.$$

Here it is useful to think of dA as being the area of an infinitesimal rectangle of length dx and height dy, so that $dA = dx \, dy$.

The Riemann sum used in the definition is just one type of Riemann sum. For a general Riemann sum the function can be evaluated anywhere in each subdivision; in fact, the subdivisions don't have to be rectangular, they can be regions of any shape. The main thing is to approximate the definite integral by making the subdivisions smaller and smaller.

The Region R

In our definition of the definite integral $\int_R f(x, y) \, dA$, the region R is a rectangle. However, the definite integral can be defined for regions of other shapes, including triangles, circles, and regions bounded by the graphs of functions.

Example 3   Let R be the rectangle $0 \le x \le 1$ and $0 \le y \le 1$. Use a Riemann sum to estimate $\int_R e^{-(x^2+y^2)} \, dA$.

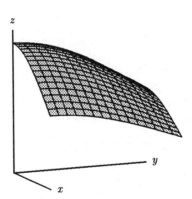

Figure 15.4: Graph of $e^{-(x^2+y^2)}$

Solution   We will divide R into 16 subrectangles by dividing each edge into 4 parts. We want to construct lower and upper sums for the integral, by getting lower and upper estimates of f on each subdivision. From Figure 15.4 we see that f decreases as we move away from the origin. Thus, to get an upper sum we evaluate f on each sub-rectangle at the corner nearest the origin. For example, in the rectangle $0 \le x \le 0.25, 0 \le y \le 0.25$, we will evaluate f at $(0, 0)$.

TABLE 15.4: *Values of $f(x, y) = e^{-(x^2+y^2)}$*

		y				
		0.0	0.25	0.50	0.75	1.00
	0.0	1	0.9394	0.7788	0.5698	0.3679
	0.25	0.9394	0.8825	0.7316	0.5353	0.3456
x	0.50	0.7788	0.7316	0.6065	0.4437	0.2865
	0.75	0.5698	0.5353	0.4437	0.3247	0.2096
	1.00	0.3679	0.3456	0.2865	0.2096	0.1353

To calculate the Riemann sum, we multiply function values by the area of each subdivision, which is $0.25^2 = 0.0625$, and add them all up. In fact it's easier to add up first and then multiply. Using Table 15.4 we find that the resulting upper sum is

$$
\begin{aligned}
[(\ 1 &+ 0.9394 + 0.7788 + 0.5698) \\
&+ (0.9394 + 0.8825 + 0.7316 + 0.5353) \\
&+ (0.7788 + 0.7316 + 0.6065 + 0.4437) \\
&+ (0.5698 + 0.5353 + 0.4437 + 0.3247)](0.0625) = 0.68.
\end{aligned}
$$

To get a lower sum, we must evaluate f at the opposite corner of each rectangle because the surface slopes down in both the x and y directions. This yields a lower estimate of 0.44. Thus,

$$
0.44 \le \int_R e^{-(x^2+y^2)} \, dA \le 0.68.
$$

As in the one-variable case, computing Riemann sums is tedious to do by hand, and is best done by a computer or calculator. In the previous example, if we keep $m = n$ and let

$$
S_n = \sum_{i,j} e^{-(x_i^2+y_j^2)} \, \Delta x \, \Delta y
$$

then for $n = 8$, 16, 32, and 64, we get from a calculator:

$$
S_8 = 0.61, \quad S_{16} = 0.59, \quad S_{32} = 0.57, \quad \text{and } S_{64} = 0.565.
$$

Each S_n as calculated here is an upper estimate for $\int_R e^{-(x^2+y^2)} \, dA$. Note that as n increases the sums S_n are decreasing. Even though this will continue, to find S_n requires n^2 computations. So as n increases the time needed to compute S_n increases by a factor of n^2. Hence, we should look for better ways to approximate the integral.

Interpretations of the Two-Variable Integral

Interpretation as a Volume

Just as the definite integral of a positive one-variable function can be interpreted as an area under the graph of the function, so the definite integral of a two-variable function can be interpreted as a volume under its graph. In the one-variable case you can see this by visualizing the Riemann sums as the total area of rectangles above the subdivisions. In the two-variable case you get solid bars instead of rectangles, because each value of f is multiplied by an area, not a length. As the number of subdivisions grows, the tops of the bars approximate the surface better and better, and the volume of the bars gets closer and closer to the volume under the surface, above the rectangle R. Figure 15.5 illustrates this process.

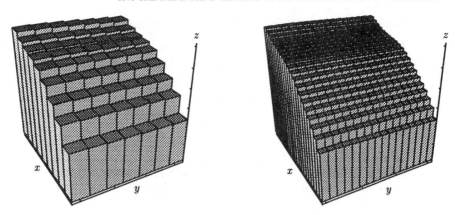

Figure 15.5: Approximating volume under a graph with finer and finer Riemann sums

Example 4   Find the volume under the graph of $f(x, y) = 2 - x^2 - y^2$ lying above the rectangle $-1 \leq x \leq 1$ and $-1 \leq y \leq 1$. (See Figure 15.6.)

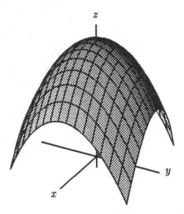

Figure 15.6: Graph of
$f(x, y) = 2 - x^2 - y^2$ above
$-1 \leq x \leq 1$ and $-1 \leq y \leq 1$

Solution   The volume we want is

$$\int_R (2 - x^2 - y^2) \, dA$$

where R is the rectangle $-1 \leq x \leq 1$, $-1 \leq y \leq 1$. Using a computer program that calculates Riemann sums, we find the values in Table 15.5. The sum S_n was calculated by subdividing the rectangle into n^2 subrectangles and evaluating f at the point with minimum x and y values. Thus, the value of the integral seems to be about 5.3.

TABLE 15.5: *Riemann sums for*
$\int_R (2 - x^2 - y^2) \, dA$

n	5	10	20	40
S_n	5.12	5.28	5.32	5.33

Interpretation of the Definite Integral as an Average Value

As in the one-variable case, the definite integral can be used to compute the average value of a function:

$$\left(\begin{array}{c} \text{Average value of } f \\ \text{on the rectangle } R \end{array} \right) = \frac{1}{\text{Area of } R} \int_R f \, dA$$

We can rewrite this as Integral = (Area of R)(Average value). Thus, if we interpret the integral as the volume under the graph of f, then we can think of the average value of f as the height of the box with the same volume that is on the same rectangular base. (See Figure 15.7.)

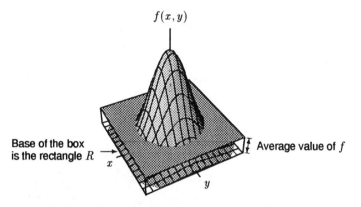

Figure 15.7: Volume and average value

One way to think of this is to imagine that the volume under the graph is made out of wax; if the wax melted and was allowed to level out within walls erected on the perimeter of R, then it would end up looking like the rectangular box, and having a constant height equal to the average value of f.

Interpretation of the Definite Integral When f is a Density Function

A function of two variables can represent a density per unit area, for example, the fox population density (in foxes per unit area), or the mass density of a thin metal plate. Then the integral $\int_R f \, dA$ represents the total population or total mass in the rectangle R.

What About the Fundamental Theorem of Calculus?

The Fundamental Theorem of Calculus was a very important tool in one-variable calculus; it enabled us to calculate definite integrals algebraically by finding antiderivatives of the integrand, instead of using Riemann sums. However, there is no exact analog for the Fundamental Theorem of Calculus in many dimensions. In one-variable calculus, the Fundamental Theorem relied on interpreting the integrand as the rate of change of something. Now we have a situation where the integrand is a function of two variables. In Chapter 20 we will see how such a function can arise (as the divergence of a vector field), and we eventually get something like the Fundamental Theorem; but it is not as useful as the one-variable Fundamental Theorem for evaluating integrals. We will have to look elsewhere to find a way of evaluating two-variable integrals algebraically. The answer is in the next section, which shows how the single-variable Fundamental Theorem of Calculus is adapted to the multivariable case.

Problems for Section 15.1 ━━━━━━━━━━━━━━━━━━━━━━━━━━━━

1. A function $f(x, y)$ has the values in the table below. Let R be the rectangle $1 \leq x \leq 1.2$, $2 \leq y \leq 2.4$. Find the Riemann sums which are reasonable over- and under-estimates for $\int_R f(x, y)\, dA$ with $\Delta x = 0.1$ and $\Delta y = 0.2$.

TABLE 15.6

		x		
		1.0	1.1	1.2
	2.0	5	7	10
y	2.2	4	6	8
	2.4	3	5	4

2. A solid is formed above the rectangle R with $0 \leq x \leq 2$, $0 \leq y \leq 4$ by the graph of $f(x, y) = 2 + xy$. Using Riemann sums with four subdivisions, find upper and lower bounds for the volume of this solid.

3. Let R be the rectangle with vertices $(0, 0)$, $(4, 0)$, $(4, 4)$, and $(0, 4)$ and let $f(x, y) = \sqrt{xy}$.

    (a) Find reasonable upper and lower bounds for $\displaystyle\int_R f\, dA$ without subdividing R.

    (b) Estimate $\displaystyle\int_R f\, dA$ by partitioning R into four subrectangles and evaluating f at its maximum and minimum values on each subrectangle.

4. Figure 15.8 shows the distribution of temperature, in °F, in a 10-foot by 10-foot heated room. Using Riemann sums, estimate the average temperature in the room.

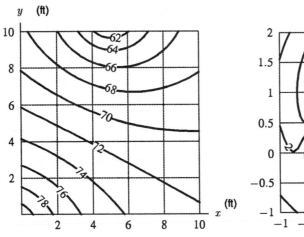

Figure 15.8: Temperature in a room

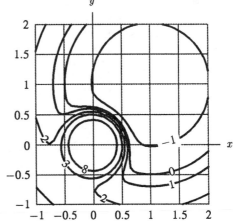

Figure 15.9: Contour diagram for Problem 5

5. Figure 15.9 shows the contour diagram of a function $z = f(x, y)$. Contours are drawn and labeled for $z = -1, 0, 1, 2, \ldots, 8$. Note that the contours for 4, 5, 6, and 7 have been omitted. Let R be the rectangle $-0.5 \leq x \leq 1$, $-0.5 \leq y \leq 1$. Is the integral $\int_R f\, dA$ positive or negative? Explain your reasoning.

6.   A biologist studying insect populations measures the population density of flies and mosquitos at various points in a rectangular study region. The graphs of the two population densities for the region are shown in Figures 15.10 and 15.11. Assuming that the units along the corresponding axes are the same in the two graphs, are there more flies or more mosquitos in the region?

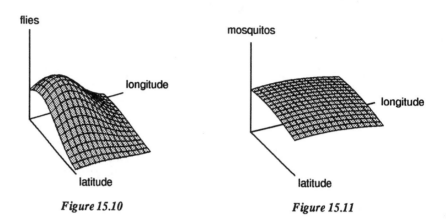

Figure 15.10   *Figure 15.11*

In Problems 7 and 8, use a computer or calculator program that finds two-dimensional Riemann sums to estimate the given integral over the given region.

7.   If R is the rectangle $1 \leq x \leq 2, 1 \leq y \leq 3$, find $\int_R (x^2 + y^2)\, dA$.

8.   If R is the rectangle $-\pi \leq x \leq 0, 0 \leq y \leq \pi/2$, find $\int_R \sin(xy)\, dA$.

9.   The hull of a certain boat has width $w(x, y)$ feet at a point x feet from the front and y feet below the water line. A table of values of w follows. Set up a definite integral that gives the volume of the hull below the waterline, and then estimate the value of the integral.

TABLE 15.7

Front of boat ⟶ Back of boat

		0	10	20	30	40	50	60
	0	2	8	13	16	17	16	10
Depth	2	1	4	8	10	11	10	8
below								
waterline	4	0	3	4	6	7	6	4
(in feet)	6	0	1	2	3	4	3	2
	8	0	0	1	1	1	1	1

10.   Let $f(x, y)$ be a function of x and y which is independent of y, that is, $f(x, y) = g(x)$ for some one-variable function g.

   (a)   What does the graph of f look like?

   (b)   Let R be the rectangle $a \leq x \leq b, c \leq y \leq d$. By interpreting the integral as a volume, and using your answer to part (a), express $\int_R f\, dA$ in terms of a one-variable integral.

11. Figure 15.12 shows contours for the annual frequency of tornados per 10,000 square miles in the U.S.[2] Each grid square is 100 miles on a side. Use the map to estimate the total number of tornados per year in (a) Texas   (b) Florida   (c) Arizona.

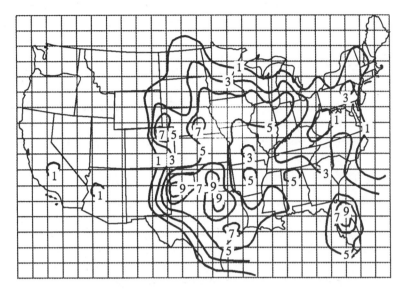

Figure 15.12

12. Figure 15.13 shows contours of annual rainfall (in centimeters) in Oregon.[3] Use it to estimate how much rain falls in Oregon in one year. Each grid square is 100 kilometers on a side.

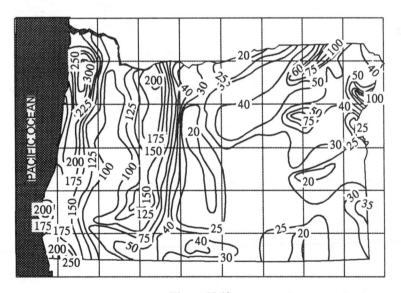

Figure 15.13

[2]From *Modern Physical Geography*, Alan H. Strahler and Arthur H. Strahler, Fourth Edition, John Wiley & Sons, New York, 1992, p. 128

[3]From *Physical Geography of the Global Environment*, H. J. de Blij and Peter O. Muller, John Wiley & Sons, New York, 1993, p. 133

13.   Let $f(x, y)$ be a function of x and y which is independent of x, that is, $f(x, y) = g(y)$ for some one-variable function g.

   (a)   What does the graph of f look like?

   (b)   Let R be the rectangle $a \leq x \leq b$, $c \leq y \leq d$. By interpreting the integral as a volume, and using your answer to part (a), express $\int_R f \, dA$ in terms of a one-variable integral.

15.2   ITERATED INTEGRALS

Although we can approximate a two-variable definite integral as closely as we like using Riemann sums, we don't yet have any way of computing one exactly. In this section we will find a way of doing this by expressing a two-variable integral in terms of one-variable integrals.

The Fox Population Again

When we estimated the fox population from the population density function $D = f(x, y)$, we found that we did not get very good estimates, even when we subdivided the area into 36 pieces. We now look at a different method of estimating the integral. We could do better by subdividing even further, but that would be a lot of work. Here is another way of estimating the population. The idea is to break the area up into horizontal strips, estimate the population in each strip, and then add up all these estimates. Figure 15.14 shows three strips.

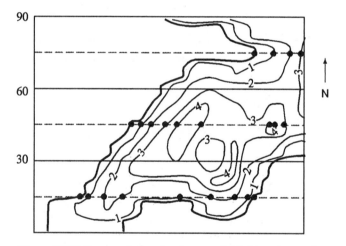

Figure 15.14: Fox population density map divided into three strips

Example 1   Estimate the fox population by estimating the population in each horizontal strip.

Solution   To estimate the population in each strip, we take a cross section of the density function through the middle of each strip. For example, look at the bottom strip in Figure 15.14. We have put a line through the middle of the strip, at $y = 15$, and marked every point where that line crosses a contour. In Figure 15.15, we have plotted the fox densities D against horizontal distance x along this cross section; the data points we obtained from the contour map are marked on the graph. Thus, Figure 15.15 is a graph of the section of f with $y = 15$.

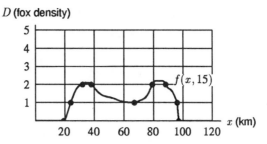

Figure 15.15: Graph of the function $D = f(x, 15)$

Now we approximate the population in this strip by supposing that the density does not vary across the width of the strip. Then the population is the width of the strip, which is 30 miles, times the integral of the one-variable density function $f(x, 15)$. Thus, we have

$$\text{Population in bottom strip} \approx 30 \int_0^{120} f(x, 15)\, dx.$$

We can estimate the integral by counting grid squares in Figure 15.15; there are about 6 grid squares, and each has an area of 20, so the integral is about 120. Thus, the fox population in the bottom strip is estimated to be about $30 \cdot 120 = 3,600$. To estimate the population in the other strips, we follow the same procedure; Figure 15.16 shows the graphs of the sections of f with $y = 45$ and $y = 75$.

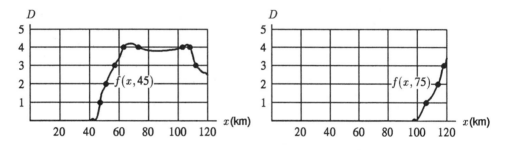

Figure 15.16: Graphs of $f(x, 45)$ and $f(x, 75)$.

From these graphs we estimate by counting grid squares that

$$\int_0^{120} f(x, 45)\, dx \approx 260 \quad \text{and} \quad \int_0^{120} f(x, 75)\, dx \approx 30.$$

Thus, the population estimates for these strips are $30 \cdot 260 = 7,800$, and $30 \cdot 30 = 900$. Combining all three estimates we get an estimate for the total population of $3600 + 7800 + 900 = 12,300$, or about 12,000 foxes. This is consistent with the upper and lower bounds we found before.

How to Express a Two-Variable Integral in Terms of One-Variable Integrals

If we want a better estimate of fox population by the method of the previous example, we can take more strips. In general, the method is to divide the diagram into horizontal strips of width Δy and take a cross-section through the middle of each strip. If y_i is the midpoint of the i-th subdivision on the y-axis, then we take the section of f with y fixed at y_i. Then we integrate with respect to x and multiply by the width Δy to get the estimate

$$\text{Population of } i\text{-th strip} \approx \left(\int_0^{120} f(x, y_i) \, dx \right) \cdot \Delta y,$$

and adding these all up we get:

$$\text{Total population} \approx \sum_{i=1}^{n} \left(\int_0^{120} f(x, y_i) \, dx \right) \cdot \Delta y.$$

This sum is itself a Riemann sum for another integral, this time with respect to y; it is the integral from $y = 0$ to $y = 90$ of the function whose value at y is the integral with respect to x

$$\int_0^{120} f(x, y) \, dx.$$

We have expressed the integral of $f(x, y)$ as an *iterated integral*

$$\int_R f \, dA = \int_0^{90} \left(\int_0^{120} f(x, y) \, dx \right) \, dy.$$

Thus, we have discovered a way of expressing two-variable integrals in terms of nested one-variable integrals:

> If R is the rectangle $a \leq x \leq b$, $c \leq y \leq d$, and if $f(x, y)$ is a continuous function of two variables, then
>
> $$\int_R f \, dA = \int_c^d \left(\int_a^b f(x, y) \, dx \right) \, dy.$$

The inside integral is performed with respect to x, holding y constant, and then the result is integrated with respect to y. The expression $\int_c^d \left(\int_a^b f(x, y) \, dx \right) \, dy$, or simply $\int_c^d \int_a^b f(x, y) \, dx \, dy$, is called an iterated integral.

Iterated Integrals from a Numerical Point of View

Here is another way to see that a two-variable integral can be computed as an iterated integral. This time we look at Riemann sums and a table of values.

Example 2   Compute $\int_R f(x, y) dA$ where R is the rectangle $0 \leq x \leq 1, 0 \leq y \leq 3$ and $f(x, y) = x + y^2$.

Solution   We compute the Riemann sum

$$\sum_{i,j} f(x_i, y_j) \, \Delta x \, \Delta y$$

and see what happens as Δx and Δy approach zero. Suppose $\Delta x = 0.25$ and $\Delta y = 0.5$. If we choose x_i and y_j each to have the two least values in each subrectangle, we get $x_i = 0$, 0.25, 0.5, 0.75 and $y_j = 0$, 0.5, 1, 1.5, 2, 2.5. The values of $f(x, y) = x + y^2$ that we need are given in Table 15.8.

TABLE 15.8: *Values for* $f(x, y) = x + y^2$

			y			
	0.0	0.5	1.0	1.5	2.0	2.5
0.0	0	0.25	1.00	2.25	4.00	6.25
0.25	0.25	0.50	1.25	2.50	4.25	6.50
0.50	0.50	0.75	1.50	2.75	4.50	6.75
0.75	0.75	1.00	1.75	3.00	4.75	7.00

(x labels the rows: 0.0, 0.25, 0.50, 0.75)

The Riemann sum we want is the sum of all entries in Table 15.8 times $\Delta x \Delta y$. We organize the sum by first adding all the values in each column, multiplying by Δx, then adding these results and multiplying by Δy. The sum for a typical column, say the column with $y = 1.5$, looks like this:

$$\sum_i (x_i + 1.5^2)\, \Delta x = (2.25 + 2.50 + 2.75 + 3.00)(0.25)$$

This is a Riemann sum for the section of f with $y = 1.5$, so as Δx approaches 0 the sum approaches the usual one-variable integral for the section of f with $y = 1.5$:

$$\int_0^1 (x + 1.5^2)\, dx$$

When we add up all the column sums to compute the full Riemann sum we get

$$\left(\sum_i (x_i + 0^2)\, \Delta x + \sum_i (x_i + 0.5^2)\, \Delta x + \cdots + \sum_i (x_i + 2.5^2)\, \Delta x \right) \Delta y$$

$$= \sum_j \left(\sum_i (x_i + y_j^2) \Delta x \right) \Delta y$$

Just as each column sum (the inner sum) approaches an integral in x as Δx approaches 0, the outer sum also approaches an integral in y as Δy approaches zero. Thus, the sum of sums, where y is fixed in each inner sum, approaches an integral of integrals, where y is fixed in each inner integral:

$$\int_0^3 \left(\int_0^1 (x + y^2)\, dx \right) dy$$

To evaluate this integral we first compute the inner integral, remembering to treat y as a constant, and then integrate the resulting expression in y:

$$\int_R (x + y^2)\, dA = \int_0^3 \left(\int_0^1 (x + y^2)\, dx \right) dy$$

$$= \int_0^3 \left(\frac{x^2}{2} + y^2 x \right) \Big|_0^1 dy = \int_0^3 \left(\frac{1}{2} + y^2 \right) dy$$

$$= \left(\frac{1}{2} y + \frac{1}{3} y^3 \right) \Big|_0^3 = \frac{21}{2}.$$

The Parallel Between Repeated Summation and Repeated Integration

It is helpful to notice how the summation and integral notation in Example 2 parallel each other. Viewing a Riemann sum as a sum of sums:

$$\sum_{i,j} f(x_i, y_j)\, \Delta x\, \Delta y = \sum_{j} \left(\sum_{i} f(x_i, y_j)\, \Delta x \right) \Delta y$$

leads us to see a two-variable integral as an integral of integrals:

$$\int_R f(x, y)\, dA = \int_c^d \left(\int_a^b f(x, y)\, dx \right) dy,$$

where R is the rectangle $a \le x \le b$ and $c \le y \le d$.

Example 3   A building is 8 feet wide and 16 feet long. It has a flat roof that is 12 feet high at one corner, and 10 feet high at each of the adjacent corners. What is the volume of the building?

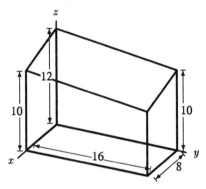

Figure 15.17: A slant-roofed hut

Solution   If we put the high corner on the z-axis, the long side along the y-axis, and the short side along the x-axis, as in Figure 15.17, then the roof is a plane with z-intercept 12, and x slope $(-2)/8 = -1/4$, and y slope $(-2)/16 = -1/8$. Hence, the equation of the roof is

$$z = 12 - \frac{1}{4}x - \frac{1}{8}y.$$

To calculate the volume, we will integrate over the rectangle $0 \le x \le 8, 0 \le y \le 16$. Setting the integral up as an iterated integral, we get

$$\int_0^8 \int_0^{16} (12 - \frac{1}{4}x - \frac{1}{8}y)\, dy\, dx.$$

The inside integral is

$$\int_0^{16} (12 - \frac{1}{4}x - \frac{1}{8}y)\, dy = \left(12y - \frac{1}{4}xy - \frac{1}{8}\frac{y^2}{2} \right) \Big|_0^{16} = 176 - 4x.$$

Then the outside integral is

$$\int_0^8 (176 - 4x)\, dx = (176x - 2x^2) \Big|_0^8 = 1280.$$

So the volume of the building is $1,280$ cubic feet.

The Order of Integration

In Example 2 we could clearly have chosen to add up the rows (fixed x) first instead of the columns. Similarly, in Example 1, we could have taken vertical strips (fixed x) instead of horizontal strips. This leads to an iterated integral where x is constant in the inner integral instead of y. Thus,

$$\int_R f(x, y)\, dA = \int_a^b \left(\int_c^d f(x, y)\, dy \right) dx$$

where R is the rectangle $a \le x \le b$ and $c \le y \le d$.

For any function you are likely to meet, it doesn't matter in which order you integrate over a rectangular region R; you get the same two-variable definite integral either way.

$$\int_R f\, dA = \int_c^d \left(\int_a^b f(x, y)\, dx \right) dy = \int_a^b \left(\int_c^d f(x, y)\, dy \right) dx$$

Example 4   Compute the integral of Example 2 as an iterated integral with x fixed in the inner integral.

Solution

$$\int_R (x + y^2)\, dA = \int_0^1 \left(\int_0^3 (x + y^2) dy \right) dx = \int_0^1 \left. \left(xy + \frac{y^3}{3} \right) \right|_0^3 dx$$

$$= \int_0^1 (3x + 9) dx = \left. \frac{3}{2}x^2 + 9x \right|_0^1 = \frac{21}{2}.$$

Iterated Integrals Over Non-Rectangular Regions

Example 5   The density at the point (x, y) of a right triangular metal plate, as shown in Figure 15.18, is $\delta(x, y)$. Express its mass as an iterated integral.

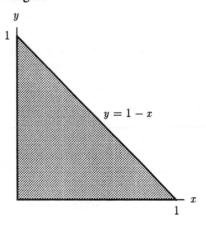

Figure 15.18: A triangular metal plate

Solution   The mass is

$$\int_R \delta(x, y)\, dA,$$

where R is the triangle. The sloping edge of the triangle is the line $y = 1 - x$. We want to compute this integral using an iterated integral. Think about how an iterated integral over a rectangle, such as

$$\int_a^b \int_c^d f(x, y)\, dy\, dx,$$

works. This integral is over the rectangle $a \leq x \leq b,\, c \leq y \leq d$. The inside integral with respect to y is along vertical strips from $y = c$ to $y = d$. There is one such integral for each x strip between $x = a$ and $x = b$. Thus, the value of the inside integral depends on x. After computing the inside integral with respect to y, we compute the outside integral with respect to x, which means putting together all the contributions from the individual vertical strips that make up the rectangle. See Figure 15.19.

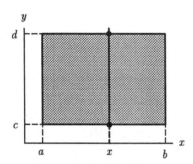

Figure 15.19: Integrating over a
rectangle

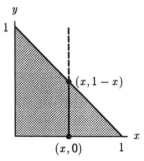

Figure 15.20: Integrating over
a triangle

For the triangular region in Figure 15.18, the idea is the same. The only difference is that the individual vertical strips no longer all go from $y = c$ to $y = d$. The vertical strip that enters the triangle at the point $(x, 0)$ leaves it at the point $(x, 1 - x)$, because the top edge of the triangle is the line $y = 1 - x$. See Figure 15.20. Thus, on this vertical strip, y goes from 0 to $1 - x$. Hence, the inside integral is

$$\int_0^{1-x} \delta(x, y)\, dy.$$

Finally, since there is one of these integrals for each x strip between 0 and 1, the outside integral goes from 0 to 1. Thus, the iterated integral we want is

$$\text{Mass} = \int_0^1 \int_0^{1-x} \delta(x, y)\, dy\, dx.$$

We could have chosen to integrate in the opposite order keeping y fixed in the inner integral instead of x. The limits are formed by looking at horizontal strips instead of vertical ones, and expressing the x-values at the end points in terms of y. A typical horizontal strip goes from $x = 0$ to $x = 1 - y$, and since y values overall range between 0 and 1, the iterated integral is

$$\text{Mass} = \int_0^1 \int_0^{1-y} \delta(x, y)\, dx\, dy.$$

Iterated integrals with variable limits on the inner integral

There are two things to remember
- The limits on the outer integral must be constants.
- If the inner integral is with respect to x, its limits should be constants or expressions in terms of y, and vice versa.

Example 6   Find the mass M of a metal plate R bounded by $y = x$ and $y = x^2$ with density given by $\delta(x, y) = 1 + xy$ kg/meter^2. (See Figure 15.21.)

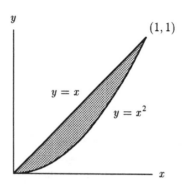

Figure 15.21: Another metal plate

Solution   The mass is

$$M = \int_R \delta(x, y) \, dA.$$

We will integrate along vertical strips first; this means we will do the y integral first, which goes from the bottom boundary $y = x^2$ to the top boundary $y = x$. Thus, the inside integral is

$$\int_{x^2}^x \delta(x, y) \, dy = \int_{x^2}^x (1 + xy) \, dy = (y + x\frac{y^2}{2})\Big|_{y=x^2}^{y=x}$$

$$= (x + \frac{x^3}{2}) - (x^2 + \frac{x^5}{2}) = x - x^2 + \frac{x^3}{2} - \frac{x^5}{2}.$$

The left edge of the region is at $x = 0$ and the right edge is at the intersection point of $y = x$ and $y = x^2$, which is $(1, 1)$. Thus, the x-coordinate of the vertical strips can vary from $x = 0$ to $x = 1$, and so the mass is

$$M = \int_0^1 (x - x^2 + \frac{x^3}{2} - \frac{x^5}{2}) \, dx = (\frac{x^2}{2} - \frac{x^3}{3} + \frac{x^4}{8} - \frac{x^6}{12})\Big|_0^1 = \frac{5}{24} \text{ kg}.$$

Example 7   A city is in the form of a semicircular region of radius 3 miles bordering on the ocean. Find the average distance from any point in the city to the ocean.

Solution   Think of the ocean as everything below the x-axis in the xy-plane and think of the city as the upper half of the circular disk of radius 3 bounded by $x^2 + y^2 = 9$. (See Figure 15.22).

The distance from any point (x, y) in the city to the ocean is the vertical distance to the x-axis, namely y. Thus, we want to compute

$$\text{Average distance} = \frac{\int_R y \, dA}{\text{area}(R)}$$

where R is the region between the upper half of the circle $x^2 + y^2 = 9$ and the x-axis. The area of R is $\pi 3^2/2 = 9\pi/2$. To compute the integral, let's try making the inner integral with respect to y. Then a typical vertical strip goes from the x-axis, namely $y = 0$, to the semicircle. The limits must be expressed in x so we solve $x^2 + y^2 = 9$ for y in terms of x to get $y = \sqrt{9 - x^2}$. Since x varies from -3 to 3 throughout the region, the integral is:

$$\int_R y \, dA = \int_{-3}^{3} \left(\int_0^{\sqrt{9-x^2}} y \, dy \right) dx = \int_{-3}^{3} \left(\frac{y^2}{2} \Big|_0^{\sqrt{9-x^2}} \right) dx$$

$$= \int_{-3}^{3} \frac{1}{2}(9 - x^2) \, dx = \frac{1}{2} \left(9x - \frac{x^3}{3} \right) \Big|_{-3}^{3} = \frac{1}{2}(18 - (-18)) = 18.$$

Therefore, the average distance is $18/(9\pi/2) = 4/\pi$ miles.

What if we choose the inner integral to be with respect to x instead? Then we get the limits by looking at horizontal strips, not vertical, and we solve $x^2 + y^2 = 9$ for x in terms of y, rather than y in terms of x. We get $x = -\sqrt{9 - y^2}$ at the left end of the strip and $x = \sqrt{9 - y^2}$ at the right. Now y varies from 0 to 3 so the integral becomes:

$$\int_0^3 \left(\int_{-\sqrt{9-y^2}}^{\sqrt{9-y^2}} y \, dx \right) dy = \int_0^3 \left(yx \Big|_{x=-\sqrt{9-y^2}}^{x=\sqrt{9-y^2}} \right) dy = \int_0^3 2y\sqrt{9 - y^2} \, dy$$

$$= -\frac{2}{3}(9 - y^2)^{3/2} \Big|_0^3 = -\frac{2}{3}(0 - 27) = 18.$$

Needless to say, we get the same average distance as before.

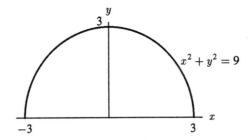

Figure 15.22: The city by the ocean

In the examples so far, the region was given and the problem was to determine the limits for an iterated integral. Sometimes the limits are given and we want to determine the region.

Example 8   Sketch the region of integration for the iterated integral $\int_0^6 \int_{x/3}^2 x\sqrt{y^3 + 1}\, dy\, dx.$

Solution   The inner integral is with respect to y, so we should imagine vertical strips crossing the region of integration. The bottom of each strip is $y = x/3$, a line through the origin, and the top is $y = 2$, a horizontal line. Since the outer integral's limits are 0 and 6, the whole region is contained between the vertical lines $x = 0$ and $x = 6$. Notice that the lines $y = 2$ and $y = x/3$ meet where $x = 6$. The region is shown in Figure 15.23 .

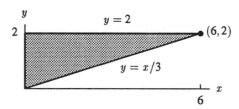

Figure 15.23: The region of integration for
Example 8

Reversing the Order of Integration

It can sometimes be helpful to reverse the order of integration in an iterated integral. Surprisingly enough, an integral which is difficult or impossible with the limits in one order can be quite straightforward in the other. The next example is such a case.

Example 9   Evaluate $\int_0^6 \int_{x/3}^2 x\sqrt{y^3 + 1}\, dy\, dx.$ This is the same integral whose region we sketched in Figure 15.23.

Solution   Since $\sqrt{y^3 + 1}$ has no elementary antiderivative, we have no algebraic way to do the inner integral. We will try reversing the order of integration. From Figure 15.23, we see that horizontal strips go from $x = 0$ to $x = 3y$. For the whole region, y varies from 0 to 2. Thus, when we change the order of integration we get:

$$\int_0^2 \int_0^{3y} x\sqrt{y^3 + 1}\, dx\, dy.$$

Now we can at least do the inner integral. What about the outer integral?

$$\int_0^2 \int_0^{3y} x\sqrt{y^3 + 1}\, dx\, dy = \int_0^2 \left(\frac{x^2}{2}\sqrt{y^3 + 1}\right)\Bigg|_{x=0}^{x=3y} dy$$

$$= \int_0^2 \frac{9y^2}{2}(y^3 + 1)^{1/2}\, dy$$

$$= (y^3 + 1)^{3/2}\Bigg|_0^2 = 27 - 1 = 26.$$

Thus, reversing the order of integration made the integral in the previous problem much easier. Notice that to reverse the order of integration it is essential first to sketch the region over which the integration is being performed.

Problems for Section 15.2

For Problems 1–3, evaluate the given integral.

1. $\int_R \sqrt{x+y}\,dA$, where R is the rectangle $0 \le x \le 1, 0 \le y \le 2$.

2. $\int_R (5x^2+1)\sin 3y\,dA$, where R is the rectangle $-1 \le x \le 1, 0 \le y \le \pi/3$.

3. $\int_R (2x+3y)^2\,dA$, where R is the triangle with vertices at $(-1,0), (0,1)$, and $(1,0)$.

For each of the regions R in Problems 4–7, set up $\int_R f\,dA$ as an iterated integral.

4.

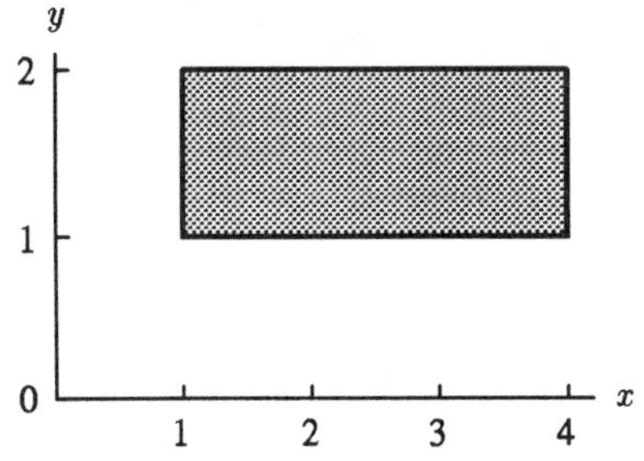

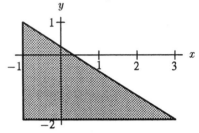

Figure 15.24

5.

Figure 15.25

6.

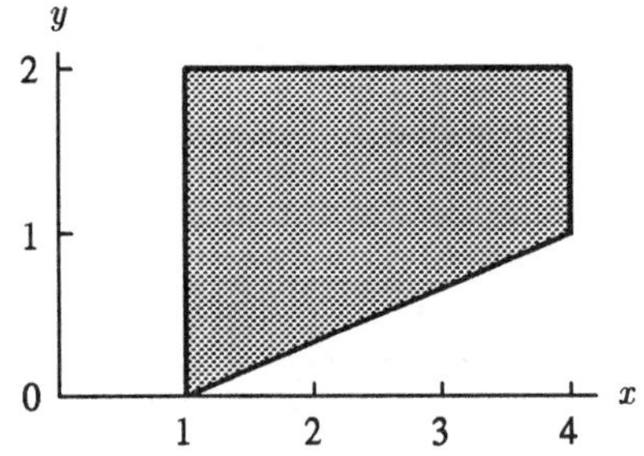

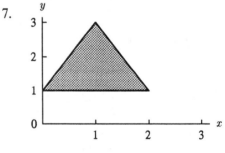

Figure 15.26:

7.

Figure 15.27

8. Do the integral in Problem 1 in reverse order.

For Problems 9–13, sketch the region of integration and evaluate the integral.

9. $\int_1^3 \int_0^4 e^{x+y}\,dy\,dx$

10. $\int_0^2 \int_0^x e^{x^2}\,dy\,dx$

11. $\int_1^5 \int_x^{2x} \sin(x)\,dy\,dx$

12. $\int_1^4 \int_{\sqrt{y}}^y x^2 y^3\,dx\,dy$

13. $\int_{-2}^0 \int_{-\sqrt{9-x^2}}^0 2xy\,dy\,dx$

14. Consider the integral $\int_0^4 \int_0^{-(y-4)/2} g(x,y)\,dx\,dy$.

(a) Sketch the region over which the integration is being performed.

(b) Write the integral with the order of the integration reversed.

Evaluate the integral in Problems 15–18 by reversing the order of integration.

15. $\displaystyle\int_0^1 \int_y^1 e^{x^2}\, dx\, dy$ 16. $\displaystyle\int_0^3 \int_{y^2}^9 y\sin(x^2)\, dx\, dy$ 17. $\displaystyle\int_0^1 \int_{\sqrt{y}}^1 \sqrt{2+x^3}\, dx\, dy$

18. $\displaystyle\int_{-4}^0 \int_0^{2x+8} f(x,y)\, dy\, dx + \int_0^4 \int_0^{-2x+8} f(x,y)\, dy\, dx$

For Problems 19–22 set up, but do not evaluate, an iterated integral for the volume of the solid.

19. Under the graph of $f(x,y) = 25 - x^2 - y^2$ and above the xy-plane.

20. Above the graph of $f(x,y) = 25 - x^2 - y^2$ and under the plane $z = 16$.

21. The three-sided pyramid whose base is the xy-plane and whose three sides are the vertical planes $y = 0$ and $y - x = 4$ and the slanted plane $2x + y + z = 4$.

22. Same as Problem 21 with the base in the plane $z = -6$.

For Problems 23–25, find the volumes of the given regions.

23. Under the graph of $f(x,y) = xy$ and above the square $0 \le x \le 2, 0 \le y \le 2$ in the xy-plane.

24. The solid between the planes $z = 3x + 2y + 1$ and $z = x + y$ and above the triangle with vertices $(1,0,0), (2,2,0)$, and $(0,1,0)$ in the xy-plane. See Figure 15.28.

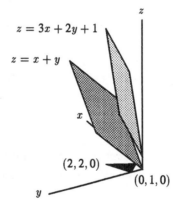

Figure 15.28

25. The region R bounded by the graph of $ax + by + cz = 1$ and the coordinate planes. Assume $a, b,$ and c are positive.

26. Find the average distance to the x-axis for points in the region bounded by the x-axis and the graph of $y = x - x^2$.

27. Prove that for a right triangle the average distance from any point in the triangle to one of the legs is one-third the length of the other leg.

28. Evaluate $\displaystyle\int_0^1 \int_y^1 \sin(x^2)\, dx\, dy$ 29. Evaluate $\displaystyle\int_0^1 \int_{e^y}^e \frac{x}{\ln x}\, dx\, dy$

15.3 THREE-VARIABLE INTEGRALS

A function of three variables can be integrated over a region R in 3-space in the same way as a function of two variables is integrated over a region in 2-space. Again, we start with a Riemann sum. First we subdivide R into smaller regions, then we multiply the volume of each region by a value of the function in that region, and then we add the results. For example, if R is the box $a \le x \le b$, $c \le y \le d$, $e \le z \le f$, then we subdivide each side into l, m, and n pieces, thereby chopping R into lmn smaller boxes, as shown in Figure 15.29.

The volume of each smaller box is

$$\Delta V = \Delta x \Delta y \Delta z,$$

where

$$\Delta x = \frac{b-a}{l}, \quad \Delta y = \frac{d-c}{m}, \quad \Delta z = \frac{f-e}{n}.$$

Using this subdivision, we construct a Riemann sum

$$\sum_{i,j,k} f(x_i, y_j, z_k)\, \Delta V.$$

As Δx, Δy, and Δz approach 0, this Riemann sum approaches the definite integral

$$\int_R f\, dV.$$

Just as in the case of two-variable integrals, we can evaluate this integral as an iterated integral:

Three-variable integral as an iterated integral or triple integral

$$\int_R f\, dV = \int_e^f \left(\int_c^d \left(\int_a^b f(x,y,z)\, dx \right) dy \right) dz.$$

where y and z are treated as constants in the innermost (dx) integral, and z is treated as a constant in the middle (dy) integral. Any other order of integration will give the same answer.

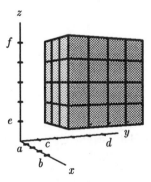

Figure 15.29: Subdividing a three-dimensional box

Example 1   A cube C is 4 centimeters on a side and is made of a material of variable density. If one corner is at the origin and the adjacent corners are on the positive x, y, and z axes, then the function giving its density (in gm/cm^3) at the point (x, y, z) is $\delta(x, y, z) = 1 + xyz$. Find the mass of the cube.

Solution   Consider a small piece ΔV of the cube, small enough so that the density remains close to constant over the piece. The mass of ΔV is then the density times the volume, i.e., it is $\delta(x, y, z) \, \Delta V$. To get the total mass, we need to put all the small pieces together in a Riemann sum and take a limit. Thus, the mass is the three-variable integral

$$M = \int_C \delta \, dV = \int_0^4 \int_0^4 \int_0^4 (1 + xyz) \, dx \, dy \, dz = \int_0^4 \int_0^4 \left[x + \frac{1}{2}x^2yz \right]_{x=0}^{x=4} dy \, dz$$

$$= \int_0^4 \int_0^4 (4 + 8yz) \, dy \, dz = \int_0^4 \left[4y + 4y^2 z \right]_{y=0}^{y=4} dz = \int_0^4 (16 + 64z) \, dz = 576 \, \text{gm}.$$

Example 2   Express the volume of the building described in Example 3 on page 236 as a triple integral.

Solution   The building is given by $0 \leq x \leq 8$, $0 \leq y \leq 16$, and $0 \leq z \leq 12 - x/4 - y/8$. (See Figure 15.30.) To find its volume, we divide it into small cubes of volume $\Delta V = \Delta x \, \Delta y \, \Delta z$ and add them up. First we form a vertical stack of cubes above the point $(x, y, 0)$. This stack goes from $z = 0$ to $z = 12 - x/4 - y/8$, and

$$\text{Volume of stack} \approx \sum_z \Delta V = \sum_z \Delta x \, \Delta y \, \Delta z = \left(\sum_z \Delta z \right) \Delta x \, \Delta y.$$

Next we line up these stacks parallel to the y-axis to form a slice from $y = 0$ to $y = 16$. So

$$\text{Volume of slice} \approx \left(\sum_y \sum_z \Delta z \, \Delta y \right) \Delta x.$$

Finally we line up the slices along the x-axis from $x = 0$ to $x = 8$ and add up their volumes, to get

$$\text{Volume} \approx \sum_x \sum_y \sum_z \Delta z \, \Delta y \, \Delta x.$$

Thus, in the limit, $\text{Volume} = \displaystyle\int_0^8 \int_0^{16} \int_0^{12-x/4-y/8} dz \, dy \, dx.$

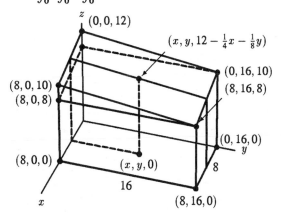

Figure 15.30: Volume of building as triple integral

Example 3   Set up an iterated integral to compute the mass of the cone bounded by $z = \sqrt{x^2 + y^2}$ and $z = 3$, if the density is given by $\delta(x, y, z) = z$.

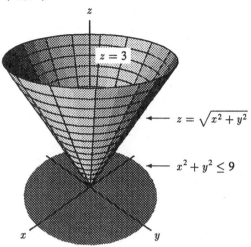

Figure 15.31

Solution   The cone is shown in Figure 15.31. Again, we break the cone into small cubes of volume $\Delta V = \Delta x \, \Delta y \, \Delta z$, on which the density is approximately constant, and approximate the mass of each piece by $\delta(x, y, z) \, \Delta x \, \Delta y \, \Delta z$. We stack the cubes vertically above the point $(x, y, 0)$, starting on the cone at height $z = \sqrt{x^2 + y^2}$ and going up to $z = 3$. Thus the inner integral is

$$\int_{\sqrt{x^2+y^2}}^{3} \delta(x, y, z) \, dz = \int_{\sqrt{x^2+y^2}}^{3} z \, dz.$$

There is a stack for every point in the xy-plane below the top of the cone, that is, for every point in the shadow cast by the cone on the xy-plane. Since the cone $z = \sqrt{x^2 + y^2}$ intersects the horizontal plane $z = 3$ in the circle $x^2 + y^2 = 9$, this means that there is a stack for all (x, y) in the region $x^2 + y^2 \leq 9$. Thus, if we line up the stacks parallel to the y-axis, we get a slice from $y = -\sqrt{9 - x^2}$ to $y = \sqrt{9 - x^2}$, for each fixed value of x. Thus, the limits on the next integral out are

$$\int_{-\sqrt{9-x^2}}^{\sqrt{9-x^2}} \int_{\sqrt{x^2+y^2}}^{3} z \, dz \, dy.$$

Finally, there is a slice for each x between -3, and 3, so the final answer is

$$\text{Mass} = \int_{-3}^{3} \int_{-\sqrt{9-x^2}}^{\sqrt{9-x^2}} \int_{\sqrt{x^2+y^2}}^{3} z \, dz \, dy \, dx.$$

Notice that setting up the limits on the two outer integrals was just like setting up the limits for a two-variable integral over the region $x^2 + y^2 \leq 9$.

There are three things to remember for a triple integral:
- The limits for the outer integral are constants.
- The limits for the middle integral can involve only one variable (that in the outer integral).
- The limits for the inner integral can involve two variables (those on the two outer integrals).

Problems for Section 15.3

In Problems 1–4, find the three-variable integrals of the given functions over the given regions.

1. $f(x, y, z) = x^2 + 5y^2 - z$, W is the rectangular box $0 \le x \le 2$, $-1 \le y \le 1$, $2 \le z \le 3$.
2. $f(x, y, z) = \sin x \cos(y + z)$, W is the cube $0 \le x \le \pi$, $0 \le y \le \pi$, $0 \le z \le \pi$.
3. $h(x, y, z) = ax + by + cz$, W is the rectangular box $0 \le x \le 1$, $0 \le y \le 1$, $0 \le z \le 2$.
4. $f(x, y, z) = e^{-x-y-z}$, W is the rectangular box with corners at $(0, 0, 0)$, $(a, 0, 0)$, $(0, b, 0)$, and $(0, 0, c)$.

For Problems 5–11 describe or sketch the region of integration for the triple integrals. If the limits do not make sense, say why.

5. $\displaystyle\int_0^6 \int_0^{3-x/2} \int_0^{6-x-2y} f(x, y, z)\, dz\, dy\, dx$
6. $\displaystyle\int_0^1 \int_0^x \int_0^x f(x, y, z)\, dz\, dy\, dx$

7. $\displaystyle\int_0^1 \int_0^z \int_0^x f(x, y, z)\, dz\, dy\, dx$
8. $\displaystyle\int_0^3 \int_{-\sqrt{9-y^2}}^0 \int_{\sqrt{x^2+y^2}}^3 f(x, y, z)\, dz\, dx\, dy$

9. $\displaystyle\int_1^3 \int_1^{x+y} \int_0^y f(x, y, z)\, dz\, dx\, dy$
10. $\displaystyle\int_0^1 \int_0^{2-x} \int_0^3 f(x, y, z)\, dz\, dy\, dx$

11.
$$\int_{-1}^1 \int_0^{\sqrt{1-x^2}} \int_0^{\sqrt{2-x^2-y^2}} f(x, y, z)\, dz\, dy\, dx$$

12. Set up $\int_R f\, dV$ as an iterated integral in all six possible orders of integration, where R is the hemisphere bounded by the upper half of $x^2 + y^2 + z^2 = 1$ and the xy-plane.
13. Find the mass of the solid bounded by the xy-plane, yz-plane, xz-plane, and the plane $(x/3) + (y/2) + (z/6) = 1$, if the density of the solid is given by $\delta(x, y, z) = x + y$.
14. Find the average value of the sum of the squares of three numbers x, y, z where each number is between 0 and 2.
15. Find the volume of the solid formed by the intersections of the cylinders $x^2 + z^2 = 1$ and $y^2 + z^2 = 1$.

15.4 NUMERICAL INTEGRATION: THE MONTE CARLO METHOD

There are many one-variable definite integrals in which the integrand has no elementary antiderivative. A familiar example is $\int_0^1 e^{-x^2}\, dx$. To evaluate such an integral, we must use some numerical technique such as the trapezoid rule or Simpson's rule. For a two or three-variable integral, we are just as likely to encounter intractable integrals. One can always use Riemann sums to approximate the integral. One can get greater accuracy by using a variant of Simpson's rule (see Problem 6). In this section, we give an alternative method called the Monte Carlo method (after the gambling casino).

A One-Variable Example

Let us consider the integral $\int_0^1 x^2\, dx$, which we know to have the value $1/3$. We now approach it probabilistically. We graph the function $y = x^2$ in the square $0 \le x \le 1$, $0 \le y \le 1$ and throw darts at the square. We expect that some darts will land above the curve, some below. The ratio of the number of darts above to those below gives a reasonable estimate of the ratio of these two areas. Since the total area plotted is the unit square of area 1, the ratio of the number of darts which land below the curve to the total number of darts thrown gives an estimate for the area under the curve. This is the basis of the Monte Carlo method.

Example 1   Approximate the value of the integral $\int_0^1 x^2\,dx$ using the Monte Carlo method.

Solution   We choose points from the unit square of Figure 15.32 at random, and we expect that the ratio of the number of points in region R, say N_R, to that of all of those picked from the square, say N, will approximate the integral:

$$\frac{N_R}{N} \approx \frac{\int_0^1 x^2\,dx}{1} = \int_0^1 x^2\,dx.$$

Since we are selecting the points at random we cannot expect to get the same ratio every time, but as the number of points increases, the approximation should improve.

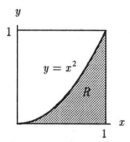

Figure 15.32: Region whose area is $\int_0^1 x^2\,dx$ as fraction of the unit square

Since we won't get the same answer every time, we try the same computation several times and compare the answers to get some idea of the accuracy. Table 15.9 shows the values of N_R/N for six different trials with $N = 50$ points.

TABLE 15.9: *Six trials for N = 50 points*

$N = 50$	1	2	3	4	5	6
N_R/N	0.2	0.24	0.52	0.36	0.38	0.28

These approximations are not particularly good. We may improve the accuracy by taking their average, giving 0.33, which has two digit accuracy. Repeating this process again for $N = 50$ gives the results found in Table 15.10.

TABLE 15.10: *Six more trials for N = 50 points*

$N = 50$	1	2	3	4	5	6
N_R/N	0.44	0.42	0.28	0.34	0.28	0.32

Notice that the average of these is 0.347. This is not as close to the true value of 1/3 as before, but remember that this is a random process. Each time it is repeated we expect a different result.

To continue with this example we now approximate $\int_0^1 x^2 \, dx$ by taking increasingly large values of N, say $N = 10, 100, 1000,$ and $10,000$. The results of one experiment are given in Table 15.11, but if you carry out a similar experiment your results will probably be slightly different.

TABLE 15.11: *Value of N_R/N as N increases*

N	10	100	1000	10000
N_R/N	0.2000	0.3400	0.3250	0.3343

It appears that as we take more points from the unit square, the ratio gets closer to the exact value of $1/3$.

The basis of the Monte Carlo method is the generation of random numbers. Fortunately, nearly all programming languages have a random number generator built in. In the previous example we used a random number generator that gives numbers between 0 and 1. Given two random numbers, say x and y, we have a point (x, y) in the unit square. We need only check to see if $y \leq x^2$. If this is true, we know the point is in the region under the parabola and count it. If this is not true, then we pick another pair of random numbers. As Table 15.11 shows, as more points are considered, our approximation usually gets better. However, it is possible for the approximation to get poorer.

When we use a Monte Carlo method, we assume that when we pick points from a region R, no point is any more likely to be chosen than any other. That is, the points that we can choose are all equally likely to be found in R. If we know the area of R, say $A(R)$, and if we choose N points from R and find that M of these N points are in S, where S is a region contained in R, then we expect, for N large, that

$$\frac{A(S)}{A(R)} \approx \frac{M}{N}$$

or

$$A(S) \approx A(R) \cdot \frac{M}{N}.$$

A Two-Variable Example

We can extend the idea of the Monte Carlo method to evaluating integrals of more than one variable.

Example 2   Use a Monte Carlo method to approximate the two-variable integral

$$\int_0^1 \int_0^1 e^{-(x^2+y^2)} \, dx \, dy.$$

Solution   This integral gives the volume of the region W above the unit square and below the graph of $z = e^{-(x^2+y^2)}$. Since the volume we are considering is contained in the cube C given by $0 \leq x \leq 1$, $0 \leq y \leq 1$, and $0 \leq z \leq 1$, we count only points of the form (x, y, z) which lie in the cube and which satisfy the condition

$$0 \leq z \leq e^{-(x^2+y^2)}.$$

If N_R of the N points chosen satisfy this condition, then, since the volume of the bounding cube is 1, we have

$$\text{Vol}(W) \approx \frac{N_R}{N}$$

So

$$\int_0^1 \int_0^1 e^{-(x^2+y^2)} \, dx \, dy = \text{Vol}(W) \approx \frac{N_R}{N}.$$

TABLE 15.12: *Ten trials with* $N = 100$

$N = 100$	1	2	3	4	5	6	7	8	9	10
N_R/N	0.54	0.60	0.57	0.60	0.51	0.53	0.59	0.56	0.56	0.57

Table 15.12 shows the value of N_R/N for ten trials of $N = 100$ points each. The average of the ten values N_R/N is 0.563. We take this as an approximate value of the integral.

Taking $N = 10,000$ gives

$$\int_0^1 \int_0^1 e^{-(x^2+y^2)} \, dx \, dy \approx N_R/N \approx 0.5654.$$

This compares favorably with estimate 0.565 that we found on page 225. Here we chose the cube C to be the smallest cube containing the volume W. This is important. Intuitively, the better "the fit" between the two volumes, the fewer random points are needed to obtain a reasonable approximation. In fact, the biggest problem with the Monte Carlo method is finding a sufficiently small rectangular box which encloses the volume.

Problems for Section 15.4

For Problems 1–2 use several runs of random numbers to decide on an approximate answer.

1. Use a Monte Carlo method to approximate the integral $\int_0^1 \sqrt{1 - x^2} \, dx$, and explain why this gives an approximation for $\pi/4$.

2. Use a Monte Carlo method to approximate the integral $\int_0^1 e^{-x^2} \, dx$.

We now give another Monte Carlo method for approximating an integral. Consider the integral $\int_R f(x, y) \, dx \, dy$ and the average value of the function given by $\dfrac{1}{\text{Area}(R)} \int_R f(x, y) \, dx \, dy$. The motivation for this is that when we divide the region R into subregions R_i, and take a point (x_i, y_i) in R_i for $i = 1, 2, \cdots, n$, then

$$\sum_i f(x_i, y_i) \text{Area}(R_i) \approx \int_R f(x, y) \, dx \, dy.$$

If all the Area (R_i)s are equal, then n Area$(R_i) = $ Area(R) and

$$\frac{\text{Area}(R)}{n} \sum_{i=1}^n f(x_i, y_i) \approx \int_R f(x, y) \, dx \, dy$$

so

$$\frac{1}{n} \sum_{i=1}^n f(x_i, y_i) \approx \frac{1}{\text{Area}(R)} \int_R f(x, y) \, dx \, dy,$$

where $\frac{1}{n} \sum_{i=1}^n f(x_i, y_i)$ is the average of the n values $f(x_i, y_i)$.

By choosing n points (x_i, y_i) at random from R, we can use the approximation:

$$\text{Area}(R) \cdot \frac{1}{n} \sum_{i=1}^n f(x_i, y_i) \approx \int_R f(x, y) \, dx \, dy.$$

Use this Monte Carlo method to approximate the integrals in Problems 3–5:

3. $\displaystyle\int_0^1 \int_0^1 e^{-xy} \, dx \, dy$ 4. $\displaystyle\int_0^1 \int_0^1 xy^{xy} \, dx \, dy$ 5. $\displaystyle\int_0^1 \int_0^1 x^{-y} \, dy \, dx$

6. Here is a way to use Simpson's rule twice to approximate a definite integral. Suppose the integral is $\int_1^5 \int_2^6 \sqrt{x^2 + y^2} \, dy \, dx$. Use Simpson's rule with $\Delta y = 1$ to approximate the inner integral when x is fixed at 1. Repeat for $x = 1.5, 2, 2.5, 3, 3.5, 4, 4.5, 5$. You now have approximations for the inner integral at nine different values of x. Now use Simpson's rule again with $\Delta x = 1$, using the nine different values for the inner integral, to approximate the outer integral (and therefore the whole double integral).

15.5 TWO-VARIABLE INTEGRALS IN POLAR COORDINATES

Integration in Polar Coordinates

We started this chapter by putting a rectangular grid on a population density map, in order to construct a Riemann sum that gave an estimate for the fox population. However, Riemann sums can be constructed from any subdivision of a region, not just a rectangular subdivision; sometimes a polar grid is more appropriate. A review of polar coordinates is in Appendix G.

Example 1   A biologist studying insect populations around a circular lake divides the area into sectors as in Figure 15.33, and measures the population density in each sector, with the results shown (in millions per square mile). Estimate the total insect population around the lake.

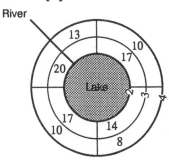

Figure 15.33: An insect infested lake

Solution   To get our estimate we will multiply the population density in each sector by the area of that sector. Unlike the rectangles in a rectangular grid, the sectors in a polar grid do not all have the same area. The inner sectors have area

$$\frac{1}{4}(\pi 3^2 - \pi 2^2) = \frac{5\pi}{4} \approx 3.93,$$

and the outer sectors have area

$$\frac{1}{4}(\pi 4^2 - \pi 3^2) = \frac{7\pi}{4} \approx 5.50,$$

so the estimated population is

$$(20)(3.93) + (17)(3.93) + (14)(3.93) + (17)(3.93) +$$
$$(13)(5.50) + (10)(5.50) + (8)(5.50) + (10)(5.50) \approx 493 \text{ million insects.}$$

What is dA in Polar Coordinates?

The subdivision in this example is the sort you get when you use a polar grid rather than a rectangular grid. A rectangular grid is constructed by putting down vertical and horizontal lines; these correspond to taking $x = k$ (a constant) and $y = l$ (another constant), respectively. In polar coordinates, setting $r = k$ gives you a circle around the origin (of radius k), and setting $\theta = l$ gives you a ray emanating from the origin (at angle l with the x-axis). A polar grid is built up out of these circles and rays, just as a rectangular grid is built up out of horizontal and vertical lines. Figure 15.34 shows a subdivision of the polar region $a \leq r \leq b$, $\alpha \leq \theta \leq \beta$, using n subdivisions each way. This is the sort of region that is naturally represented in polar coordinates; a sort of circular rectangle bent around the origin.

In general, if you divide R as shown in Figure 15.34, you will get a Riemann sum similar to the ones coming from rectangular grids:

$$\sum_{i,j} f(r_i, \theta_j) \, \Delta A.$$

However, calculating the area ΔA is more complicated. Figure 15.35 shows ΔA. For Δr, $\Delta \theta$ small, the shaded region is approximately rectangular, with sides $r \, \Delta \theta$ and Δr, so

$$\Delta A \approx r \Delta \theta \Delta r.$$

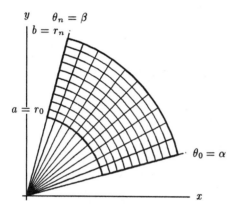

Figure 15.34: How to divide up a region using polar coordinates

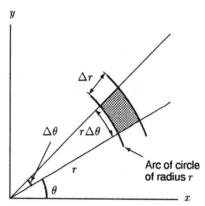

Figure 15.35: Area ΔA in polar coordinates

Thus, the Riemann sum is approximately

$$\sum_{i,j} f(r_i, \theta_j) \, r_i \, \Delta \theta \, \Delta r.$$

If we take the limit as Δr and $\Delta \theta$ approach 0, we end up with

$$\int_R f \, dA = \int_\alpha^\beta \int_a^b f(r, \theta) \, r \, dr \, d\theta.$$

> When computing integrals in polar coordinates, put $dA = r \, dr \, d\theta$ or $dA = r \, d\theta \, dr$.

Example 2   Compute the integral of $f(x, y) = 1/(x^2 + y^2)^{3/2}$ over the region R shown in Figure 15.36.

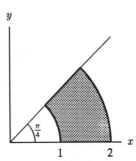

Figure 15.36: A polar
region

Solution   The region can be described in polar coordinates by the inequalities $1 \le r \le 2, 0 \le \theta \le \pi/4$. In
polar coordinates, since $r = \sqrt{x^2 + y^2}$, we can write f as

$$f(r, \theta) = \frac{1}{r^3},$$

Then

$$\int_R f \, dA = \int_0^{\pi/4} \int_1^2 \frac{1}{r^3} r \, dr \, d\theta = \int_0^{\pi/4} \left(\int_1^2 r^{-2} \, dr \right) d\theta$$

$$= \int_0^{\pi/4} \left[-\frac{1}{r} \right]_{r=1}^{r=2} d\theta = \int_0^{\pi/4} \frac{1}{2} \, d\theta = \frac{\pi}{8}.$$

Example 3   On the basis of the shape of each region in Figure 15.37, decide whether to integrate using polar or
Cartesian coordinates. Set up the integral of an arbitrary function $z = f(x, y)$ over the region.

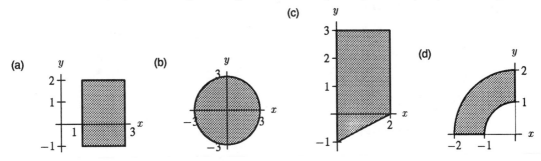

Figure 15.37: Four regions

Solution   (a)   Since this is a rectangular region, Cartesian coordinates are the right choice. The rectangle is
described by the inequalities $1 \le x \le 3$ and $-1 \le y \le 2$, so the integral is

$$\int_{-1}^2 \int_1^3 f(x, y) \, dx \, dy.$$

(b)   A circle is best described in polar coordinates. The radius is 3, so r goes from 0 to 3, and the
range of θ is 0 to 2π to describe the whole circle, so the integral is

$$\int_0^{2\pi} \int_0^3 f(r \cos \theta, r \sin \theta) r \, dr \, d\theta.$$

(c)   The bottom boundary of this trapezoid is the line $y = (x/2) - 1$, and the top is the line $y = 3$, so use Cartesian coordinates. If we integrate with respect to y first we get $(x/2) - 1$ for the lower limit of the integral and 3 for the upper limit. After that we integrate with respect to x from $x = 0$ to $x = 2$. So the integral is

$$\int_0^2 \int_{(x/2)-1}^3 f(x, y)\, dy\, dx.$$

(d)   This is another polar region: it is a piece of a ring with inner radius 1 and outer radius 2, so r goes from 1 to 2. Its position in the second quadrant shows that θ goes from $\pi/2$ to π. The integral is

$$\int_{\pi/2}^{\pi} \int_1^2 f(r \cos \theta, r \sin \theta)\, r\, dr\, d\theta.$$

Problems for Section 15.5

For each of the regions R in Problems 1-4, set up $\displaystyle\int_R f\, dA$ as an iterated integral in polar coordinates.

1.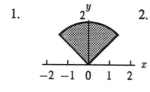

Figure 15.38

2.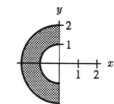

Figure 15.39

3.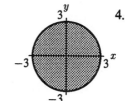

Figure 15.40

4.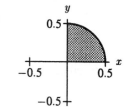

Figure 15.41

Sketch the region over which the integrals in Problems 5–11 are computed.

5. $\displaystyle\int_0^{2\pi} \int_1^2 f(r, \theta)\, r\, dr\, d\theta.$

6. $\displaystyle\int_{\pi/2}^{\pi} \int_0^1 f(r, \theta)\, r\, dr\, d\theta.$

7. $\displaystyle\int_{\pi/6}^{\pi/3} \int_0^1 f(r, \theta)\, r\, dr\, d\theta.$

8. $\displaystyle\int_3^4 \int_{3\pi/4}^{3\pi/2} f(r, \theta)\, r\, d\theta\, dr.$

9. $\displaystyle\int_0^{\pi/4} \int_0^{1/\cos \theta} f(r, \theta)\, r\, dr\, d\theta.$

10. $\displaystyle\int_{\pi/4}^{\pi/2} \int_0^{2/\sin \theta} f(r, \theta)\, r\, dr\, d\theta.$

11. $\displaystyle\int_0^4 \int_{-\pi/2}^{\pi/2} f(r, \theta)\, r\, d\theta\, dr.$

Evaluate the integral in Problems 12–13 over the region indicated.

12. $\displaystyle\int_R \sin(x^2 + y^2)\, dA$, where R is the disc of radius 2 centered at the origin.

13. $\int_R (x^2 - y^2)\, dA$, where R is the first quadrant region between the circles of radius 1 and radius 2.

14. Consider the integral $\int_0^3 \int_{x/3}^1 f(x, y)\, dy\, dx$.

   (a) Sketch the region R over which the integration is being performed.
   (b) Rewrite the integral with the order of integration reversed.
   (c) Rewrite the integral in polar coordinates.

Change the integrals in Problems 15–17 to polar coordinates and evaluate.

15. $\int_{-1}^0 \int_{-\sqrt{1-x^2}}^{\sqrt{1-x^2}} x\, dy\, dx$ 16. $\int_0^{\sqrt{2}} \int_y^{\sqrt{4-y^2}} xy\, dx\, dy$ 17. $\int_0^3 \int_{-x}^x \frac{x}{y^2}\, dy\, dx$

18. Find the volume of the region between the graph of $f(x, y) = 25 - x^2 - y^2$ and the xy plane.

19. An ice cream cone can be modeled by the region bounded by the hemisphere $z = \sqrt{8 - x^2 - y^2}$ and the cone $z = \sqrt{x^2 + y^2}$. Find its volume.

20. A disk of radius 5 cm has density 10 g/cm² at its center, has density 0 at its edge, and its density is a linear function of the distance from the center. Find the mass of the disk.

21. A city by the ocean surrounds a bay, and has the shape shown in Figure 15.42.

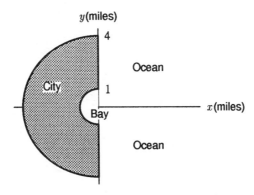

Figure 15.42

   (a) The population density of the city (in thousands of people per square mile) is given by the function $\delta(r, \theta)$, where r and θ are polar coordinates with respect to the x and y-axes shown, and the distances indicated on the y-axis are in miles. Set up a double integral in polar coordinates that gives the total population of the city.
   (b) The population density decreases as you move away from the shoreline of the bay, and also decreases the further you have to drive to get to the ocean. Which of the following functions best describes this situation?
   (i) $\delta(r, \theta) = (4 - r)(2 + \cos \theta)$
   (ii) $\delta(r, \theta) = (4 - r)(2 + \sin \theta)$
   (iii) $\delta(r, \theta) = (r + 4)(2 + \cos \theta)$
   (c) Evaluate the integral you set up in (a) with the function you chose in (b), and give the resulting estimate for the population.

15.6 INTEGRALS IN CYLINDRICAL AND SPHERICAL COORDINATES

In our work on two-variable integrals we saw that it was easier to evaluate some integrals using polar, rather than Cartesian coordinates. Similarly, in trying to evaluate three-variable integrals, there are instances when coordinates other than rectangular ones will make the evaluation of the integral much easier.

Cylindrical Coordinates

The cylindrical coordinates of a point (x, y, z) in 3-space are obtained by representing the x and y coordinates in polar coordinates and letting the z-coordinate be the z-coordinate of the Cartesian coordinate system. (See Figure 15.43.)

Relation between Cartesian and cylindrical coordinates

$$x = r \cos \theta,$$
$$y = r \sin \theta,$$
$$z = z.$$

As with polar coordinates in the plane, note that $x^2 + y^2 = r^2$.

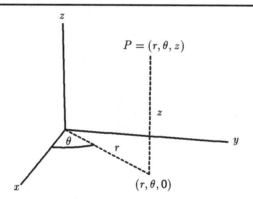

Figure 15.43: Cylindrical coordinates: (r, θ, z)

A useful way to visualize how cylindrical coordinates work is to sketch the fundamental surfaces obtained by setting one of the coordinates equal to a constant. See Figures 15.44–15.46.

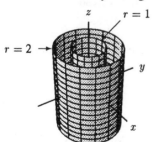

Figure 15.44: The surfaces $r = 1$ and $r = 2$

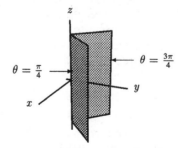

Figure 15.45: The surfaces $\theta = \pi/4$ and $\theta = 3\pi/4$

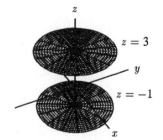

Figure 15.46: The surfaces $z = -1$ and $z = 3$

Setting r = constant gives a cylinder around the z-axis whose radius is equal to the constant; setting θ = constant gives a half-plane perpendicular to the xy plane, with one edge along the z-axis, making an angle θ with the x-axis. Finally, setting z = constant gives a horizontal plane z units from the xy-plane.

The sorts of regions that can easily be described in cylindrical coordinates are the regions whose boundaries are formed by parts of these fundamental surfaces. (For example, vertical cylinders, or wedge-shaped parts of vertical cylinders.)

Example 1   Describe in cylindrical coordinates a wedge of cheese cut from a cylinder of cheese 4 cm high and 6 cm in radius if this wedge subtends an angle of $\pi/6$ at the center.

Solution   If we put the wedge in a coordinate system as shown in Figure 15.47, then it is described by the inequalities $0 \le r \le 6$, and $0 \le z \le 4$, and $0 \le \theta \le \pi/6$.

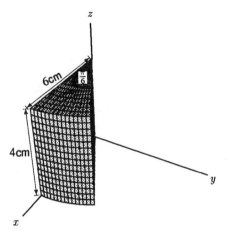

Figure 15.47: A wedge of cheese

Integration in Cylindrical Coordinates

When we integrated in polar coordinates, we had to express the area element dA in terms of polar coordinates: $dA = r\, dr\, d\theta$. To evaluate a triple integral $\int_W f\, dV$ in cylindrical coordinates, we need to express the volume element dV in cylindrical coordinates.

Consider the element of volume ΔV as shown in Figure 15.48. It is bounded by the fundamental surfaces described earlier; hence, it is a piece of a wedge out of a cylinder. As we saw in the section on polar coordinates, the area of the base of the volume element is given approximately by $r\Delta r\Delta\theta$. Since the height is Δz, the element of volume is given approximately by $\Delta V \approx r\,\Delta r\,\Delta\theta\,\Delta z$.

When computing integrals in cylindrical coordinates, put $dV = r\, dr\, d\theta\, dz$. Other orders of integration are also possible.

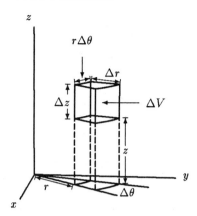

Figure 15.48: Volume element in
cylindrical coordinates

Example 2   Find the mass of the wedge of cheese in Example 1, if its density is 0.8 grams/cm^3.

Solution   Call the wedge W. Then its mass is

$$\int_W 0.8 \, dV.$$

In cylindrical coordinates this integral is

$$\int_0^4 \int_0^{\pi/6} \int_0^6 0.8 \, r \, dr \, d\theta \, dz = \int_0^4 \int_0^{\pi/6} 0.4 r^2 \Big|_0^6 \, d\theta \, dz = 14.4 \int_0^4 \int_0^{\pi/6} d\theta \, dz$$

$$= 14.4 \left(\frac{\pi}{6}\right) \int_0^4 dz = 14.4 \left(\frac{\pi}{6}\right) 4 \approx 30.16 \text{ grams.}$$

Example 3   A water tank in the shape of a hemisphere has radius a; its base is its plane face. Find the volume, V, of water in the tank as a function of h, the depth of the water.

Solution   In Cartesian coordinates a sphere of radius a has the equation $x^2 + y^2 + z^2 = a^2$. Since $r^2 = x^2 + y^2$, this becomes

$$r^2 + z^2 = a^2$$

in cylindrical coordinates. Thus, if we want to describe the amount of water in the tank in cylindrical coordinates, we let r go from 0 to $\sqrt{a^2 - z^2}$, we let θ go from 0 to 2π, and we let z go from 0 to h. So its volume is

$$\int_W dV = \int_0^{2\pi} \int_0^h \int_0^{\sqrt{a^2 - z^2}} r \, dr \, dz \, d\theta = \int_0^{2\pi} \int_0^h \frac{r^2}{2} \Big|_{r=0}^{r=\sqrt{a^2 - z^2}} dz \, d\theta$$

$$= \int_0^{2\pi} \int_0^h \frac{1}{2}(a^2 - z^2) \, dz \, d\theta = \int_0^{2\pi} \frac{1}{2}\left(a^2 z - \frac{z^3}{3}\right) \Big|_{z=0}^{z=h} d\theta$$

$$= \int_0^{2\pi} \frac{1}{2}\left(a^2 h - \frac{h^3}{3}\right) d\theta = \pi\left(a^2 h - \frac{h^3}{3}\right).$$

Spherical Coordinates

We define another 3-dimensional system of coordinates called spherical coordinates as follows. In Figure 15.49, the point P has coordinates (x, y, z) in the rectangular coordinate system. We define spherical coordinates ρ, ϕ, and θ for P as follows: $\rho = \sqrt{x^2 + y^2 + z^2}$ is the distance of P from the origin; ϕ is the angle between the positive z axis and the line through the origin and the point P; and θ is the same as in cylindrical coordinates.

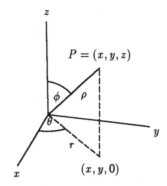

Figure 15.49: Spherical coordinates

In cylindrical coordinates,

$$x = r \cos \theta,$$
$$y = r \sin \theta,$$
$$z = z,$$

but from the figure we have $z = \rho \cos \phi$ and $r = \rho \sin \phi$, so that we have the following:

Relation between Cartesian and Spherical coordinates

$$x = \rho \sin \phi \cos \theta$$
$$y = \rho \sin \phi \sin \theta$$
$$z = \rho \cos \phi.$$

Also, as noted above,

$$\rho^2 = x^2 + y^2 + z^2.$$

Thus, each point in 3-space is represented using spherical coordinates where $0 \leq \rho < \infty, 0 \leq \phi \leq \pi$, and $0 \leq \theta \leq 2\pi$. As the name indicates, this system of coordinates is useful when there is spherical symmetry with respect to the origin, either in the region of integration or in the integrand.

You might note that the angles θ and ϕ in spherical coordinates are similar to those of longitude and latitude on the globe. The angle θ is the longitude measured east of Greenwich. The angle $\frac{\pi}{2} - \phi$ is what map-makers call the latitude, the angle measured from the equator, while ϕ itself would be measured from the north pole of the globe.

The fundamental surfaces in spherical coordinates are $\rho = k$ (a constant), which is a sphere of radius k centered at the origin, $\theta = k$ (a constant), which is the half-plane with its edge along the z-axis, and $\phi = k$ (a constant), which is a cone if $k \neq \pi/2$ and the xy-plane if $k = \frac{\pi}{2}$. (See Figures 15.50-15.52.)

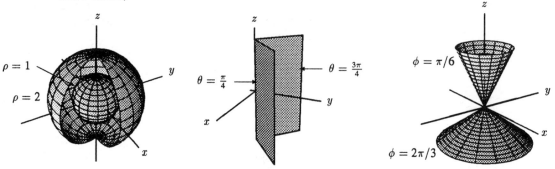

Figure 15.50: The surfaces $\rho = 1$ and $\rho = 2$

Figure 15.51: The surfaces $\theta = \pi/4$ and $\theta = 3\pi/4$

Figure 15.52: The surfaces $\phi = \pi/3$ and $\phi = 2\pi/3$

Integration in Spherical Coordinates

To use spherical coordinates in triple integrals we need to determine the volume element for $\int_W f \, dV$ where W is a region in three space and f is a function defined on W. From Figure 15.53, we see that the volume element can be approximated by a rectangular box. One edge has length $\Delta\rho$. The edge parallel to the xy-plane is an arc of a circle made from rotating the cylindrical radius $r \, (= \rho \sin \phi)$ through an angle $\Delta\theta$, and so has length $\rho \sin \phi \, \Delta\theta$. The remaining side comes from rotating the radius ρ through an angle $\Delta\phi$, and so has length $\rho \, \Delta\phi$. Therefore,

$$\Delta V \approx \Delta\rho (\rho \, \Delta\phi)(\rho \sin \phi \, \Delta\theta) = \rho^2 \sin \phi \, \Delta\rho \, \Delta\phi \, \Delta\theta$$

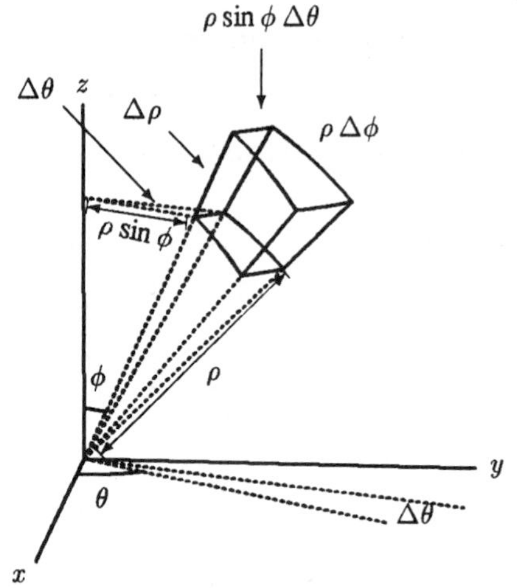

Figure 15.53: The volume element in spherical coordinates

Thus,

When computing integrals in spherical coordinates, put $dV = \rho^2 \sin \phi \, d\rho \, d\phi \, d\theta$. Other orders of integration are also possible.

Example 4   Use spherical coordinates to derive the formula for the volume of a solid sphere of radius a.

Solution   In spherical coordinates, a sphere of radius a is described by the inequalities $0 \leq \rho \leq a, 0 \leq \theta \leq 2\pi$, and $0 \leq \phi \leq \pi$. Note that θ goes all the way around the circle (just like longitude), whereas ϕ only goes from 0 to π. We can find the volume by integrating the constant density function 1 over the sphere:

$$\text{Volume} = \int_R dV = \int_0^{2\pi} \int_0^{\pi} \int_0^a \rho^2 \sin\phi \, d\rho \, d\phi \, d\theta = \int_0^{2\pi} \int_0^{\pi} \frac{1}{3} a^3 \sin\phi \, d\phi \, d\theta$$

$$= \frac{1}{3} a^3 \int_0^{2\pi} -\cos\phi \Big|_0^{\pi} \, d\theta = \frac{2}{3} a^3 \int_0^{2\pi} d\theta = \frac{4\pi a^3}{3}.$$

Example 5   Find the magnitude of the gravitational force exerted by a solid hemisphere of radius a and constant density δ on a unit mass located at the center of the base of the hemisphere.

Solution   Assume the base of the hemisphere rests on the xy-plane with center at the origin. Because the hemisphere is symmetric about the z-axis, the force on the mass in the horizontal x or y direction is 0; the horizontal force exerted on the unit mass by the part of the hemisphere at (x, y, z) cancels with that at $(-x, -y, z)$. Thus, we need only compute the vertical component of the force. If we use spherical coordinates, a piece of the hemisphere of volume dV located at (ρ, θ, ϕ) exerts on the unit mass a force of magnitude $G(\delta \, dV)/\rho^2$, since the magnitude of the force is $G \cdot \text{mass}/(\text{distance})^2$. The magnitude of the vertical component of this force is

$$(G(\delta \, dV)/\rho^2) \cos\phi.$$

Adding up all the contributions of all the small pieces of volume, we get a vertical force with magnitude

$$F = \int_0^{2\pi} \int_0^{\pi/2} \int_0^a (G\delta/\rho^2)(\cos\phi)\rho^2 \sin\phi \, d\rho \, d\phi \, d\theta = \int_0^{2\pi} \int_0^{\pi/2} G\delta(\cos\phi \sin\phi)\rho \Big|_{\rho=0}^{\rho=a} d\phi \, d\theta$$

$$= \int_0^{2\pi} \int_0^{\pi/2} G\delta a \cos\phi \sin\phi \, d\phi \, d\theta = \int_0^{2\pi} G\delta a \left(-\frac{(\cos\phi)^2}{2}\right) \Big|_{\phi=0}^{\phi=\pi/2} d\theta$$

$$= \int_0^{2\pi} G\delta a \left(\frac{1}{2}\right) d\theta = G\delta a \pi.$$

Problems for Section 15.6

In Problems 1–2, evaluate the indicated three-variable integrals in cylindrical coordinates.

1. $f(x, y, z) = x^2 + y^2 + z^2$, W is the region $0 \leq r \leq 4$, $\pi/4 \leq \theta \leq 3\pi/4$, $-1 \leq z \leq 1$.

2. $f(x, y, z) = \sin(x^2 + y^2)$, W is the solid cylinder with height 4 and with base of radius 1 centered on the z axis at $z = -1$.

In Problems 3–4, evaluate the indicated three-variable integrals in spherical coordinates.

3. $f(x, y, z) = 1/(x^2 + y^2 + z^2)^{1/2}$ over the bottom half of the sphere of radius 5 centered at the origin.

4. $f(\rho, \theta, \phi) = \sin \phi$, over the region $0 \le \theta \le 2\pi$, $0 \le \phi \le \pi/4$, $1 \le \rho \le 2$.

For Problems 5–9, choose a set of coordinate axes, and then set up the three-variable integral in an appropriate coordinate system for integrating a density function δ over the given region.

5.

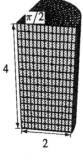

Figure 15.54

6.

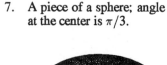

Figure 15.55

7. A piece of a sphere; angle at the center is $\pi/3$.

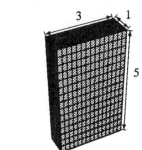

Figure 15.56

8.

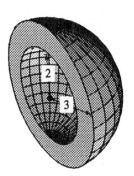

Figure 15.57

9.

Figure 15.58

10. Sketch the region R over which the integration is being performed:

$$\int_0^{\pi/2} \int_{\pi/2}^{\pi} \int_0^1 f(\rho, \phi, \theta)\rho^2 \sin \phi \, d\rho \, d\phi \, d\theta.$$

Evaluate the integrals in Problems 11–12.

11. $\displaystyle \int_0^1 \int_{-\sqrt{1-x^2}}^{\sqrt{1-x^2}} \int_{-\sqrt{1-x^2-z^2}}^{\sqrt{1-x^2-z^2}} \frac{1}{(x^2 + y^2 + z^2)^{1/2}} \, dy \, dz \, dx$

12. $\displaystyle \int_0^1 \int_{-1}^1 \int_{-\sqrt{1-x^2}}^{\sqrt{1-x^2}} \frac{1}{(x^2 + y^2)^{1/2}} \, dy \, dx \, dz$

13. Write a triple integral representing the volume of a slice of the cylindrical cake of height 2 and radius 5 between the planes $\theta = \pi/6$ and $\theta = \pi/3$. Evaluate this integral.

14. Find the mass M of the solid region W given in spherical polar coordinates by

    $$W = \{(\rho, \phi, \theta) : 0 \leq \rho \leq 3, 0 \leq \theta < 2\pi, 0 \leq \phi \leq \pi/4\}$$

    if the density at any point P is given by $\delta(P) =$ distance of P from the origin.

15. A particular spherical cloud of gas of radius 3 km is more dense at the center than towards the edge. The density, D, of the gas at a distance ρ km from the center is given by $D(\rho) = 3 - \rho$. Write an integral representing the total mass of the cloud of gas, and evaluate it.

16. Find the volume that remains after a cylindrical hole of radius R is bored through a sphere of radius a, where $0 < R < a$, passing through the center of the sphere along the pole.

17. Use appropriate coordinates to find the average distance to the origin for points in the ice cream cone region bounded by the hemisphere $z = \sqrt{8 - x^2 - y^2}$ and the cone $z = \sqrt{x^2 + y^2}$. (Hint: The volume of this region is computed in Problem 19 on page 255.)

18. Compute the force of gravity exerted by a solid cylinder of radius R, height H, and constant density δ on a unit mass at the center of the base of the cylinder.

19. Let W_1 be a solid cylinder of height 4, radius 1, and with the origin at the center of its base. Let W_2 be the top half of the sphere of radius 5 centered at the origin. Calculate the integrals of both $f(x, y, z) = x^2 + y^2 + z^2$ and $g(x, y, z) = z$ over W_1 and over W_2.

15.7 APPLICATIONS OF INTEGRATION TO PROBABILITY

To understand how a quality such as height or weight is distributed throughout a population, we use a single variable density function. A review of single variable density functions is in Appendix F. Often we want to study two or more qualities at the same time, and how they are related. In this section we will see how to use a multivariable density function to represent multivariable data.

Density Functions

Distribution of Weight and Height in Expectant Mothers

Table 15.13 shows the distribution of weight and height in a survey of expectant mothers. We represent this data by a histogram in Figure 15.59 in such a way that the volume of each bar represents the percentage in the corresponding weight and height range. For example, the bar representing the mothers who weighed 60–70 kg and were 160–165 cm tall has base of area $10 \text{ kg} \cdot 5 \text{ cm} = 50 \text{ kg cm}$. We want the volume of this bar to be 12%, so its height is $12\%/50 \text{ kg cm} = .24\%/ \text{kg cm}$. The total volume is $100\% = 1$.

TABLE 15.13: *Distribution of weight and height in a survey of expectant mothers*

	45–50Kg	50–60Kg	60–70Kg	70–80Kg	80–105Kg	Totals by height
150–155cm	2	4	4	2	1	13
155–160cm	0	12	8	2	1	23
160–165cm	1	7	12	4	3	27
165–170cm	0	8	12	6	2	28
170–180cm	0	1	3	4	1	9
Totals by weight	3	32	39	18	8	100

Notice that the units on the vertical axis are %/ kg cm. Thus volumes under the histogram are in units of %.

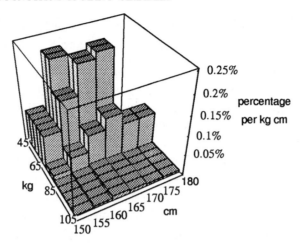

Figure 15.59: Histogram representing the data in Table 15.13.

Example 1   Find the percentage of mothers in the survey with height between 170 and 180 cms.

Solution   We add the percentages across the row corresponding to the 170–180 cm height range:

$$0 + 1 + 3 + 4 + 1 = 9\%.$$

This is equivalent to finding the volumes of the corresponding rectangular solids in the histogram.

Smoothing the Histogram

If we had smaller weight and height groups (and a larger sample), we would be able to draw a smoother histogram from which we could get finer estimates. In the limit, we could replace the histogram with a smooth surface, in such a way that the volume under the surface over the rectangle representing a given weight/height group would be the percentage of mothers in the group. We define a *density* function, $p(w, h)$, to be the function whose graph is the smooth surface. It has the property that

$$\begin{pmatrix} \text{\% of sample with} \\ \text{weight between } a \text{ and } b \text{ and} \\ \text{height between } c \text{ and } d \end{pmatrix} = \begin{pmatrix} \text{Volume under graph of } p \\ \text{over the rectangle} \\ a \leq w \leq b, c \leq h \leq d \end{pmatrix} = \int_a^b \int_c^d p(w, h)\, dh\, dw.$$

Two-Variable Density Functions

We can generalize this idea to look at how any two characteristics, x and y, are distributed throughout a population.

A function $p(x, y)$ is called a *joint density function* for x and y if

$$\begin{pmatrix} \text{\% of population with} \\ x \text{ between } a \text{ and } b \text{ and} \\ y \text{ between } c \text{ and } d \end{pmatrix} = \begin{pmatrix} \text{Volume under graph of } p \\ \text{above the rectangle} \\ a \leq x \leq b, c \leq y \leq d \end{pmatrix} = \int_a^b \int_c^d p(x, y)\, dy\, dx$$

where

$$\int_{-\infty}^{\infty} \int_{-\infty}^{\infty} p(x, y)\, dy\, dx = 1 \quad \text{and} \quad p(x, y) \geq 0 \text{ for all } x \text{ and } y.$$

Example 2   Let $p(x, y)$ be defined on the square $0 \leq x \leq 1, 0 \leq y \leq 1$ by $p(x, y) = x + y$, and let $p(x, y) = 0$ if (x, y) is outside this square. Verify that p is a joint density function. What does it mean in terms of the distribution of x and y in the population that $p(x, y) = 0$ outside the rectangle?

Solution   First, we have $p(x, y) \geq 0$ for all x and y. To verify that p is a joint density function, we must check that the total volume under the graph is 1:

$$\begin{aligned} \int_{-\infty}^{\infty} \int_{-\infty}^{\infty} p(x, y)\, dy\, dx &= \int_0^1 \int_0^1 (x + y)\, dy\, dx \\ &= \int_0^1 \left(xy + \frac{y^2}{2} \right) \Big|_0^1 dx \\ &= \int_0^1 \left(x + \frac{1}{2} \right) dx \\ &= \left(\frac{x^2}{2} + \frac{x}{2} \right) \Big|_0^1 = 1 \end{aligned}$$

The fact that $p(x, y) = 0$ outside the square means that the variables never take values outside that range, that is, the value of x and y for any individual in the population is always between 0 and 1.

Example 3   Suppose two variables x and y are distributed in a population according to the density function of Example 2. Find the fraction of the population with $x \leq 1/2$, the fraction with $y \leq 1/2$, and the fraction with both $x \leq 1/2$ and $y \leq 1/2$.

Solution   The fraction with $x \leq 1/2$ is the volume under the graph to the left of the line $x = 1/2$:

$$\begin{aligned} \int_0^{1/2} \int_0^1 (x + y)\, dy\, dx &= \int_0^{1/2} \left(xy + \frac{y^2}{2} \right) \Big|_0^1 dx \\ &= \int_0^{1/2} \left(x + \frac{1}{2} \right) dx \\ &= \left(\frac{x^2}{2} + \frac{x}{2} \right) \Big|_0^{1/2} = \frac{1}{8} + \frac{1}{4} = \frac{3}{8}. \end{aligned}$$

Since the function is symmetric in x and y, the fraction with $y \leq 1/2$ is also 3/8. Finally, the fraction with both $x \leq 1/2$ and $y \leq 1/2$ is

$$\int_0^{1/2} \int_0^{1/2} (x + y)\, dy\, dx = \int_0^{1/2} \left(xy + \frac{y^2}{2}\right) \Big|_0^{1/2} dx$$

$$= \int_0^{1/2} \left(\frac{1}{2}x + \frac{1}{8}\right) dx$$

$$= \left(\frac{1}{4}x^2 + \frac{1}{8}x\right) \Big|_0^{1/2} = \frac{1}{16} + \frac{1}{16} = \frac{1}{8}$$

Recall that a one-variable density function $p(x)$ is a function such that $p(x) \geq 0$ for all x, and $\int_{-\infty}^{\infty} p(x)\, dx = 1$. (See Appendix F.)

Example 4   Let p_1 be a one-variable density function for x and p_2 a one-variable density function for y. Verify that

$$p(x, y) = p_1(x)p_2(y)$$

is a joint density function.

Solution   Since both p_1 and p_2 are density functions, and therefore non-negative everywhere, their product $p = p_1 p_2$ is non-negative everywhere.

Now we must check that the volume under the graph of f is 1. Since $\int_{-\infty}^{\infty} p_2(y)\, dy = 1$ and $\int_{-\infty}^{\infty} p_1(x)\, dx = 1$,

$$\int_{-\infty}^{\infty} \int_{-\infty}^{\infty} p(x, y)\, dy\, dx = \int_{-\infty}^{\infty} \int_{-\infty}^{\infty} (p_1(x)p_2(y))\, dy\, dx$$

$$= \int_{-\infty}^{\infty} p_1(x) \left(\int_{-\infty}^{\infty} p_2(y)\, dy\right) dx$$

$$= \int_{-\infty}^{\infty} p_1(x)(1)\, dx$$

$$= \int_{-\infty}^{\infty} p_1(x)\, dx = 1.$$

Density Functions and Probability

What is the probability that an expectant mother weighs 60–70 kg and is 155–160 cm tall? From Table 15.13 we see that 8% of mothers fall in this group, so the probability that a randomly chosen mother would fall in this group is .08.

Thus we can interpret the volume under the joint density function as a probability. Consider the joint density function $p(w, h)$ for the weight and height of expectant mothers.

Probability that a mother has weight between a and b and height between c and d	$=$	Volume under graph of p over the rectangle $a \leq w \leq b, c \leq h \leq d$	$= \int_a^b \int_c^d p(w, h) \, dh \, dw.$

For a general joint density function $p(x, y)$, the probability that x falls in an interval of width Δx around x_0 and y falls in an interval of width Δy around y_0 is approximately $p(x_0, y_0) \Delta x \Delta y$. The next example shows how a probability density function is used.

Example 5 A machine in a factory is set to produce components 10 inches long and 5 inches in diameter. In fact there is a slight variation from one component to the next. A component is usable if its length and diameter deviate from the correct values by less than 0.1 in. If the length is x inches and the diameter is y inches, the probability density function for the variation in x and y is

$$p(x, y) = \frac{50\sqrt{2}}{\pi} e^{-100(x-10)^2} e^{-50(y-5)^2}.$$

What is the probability that a component will be usable?

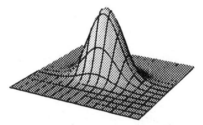

Figure 15.60: The density function
$p(x, y) = \frac{50\sqrt{2}}{\pi} e^{-100(x-10)^2} e^{-50(y-5)^2}$.

Solution The probability that a particular component has length x in the range $x_0 - \Delta x \leq x \leq x_0 + \Delta x$ and diameter y in the range $y_0 - \Delta y \leq y \leq y_0 + \Delta y$ is the definite integral

$$\frac{50\sqrt{2}}{\pi} \int_{y_0 - \Delta y}^{y_0 + \Delta y} \int_{x_0 - \Delta x}^{x_0 + \Delta x} e^{-100(x-10)^2} e^{-50(y-5)^2} \, dx \, dy.$$

Thus,

$$\left(\begin{array}{c} \text{Probability that} \\ \text{component is usable} \end{array} \right) = \frac{50\sqrt{2}}{\pi} \int_{4.9}^{5.1} \int_{9.9}^{10.1} e^{-100(x-10)^2} e^{-50(y-5)^2} \, dx \, dy.$$

The double integral cannot be evaluated using elementary functions, so we evaluate it using a numerical integration method. This yields a value of 0.02556 and so the required probability is

$$\frac{50\sqrt{2}}{\pi} (0.02556) \approx 0.57530.$$

Thus there is a 57.5% chance that the component will be usable.

Dependence and Independence of Variables

If we studied the distribution of height and weight in a population, we would expect to see some relation between them. All other things being equal, taller people are more likely to be heavy than short people. On the other hand, if we studied the distribution of height and annual income, we would not expect to see much of a relationship; tall people and short people probably earn the same, on average.

How Can We Determine Dependence from the Joint Density Function?

Let's see how the dependence between weight and height shows up in the data in Table 15.13. First, look at the column corresponding to expectant mothers weighing 70–80kg. This group is 18% of the whole sample. The subset of this group with height between 170 and 180 cm is 4% of the whole sample. So the probability that a woman in this weight group is 170–180 cm tall is

$$\frac{\text{Probability that height is 170–180 cm and weight 70–80kg}}{\text{Probability that weight is 70–80 kg}} = 4/18 = .22.$$

Now let us look at a lighter group, the women who weigh 60–70 kg. This group forms 39% of the total, and the subset with height 170–180 cm forms 3%. So the probability that a woman in this group is 170–180 cm high is

$$\frac{\text{Probability that height is 170–180 cm and weight 60–70kg}}{\text{Probability that weight is 60–70 kg}} = 3/39 = .08.$$

This is a lower probability than what we obtained for the 70–80 kg group. This makes sense, since it is less likely for a light woman to be tall than it is for a heavy woman. In this situation, we say that the two variables w and h are *dependent*, because to a certain extent they depend on each other.

Conditional Probability

We can generalize these ideas to any joint density function. We want to calculate the probability that y falls in a certain range, given that x falls in a certain range.

If $p(x, y)$ is a probability density function, we define the **conditional probability** that $a \leq x \leq b$ given $c \leq y \leq d$ by

$$\frac{\begin{array}{c}\text{Probability that}\\ a \leq x \leq b \text{ and } c \leq y \leq d\\ \hline \text{Probability that}\\ c \leq y \leq d\end{array}}{} = \frac{\int_a^b \int_c^d p(x,y)\, dy\, dx}{\int_{-\infty}^{\infty} \int_c^d p(x,y)\, dy\, dx}.$$

Example 6   For the probability density function in Example 5, calculate the probability that:
   (a)   The length is between 9.9 and 10.1 in, given that the diameter is between 4.9 and 5.1 in.
   (b)   The length is between 9.9 and 10.1 in, given that the diameter is between 5.3 and 5.5.

Solution   (a)   We have that

$$\frac{\text{Probability that } 9.9 \leq x \leq 10.1}{\text{given } 4.9 \leq y \leq 5.1} = \frac{\text{Probability that } 9.9 \leq x \leq 10.1 \text{ and } 4.9 \leq y \leq 5.1}{\text{Probability that } 4.9 \leq y \leq 5.1}$$

$$= \frac{\dfrac{50\sqrt{2}}{\pi} \displaystyle\int_{9.9}^{10.1} \int_{4.9}^{5.1} e^{-100(x-10)^2} e^{-50(y-5)^2} \, dy \, dx}{\dfrac{50\sqrt{2}}{\pi} \displaystyle\int_{-\infty}^{\infty} \int_{4.9}^{5.1} e^{-100(x-10)^2} e^{-50(y-5)^2} \, dy \, dx}$$

$$\approx \frac{0.57}{0.68} \approx 0.84.$$

(b)   Similarly, we find that

$$\frac{\text{Probability that } 9.9 \leq x \leq 10.1}{\text{given } 5.3 \leq y \leq 5.5} = \frac{\text{Probability that } 9.9 \leq x \leq 10.1 \text{ and } 5.3 \leq y \leq 5.5}{\text{Probability that } 5.3 \leq y \leq 5.5}$$

$$= \frac{\dfrac{50\sqrt{2}}{\pi} \displaystyle\int_{9.9}^{10.1} \int_{5.3}^{5.5} e^{-100(x-10)^2} e^{-50(y-5)^2} \, dy \, dx}{\dfrac{50\sqrt{2}}{\pi} \displaystyle\int_{-\infty}^{\infty} \int_{5.3}^{5.5} e^{-100(x-10)^2} e^{-50(y-5)^2} \, dy \, dx}$$

$$\approx \frac{0.00114}{0.00135} \approx 0.84.$$

Look at the denominators in the ratios used to calculate the conditional probabilities. These are the probabilities of the condition on y. Notice that it much less likely that $5.3 \leq y \leq 5.5$ than that $4.9 \leq y \leq 5.1$ (a probability of .00135 in the first case as opposed to .68 in the second). However, given the condition, the probability for the length x to fall in the range $9.9 \leq x \leq 10.1$ is the same in both cases, about .84. Thus it seems that the variation in the length is independent of the variation in the diameter. In this situation we say that the variables x and y are *independent*.

Example 7   For the density in Example 2, find the probability that $x \leq 1/2$ given $y \leq 1/2$.

Solution

$$\frac{\text{Probability that}}{x \leq \tfrac{1}{2} \text{ given } y \leq \tfrac{1}{2}} = \frac{\text{Probability that } x \leq \tfrac{1}{2} \text{ and } y \leq \tfrac{1}{2}}{\text{Probability that } y \leq \tfrac{1}{2}} = \frac{1/8}{3/8} = \frac{1}{3}.$$

This is less than $3/8 = $ Probability that $x \leq 1/2$. Thus having $y \leq 1/2$ makes $x \leq 1/2$ less likely. In this case the variables do not seem to be independent.

How Can we Tell if Two Variables are Independent?

Two events are said to be independent if the probability that they both happen is simply the product of the probabilities that they would happen individually. For example, if you throw two dice, the probability of a double four is $(1/6) \cdot (1/6) = 1/36$. This is because the face showing on the first die is independent of the face showing on the second, and the probability of a four on either die is

1/6. Let's use this idea to find the joint probability distribution function for two qualities x and y that vary independently in a population.

Suppose x has density function $p_1(x)$. Consider a small range of x-values of length Δx, say $x_0 \leq x \leq x_0 + \Delta x$. The probability that x falls in this range is the area under the graph of p_1 between x_0 and $x_0 + \Delta x$, which is approximately $p_1(x_0) \Delta x$ if Δx is small. Similarly, if y has density function $p_2(y)$, the probability that $y_0 \leq y \leq y_0 + \Delta y$ is about $p_2(y_0) \Delta y$. If x and y are independent, then we would expect

$$\left(\begin{array}{c} \text{Probability that} \\ x_0 \leq x \leq x_0 + \Delta x \\ \text{and } y_0 \leq y \leq y_0 + \Delta y \end{array} \right) = \left(\begin{array}{c} \text{Probability that} \\ x_0 \leq x \leq x_0 + \Delta x \end{array} \right) \cdot \left(\begin{array}{c} \text{Probability that} \\ y_0 \leq y \leq y_0 + \Delta y \end{array} \right)$$

$$\approx (p_1(x_0) \Delta x) \cdot (p_2(y_0) \Delta y) = p_1(x_0) p_2(y_0) \Delta x \, \Delta y.$$

On the other hand, if $p(x, y)$ is the joint density function for x and y,

$$\left(\begin{array}{c} \text{Probability that} \\ x_0 \leq x \leq x_0 + \Delta x \\ \text{and } y_0 \leq y \leq y_0 + \Delta y \end{array} \right) = \left(\begin{array}{c} \text{Volume under the graph of } p \\ \text{above } x_0 \leq x \leq x_0 + \Delta x, y_0 \leq y \leq y_0 + \Delta y \end{array} \right)$$

$$\approx p(x_0, y_0) \Delta x \, \Delta y.$$

Thus,

$$p_1(x_0) p_2(y_0) \Delta x \, \Delta y \approx p(x_0, y_0) \Delta x \, \Delta y.$$

Dividing out by $\Delta x \, \Delta y$, we arrive at the conclusion that:

> If x has probability density $p_1(x)$ and y has probability density $p_2(y)$, and if x and y are independent, then the joint density for x and y is $p(x, y) = p_1(x) p_2(y)$.

Conversely, if the joint density function $p(x, y)$ can be written as a product of one-variable density functions $p_1(x)$ and $p_2(y)$, then

$$\begin{array}{rl} \begin{array}{c} \text{Probability that} \\ a \leq x \leq b \text{ and } c \leq y \leq d \end{array} & = \int_a^b \int_c^d (p_1(x) p_2(y)) \, dy \, dx \\[2mm] & = \int_a^b p_1(x) \left(\int_c^d p_2(y) \, dy \right) dx \\[2mm] & = \int_a^b p_1(x) \, dx \cdot (\text{Probability that } c \leq y \leq d) \\[2mm] & = (\text{Probability that } a \leq x \leq b) \cdot (\text{Probability that } c \leq y \leq d). \end{array}$$

Hence the variables are independent. Thus,

> If the joint density function p of x and y can be expressed as a product
> $$p(x, y) = p_1(x) p_2(y),$$
> where p_1 and p_2 are density functions, then x and y are independent.

The one-variable *normal* probability density function with mean μ and standard deviation σ is defined by

$$p(x) = \frac{1}{\sigma\sqrt{2\pi}}e^{-(x-\mu)^2/(2\sigma^2)}.$$

(See Appendix F.) The normal density function arises very frequently in applications and is one of the most widely used probability density functions.

Example 8 Show that the length and radius of the components produced by the machine in Example 5 are independent.

Solution The joint density function may be written as

$$\frac{50\sqrt{2}}{\pi}e^{-100(x-10)^2}e^{-50(y-5)^2} = \left(\frac{10\sqrt{2}}{\sqrt{2\pi}}e^{-(x-10)^2/(2(\frac{1}{10\sqrt{2}})^2)}\right)\left(\frac{10}{\sqrt{2\pi}}e^{-(y-5)^2/(2(\frac{1}{10})^2)}\right),$$

which is a product of a normal distribution in x with mean 10 and standard deviation $1/(10\sqrt{2})$ and a normal distribution in y with mean 5 and standard deviation $1/10$.

Problems for Section 15.7

1. Let x and y have joint density function

$$p(x, y) = \begin{cases} \frac{2}{3}(x + 2y) & \text{for } 0 \le x \le 1, 0 \le y \le 1. \\ 0 & \text{otherwise.} \end{cases}$$

   Find

   (a) The probability that $x > 1/3$.
   (b) The probability that $x < (1/3) + y$.

2. Table 2 gives some values of the joint density function for two variables x and y. We assume x can take the values 1, 2, 3 and 4 and y can take the values 1, 2 and 3.

TABLE 15.14

		y		
		1	2	3
x	1	0.3	0.2	0.1
	2	0.2	0.1	0
	3	0.1	0	0
	4	0	0	0

   (a) Explain why this table defines a joint density function.
   (b) What is the probability that $x = 2$?
   (c) Find the probability that $y \le 2$.
   (d) Find the probability that $x \le 3$ and $y \le 2$.

3.   A joint density function is given by

$$f(x, y) = \begin{cases} kx^2 & \text{for } 0 \le x \le 2 \text{ and } 0 \le y \le 1, \\ 0 & \text{otherwise.} \end{cases}$$

   (a)   Find the value of the constant k.
   (b)   Find the probability that a point (x, y) satisfies $x + y \le 2$.
   (c)   Find the probability that a point (x, y) satisfies $x \le 1$ and $y \le 1/2$

4.   Assume that the joint density function for x, y is given by

$$f(x, y) = \begin{cases} kxy & \text{for } 0 \le x \le y \le 1, \\ 0 & \text{otherwise.} \end{cases}$$

   (a)   Determine the value of k.
   (b)   Find the probability that (x, y) lies in the shaded region in Figure 15.61.

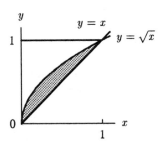

Figure 15.61

5.   A health insurance company wants to know what proportion of its policies are going to cost them a lot of money because the insured people are over 65 and sick. In order to compute this proportion, the company defines a *disability index*, x, with $0 \le x \le 1$, where $x = 0$ represents perfect health and $x = 1$ represents total disability. In addition, the company uses a density function, $f(x, y)$, defined in such a way that the quantity

$$f(x, y)\, \Delta x\, \Delta y$$

   approximates the fraction of the population with disability index between x and $x + \Delta x$, and aged between y and $y + \Delta y$. The company knows from experience that a policy no longer covers its costs if the insured person is over 65 and has a disability index exceeding 0.8. Write an expression for the fraction of the company's policies held by people meeting these criteria.

6.   Assume that a point is chosen at random from the region S in the xy-plane containing all points (x, y) such that $-1 \le x \le 1, -2 \le y \le 2$ and $x - y \ge 0$ (at random means that the density function is constant on S).
   (a)   Determine the joint density function for x and y.
   (b)   If T is a subset of S with area α, then find the probability that a point (x, y) is in T.

7.   Give the joint density of x and y where x and y are independent, x has a normal distribution with mean 5 and standard deviation $1/10$, and y has a normal distribution with mean 15 and standard deviation $1/6$.

8. The probability that a radioactive substance will decay at time t is modeled by the density function
$$p(t) = \lambda e^{-\lambda t}$$
for $t \geq 0$, and $p(t) = 0$ for $t < 0$. The positive constant λ depends on the material, and is called the decay rate.

(a) Verify that p is a density function.

(b) Consider two materials with decay rates λ and μ, which decay independently of each other. Write the joint density function for the probability that the first material decays at time t and the second at time s.

(c) Find the probability that the first substance decays before the second.

15.8   NOTES ON CHANGE OF VARIABLES IN A MULTIPLE INTEGRAL

In the previous sections, we used polar, cylindrical, and spherical coordinates to simplify iterated integrals. In this section, we discuss more general changes of variable. In the process, we will see where the extra factor of r comes from when we change from Cartesian to polar coordinates and the factor $\rho^2 \sin \phi$ when we change from Cartesian to spherical coordinates.

Polar Change of Variables Revisited

Suppose you have to evaluate the integral $\int_R (x + y)\,dA$ where R is the region in the first quadrant bounded by the circle $x^2 + y^2 = 16$ and the x and y-axes. Writing the integral in Cartesian and polar coordinates we have

$$\int_R (x + y)\,dA = \int_0^4 \int_0^{\sqrt{16-x^2}} (x + y)\,dy\,dx = \int_0^{\pi/2} \int_0^4 (r \cos \theta + r \sin \theta) r\,dr d\theta.$$

This is an integral over the rectangle in the $r\theta$-space given by $0 \leq r \leq 4$, $0 \leq \theta \leq \pi/2$. The conversion from polar to Cartesian coordinates changes this rectangle into a quarter-disc. Figure 15.62 shows how a typical rectangle (shaded) in the $r\theta$-plane with sides of length Δr and $\Delta \theta$ corresponds to a curved rectangle in the xy-plane with sides of length Δr and $r\Delta \theta$. The extra r is needed because the correspondence between r, θ and x, y not only curves the lines $r = 1, 2, 3 \ldots$ into circles, it also stretches those lines around larger and larger circles.

In general, to convert an integral from Cartesian to polar coordinates we make three changes:

1. Substitute $x = r \cos \theta$, $y = r \sin \theta$ in the integrand,

2. Change the limits of integration,

3. Replace dA by $r\,dr\,d\theta$.

Any other change of variable will have three similar steps. This section shows a more general way of computing the factor r in the area element $r\,dr\,d\theta$.

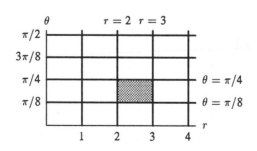 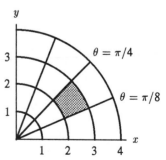

Figure 15.62: A grid in the $r\theta$-plane and the corresponding curved grid in the xy-plane

General Change of Variables

We now consider a general change of variable, where x, y coordinates are related to some other u, v coordinates by the functions

$$x = x(u, v) \quad y = y(u, v).$$

Just as circular regions in the xy-plane correspond to rectangular regions in the $r\theta$-plane, we look at a curved region R in the xy-plane corresponding to a rectangular region T in the uv-plane. We want to know what happens when we divide T into small rectangles $T_{i,j}$. In Figure 15.63 is a typical rectangle $T_{i,j}$ with sides of length Δu and Δv and lower left corner at (u, v).

The corresponding piece $R_{i,j}$ of the xy-plane is a quadrilateral with curved sides. If we choose Δu and Δv very small, by local linearity, $R_{i,j}$ is approximately a parallelogram. The edges no longer have length Δu and Δv, and the area of $R_{i,j}$ is not $\Delta u \Delta v$.

Recall from Chapter 12 that the area of the parallelogram with sides $\vec{a}$ and $\vec{b}$ is $\|\vec{a} \times \vec{b}\|$. Thus, we need to find the sides of $R_{i,j}$ as vectors. The side of $R_{i,j}$ corresponding to the bottom side of $T_{i,j}$ has end points $(x(u, v), y(u, v))$ and $(x(u + \Delta u, v), y(u + \Delta u, v))$, so in vector form that side is

$$\vec{a} = (x(u+\Delta u, v)-x(u, v))\vec{i} +(y(u+\Delta u, v)-y(u, v))\vec{j} +0\vec{k} \approx \left(\frac{\partial x}{\partial u}\Delta u\right)\vec{i} +\left(\frac{\partial y}{\partial u}\Delta u\right)\vec{j} +0\vec{k}.$$

Similarly, the side of $R_{i,j}$ corresponding to the left edge of $T_{i,j}$ is given by

$$\vec{b} \approx \left(\frac{\partial x}{\partial v}\Delta v\right)\vec{i} + \left(\frac{\partial y}{\partial v}\Delta v\right)\vec{j} + 0\vec{k}.$$

Computing the cross product, we get

$$\text{Area } R_{i,j} \approx \|\vec{a} \times \vec{b}\| \approx \left|\left(\frac{\partial x}{\partial u}\Delta u\right)\left(\frac{\partial y}{\partial v}\Delta v\right) - \left(\frac{\partial x}{\partial v}\Delta v\right)\left(\frac{\partial y}{\partial u}\Delta u\right)\right|$$

$$= \left|\frac{\partial x}{\partial u}\cdot\frac{\partial y}{\partial v} - \frac{\partial x}{\partial v}\cdot\frac{\partial y}{\partial u}\right|\Delta u \Delta v.$$

Using determinant notation, we define the *Jacobian*, $\dfrac{\partial(x, y)}{\partial(u, v)}$, as follows

$$\frac{\partial x}{\partial u}\cdot\frac{\partial y}{\partial v} - \frac{\partial x}{\partial v}\cdot\frac{\partial y}{\partial u} = \begin{vmatrix} \frac{\partial x}{\partial u} & \frac{\partial y}{\partial u} \\ \frac{\partial x}{\partial v} & \frac{\partial y}{\partial v} \end{vmatrix} = \frac{\partial(x, y)}{\partial(u, v)}.$$

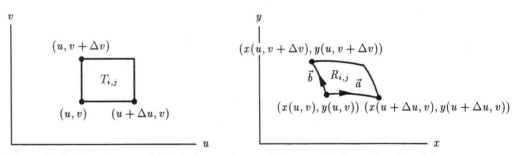

Figure 15.63: A small rectangle $T_{i,j}$ in the uv-plane and the corresponding region $R_{i,j}$ of the xy-plane

Thus, we can write

$$\text{Area } R_{i,j} \approx \left| \frac{\partial(x, y)}{\partial(u, v)} \right| \Delta u \, \Delta v.$$

To compute $\int_T f(x, y) \, dA$, we look at the Riemann sum obtained by dividing the region R into the little curved regions $R_{i,j}$, giving

$$\int_R f(x, y) \, dA \approx \sum_{i,j} f(x_i, y_j) \cdot (\text{Area of } R_{i,j})$$

$$\approx \sum_{i,j} f(x_i, y_j) \left| \frac{\partial(x, y)}{\partial(u, v)} \right| \Delta u \, \Delta v.$$

Each point (x_i, y_j) corresponds to a point (u_i, v_j) so the sum can be written in terms of u and v

$$\sum_{i,j} f(x(u_i, v_j), y(u_i, v_j)) \left| \frac{\partial(x, y)}{\partial(u, v)} \right| \Delta u \, \Delta v.$$

This is a Riemann sum in terms of u and v, so as Δu and Δv approach 0, we get

$$\int_R f(x, y) \, dA = \int_T f(x(u, v), y(u, v)) \left| \frac{\partial(x, y)}{\partial(u, v)} \right| du \, dv.$$

To convert an integral from x, y to u, v coordinates we make three changes:
1. Substitute for x and y in the integrand in terms of u and v.
2. Change the xy region R into a uv region T.
3. Introduce the absolute value of the Jacobian $\left| \dfrac{\partial(x, y)}{\partial(u, v)} \right|$, representing the change in the area element.

Example 1   Verify that the Jacobian $\dfrac{\partial(x, y)}{\partial(u, v)} = r$ for polar coordinates $x = r \cos \theta$, $y = r \sin \theta$.

Solution $\dfrac{\partial(x, y)}{\partial(r, \theta)} = \begin{vmatrix} \frac{\partial x}{\partial r} & \frac{\partial y}{\partial r} \\ \frac{\partial x}{\partial \theta} & \frac{\partial y}{\partial \theta} \end{vmatrix} = \begin{vmatrix} \cos \theta & \sin \theta \\ -r \sin \theta & r \cos \theta \end{vmatrix} = r \cos^2 \theta + r \sin^2 \theta = r.$

Example 2   Find the area of the ellipse $\dfrac{x^2}{a^2} + \dfrac{y^2}{b^2} = 1$

Solution   Let $x = au$, $y = bv$. Then the ellipse $x^2/a^2 + y^2/b^2 = 1$ in the xy-plane corresponds to the circle $u^2 + v^2 = 1$ in the uv-plane. The Jacobian is $\begin{vmatrix} a & 0 \\ 0 & b \end{vmatrix} = ab$. Thus, if we let R be the ellipse in the xy-plane and T the circle in the uv-plane, we get

$$\text{Area of } xy\text{-ellipse} = \int_R 1 \, dA = \int_T 1ab \, du \, dv = ab \int_T du \, dv = ab(\text{Area of } uv\text{-circle}) = \pi ab.$$

Change of Variables in Three-Variable Integrals

For three-variable integrals there is a similar formula. If the functions

$$x = x(u, v, w), \quad y = y(u, v, w), \quad z = z(u, v, w)$$

define a change of variables from a region S in uvw-space to a region W in xyz-space, the Jacobian of this change of variables is given by the determinant

$$\frac{\partial(x, y, z)}{\partial(u, v, w)} = \begin{vmatrix} \frac{\partial x}{\partial u} & \frac{\partial y}{\partial u} & \frac{\partial z}{\partial u} \\ \frac{\partial x}{\partial v} & \frac{\partial y}{\partial v} & \frac{\partial z}{\partial v} \\ \frac{\partial x}{\partial w} & \frac{\partial y}{\partial w} & \frac{\partial z}{\partial w} \end{vmatrix}.$$

By analogy with two dimensions, this Jacobian represents the change in the volume element. Thus, we have

$$\int_W f(x, y, z)\, dx\, dy\, dz = \int_S f(x(u, v, w), y(u, v, w), z(u, v, w)) \left| \frac{\partial(x, y, z)}{\partial(u, v, w)} \right| du\, dv\, dw.$$

Problem 3 at the end of this section asks you to verify that the Jacobian for the change of variables for spherical coordinates is exactly the $\rho^2 \sin \phi$ factor you already use. The next example generalizes from Example 2 to ellipsoids.

Example 3   Find the volume of the ellipsoid $\dfrac{x^2}{a^2} + \dfrac{y^2}{b^2} + \dfrac{z^2}{c^2} = 1$.

Solution   Let $x = au$, $y = bv$, $z = cw$. The Jacobian is easily computed to be abc. The xyz-ellipsoid corresponds to the uvw-sphere $u^2 + v^2 + w^2 = 1$. Thus, as in Example 2,

$$\text{Volume of } xyz\text{-ellipsoid} = (abc)(\text{Volume of } uvw\text{-sphere}) = abc\frac{4}{3}\pi = \frac{4}{3}\pi abc.$$

Problems for Section 15.8

1. Find the region R in the xy-plane corresponding to the region $T = \{(u, v) \mid 0 \le u \le 3, 0 \le v \le 2\}$ under the change of variables $x = 2u - 3v$, $y = u - 2v$. Check that

$$\int_R dx\, dy = \int_T \left| \frac{\partial(x, y)}{\partial(u, v)} \right| du\, dv.$$

2. Find the region R in the xy-plane corresponding to the region $T = \{(u, v) \mid 0 \le u \le 2, u \le v \le 2\}$ under the change of variables $x = u^2$, $y = v$. Check that

$$\int_R dx\, dy = \int_T \left| \frac{\partial(x, y)}{\partial(u, v)} \right| du\, dv.$$

3. Compute the Jacobian for the change of variables:

$$x = \rho \sin \varphi \cos \theta, \quad y = \rho \sin \varphi \sin \theta, \quad z = \rho \cos \varphi.$$

4. For the change of variables $x = 3u - 4v$, $y = 5u + 2v$, show that

$$\frac{\partial(x,y)}{\partial(u,v)} \cdot \frac{\partial(u,v)}{\partial(x,y)} = 1$$

5. Use the change of variables $x = 2u + v$, $y = u - v$ to compute the integral $\int_R (x + y)\, dA$, where R is the parallelogram formed by $(0,0)$, $(3,-3)$, $(5,-2)$, and $(2,1)$.

6. Use the change of variables $x = \frac{1}{2}u$, $y = \frac{1}{3}v$ to compute the integral $\int_R (x^2 + y^2)\, dA$, where R is the region bounded by the curve $4x^2 + 9y^2 = 36$.

7. Use the change of variables $u = xy$, $v = xy^2$ to compute $\int_R xy^2\, dA$, where R is the region bounded by $xy = 1$, $xy = 4$, $xy^2 = 1$, $xy^2 = 4$.

8. Evaluate the integral $\int_R \cos\left(\frac{x-y}{x+y}\right)\, dx\, dy$ where R is the triangle bounded by $x + y = 1$, $x = 0$, and $y = 0$.

REVIEW PROBLEMS FOR CHAPTER FIFTEEN

1. Figure 15.64 shows contours of average annual rainfall (in inches) in South America.[4] Each grid square is 500 miles on a side. Estimate the total volume of rain that falls on the considered area in a year.

2. Figure 15.65 gives isotherms for low winter temperature in Washington, D.C.[5] The grid squares are one mile on a side. Find the average low temperature over the whole city (the city is the shaded region).

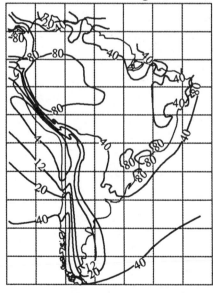

Figure 15.64

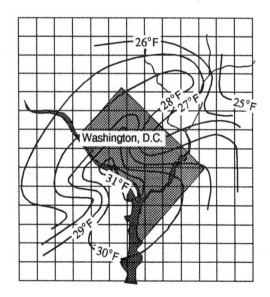

Figure 15.65

[4]From *Modern Physical Geography*, Alan H. Strahler and Arthur H. Strahler, Fourth Edition, John Wiley & Sons, New York, 1992, p. 144

[5]From *Physical Geography of the Global Environment*, H. J. de Blij and Peter O. Muller, John Wiley & Sons, New York, 1993, p. 220

Sketch the regions over which the integrals in Problems 3–6 are being performed.

3. $\int_{1}^{4} \int_{-\sqrt{y}}^{\sqrt{y}} f(x, y)\, dx\, dy.$

4. $\int_{0}^{1} \int_{0}^{\sin^{-1} y} f(x, y)\, dx\, dy.$

5. $\int_{-1}^{1} \int_{-\sqrt{1-x^2}}^{\sqrt{1-x^2}} f(x, y)\, dy\, dx.$

6. $\int_{0}^{2} \int_{-\sqrt{4-y^2}}^{0} f(x, y)\, dx\, dy.$

Calculate exactly the integrals in Problems 7–11. (Your answer may contain e, π, $\sqrt{2}$, and so on).

7. $\int_{0}^{1} \int_{0}^{z} \int_{0}^{2} (y + z)^7\, dx\, dy\, dz.$

8. $\int_{0}^{1} \int_{3}^{4} (\sin(2 - y)) \cos(3x - 7)\, dx\, dy.$

9. $\int_{0}^{10} \int_{0}^{0.1} x e^{xy}\, dy\, dx.$

10. $\int_{0}^{1} \int_{0}^{y} (\sin^3 x)(\cos x)(\cos y)\, dx\, dy.$

11. $\int_{3}^{4} \int_{0}^{1} x^2 y \cos(xy)\, dy\, dx.$

12. Write $\int_{R} f(x, y)\, dA$ as an iterated integral if R is the region in Figure 15.66.

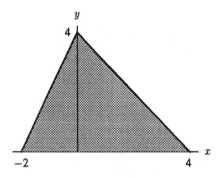

Figure 15.66

13. Evaluate $\int_{R} \sqrt{x^2 + y^2}\, dA$ where R is the region in Figure 15.67.

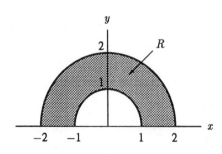

Figure 15.67

Evaluate the integrals in Problems 14–16 by changing them to cylindrical or spherical coordinates as appropriate.

14. $\displaystyle\int_{-\sqrt{3}}^{\sqrt{3}}\int_{-\sqrt{3-x^2}}^{\sqrt{3-x^2}}\int_{1}^{4-x^2-y^2}\frac{1}{z^2}\,dz\,dy\,dx$

15. $\displaystyle\int_{0}^{1}\int_{0}^{\sqrt{1-x^2}}\int_{0}^{\sqrt{x^2+y^2}}(z+\sqrt{x^2+y^2})\,dz\,dy\,dx$

16. $\displaystyle\int_{0}^{3}\int_{-\sqrt{9-z^2}}^{\sqrt{9-z^2}}\int_{0}^{\sqrt{9-y^2-z^2}}x^2\,dx\,dy\,dz$

17. For $R=\{(x,y,z):1\le x^2+y^2\le 4, 0\le z\le 4\}$ evaluate the integral
$$\int_{R}\frac{z}{(x^2+y^2)^{3/2}}\,dV.$$

18. Write an integral representing the mass of a sphere of radius 3 if the density of the sphere at any point is twice the distance of that point from the center of the sphere.

19. Compute the integral $\int_{0}^{1}\int_{-\sqrt{1-x^2}}^{\sqrt{1-x^2}}e^{-(x^2+y^2)}\,dy\,dx$.

20. A forest next to a road has the shape in Figure 15.68. The population density of rabbits is proportional to the distance from the road. It is 0 at the road, and 10 rabbits per square mile at the opposite edge of the forest. Find the total rabbit population in the forest.

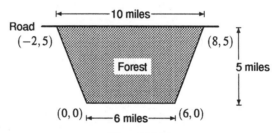

Figure 15.68

21. Consider a solid body W with an axis L passing through it, and suppose that W is spinning around on L. The *moment of inertia* of W tells you how great will be the angular acceleration of the body for a given torque (a force twisting the body). The moment of inertia may be calculated by the following triple integral:
$$\int_{W}S(x,y,z)d(x,y,z)^2\,dV,$$
where $S(x,y,z)$ is the density of the body at the point (x,y,z), and $d(x,y,z)$ is the distance of the point from the axis L. Thus, for example, if L is the z-axis, then $d(x,y,z)^2=x^2+y^2$. Consider a rectangular brick with length 5, width 3, and height 1, and of uniform density 1. Compute the moment of inertia about each of the three axes passing through the center of the brick, perpendicular to one of the sides.

22. Compute the moment of inertia of a sphere of radius R about an axis passing through its center. Assume that the sphere has a constant density of 1.

CHAPTER SIXTEEN

PARAMETRIC CURVES AND SURFACES

In single-variable calculus, we study the motion of a particle along a straight line. For example, we represent the motion of an object thrown straight up into the air by a single function $h(t)$, the height of the object above the ground at time t.

To study the motion of a particle in 2-dimensional or 3-dimensional space, we must express all the coordinates of the particle in terms of t: $x(t)$, $y(t)$, and $z(t)$ if the motion is in 3-space.

This leads us to the parametric representation of the path of motion, which is a curve. We can study surfaces in 3-dimensional space in an analogous way.

16.1 MOTION IN SPACE

How do we Represent Motion?

To represent the motion of a particle in the xy-plane we use two equations, one for the x-coordinate of the particle, $x = f(t)$, and another for the y-coordinate, $y = g(t)$. Thus at time t the particle is at the point $(f(t), g(t))$. The equation for x describes the right-left motion; the equation for y describes the up-down motion. The two equations for x and y are called *parametric equations* with *parameter* t.

Example 1   Describe the motion of the particle whose coordinates at time t are

$$x = \cos t, \quad y = \sin t.$$

Solution   Since $(\cos t)^2 + (\sin t)^2 = 1$, we have $x^2 + y^2 = 1$. That is, at any time t the particle is at a point (x, y) on the unit circle $x^2 + y^2 = 1$. We plot points at different times to see how the particle moves on the circle. See Figure 16.1 and Table 16.1. The particle moves at a uniform speed, completing one full trip counterclockwise around the circle every 2π units of time. Notice how the x-coordinate goes back and forth from 1 to 0 to -1 to 0 to 1 while the y-coordinate goes up and down from 0 to 1 to 0 to -1 to 0. The two motions combine to trace out a circle.

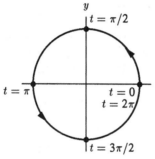

Figure 16.1: The circle parameterized by $x = \cos t, y = \sin t$

TABLE 16.1: *Points on the circle with $x = \cos t$, $y = \sin t$*

t	x	y
0	1	0
$\pi/2$	0	1
π	-1	0
$3\pi/2$	0	-1
2π	1	0

Example 2   Figure 16.2 shows the graphs of two functions, $f(t)$ and $g(t)$. Describe the motion of the particle whose coordinates at time t are

$$x = f(t), \quad y = g(t).$$

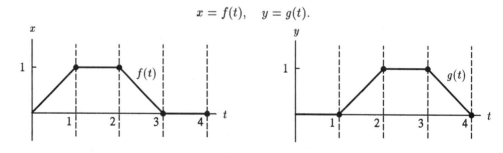

Figure 16.2: Graphs of $x = f(t)$ and $y = g(t)$

Solution   Between times $t = 0$ and $t = 1$, the x-coordinate goes from 0 to 1, while the y-coordinate stays fixed at 0. So the particle moves along the x-axis from $(0,0)$ to $(1,0)$. Then, between time $t = 1$ and $t = 2$, the x-coordinate stays fixed at $x = 1$, while the y-coordinate goes from 0 to 1. Thus the particle moves along the vertical line from $(1,0)$ to $(1,1)$. Similarly, between times $t = 2$ and $t = 3$, it moves horizontally back to $(0,1)$, and between time $t = 3$ and $t = 4$ it moves down the y-axis to $(0,0)$. Thus it traces out a square. See Figure 16.3.

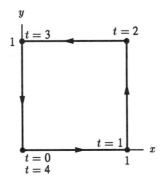

Figure 16.3: A square

Parametric Equations in Three Dimensions

To describe a motion in 3-dimensional space parametrically, we need an extra equation giving z in terms of t.

Example 3   Describe the motion given parametrically by

$$x = \cos t, \quad y = \sin t, \quad z = t.$$

Solution   The particle's x- and y-coordinates are the same as in Example 1, which gives circular motion in the xy-plane, while the z-coordinate steadily increases. Thus the particle traces out a rising spiral, like a coil spring. See Figure 16.4. This curve is called a *helix*.

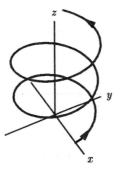

Figure 16.4: A helix.

Different Motions Along the Same Path

Example 4   Describe the motion of the particle whose x and y coordinates at time t are given by the equations

$$x = \cos 3t, \quad y = \sin 3t.$$

Solution   Since $(\cos 3t)^2 + (\sin 3t)^2 = 1$, so $x^2 + y^2 = 1$, we again have motion on the circle. But if we plot points at different times, we see that the particle is moving three times as fast as in Example 1 on page 282. (See Figure 16.5 and Table 16.2.)

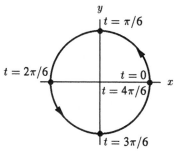

Figure 16.5: The circle parameterized by $x = \cos 3t$, $y = \sin 3t$

TABLE 16.2: *Points on circle with $x = \cos 3t$, $y = \sin 3t$*

t	x	y
0	1	0
$\pi/6$	0	1
$2\pi/6$	-1	0
$3\pi/6$	0	-1
$4\pi/6$	1	0

Example 4 is the same as Example 1 except that t is replaced by $3t$. Replacing t by a function of t is called a *change in parameter*.   If we make a change in parameter, the particle follows the same curve but it traces out the curve at a different speed or in a different direction, or traces only a portion of the curve. In Section 16.3 we will discuss the speed of a moving particle in more detail.

Example 5   Describe the motion of the particle whose x and y coordinates at time t are

$$x = \cos(e^{-t^2}), \quad y = \sin(e^{-t^2}).$$

Solution   As in Examples 1 and 4, we still have $x^2 + y^2 = 1$ so the motion lies on the unit circle. As time t goes from $-\infty$ (way back in the past) to 0 (the present) to ∞ (way off in the future), e^{-t^2} goes from 0 to 1 back to 0. So $(x, y) = (\cos(e^{-t^2}), \sin(e^{-t^2}))$ goes from $(1, 0)$ to $(\cos(1), \sin(1))$ and back to $(1, 0)$. The particle does not actually reach the point $(1, 0)$ at any finite time. (See Figure 16.6 and Table 16.3.)

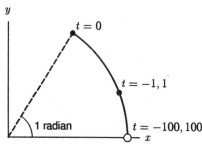

Figure 16.6: The circle parameterized by $x = \cos(e^{-t^2})$, $y = \sin(e^{-t^2})$

TABLE 16.3: *Points on circle with $x = \cos(e^{-t^2})$, $y = \sin(e^{-t^2})$*

t	x	y
-100	~ 1	~ 0
-1	0.93	0.36
0	0.54	0.84
1	0.93	0.36
100	~ 1	~ 0

Complicated Curves

Using parametric equations, we can plot curves that we do not usually see when graphing equations in x and y.

Example 6   Sketch the curve traced out by the particle whose motion is given by

$$x = \cos 3t, \quad y = \sin 5t.$$

Solution   The x-coordinate oscillates back and forth between 1 and -1, completing 3 oscillations every 2π seconds (assuming t is time in seconds). The y-coordinate oscillates up and down between 1 and -1, completing 5 oscillations every 2π seconds. Since both the x and y coordinates return to their original values every 2π seconds, the curve is retraced every 2π seconds. The result is a pattern called a Lissajous figure . See Figure 16.7. It is difficult to sketch the graph by hand, but a graphing calculator set on parametric graphing mode can draw the figure quickly. Problems 14–17 experiment with other Lissajous figures $x = \cos at$, $y = \sin bt$, for different values of a and b.

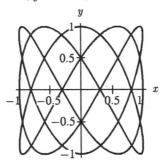

Figure 16.7: A Lissajous figure.

Example 7   Describe the curve given by the parametric equations

$$x = (4 + \sin 3t) \cos 2t, \qquad y = (4 + \sin 3t) \sin 2t, \qquad z = \cos 3t$$

Solution   Since the equations are periodic with period 2π, the entire curve is traced out as t ranges from 0 to 2π. To picture the curve, first look at the x and y equations only. If we ignore the $\sin 3t$ term we get $x = 4 \cos 2t$, $y = 4 \sin 2t$, which goes twice around a circle of radius 4. If we now include the $\sin 3t$ term, the radius $(4 + \sin 3t)$ oscillates three times between 3 and 5 as the circle goes twice around the origin. Meanwhile, the z-coordinate oscillates three times between -1 and 1. The result is a curve that goes two times around the z-axis, meanwhile going through three full twists. Figure 16.8 shows the curve in xyz-space and the shadow it casts in the xy-plane (by a light very high up on the z-axis).

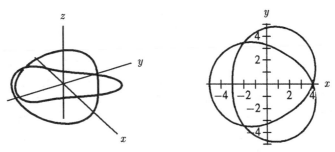

Figure 16.8: A complicated curve and its shadow in the xy-plane

Problems for Section 16.1

For Problems 1–4, describe the motion of a particle whose position at time t is $x = f(t)$, $y = g(t)$, where the graphs of f and g are as shown.

1.
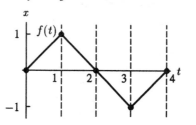
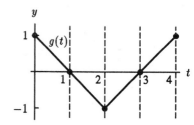

Figure 16.9

2.
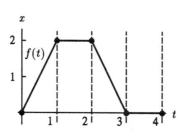
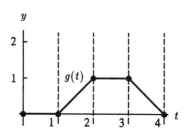

Figure 16.10

3.
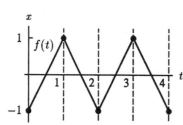
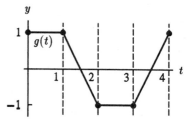

Figure 16.11

4.
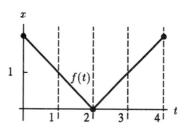
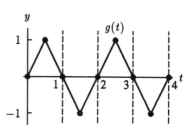

Figure 16.12

In Problems 5–7, the curve being traced out is a circle. Describe in words how the circle is traced out, including when and where the particle is moving clockwise and when and where the particle is moving counterclockwise.

5. $x = \cos(t^3 - t)$, $y = \sin(t^3 - t)$ 6. $x = \cos(\ln t)$, $y = \sin(\ln t)$

7. $x = \cos(\cos t)$, $y = \sin(\cos t)$

8. Describe the similarities and differences among the motions in the plane given by the three following pairs of parametric equations:
    (a) $x = t$, $y = t^2$ (b) $x = t^2$, $y = t^4$ (c) $x = t^3$, $y = t^6$.

On a graphing calculator or a computer, plot the curves given by the equations in Problems 9–13.

9. $x = (\cos 3t)(\cos t)$, $y = (\cos 3t)(\sin t)$ 10. $x = t^2 - t$, $y = t^3 + t^2$

11. $x = 3\cos t + \cos 3t$, $y = 3\sin t + \sin 3t$ 12. $x = 3e^t + 5e^{2t}$, $y = 2e^t - 7e^{2t}$

13. $x = t + 2\sin t$, $y = 2\cos t$

Graph the Lissajous figures in Problems 14–17 using a computer or graphing calculator.

14. $x = \cos 2t$, $y = \sin 5t$ 15. $x = \cos 3t$, $y = \sin 7t$

16. $x = \cos 2t$, $y = \sin 4t$ 17. $x = \cos 2t$, $y = \sin \sqrt{3}t$

For Problems 18–21, use a computer to draw the given curve.

18. $x = t^2 + t$, $y = t^2 - t$, $z = t^2$ 19. $x = t\cos t$, $y = t\sin t$, $z = t$

20. $x = \cos 3t$, $y = \sin 5t$, $z = \cos(2t - \pi/6)$
21. $x = (4 + \sin 5t)\cos 3t$, $y = (4 + \sin 5t)\sin 3t$, $z = \cos 3t$.
22. Motion along a straight line is given by a single equation, say, $x = t^3 - t$ where x is distance along the line. It is difficult to see the motion from a plot; it just traces out the x-line, as in Figure 16.13. To visualize the motion, we introduce a y-coordinate and let it slowly increase, giving Figure 16.14. Try the following on a graphing calculator or computer. Let $y = t$. Now plot the pair of parametric equations $x = t^3 - t$, $y = t$ for, say, $-2 \leq t \leq 3$. What does the plot tell you about the particle's motion?

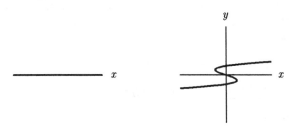

Figure 16.13   *Figure 16.14*

For Problems 23–25, graph the motion along the x-line by the method of Problem 22. What does the path tell you about the particle's motion?

23. $x = \cos t$, $-10 \leq t \leq 10$ 24. $x = t^4 - 2t^2 + 3t - 7$, $-3 \leq t \leq 2$

25. $x = t\ln t$, $0.01 \leq t \leq 10$
26. Imagine a light shining on the helix of Example 3 on page 283 from far down each of the axes. Sketch the shadow cast by the helix on each of the coordinate planes: xy, xz, and yz.

16.2 PARAMETERIZED CURVES

Parametric Representations of Curves in the Plane

In the previous section we concentrated on the motion of a particle. Sometimes we are more interested in the curve traced out by the particle than we are in the motion itself. In that case we will call the parametric equations a *parameterization* of the curve. Though the parameter may not have physical meaning, it is still helpful to think of it as time t.

Example 1   Give a parameterization of the semicircle of radius 1, shown in Figure 16.15.

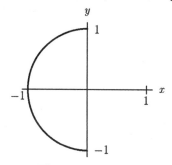

Figure 16.15: Find a parameterization of this semicircle

Solution   We can use the equations $x = \cos t$ and $y = \sin t$ for counterclockwise motion in a circle, from Example 1 on page 282. The particle passes $(0, 1)$ at $t = \pi/2$, moves counterclockwise around the circle, and reaches $(0, -1)$ at $t = 3\pi/2$. So the parameterization is

$$x = \cos t, \; y = \sin t, \quad \frac{\pi}{2} \le t \le \frac{3\pi}{2}.$$

When parameterizing a curve, we usually require that the parameterization go from one end of the curve to the other, without tracing the same portion of the curve twice. This is different from parameterizing the motion of a particle, where, for example, we could have a particle move around the same circle many times.

How Do We Find a Parameterization?

Parametric Representation of Graphs of Functions

The graph of any function $y = f(x)$ can be given parametrically simply by letting the parameter t be x:

$$x = t, \quad y = f(t).$$

Example 2   Give parametric equations for the curve $y = x^3 - x$.

Solution   Let $x = t$, $y = t^3 - t$. Imagine a particle whose motion is described by these equations. At any time t, the particle's position (x, y) satisfies $y = t^3 - t = x^3 - x$. Since $x = t$, as time increases the x-coordinate moves from left to right, thus, the particle traces out the curve $y = x^3 - x$ from left to right.

Using Position Vectors to Write Parameterized Curves as Vector Functions

Recall that a point in the plane with coordinates (x, y) can be represented by a position vector $\vec{r}$ with its tail at the origin and its tip at the point (x, y) as shown in Figure 16.16.

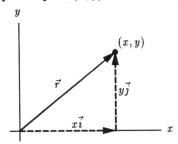

Figure 16.16: Position vector $\vec{r}$ for
the point (x, y)

Since $\vec{r} = x\vec{i} + y\vec{j}$, we can write a pair of parametric equations $x = f(t)$, $y = g(t)$ as a single vector equation $\vec{r} = \vec{F}(t)$, where $\vec{F}(t) = f(t)\vec{i} + g(t)\vec{j}$. For example, the circular motion $x = \cos t$, $y = \sin t$ can be written as $\vec{r} = (\cos t)\vec{i} + (\sin t)\vec{j}$. Similarly, in 3-space we use $\vec{r} = x\vec{i} + y\vec{j} + z\vec{k}$. For example, the helix $x = \cos t$, $y = \sin t$, $z = t$ can be written as $\vec{r} = (\cos t)\vec{i} + (\sin t)\vec{j} + t\vec{k}$.

Example 3   Give a parameterization for the circle of radius 3 centered at $(-1, 2)$.

Solution   The circle of radius 1 centered at the origin is parameterized by the vector equation

$$\vec{r} = \cos t\vec{i} + \sin t\vec{j}, \quad 0 \le t \le 2\pi.$$

We get the circle of radius 3 centered at $(-1, 2)$ by adding $3\vec{r}$ to the position vector $\vec{r}_0$ of $(-1, 2)$, so

$$\vec{s} = x\vec{i} + y\vec{j} = \vec{r}_0 + 3\vec{r} = -\vec{i} + 2\vec{j} + 3(\cos t\vec{i} + \sin t)\vec{j}$$
$$= (-1 + 3\cos t)\vec{i} + (2 + 3\sin t)\vec{j},$$

or, equivalently,

$$x = -1 + 3\cos t, \quad y = 2 + 3\sin t, \quad 0 \le t \le 2\pi.$$

See Figures 16.17 and 16.18.

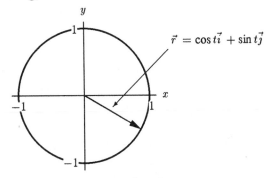

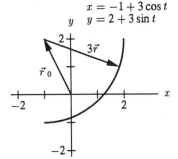

Figure 16.17: The circle $x^2 + y^2 = 1$ parameterized by $\vec{r} = \cos t\vec{i} + \sin t\vec{j}$

Figure 16.18: The circle of radius 3 centered at $(-1, 2)$

Parametric Equations of Lines

Consider a straight line in the direction of a vector $\vec{v}$ passing through the point with position vector $\vec{r}_0$. You can get to all points on the line by starting at $\vec{r}_0$ and moving up and down the line, adding different multiples of $\vec{v}$ to $\vec{r}_0$. See Figure 16.19.

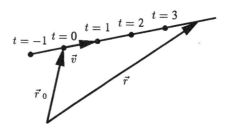

Figure 16.19: The line $\vec{r}(t) = \vec{r}_0 + t\vec{v}$ where
$$\vec{r}_0 = x_0\vec{i} + y_0\vec{j} + z_0\vec{k}$$

In this way, every point on the line can be written as $\vec{r}_0 + t\vec{v}$, which yields:

Parametric equation for a line

The parametric equation of a line through the point with position vector $\vec{r}_0 = x_0\vec{i} + y_0\vec{j} + z_0\vec{k}$ in the direction of the vector $\vec{v} = a\vec{i} + b\vec{j} + c\vec{k}$ is:

$$\vec{r} = \vec{r}_0 + t\vec{v} = x_0\vec{i} + y_0\vec{j} + z_0\vec{k} + t(a\vec{i} + b\vec{j} + c\vec{k}),$$

or equivalently,

$$x = x_0 + at, \quad y = y_0 + bt, \quad z = z_0 + ct.$$

Notice that the parameterization of a line given above expresses the coordinates x, y, and z as linear functions of the parameter t.

Example 4   Find the parametric equation for:

  (i)   The line passing through the points $(2, -1, 3)$ and $(-1, 5, 4)$.
  (ii) The line segment from $(2, -1, 3)$ to $(-1, 5, 4)$.

Solution   (i)   The line passes through $(2, -1, 3)$ and is parallel to the displacement vector $\vec{v} = -3\vec{i} + 6\vec{j} + \vec{k}$ from $(2, -1, 3)$ to $(-1, 5, 4)$. Thus the parametric equation is

$$\vec{r} = 2\vec{i} - \vec{j} + 3\vec{k} + t(-3\vec{i} + 6\vec{j} + \vec{k}).$$

  (ii) Note that in this parameterization, $t = 0$ corresponds to the point $(2, -1, 3)$ and $t = 1$ corresponds to the point $(-1, 5, 4)$. So the parameterization is

$$\vec{r} = 2\vec{i} - \vec{j} + 3\vec{k} + t(-3\vec{i} + 6\vec{j} + \vec{k}), \qquad 0 \le t \le 1.$$

Implicit, Explicit, and Parametric Representations in 2-Space

The circle of radius 1 centered at the origin can be represented implicitly by the equation

$$x^2 + y^2 = 1,$$

explicitly by the equations

$$y = \sqrt{1 - x^2} \quad \text{and} \quad y = -\sqrt{1 - x^2},$$

or parametrically by

$$x = \cos t, \; y = \sin t, \quad 0 \le t \le 2\pi.$$

In general

- An **implicit** representation of a curve in the xy-plane is given by a single equation in x and y, $f(x, y) = 0$.
- An **explicit** representation of a curve in the xy-plane is given by equations expressing y in terms of x or x in terms of y of the form $y = g(x)$ or $x = h(y)$.
- A **parametric** representation of a curve in the xy-plane is given by a pair of equations expressing x and y in terms of a third variable, often denoted t.

There can be many different implicit or parametric representations of a given curve.

Example 5 Give implicit, explicit, and parametric representations of the line passing through the points $(3, 0)$ and $(0, 5)$.

Solution An implicit representation is $x/3 + y/5 - 1 = 0$ (check that the x-intercept is 3 and y-intercept is 5). An explicit representation is $y = 5 - (5/3)x$. A parametric representation is $x = 3t$, $y = 5 - 5t$.

Example 6 Give implicit and explicit representations of the curve having the parametric representation

$$x = 3 + 5 \sin t, \quad y = 1 + 2 \cos t, \quad 0 \le t \le 2\pi.$$

Solution We need to eliminate the parameter t. Solving for $\sin t$ and $\cos t$, we get $(x - 3)/5 = \sin t$, $(y - 1)/2 = \cos t$. Since $\sin^2 t + \cos^2 t = 1$, we have

$$\left(\frac{x - 3}{5}\right)^2 + \left(\frac{y - 1}{2}\right)^2 = 1,$$

which is an implicit representation for an ellipse centered at the point $(3, 1)$. To get an explicit representation, we solve for one variable in terms of the other

$$\left(\frac{y - 1}{2}\right)^2 = 1 - \left(\frac{x - 3}{5}\right)^2$$

$$\frac{y - 1}{2} = \pm\sqrt{1 - \left(\frac{x - 3}{5}\right)^2}$$

$$y = 1 \pm 2\sqrt{1 - \frac{(x - 3)^2}{25}}.$$

We do not get one explicit representation for the whole ellipse; rather, we get one for the upper half (the positive square root) and one for the lower half (the negative square root).

Explicit and parametric equations are easier to plot than implicit equations. For example, to sketch $y = f(x)$, we evaluate $f(x)$ for various values of x and plot points. Similarly, for a curve given parametrically, we evaluate x and y for various values of t and plot points. For an implicit representation, however, we can try values for x, but then we must solve the implicit equation for y. There may be many or no values for y for a given value of x. Moreover, it may be impossible to solve the equation for y algebraically.

Where Does a Curve Pierce a Surface?

Parametric equations for a curve are ideally suited for determining where the curve intersects a given surface.

Example 7   Find the points at which the line $x = t$, $y = 2t$, $z = 1 + t$ pierces the sphere of radius 10 centered at the origin.

Solution   The equation for the sphere is

$$x^2 + y^2 + z^2 = 100.$$

To find the values of the parameter t that correspond to points of intersection of the line and the sphere, we can substitute the parametric equations of the line into the equation of the sphere, getting

$$t^2 + 4t^2 + (1 + t)^2 = 100,$$

so

$$6t^2 + 2t - 99 = 0,$$

which has the two solutions $t = -4.23$ and $t = 3.90$. Thus the line touches the sphere at the two points:

$$(x, y, z) = (-4.23, 2(-4.23), 1 + (-4.23)) = (-4.23, -8.46, -3.23)$$

and

$$(x, y, z) = (3.90, 2(3.90), 1 + 3.90) = (3.90, 7.80, 4.90)$$

Problems for Section 16.2

Write a parameterization for each of the curves in the xy-plane in Problems 1–7.

1. A circle of radius 3 centered at the origin and traced out clockwise.
2. A line through the point $(1, 3)$ and parallel to the vector $-2\vec{i} - 4\vec{j}$.
3. A vertical line through the point $(-2, -3)$.
4. A circle of radius 5 centered at the point $(2, 1)$ and traced out counterclockwise.
5. A circle of radius 2 centered at the origin traced out clockwise starting at the point $(-2, 0)$ when $t = 0$.
6. An ellipse centered at the origin and crossing the x-axis at ±5 and the y-axis at ±7.
7. An ellipse centered at the origin, crossing the x-axis at ±3 and the y-axis at ±7. The parameterization should start at the point $(-3, 0)$ and trace out the ellipse counterclockwise.
8. True or false? The equations $x = \cos(t)$, $y = \cos(t)$, $0 \le t \le \pi$, parameterize a line segment. Explain.

9. If t is allowed to take on all real values, the parametric equations

$$x = 2 + 3t, \quad y = 4 + 7t$$

describe a line in the plane.

(a) What part of the line is obtained by restricting t to nonnegative numbers?
(b) What part of the line is obtained if t is restricted to $-1 \le t \le 0$?
(c) How should t be restricted to give the part of the line to the left of the y-axis?

10. (a) Explain why the two pairs of equations

$$x\vec{i} + y\vec{j} = (2 + t)\vec{i} + (4 + 3t)\vec{j},$$
$$x\vec{i} + y\vec{j} = (1 - 2t)\vec{i} + (1 - 6t)\vec{j}$$

parameterize the same line.

(b) What are the slope and y intercept of this line?

11. Two particles are traveling through space. At time t the first particle is at the point $(-1 + t, 4 - t, -1 + 2t)$ and the second is at $(-7 + 2t, -6 + 2t, -1 + t)$.

(a) Describe the two paths.
(b) Will the two particles collide, and if so when and where?
(c) Do the paths of the two particles cross, and if so where?

12. Suppose $a, b, c, d, m, n, p, q > 0$ and match each pair of parametric equations below with one of the lines l_1, l_2, l_3, l_4 in Figure 16.20.

I. $\begin{cases} x = a + ct, \\ y = -b + dt. \end{cases}$ II. $\begin{cases} x = m + pt, \\ y = n - qt. \end{cases}$

13. What can you say about the values of a, b and k if the equations

$$x = a + k \cos t, \quad y = b + k \sin t, \quad 0 \le t \le 2\pi.$$

trace out each of the circles in Figure 16.21: (a) C_1? (b) C_2? (c) C_3?

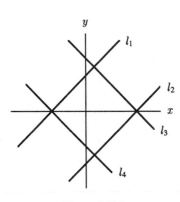

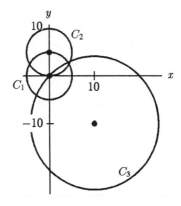

Figure 16.20   Figure 16.21:

What curves do the parametric equations in Problems 14–16 trace out? Find an implicit or explicit equation for each curve.

14. $x = 2 + \cos t, \ y = 2 - \sin t$ 15. $x = 2 + \cos t, \ y = 2 - \cos t$

16. $x = 2 + \cos t, \ y = \cos^2 t$

17. Describe the curve represented by the parametric equations

$$x = 3 + t^3, \quad y = 5 - t^3, \quad z = 7 + 2t^3.$$

State whether the equations in Problems 18–20 represent a curve parametrically, implicitly, or explicitly. Give the two other types of representations for the same curve.

18. $xy = 1$ for $x > 0$ 19. $x^2 - 2x + y^2 = 0$ for $y < 0$

20. $x = e^t, \quad y = e^{2t}$ for all t

Write a parameterization for each of the curves in Problems 21–25.

21. The circle of radius 2 in the xz-plane, centered at the origin.
22. The circle of radius 3 centered at the point $(0, 0, 2)$ parallel to the xy-plane.
23. The line through the points $(2, -1, 4)$ and $(1, 2, 5)$.
24. The line through the point $(1, 3, 2)$ perpendicular to the xz-plane.
25. The line through the point $(1, 1, 1)$ perpendicular to the plane $2x - 3y + 5z = 4$.
26. The equation $\vec{r} = 10\vec{k} + t(\vec{i} + 2\vec{j} + 3\vec{k})$ parameterizes a line.

(a) Suppose we restrict ourselves to the part of the line where $t < 0$. What do we get?
(b) Suppose we restrict ourselves to the part of the line where $0 \leq t \leq 1$. What do we get?

27. (a) Explain why the line of intersection of two planes must be parallel to the cross product of a normal vector to the first plane and a normal vector to the second.
(b) Find a vector parallel to the line of intersection of the two planes $x + 2y - 3z = 7$ and $3x - y + z = 0$.
(c) Find parametric equations for the line in part (b).

For Problems 28–32, give parametric equations for the given line.

28. The line through the points $(2, 3, -1)$ and $(5, 2, 0)$.
29. The line pointing in the direction of the vector $3\vec{i} - 3\vec{j} + \vec{k}$ and through the point $(1, 2, 3)$.
30. The line parallel to the z-axis passing through the point $(1, 0, 0)$.
31. The line of intersection of the planes $x - y + z = 3$ and $2x + y - z = 5$.
32. The line perpendicular to the surface $z = x^2 + y^2$ at the point $(1, 2, 5)$.
33. Do the lines in Problems 28 and 29 intersect?
34. Is the point $(-3, -4, 2)$ visible from the point $(4, 5, 0)$ if there is an opaque ball of radius 1 centered at the origin?
35. Consider the parameterized curve

$$x = \frac{a \cos \theta}{\sqrt{1 + (A \cos \theta + B \sin \theta)^2}}$$

$$y = \frac{a \sin \theta}{\sqrt{1 + (A \cos \theta + B \sin \theta)^2}}$$

$$z = \frac{a(A \cos \theta + B \sin \theta)}{\sqrt{1 + (A \cos \theta + B \sin \theta)^2}}.$$

(a) Show that the curve lies entirely on a sphere centered at the origin.

(b) Show that the curve lies entirely on a plane through the origin.

(c) Describe the curve in words.

36. An important application of the parametric representation of a line in three dimensions is to find where the line between two points intersects a given plane. For example, the pictures of curves and surfaces in three-dimensional space in this book are drawn by computer. To do this, the computer first calculates the xyz-coordinates of some points on the curve or surface. For each such point, it then computes the line of sight from that point to the eye of an imaginary viewer and determines where that line intersects an imaginary window (the plane of the computer screen) lying between the point and the viewer's eye. The two-dimensional screen coordinates of that point of intersection are computed so the point can be plotted on the screen. Find formulas for the coordinates of the point of intersection of the plane $Ax + By + Cz = D$ with the line from the point (a, b, c) to a viewer at the point (A, B, C).

37. In Problem 36, the xyz-coordinates are computed for the point where the line of sight from a viewer at (A, B, C) to a point (a, b, c) meets a viewing plane (the screen) $Ax + By + Cz = D$. The computer needs to compute screen coordinates of this point, not the xyz-space coordinates. To do this we take two vectors $\vec{u}$ and $\vec{v}$ at right angles to each other beginning at the screen origin and lying in the plane of the screen. Then any point on the screen can be written as $r\vec{u} + s\vec{v}$ for some numbers r, s; these numbers are the screen coordinates. We choose for the screen origin the point Q where the line of sight from the viewer (A, B, C) to the xyz-origin $(0, 0, 0)$ intersects the viewing plane $Ax + By + Cz = D$. We choose $\vec{u}$ to be a unit vector parallel to the xy-plane pointing to the viewer's right and $\vec{v}$ to be a unit vector at right angles to $\vec{u}$ and pointing up (its z-component is positive). The screen coordinates are found by taking the dot product with $\vec{u}$ and $\vec{v}$ of the vector from the screen origin Q to the point of intersection computed in Problem 36.

(a) Find the xyz-coordinates of the screen origin Q in terms of A, B, C, D.

(b) Find the vector $\vec{u}$ in terms of A, B, C.

(c) Find the vector $\vec{v}$ in terms of A, B, C.

(d) Find the coordinates of the point of intersection computed in Problem 36..

(e) Find the screen coordinates r and s of the point computed in Problem 36. That is, find $r = \vec{u} \cdot (\vec{P} - \vec{Q})$ and $s = \vec{v} \cdot (\vec{P} - \vec{Q})$. [Hint: Use the fact that $\vec{u} \cdot (A\vec{i} + B\vec{j} + C\vec{k}) = 0$ and $\vec{v} \cdot (A\vec{i} + B\vec{j} + C\vec{k}) = 0$.]

16.3 VELOCITY AND ACCELERATION VECTORS

We can represent the velocity of a moving particle by a vector with the following properties:

The Velocity Vector

The velocity vector of a moving object is a vector $\vec{v}$ such that:
- The magnitude of $\vec{v}$ is the speed of the object.
- The direction of $\vec{v}$ is the direction of motion.

Thus $\|\vec{v}\|$ = speed of object and the velocity vector is tangent to the object's path.

Example 1   A child is sitting on the edge of a merry-go-round of diameter 10 ft moving counterclockwise at 15 revolutions per minute. Draw two velocity vectors for the child at different times.

Solution   The child moves at a constant speed around a circle of radius 5, completing one revolution counterclockwise every 4 seconds. One revolution around a circle of radius 5 is a distance of 10π, so the child's speed is $10\pi/4 = 5\pi/2 \approx 7.9$ ft/sec. Hence, the magnitude of the velocity vector is 7.9. The direction of motion is tangent to the circle, and hence perpendicular to the radius at that point. Figure 16.22 shows the vector at two different points on the circle.

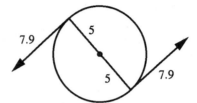

Figure 16.22: The velocity vector of a child on a merry-go-round.

Example 2   An object moving with constant velocity in three-space (with coordinates in meters) passes the point $(1, 1, 1)$, and then passes the point $(2, -1, 3)$ five seconds later. What is its velocity vector?

Solution   The displacement vector from $(1, 1, 1)$ to $(2, -1, 3)$ in 5 seconds is $\vec{d} = (2\vec{i} - \vec{j} + 3\vec{k}) - (\vec{i} + \vec{j} + \vec{k}) = \vec{i} - 2\vec{j} + 2\vec{k}$ meters. The velocity vector has the same direction as $\vec{d}$, and is given by

$$\vec{v} = \frac{\vec{d}}{5} = 0.2\vec{i} - 0.4\vec{j} + 0.4\vec{k}.$$

Example 3   A particle that passes the point $P = (5, 4, 3)$ at time $t = 7$ is moving with constant velocity $\vec{v} = 3\vec{i} + \vec{j} + 2\vec{k}$. Find the equations for its position at time t.

Solution   At time t the particle is $s = t - 7$ seconds from P, so the displacement vector from the point P to the particle is $\vec{d} = s\vec{v}$. To find the position vector of the particle at time t, we add this to the position vector for P, $\vec{r}_0 = 5\vec{i} + 4\vec{j} + 3\vec{k}$. Thus the vector equation for the motion is:

$$\vec{r} = \vec{r}_0 + s\vec{v}$$
$$= (5\vec{i} + 4\vec{j} + 3\vec{k}) + (t - 7)(3\vec{i} + \vec{j} + 2\vec{k}),$$

or equivalently,

$$x = 5 + 3(t - 7)$$
$$y = 4 + 1(t - 7)$$
$$z = 3 + 2(t - 7).$$

Notice that these equations are linear. They describe motion on a straight line through the point $(5, 4, 3)$ that is parallel to the velocity vector $\vec{v} = 3\vec{i} + \vec{j} + 2\vec{k}$.

As suggested by Example 3, motion with constant velocity is always described by linear parametric equations.

How to Compute the Velocity

If the velocity of an object is not constant, we find the velocity as in one-variable calculus: by taking a limit. If the position vector of the particle is $\vec{r}(t)$ at time t, then the displacement vector between its positions at times t and $t + \Delta t$ is $\Delta \vec{r} = \vec{r}(t + \Delta t) - \vec{r}(t)$. See Figure 16.23. Thus,

$$\vec{v}(t) \approx \frac{\Delta \vec{r}}{\Delta t},$$

and, taking the limit as Δt goes to zero, we have the following result:

> The velocity vector of a moving object with position vector $\vec{r}(t)$ at time t is
> $$\vec{v}(t) = \lim_{\Delta t \to 0} \frac{\Delta \vec{r}}{\Delta t} = \lim_{\Delta t \to 0} \frac{\vec{r}(t + \Delta t) - \vec{r}(t)}{\Delta t},$$
> whenever the limit exists. We use the notation $\vec{v} = \dfrac{d\vec{r}}{dt}$.

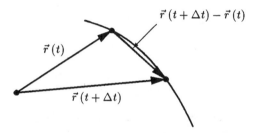

Figure 16.23: The change in the position vector for a particle moving on a curve.

The Components of the Velocity Vector

If we represent a curve parametrically by

$$x = f(t), \quad y = g(t), \quad z = h(t),$$

then we can write the components of its position vector: $\vec{r}(t) = f(t)\vec{i} + g(t)\vec{j} + h(t)\vec{k}$. This enables us to compute the components of the velocity vector:

$$
\begin{aligned}
\vec{v}(t) &= \lim_{\Delta t \to 0} \frac{\vec{r}(t + \Delta t) - \vec{r}(t)}{\Delta t} \\
&= \lim_{\Delta t \to 0} \frac{(f(t + \Delta t)\vec{i} + g(t + \Delta t)\vec{j} + h(t + \Delta t)\vec{k}) - (f(t)\vec{i} + g(t)\vec{j} + h(t)\vec{k})}{\Delta t} \\
&= \lim_{\Delta t \to 0} \left(\frac{f(t + \Delta t) - f(t)}{\Delta t}\vec{i} + \frac{g(t + \Delta t) - g(t)}{\Delta t}\vec{j} + \frac{h(t + \Delta t) - h(t)}{\Delta t}\vec{k} \right) \\
&= f'(t)\vec{i} + g'(t)\vec{j} + h'(t)\vec{k} \\
&= \frac{dx}{dt}\vec{i} + \frac{dy}{dt}\vec{j} + \frac{dz}{dt}\vec{k}.
\end{aligned}
$$

Thus we have the following result:

The Components of the Velocity Vector

The velocity vector at time t of a particle moving in space is

$$\vec{v} = \frac{dx}{dt}\vec{i} + \frac{dy}{dt}\vec{j} + \frac{dz}{dt}\vec{k}.$$

Example 4   Find the components of the velocity vector for the child on the merry-go-round in Example 1 using a coordinate system which has its origin at the center of the merry-go-round.

Solution   The merry-go-round has radius 5 and completes 1 revolution counterclockwise every 4 seconds. Thus the motion is described by the equation

$$\vec{r} = 5\cos(\frac{2\pi}{4}t)\vec{i} + 5\sin(\frac{2\pi}{4}t)\vec{j} = 5\cos(\frac{\pi}{2}t)\vec{i} + 5\sin(\frac{\pi}{2}t)\vec{j},$$

where t is in seconds. So

$$\vec{v} = \frac{dx}{dt}\vec{i} + \frac{dy}{dt}\vec{j} = -\frac{5\pi}{2}\sin(\frac{\pi}{2}t)\vec{i} + \frac{5\pi}{2}\cos(\frac{\pi}{2}t)\vec{j}.$$

We observe that the magnitude of $\vec{v}$ is

$$\|\vec{v}\| = \sqrt{(\frac{5\pi}{2})^2 \sin^2(\frac{\pi}{2}t) + (\frac{5\pi}{2})^2 \cos^2(\frac{\pi}{2}t)}$$

$$= \frac{5\pi}{2}\sqrt{\sin^2\frac{\pi}{2}t + \cos^2\frac{\pi}{2}t} = \frac{5\pi}{2} \approx 7.9,$$

which agrees with the speed we calculated in Example 1. To see that the direction is correct, we should show that the vector $\vec{v}$ at any time t is perpendicular to the position vector of the particle at time t, using the dot product of $\vec{v}$ and $\vec{r}$:

$$\vec{v} \cdot \vec{r} = (-\frac{5\pi}{2}\sin(\frac{\pi}{2}t)\vec{i} + \frac{5\pi}{2}\cos(\frac{\pi}{2}t)\vec{j}) \cdot (5\cos(\frac{\pi}{2}t)\vec{i} + 5\sin(\frac{\pi}{2}t)\vec{j})$$

$$= -\frac{25\pi}{2}\sin(\frac{\pi}{2}t)\cos(\frac{\pi}{2}t) + \frac{25\pi}{2}\cos(\frac{\pi}{2}t)\sin(\frac{\pi}{2}t) = 0.$$

Thus, the velocity vector is perpendicular to $\vec{r}$ and hence tangent to the circle, as expected.

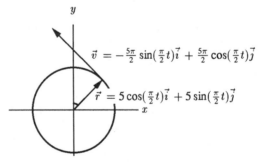

Figure 16.24: Velocity and radius vector of motion around a circle

Velocity Vectors and Tangent Lines

Since the velocity vector is tangent to the path of motion, it can be used to find parametric equations for the tangent line, if there is one.

Example 5 Find the tangent line to the curve defined by the parametric equation

$$\vec{r} = t^2\vec{i} + t^3\vec{j} + 2t\vec{k}$$

at the point $(1, 1, 2)$.

Solution At time $t = 1$ the particle is at the point $(1, 1, 2)$ with position vector $\vec{r}_0 = \vec{i} + \vec{j} + 2\vec{k}$. The velocity vector at time t is $\vec{v}(t) = 2t\vec{i} + 3t^2\vec{j} + 2\vec{k}$, so at time $t = 1$ it is $\vec{v} = 2\vec{i} + 3\vec{j} + 2\vec{k}$. The tangent line passes through $(1, 1, 2)$ in the direction of $\vec{v}$, so it has equation

$$\vec{r} \approx \vec{r}_0 + t\vec{v}$$
$$= (\vec{i} + \vec{j} + 2\vec{k}) + t(2\vec{i} + 3\vec{j} + 2\vec{k}),$$

or equivalently,

$$x = 1 + 2t, \quad y = 1 + 3t, \quad z = 2 + 2t.$$

Different parametrizations of the same curve might give different velocity vectors at a given point, so we might get different parametric equations for the (same) tangent line through the point. Also, a curve which crosses itself at a point might not have a tangent line there, even though the motion has a velocity vector each time it passes that point.

The Length of a Curve

The speed of a particle is the magnitude of its velocity vector:

$$\text{Speed} = \|\vec{v}\| = \sqrt{\left(\frac{dx}{dt}\right)^2 + \left(\frac{dy}{dt}\right)^2 + \left(\frac{dz}{dt}\right)^2}.$$

Just as we can find the total distance traveled by a particle moving along a straight line by integrating the absolute value of its velocity, we can find the distance traveled along a curve by integrating the magnitude of its velocity vector. Thus,

$$\text{Distance traveled} = \int_a^b \|\vec{v}(t)\|\, dt.$$

If the particle never stops and reverses its direction as it moves along the curve, the distance it travels will be the same as the length of the curve. This suggests the following formula, which is justified in Problem 33:

If the curve C in the plane is given parametrically for $a \leq t \leq b$ and if the velocity vector $\vec{v}$ of the parameterization exists and is not $\vec{0}$ for $a \leq t \leq b$ then

$$\text{Length of } C = \int_a^b \|\vec{v}\|\, dt.$$

Example 6   Find the circumference of the ellipse given by the parametric equations

$$x = 2\cos t, \quad y = \sin t, \quad 0 \le t \le 2\pi.$$

Solution   The length of this curve is given by the integral

$$\int_0^{2\pi} \sqrt{\left(\frac{dx}{dt}\right)^2 + \left(\frac{dy}{dt}\right)^2}\, dt = \int_0^{2\pi} \sqrt{(-2\sin t)^2 + (\cos t)^2}\, dt$$

$$= \int_0^{2\pi} \sqrt{4\sin^2 t + \cos^2 t}\, dt.$$

Simpson's rule with $n = 20$ gives a value of 9.688449 and with $n = 40$ gives a value of 9.688448. Thus, the length of the curve is approximately 9.68845. Since the given ellipse is inscribed in a circle of radius 2 and circumscribes a circle of radius 1, we would expect the length of the ellipse to be between $2\pi(2) \cong 12.57$ and $2\pi(1) \cong 6.28$ so the value of 9.68845 is reasonable.

Just as the velocity of a particle moving in 2-space or 3-space is a vector quantity, so is the rate of change of the velocity of the particle, namely its acceleration.

Limit Definition of the Acceleration Vector

Acceleration is the rate of change of velocity. Figure 16.25 shows a particle at time t with velocity vector $\vec{v}(t)$ and then at time $t + \Delta t$ a little later. The vector $\Delta\vec{v} = \vec{v}(t + \Delta t) - \vec{v}(t)$ is the change in velocity and points approximately in the direction of the acceleration. In the limit as $\Delta t \to 0$, we have the instantaneous acceleration at time t:

Acceleration Vector

The acceleration vector of an object moving with velocity $\vec{v}(t)$ at time t is

$$\vec{a}(t) = \lim \frac{\Delta\vec{v}}{\Delta t} = \lim_{\Delta t \to 0} \frac{\vec{v}(t + \Delta t) - \vec{v}(t)}{\Delta t},$$

if the limit exists.

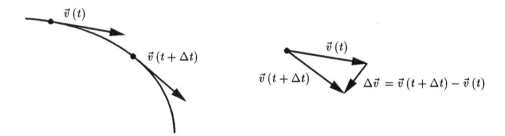

Figure 16.25: Computing the difference between two velocity vectors

We also use the notation $\vec{a} = d\vec{v}/dt = d^2\vec{r}/dt^2$.

Components of the Acceleration Vector

If we represent a curve in space parametrically by $x = f(t)$, $y = g(t)$, $z = h(t)$, we can express the acceleration in components. The velocity vector $\vec{v}(t)$ has components

$$\vec{v}(t) = f'(t)\vec{i} + g'(t)\vec{j} + h'(t)\vec{k}.$$

From the definition of the acceleration vector, we have

$$\vec{a}(t) = \lim_{\Delta t \to 0} \frac{\vec{v}(t + \Delta t) - \vec{v}(t)}{\Delta t} = \frac{d\vec{v}}{dt}.$$

Thus:

Components of the Acceleration Vector

The acceleration vector, $\vec{a}(t)$, at time t of a particle moving in space with motion described parametrically by $\vec{r}(t) = f(t)\vec{i} + g(t)\vec{j} + h(t)\vec{k}$ is

$$\vec{a}(t) = f''(t)\vec{i} + g''(t)\vec{j} + h''(t)\vec{k} = \frac{d^2 x}{dt^2}\vec{i} + \frac{d^2 y}{dt^2}\vec{j} + \frac{d^2 z}{dt^2}\vec{k}.$$

Acceleration of Motion in a Circle and in a Line

Example 7   Find the acceleration vector of the child on the merry-go-round in Example 1.

Solution   In Example 4 we saw that the velocity vector was

$$\vec{v}(t) = \frac{dx}{dt}\vec{i} + \frac{dy}{dt}\vec{j} = -\frac{5\pi}{2}\sin(\frac{\pi}{2}t)\vec{i} + \frac{5\pi}{2}\cos(\frac{\pi}{2}t)\vec{j}.$$

So the acceleration vector is

$$\begin{aligned}
\vec{a} &= \frac{d^2 x}{dt^2}\vec{i} + \frac{d^2 y}{dt^2}\vec{j} \\
&= -(\frac{5\pi}{2} \cdot \frac{\pi}{2} \cos \frac{\pi}{2}t)\vec{i} - (\frac{5\pi}{2} \cdot \frac{\pi}{2} \sin \frac{\pi}{2}t)\vec{j} \\
&= -\frac{5\pi^2}{4} \cos(\frac{\pi}{2}t)\vec{i} - \frac{5\pi^2}{4} \sin(\frac{\pi}{2}t)\vec{j}
\end{aligned}$$

Observe that the acceleration vector is $\pi^2/4$ times the negative of the position vector $x\vec{i} + y\vec{j} = 5\cos(\frac{\pi}{2}t)\vec{i} + 5\sin(\frac{\pi}{2}t)\vec{j}$ and thus points toward the origin.

In general, we have the following result:

A particle traveling around a circle of radius R with constant speed $\|\vec{v}\|$ has its acceleration vector pointing to the center of the circle.

In uniform circular motion, the acceleration vector reflects the fact that the velocity vector does not change in magnitude, only in direction. We now look at straight line motion where the velocity vector always has the same direction but the magnitude changes. We expect that the acceleration vector will point in the same direction as the velocity vector if the speed is increasing and in the opposite direction as the velocity vector if the speed is decreasing.

Example 8   Consider the motion given by the vector equation

$$\vec{r} = 2\vec{i} + 6\vec{j} + (t^3 + t)(4\vec{i} + 3\vec{j} + \vec{k}).$$

Show that this is straight line motion in the direction of the vector $4\vec{i} + 3\vec{j} + \vec{k}$ and relate the acceleration vector to the velocity vector.

Solution   The velocity vector is

$$\vec{v} = (3t^2 + 1)(4\vec{i} + 3\vec{j} + \vec{k}).$$

Since $(3t^2 + 1)$ is a positive scalar, the velocity vector $\vec{v}$ always points in the direction of the vector $4\vec{i} + 3\vec{j} + \vec{k}$. In addition,

$$\text{Speed} = \|\vec{v}\| = (3t^2 + 1)\sqrt{4^2 + 3^2 + 1^2} = \sqrt{26}(3t^2 + 1)$$

Notice that the speed is decreasing until $t = 0$, then starts increasing. The acceleration vector is

$$\vec{a} = 6t(4\vec{i} + 3\vec{j} + \vec{k}).$$

The acceleration vector for $t > 0$ points in the same direction as $4\vec{i} + 3\vec{j} + \vec{k}$, which is the same direction as $\vec{v}$. This makes sense because the object is speeding up. For $t < 0$, the acceleration vector $6t(4\vec{i} + 3\vec{j} + \vec{k})$ points in the opposite direction of the velocity vector because the object is slowing down. Finally, the magnitude of the acceleration vector is

$$\|\vec{a}\| = |6t|\sqrt{4^2 + 3^2 + 1^2} = 6\sqrt{26}\,|t|,$$

which equals the absolute value of the rate of change, $6\sqrt{26}t$, of the speed $\|\vec{v}\|$.

Problems for Section 16.3

1. Table 16.4 gives x and y coordinates of a particle in the plane at time t. Assuming the path is smooth, estimate the following quantities:

    (a) The velocity vector and speed at time $t = 2$.
    (b) Any times when the particle is moving straight up or down (parallel to the y-axis).
    (c) Any times when the particle has come to a stop.

    TABLE 16.4

t	0	0.5	1.0	1.5	2.0	2.5	3.0	3.5	4.0
x	1	4	6	7	6	3	2	3	5
y	3	2	3	5	8	10	11	10	9

2. Give a table of values near $t = 1$ for the circular motion with position vector

    $$\vec{r}(t) = (\cos t)\vec{i} + (\sin t)\vec{j}$$

    and interpret the table in terms of the velocity vector at time $t = 1$.

3. (a) Sketch the parameterized curve $x = t \cos t$, $y = t \sin t$ for $0 \le t \le 4\pi$.
   (b) Use difference quotients to approximate the velocity vectors $\vec{v}(t)$ for $t = 2, 4$, and 6.
   (c) Compute exactly the velocity vectors $\vec{v}(t)$ for $t = 2, 4$, and 6, and sketch them onto the graph of the curve.

For Problems 4–9, find the velocity vector $\vec{v}(t)$ for the motion of a particle given by the parametric equations. In each case, also give the speed $\|\vec{v}(t)\|$ and any times when the particle comes to a stop.

4. $x = t^2$, $y = t^3$

5. $x = \cos(t^2)$, $y = \sin(t^2)$

6. $x = \cos 2t$, $y = \sin t$

7. $x = \cos 3t$, $y = \sin 5t$

8. $x = t$, $y = t^2$, $z = t^3$

9. $x = t^2 - 2t$, $y = t^3 - 3t$, $z = 3t^4 - 4t^3$

10. Find parametric equations for the tangent line at $t = 2$ for Problem 4.
11. Find parametric equations for the tangent line at $t = 2$ for Problem 8.

Find the length of the curves in Problems 12–14.
12. $x = 3 + 5t$, $y = 1 + 4t$, $z = 3 - t$ for $1 \le t \le 2$. Explain why your answer is reasonable.
13. $x = \cos(e^t)$, $y = \sin(e^t)$ for $0 \le t \le 1$. Explain why your answer is reasonable.
14. $x = \cos 3t$, $y = \sin 5t$ for $0 \le t \le 2\pi$.
15. Consider the motion of the particle given by the parametric equations

$$x = t^3 - 3t, \quad y = t^2 - 2t,$$

where the y-axis is vertical and the x-axis is horizontal.
   (a) Does the particle ever come to a stop? If so, when and where?
   (b) Is the particle ever moving straight up or down? If so, when and where?
   (c) Is the particle ever moving straight horizontally right or left? Is so, when and where?

16. Suppose $\vec{r} = \cos t\,\vec{i} + \sin t\,\vec{j} + 2t\,\vec{k}$ represents the position of a particle on a helix, where z is the height of the particle above the ground.
   (a) Is the particle ever moving downwards? When?
   (b) When does the particle reach a point 10 units above the ground?
   (c) What is the velocity of the particle when it is 10 units above the ground?
   (d) Suppose the particle leaves the helix and moves along the tangent line to the spiral at this point. What is the equation of the tangent line?

17. A particle that passes the point $P = (5, 4, -2)$ at time $t = 4$ is moving with constant velocity $\vec{v} = 2\vec{i} - 3\vec{j} + \vec{k}$. Find the parametric equations for its motion.

18. Mr. Skywalker is traveling along the curve given by:

$$\vec{r}(t) = -2e^{3t}\vec{i} + 5\cos t\,\vec{j} - 3\sin(2t)\vec{k}.$$

If the power thrusters are turned off, his ship flies off on a tangent line to $\vec{r}(t)$. He is almost out of power when he notices that a station on Xardon is open at the point with coordinates $(1.5, 5, 3.5)$. Quickly calculating his position, he turns off the thrusters at $t = 0$. Does he make it to the Xardon station? Explain.

19. Suppose $F(x, y) = 1/(x^2 + y^2 + 1)$ gives the temperature at the point (x, y) in the plane. Suppose a ladybug moves along a parabola according to the parametric equations

$$x = t, \quad y = t^2.$$

Find the rate of change in the temperature of the ladybug at time t.

20. This problem generalizes the result of Problem 19. Suppose $F(x, y)$ gives the temperature at any point (x, y) in the plane and that a ladybug moves in the plane with position vector at time t given by $\vec{r}(t) = x(t)\vec{i} + y(t)\vec{j}$ and velocity vector $\vec{v}(t)$. Use the chain rule to show that

Rate of change in the temperature of the bug at time $t = \nabla F(x(t), y(t)) \cdot \vec{v}(t).$

21. Emily is standing on the outer edge of a merry-go-round, 30 feet from the center. The merry-go-round completes one full revolution every 15 seconds. As Emily passes over a point P on the ground, she drops a ball from 9 feet above the ground.

    (a) How fast is Emily going?
    (b) How far from P does the ball hit the ground? (The acceleration of gravity is 32 ft/sec^2; you will need to compute how long it takes for the ball to fall 9 feet).
    (c) How far from Emily does the ball hit the ground?

22. A lighthouse L is located on an island in the middle of a lake as shown in Figure 16.26. Consider the motion of the point where the light beam from L hits the shore of the lake.

    (a) Suppose the beam rotates counterclockwise about L at a constant angular velocity. At which of A, B, C, D, or E is the speed of the point greatest, and at which point smallest?
    (b) Repeat (a), supposing the beam rotates counterclockwise so that it sweeps out equal areas of the lake in equal times.
    (c) What happens if you place the lighthouse at different points in the lake. Can the speed of the point on the shore ever be infinite for part (a)?
    (d) Suppose now that the lake is rectangular instead. What happens to the velocity vector at the corners? For part (b) show that the speed is constant along each side (possibly a different constant on each side).

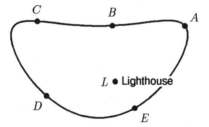

Figure 16.26: The lighthouse in the lake

23. A hypothetical moon orbits a planet which in turn orbits a star. Suppose that the orbits are circular and that the moon orbits the planet 12 times in the time it takes the planet to orbit the star once. In this problem we will investigate whether the moon could come to a stop at some instant. (See Figure 16.27).

    (a) Suppose the radius of the moon's orbit around the planet is 1 unit and the radius of the planet's orbit around the star is R units. Explain why the motion of the moon relative to the star can be described by the parametric equations:

$$x = R\cos t + \cos 12t, \quad y = R\sin t + \sin 12t.$$

(b) Find a value for R and t such that the moon stops relative to the star at time t.
(c) On a graphing calculator, plot the path of the moon for the value of R you obtained in part (b). Experiment with other values for R.

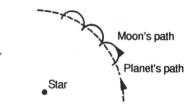

Figure 16.27: The motion of the moon

For Problems 24–28, find the velocity and acceleration vectors for the parameterized motions.

24. $x = 3\cos t, y = 4\sin t$

25. $x = t, y = t^3 - t$

26. $x = 2 + 3t, y = 4 + t, z = 1 - t$

27. $x = 2 + 3t^2, y = 4 + t^2, z = 1 - t^2$

28. $x = t, y = t^2, z = t^3$

29. The moon is revolving around the earth which is itself revolving around the sun. As a reasonable approximation, we can write the position of the moon relative to the sun parametrically as

$$\vec{r} = x\vec{i} + y\vec{j} = (93\cos(2\pi t) + 0.24\cos(24\pi t))\vec{i} + (93\sin(2\pi t) + 0.24\sin(24\pi t))\vec{j}$$

where x and y are in millions of miles.

(a) What are the units of t? What is the period of each of the two motions?
(b) How far is it from the earth to the sun and from the earth to the moon?
(c) What are the longest and shortest distances between the moon and the sun?
(d) What is the velocity, speed, and acceleration of the moon at $t = 1/3$?

30. The moon takes about 27 days to orbit the earth once at a distance of about 240,000 miles from the center of the earth. Assuming the orbit of the moon around the earth is a circle and the speed of the moon is constant, compute the magnitude of the acceleration of the moon in ft/sec^2. Newton used this computation to verify his law of gravitation, which says the magnitude of the acceleration caused by the gravitational pull of the earth should be proportional to the inverse of the square of the distance from the center of the earth. Compare the magnitude of the acceleration of the moon with the magnitude of the acceleration due to gravity at the surface of earth, namely 32 ft/sec^2 (the surface of the earth is about 4,000 miles from the center of the earth). Do you have rough agreement with Newton's law?

31. Determine the position vector $\vec{r}(t)$ for a rocket which is launched from the origin at time $t = 0$ seconds, reaches its highest point of $(1000, 3000, 10,000)$ ft, and is only subject to the force of gravity after the launch.

32. Find the velocity and acceleration vectors for the helical motion:

$$x = 3\cos(t^2), \quad y = 3\sin(t^2), \quad z = t^2.$$

33.   In this exercise we will justify the formula for the length of a curve given on page 299. Suppose we want to find the length of a curve C given by differentiable parametric equations $x = x(t)$, $y = y(t)$, $z = z(t)$ for $a \le t \le b$. By dividing the parameter interval $a \le t \le b$ at points $t_0 = a, t_1, \ldots, t_n = b$ into small segments of length Δt, we get a corresponding division of the curve C into small pieces. See Figure 16.28, where the points $P_i = (x(t_i), y(t_i), z(t_i))$ on the curve C correspond to parameter values $t = t_i$.

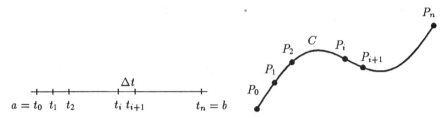

Figure 16.28: A subdivision of the parameter interval and the corresponding subdivision of the curve C

Let C_i be the portion of the curve C between P_i and P_{i+1}.

(a)   Use local linearity to show

$$\text{Length of } C_i \approx \sqrt{x'(t_i)^2 + y'(t_i)^2 + z'(t_i)^2} \, \Delta t.$$

(b)   Use (a) and a Riemann sum to explain why

$$\text{Length of } C = \int_a^b \sqrt{x'(t)^2 + y'(t)^2 + z'(t)^2} \, dt.$$

16.4 PARAMETERIZED SURFACES

How Do We Parameterize a Surface?

In Section 16.1 we saw how to parameterize a circle in 2-space using the equations

$$x = \cos t, \quad y = \sin t.$$

In 3-space, the same circle in the xy-plane has parametric equations

$$x = \cos t, \quad y = \sin t, \quad z = 0.$$

We must add the equation $z = 0$ to specify that the circle is in the xy-plane; if we wanted a circle in the plane $z = 3$, we would use the equations

$$x = \cos t, \quad y = \sin t, \quad z = 3.$$

Suppose now we let z vary freely, as well as t. Then we would get circles in every horizontal plane, forming the cylinder in Figure 16.29. Thus we need two parameters, t and z, to parameterize the cylinder.

This is true in general. A curve, though it may live in two or three dimensions, is itself one-dimensional; if you move along it you can only move backwards and forwards in one direction. Thus it only requires one parameter to trace out a curve.

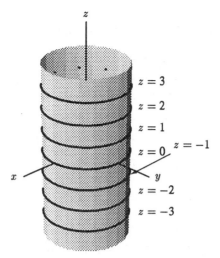

Figure 16.29: The cylinder
$x = \cos t,\ y = \sin t,\ z = z$.

A surface is two-dimensional; at any given point there are two independent directions you can move in. For example, on the cylinder you can move vertically, or you can move around the z-axis horizontally. So we need *two* parameters to describe it. We can think of the parameters as map coordinates, like longitude and latitude on the surface of the Earth. To put coordinates on the surface we specify a point (x, y, z) on it in terms of two independent parameters, p and q:

$$x = f_1(p, q), \quad y = f_2(p, q), \quad z = f_3(p, q).$$

The three equations for x, y and z are called parametric equations for the surface with *parameters* p and q. We say that we have a *parameterized surface*. In the case of the cylinder our parameters are t and z, so the parameterization is

$$x = \cos t, \quad y = \sin t, \quad z = z.$$

The last equation looks strange, but it reminds us that we are in three dimensions, not two, and that the z-coordinate on our surface is allowed to vary freely.

Using Position Vectors

Just as for parameterized curves, we can use the position vector $\vec{r} = x\vec{i} + y\vec{j} + z\vec{k}$ to combine the three parametric equations for a surface into a single vector equation. For example, the parameterization of the cylinder $x = \cos t, y = \sin t, z = z$ can be written as

$$\vec{r} = \cos t\vec{i} + \sin t\vec{j} + z\vec{k}$$

How do we Find a Parameterization?

Parameterizing a Surface of the Form $z = f(x, y)$

The graph of a function $z = f(x, y)$ can be given parametrically simply by letting the parameters p and q be x and y:

$$x = p, \quad y = q, \quad z = f(p, q).$$

Example 1   Give a parametric description of the lower hemisphere of the sphere $x^2 + y^2 + z^2 = 1$.

Solution   The surface is the graph of the function $z = -\sqrt{1 - x^2 - y^2}$ over the region $x^2 + y^2 \leq 1$ in the plane. Then parametric equations are $x = p$, $y = q$, $z = -\sqrt{1 - p^2 - q^2}$, where the parameters p and q vary inside the unit circle.

In practice we usually regard the two independent variables x and y as parameters rather than to introduce new variables p and q. Thus, we write $x = x$, $y = y$, $z = f(x, y)$, even though the first two equations do not really add anything.

Parameterizing Planes

Consider a plane containing two nonparallel vectors $\vec{v}_1$ and $\vec{v}_2$ and a point $P_0 = (x_0, y_0, z_0)$ with position vector $\vec{r}_0$. You can get to all points of the plane by starting at P_0 and moving parallel to $\vec{v}_1$ and $\vec{v}_2$, adding multiples of them to $\vec{r}_0$. See Figure 16.30.

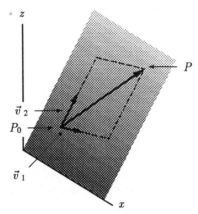

Figure 16.30: How to get from P_0 to P moving only in directions parallel to $\vec{v}_1$ and $\vec{v}_2$.

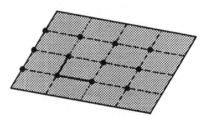

Figure 16.31: The plane $\vec{r}(p, q) = \vec{r}_0 + p\vec{v}_1 + q\vec{v}_2$ and some points corresponding to various choices of (p, q)

This yields

Parametric Equations for a Plane

The plane through the point $\vec{r}_0 = (x_0, y_0, z_0)$ and containing the two non-parallel vectors $\vec{v}_1$ and $\vec{v}_2$ has parametric equations:

$$\vec{r}(p, q) = \vec{r}_0 + p\vec{v}_1 + q\vec{v}_2.$$

If $\vec{v}_1 = a_1\vec{i} + a_2\vec{j} + a_3\vec{k}$, and $\vec{v}_2 = b_1\vec{i} + b_2\vec{j} + b_3\vec{k}$, then the equations of the plane can be written in the form:

$$x = x_0 + pa_1 + qb_1$$
$$y = y_0 + pa_2 + qb_2$$
$$z = z_0 + pa_3 + qb_3.$$

See Figure 16.31. Notice that the parameterization of the plane expresses the coordinates x, y, and z as linear functions of the parameters p and q.

Example 2 Write parametric equations for the plane through the point $(2, -1, 3)$ and containing the vectors $\vec{v}_1 = 2\vec{i} + 3\vec{j} - \vec{k}$ and $\vec{v}_2 = \vec{i} - 4\vec{j} + 5\vec{k}$.

Solution The parametric equations are

$$
\begin{aligned}
\vec{r}(p, q) &= \vec{r}_0 + p\vec{v}_1 + q\vec{v}_2 \\
&= 2\vec{i} - \vec{j} + 3\vec{k} + p(2\vec{i} + 3\vec{j} - \vec{k}) + q(\vec{i} - 4\vec{j} + 5\vec{k}) \\
&= (2 + 2p + q)\vec{i} + (-1 + 3p - 4q)\vec{j} + (3 - p + 5q)\vec{k},
\end{aligned}
$$

or equivalently,

$$
x = 2 + 2p + q, \quad y = -1 + 3p - 4q, \quad z = 3 - p + 5q.
$$

Parameterizing Surfaces of Revolution

Many surfaces that come up in applications have an axis of rotational symmetry and circular cross-sections perpendicular to that axis. These surfaces are usually referred to as *surfaces of revolution*.

Example 3 Find a parameterization of the cone whose base is the circle $x^2 + y^2 = a^2$ in the xy-plane and whose vertex is at a height h above the xy-plane. See Figure 16.32.

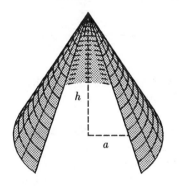
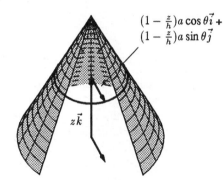

$(1 - \frac{z}{h})a \cos\theta \vec{i} +$
$(1 - \frac{z}{h})a \sin\theta \vec{j}$

$z\vec{k}$

Figure 16.32

Solution The radius of the circular cross section decreases linearly as we move up the axis, from a when $z = 0$ to 0 when $z = h$. At a height z the radius of the circle is $(1 - z/h)a$ and so this circle is traced out by the vector $(1 - z/h)a \cos\theta \vec{i} + (1 - z/h)a \sin\theta \vec{j}$ as θ goes from 0 to 2π. We get to a point on the cone by adding this to the vector $z\vec{k}$ going up the axis. This gives

$$
x\vec{i} + y\vec{j} + z\vec{k} = a\left(1 - \frac{z}{h}\right)\cos\theta \vec{i} + a\left(1 - \frac{z}{h}\right)\sin\theta \vec{j} + z\vec{k},
$$

which can be written as

$$
x = \left(1 - \frac{z}{h}\right)a \cos\theta, \quad y = \left(1 - \frac{z}{h}\right)a \sin\theta, \quad z = z,
$$

for $0 \le z \le h$ and $0 \le \theta \le 2\pi$.

Example 4   Consider the bell of a trumpet. A reasonable model for the radius $z = f(x)$ of the horn (in cm) at a distance x cm from the large open end is given by the function

$$f(x) = \frac{6}{(x+1)^{0.7}}.$$

The bell is obtained by rotating the graph of f about the x-axis. Find a parameterization for the first 24 cm of the bell. See Figure 16.33.

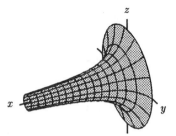

Figure 16.33: The bell of a trumpet

Solution   At distance x from the open end of the horn, the cross section parallel to the yz-plane is a circle of radius $f(x)$, with center on the x-axis. Such a circle can be parameterized by $y = f(x)\cos\theta$, $z = f(x)\sin\theta$. Thus we have the parameterization

$$x = x, \quad y = \left(\frac{6}{(x+1)^{0.7}}\right)\cos\theta, \quad z = \left(\frac{6}{(x+1)^{0.7}}\right)\sin\theta,$$

for $0 \le x \le 24$ and $0 \le \theta < 2\pi$.

Another Look at Spherical Coordinates

Recall the spherical coordinates ρ, ϕ, and θ introduced on page 259 of Chapter 15. On a sphere of fixed radius $\rho = a$ we can use ϕ and θ as coordinates, similar to latitude and longitude on the surface of the earth. See Figure 16.34.

The latitude and ϕ are related as follows: If $0 \le \phi \le \pi/2$, then the latitude is $\pi/2 - \phi$ North, and if $\pi/2 \le \phi \le \pi$, then it is $\phi - \pi/2$ South. The north pole has $\phi = 0$, the equator has $\phi = \pi/2$ and the south pole has $\phi = \pi$. Assuming the Greenwich meridian passes through the positive x-axis, the longitude and θ are related as follows: If $0 \le \theta \le \pi$, then the longitude is θ East, and if $\pi \le \theta \le 2\pi$, then it is $2\pi - \theta$ West.

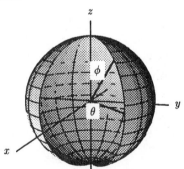

Figure 16.34: Parameterizing the sphere by ϕ and θ.

Example 5   You are at a point on a sphere with $\phi = 3\pi/4$. Are you in the northern or southern hemisphere? If ϕ decreases, do you move closer to or farther from the equator?

Solution   The equator has $\phi = \pi/2$. Since $3\pi/4 > \pi/2$, you are in the southern hemisphere. If your ϕ decreases you move closer to the equator.

Example 6   On a sphere, you are standing at a point with coordinates ϕ and θ. Your *antipodal* point is the point on the other side of the sphere on a line through you and the center. What are the coordinates of your antipodal point?

Solution   See Figure 16.35. The coordinates are $\pi + \theta$ and $\pi - \phi$. Notice that if you are on the equator, then so is your antipodal point.

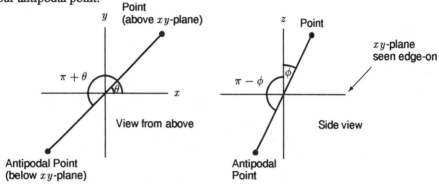

Figure 16.35: Two views of the xyz-coordinate system showing coordinates of antipodal points.

Parametrizing a Sphere Using Spherical Coordinates

The sphere with radius 1 centered at the origin is parametrized by

$$x = \sin \phi \cos \theta$$
$$y = \sin \phi \sin \theta$$
$$z = \cos \phi.$$

See Figure 16.36.

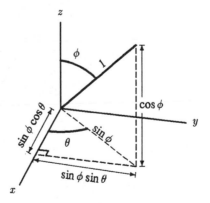

Figure 16.36: The relationship between x, y, z and ϕ, θ on a sphere of radius 1.

As a check, verify that $x^2 + y^2 + z^2 = \sin^2 \phi (\cos^2 \theta + \sin^2 \theta) + \cos^2 \phi = \sin^2 \phi + \cos^2 \phi = 1$. Notice that the z-coordinate depends only on the parameter ϕ. Geometrically, this means that all points on the same latitude have the same z-coordinate.

We can also write these equations in vector form:

$$\vec{r}(\theta, \phi) = \sin \phi \cos \theta \, \vec{i} + \sin \phi \sin \theta \, \vec{j} + \cos \phi \, \vec{k} \, .$$

Example 7   Find parametric equations for the sphere:
  (a) Centered at the origin and having radius 2.
  (b) Centered at the point $(2, -1, 3)$ and having radius 2.

Solution   (a) We must scale the distance from the origin by 2. Thus, we have

$$x = 2 \sin \phi \cos \theta$$
$$y = 2 \sin \phi \sin \theta$$
$$z = 2 \cos \phi,$$

or, in vector form,
$$\vec{r} = 2 \sin \phi \cos \theta \vec{i} + 2 \sin \phi \sin \theta \vec{j} + 2 \cos \phi \vec{k} \, .$$

  (b) To shift the center of the sphere from the origin to the point $(2, -1, 3)$, add the vector parameterization we found in part (a) to the position vector of $(2, -1, 3)$. See Figure 16.37. This gives

$$\vec{r} = 2\vec{i} - \vec{j} + 3\vec{k} + (2 \sin \phi \cos \theta \vec{i} + 2 \sin \phi \sin \theta \vec{j} + 2 \cos \phi \vec{k})$$
$$= (2 + 2 \sin \phi \cos \theta)\vec{i} + (-1 + 2 \sin \phi \sin \theta)\vec{j} + (3 + 2 \cos \phi)\vec{k} \, ,$$

or

$$x = 2 + 2 \sin \phi \cos \theta$$
$$y = -1 + 2 \sin \phi \sin \theta$$
$$z = 3 + 2 \cos \phi.$$

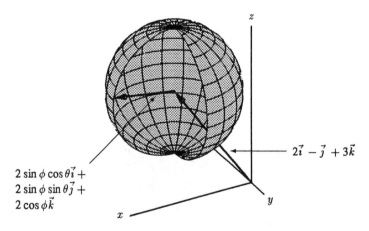

Figure 16.37: Sphere centered at the point $(2, -1, 3)$ and with radius 2

Note that the same point can have more than one value for θ or ϕ. For example, points with $\theta = 0$ also have $\theta = 2\pi$. Also, the north pole, at $\phi = 0$, and the south pole, at $\phi = \pi$, can have any value of θ.

Parameter Curves

On a parameterized surface, the curve obtained by setting one of the parameters equal to a constant and letting the other vary is called a *parameter curve*.

Example 8   Consider the vertical cylinder

$$x = \cos t, \quad y = \sin t, \quad z = z.$$

(a) Describe the two parameter curves through the point $(0, 1, 1)$.
(b) Describe the family of parameter curves with $t =$ constant, and the family with $z =$ constant.

Solution   (a)   Since the point $(0, 1, 1)$ corresponds to the parameter values $t = \pi/2$ and $z = 1$, there will be two parameter curves, one with $t = \pi/2$ and the other with $z = 1$. The parameter curve with $t = \pi/2$ is given by the parametric equations

$$x = \cos(\frac{\pi}{2}) = 0, \quad y = \sin(\frac{\pi}{2}) = 1, \quad z = z,$$

with parameter z. This is a straight line through the point $(0, 1, 1)$ parallel to the z-axis.
    The parameter curve with $z = 1$ is given by the parametric equations

$$x = \cos t, \quad y = \sin t, \quad z = 1,$$

with parameter t. This is the unit circle parallel to and one unit above the xy-plane centered on the z-axis.

(b)   First, take fixed values $t = t_0$ for t and let z vary. Then we get curves

$$x = \cos t_0, \quad y = \sin t_0, \quad z = z,$$

with parameter z. These are vertical lines on the cylinder parallel to the z-axis.
    The other family is obtained by taking fixed values $z = z_0$ for z and varying t. Curves in this family are parameterized by

$$x = \cos t, \quad y = \sin t, \quad z = z_0,$$

with parameter t. They are circles of radius 1 parallel to the xy-plane centered on the z-axis. See Figure 16.38.

 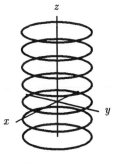

Figure 16.38: The two families of parameter curves for the cylinder $x = \cos t, y = \sin t, z = z$

Example 9   Describe the families of parameter curves with $\theta =$ constant and $\phi =$ constant for the sphere

$$x = \sin \phi \cos \theta$$
$$y = \sin \phi \sin \theta$$
$$z = \cos \phi.$$

Solution   Since ϕ measures latitude, the family with ϕ constant consists of the circles of constant latitude. Similarly, the family with θ constant consists of the meridians (semicircles) running between the north and south poles. See Figure 16.39.

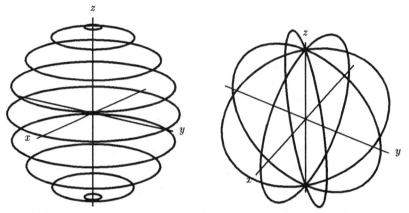

Figure 16.39: The two families of parameter curves for a sphere parameterized by latitude ϕ and longitude θ.

We have seen parameter curves before. The sections with $x = a$ and $y = b$ on a surface $z = f(x, y)$ are examples of parameter curves. So are the grid lines on a computer sketch of a surface. The small regions shaped somewhat like parallelograms surrounded by nearby pairs of parameter curves are called *parameter rectangles*. See Figure 16.40.

Parameter curve: with $q =$ constant

Parameter curve: with $p =$ constant

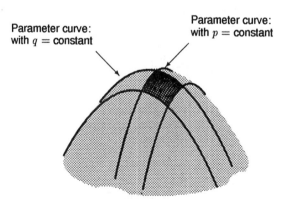

Figure 16.40: Shaded region is parameter rectangle

Problems for Section 16.4

1. Describe in words the curve $\phi = \pi/4$ on the surface of the globe.
2. Describe in words the curve $\theta = \pi/4$ on the surface of the globe.
3. You are at a point on the earth with longitude 80° West of Greenwich, England, and latitude 40° North of the equator.
    (a) If your latitude decreases have you moved nearer to or farther from the equator?
    (b) If your latitude decreases, have you moved nearer to or farther from the north pole?
    (c) If your longitude increases (say, to 90° West), have you moved nearer to or farther from Greenwich?
4. If the sphere is parameterized using the spherical coordinates θ and ϕ, describe in words the part of the sphere given by the following restrictions:

    (a) $\pi \leq \theta < 2\pi, \quad 0 \leq \phi \leq \pi/2$
    (c) $\pi/4 \leq \theta < \pi/3, \quad 0 \leq \phi \leq \pi$
    (b) $0 \leq \theta < 2\pi, \quad 0 \leq \phi \leq \pi$
    (d) $0 \leq \theta \leq \pi, \quad \pi/4 \leq \phi < \pi/3$

5. A city is described parametrically by the equation
$$x\vec{i} + y\vec{j} + z\vec{k} = (x_0\vec{i} + y_0\vec{j} + z_0\vec{k}) + p\vec{v_1} + q\vec{v_2}$$
where $\vec{v}_1 = 2\vec{i} - 3\vec{j} + 2\vec{k}$ and $\vec{v}_2 = \vec{i} + 4\vec{j} + 5\vec{k}$. A city block is a rectangle determined by $\vec{v}_1$ and $\vec{v}_2$. East is in the direction of $\vec{v}_1$ and north is in the direction of $\vec{v}_2$. Starting at the point (x_0, y_0, z_0) you walk 5 blocks east, 4 blocks north, 1 block west and 2 blocks south. What are the parameters of the point where you end up? What are your x, y and z coordinates at that point?

6. Find a parameterization for the plane through $(1, 3, 4)$ and orthogonal to $\vec{n} = 2\vec{i} + \vec{j} - \vec{k}$.
7. Find parametric equations for the sphere centered at the origin and having radius 5.
8. Find parametric equations for the sphere centered at the point $(2, -1, 3)$ and with radius 5.
9. Find parametric equations for the sphere $(x - a)^2 + (y - b)^2 + (z - c)^2 = d^2$.
10. Find parametric equations for the cone $x^2 + y^2 = z^2$.
11. Find a parameterization of a circular cylinder of radius a whose axis is along the z-axis, from $z = 0$ to a height $z = h$. See Figure 16.41.

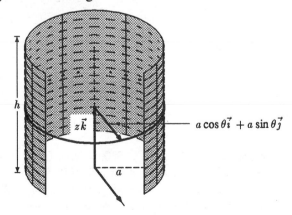

Figure 16.41: Parameterization of a cylinder

12.   Parameterize the cone in Example 3 on page 309 in terms of r and θ instead of z and θ.

13.   Parameterize a cone of height h and maximum radius a with vertex at the origin and opening upward. Do this in two ways :

   (a)   Using r and θ and

   (b)   Using z and θ.

   In each case, give the range of values for each parameter.

14.   Adapt the parameterization for the sphere to find a parameterization for the ellipsoid

$$\frac{x^2}{a^2} + \frac{y^2}{b^2} + \frac{z^2}{c^2} = 1.$$

15.   Parameterize the paraboloid $z = x^2 + y^2$ using cylindrical coordinates.

16.   Parameterize a vase formed by rotating the curve $z = 10\sqrt{x - 1}$, $1 \le x \le 2$, around the z-axis. Sketch the vase.

17.   Suppose you are standing at a point on the equator of a sphere, parameterized by spherical coordinates θ, and ϕ. If you go halfway around the equator and halfway up toward the north pole along a longitude, what will be your new θ and ϕ in terms of your old θ and ϕ?

18.   Find a parameterization for the following surfaces.

   (a)   The spherical cap $x^2 + y^2 + z^2 = 1$, with $1/2 \le z \le 1$

   (b)   The spherical cap $x^2 + y^2 + z^2 = 1$, with $-1 \le x \le -1/3$

   (c)   The intersection of the spherical caps in part (a) and part (b)

19.   Figure 16.42 is a picture of the parametric surface

$$x = a(p + q), \quad y = b(p - q), \quad z = 4cq^2,$$

   for $a = 1$, $b = 1$ and $c = 1$. What happens if you increase a? Increase b? Increase c? How could we flip the surface upside down?

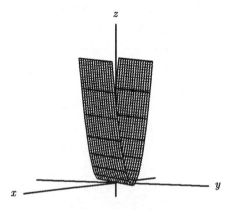

Figure 16.42

20.   You are standing at the point $(-5, 0, 5)$, your friend Jane is standing at the point $(0, -5, -5)$ and her friend Jo is standing at the point $(10, 5, 0)$. Find parametric equations for the plane you are all standing on so that your parameters are $(0, 0)$, Jane's are $(1, 0)$ and Jo's are $(0, 1)$. Find a parameterization so Jane's and Jo's parameters are switched. Find a parameterization so that your parameters and Jane's are switched.

21. There is a famous way to parameterize a sphere called *stereographic projection*. We work with the sphere $x^2 + y^2 + z^2 = 1$. Draw a line from a point (a, b) in the xy-plane to the north pole $(0, 0, 1)$. This line intersects the sphere in a point (x, y, z). This gives a parameterization of the sphere by points in the plane.

    (a) Which point corresponds to the south pole?

    (b) Which points correspond to the equator?

    (c) Do we get all the points of the sphere by this parameterization?

    (d) Which points correspond to the upper hemisphere?

    (e) Which points correspond to the lower hemisphere?

22. Does the plane $\vec{r} = (2 + p)\vec{i} + (3 + p + q)\vec{j} + 4q\vec{k}$ contain the following points?

    (a) $(4, 8, 12)$ (b) $(1, 2, 3)$

23. Find three vectors parallel to the plane

$$x = 3 + 5p - 2q, \quad y = -4 + p + 2q, \quad z = 9p + q.$$

24. Find parametric equations for the plane $3x + 4y + 5z = 10$.

25. (a) Find a vector normal to the plane $\vec{r} = (3 - 5p + 2q)\vec{i} + (1 + p + 3q)\vec{j} + (p - q)\vec{k}$.

    (b) Find an implicit equation for the plane.

26. Are the planes

$$x = 2 + p + q, \quad y = 4 + p - q, \quad z = 1 + 2p,$$

and

$$x = 2 + p + 2q, \quad y = q, \quad z = p - q,$$

parallel?

27. Obtain a parameterization of the surface obtained by rotating the curve $x^2 z = 1$, for $x > 0$, in the xz-plane (a) about the x-axis.   (b) about the z-axis.

    See Figure 16.43

Figure 16.43: $x^2 z = 1$
for $x > 0$

28. Many brass instruments are approximately Bessel horns, which are surfaces of revolution about the x-axis of

$$z = f(x) = \frac{b}{(x + a)^m}$$

for positive constants a, b, and m. Thus $f(x)$ is the radius of the bell at a distance x from the large open end. Usually m is in the range $0.5 \leq m \leq 1$.

    Determine a and b and sketch the graph of f for $0 \leq x \leq 20$ if the radius at $x = 0$ is 15 cm and the radius at $x = 20$ cm is 1 cm for each of the following three cases. Can you see why m is called the flare parameter?   (a) $m = 0.5$ (b) $m = 0.7$ (c) $m = 1$

29. Give a parameterization of the bell of the horn of Problem 28(a).

30. Describe the surface
$$\vec{r} = p^2 \cos\theta \vec{i} + p\vec{j} + p^2 \sin\theta \vec{k} \,.$$

31. (a) Describe the surface given parametrically by the equations
$$x = \cos(p - q) \quad y = \sin(p - q) \quad z = p + q.$$

(b) Describe the two families of parameter curves on the surface.

32. Give a parameterization of the circle of radius a centered at the point (x_0, y_0, z_0) and in the plane parallel to two given unit vectors $\vec{u}$ and $\vec{v}$ such that $\vec{u} \cdot \vec{v} = 0$.

33. A torus (doughnut) is constructed by rotating a small circle of radius a in a large circle of radius b about the origin. The small circle is in a (rotating) vertical plane through the origin and the large circle is in the xy-plane. See Figure 16.44. Parameterize the torus by the following method.

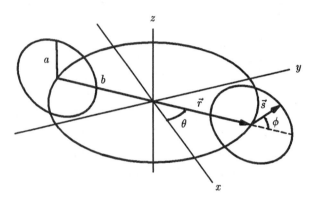

Figure 16.44

(a) Parameterize the large circle.

(b) For a typical point on the large circle, find two unit vectors which are perpendicular to one another and in the plane of the small circle at that point. Use these vectors to parameterize the small circle relative to its center.

(c) Combine your answers to parts (b) and (c) to parameterize the torus.

For Problems 34–37,

(a) Write down an equation in x, y, z for the given parametric surface and identify the surface.

(b) Draw a picture of the surface.

34. $x = 2p \qquad\quad 0 \le p \le 1$
$y = p + q \qquad 0 \le q \le 1$
$z = 1 + p - q$

35. $x = p + q \qquad 0 \le p \le 1$
$y = p - q \qquad 0 \le q \le 1$
$z = p^2 + q^2$

36. $x = 3\sin p \quad 0 \le p \le \pi$
$y = 3\cos p \quad 0 \le q \le 1$
$z = q + 1$

37. $x = p \qquad\qquad p^2 + q^2 \le 1$
$y = q \qquad\qquad p, q \ge 0$
$z = \sqrt{1 - p^2 - q^2}$

REVIEW PROBLEMS FOR CHAPTER SIXTEEN

Write a parameterization for each of the curves in Problems 1–7

1. The horizontal line through the point $(0, 5)$.
2. The circle of radius 2 centered at the origin starting at the point $(0, 2)$ when $t = 0$.
3. The circle of radius 4 centered at the point $(4, 4)$ starting on the x-axis when $t = 0$.
4. The circle of radius 1 in the xy-plane centered at the origin, traversed counterclockwise when viewed from above.
5. The circle of radius 2 in 3-space parallel to the xy-plane, centered at the point $(0, 0, 1)$, and traversed counterclockwise when viewed from below.
6. The circle of radius 3 parallel to the xz-plane, and centered at the point $(0, 5, 0)$, traversed counterclockwise when viewed from $(0, 10, 0)$.
7. The circle of radius 2 centered at $(0, 1, 0)$ lying in the plane $x + z = 0$.
8. Consider the parametric equations below for $0 \leq t \leq \pi$.

   (I) $\vec{r} = \cos(2t)\vec{i} + \sin(2t)\vec{j}$ (II) $\vec{r} = 2\cos t\,\vec{i} + 2\sin t\,\vec{j}$

   (III) $\vec{r} = \cos(t/2)\vec{i} + \sin(t/2)\vec{j}$ (IV) $\vec{r} = 2\cos t\,\vec{i} - 2\sin t\,\vec{j}$

   (a) Match the equations above with four of the curves C_1, C_2, C_3, C_4, C_5 and C_6 in Figure 16.45. (Each curve is part of a circle.)

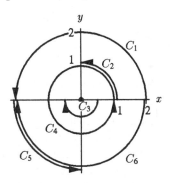

Figure 16.45

   (b) Give parametric equations for the curves which have not been matched, again assuming $0 \leq t \leq \pi$.

Describe in words the objects parameterized by the equations in Problems 9–16.

9. $x = r \cos \theta$ $0 \leq r \leq 5$
   $y = r \sin \theta$ $0 \leq \theta \leq 2\pi$
   $z = 7$

10. $x = 5 \cos \theta$ $0 \leq \theta \leq 2\pi$
    $y = 5 \sin \theta$
    $z = 7$

11. $x = 5 \cos \theta$ $0 \leq \theta \leq 2\pi$
    $y = 5 \sin \theta$ $0 \leq z \leq 7$
    $z = z$

12. $x = 5 \cos \theta$ $0 \leq \theta \leq 2\pi$
    $y = 5 \sin \theta$
    $z = 5\theta$

13. $x = r \cos \theta \quad 0 \leq r \leq 5$
$y = r \cos \theta \quad 0 \leq \theta \leq 2\pi$
$z = r$

14. $x = 2z \cos \theta \quad 0 \leq z \leq 7$
$y = 2z \sin \theta \quad 0 \leq \theta \leq 2\pi$
$z = z$

15. $x = 3 \cos \theta \quad 0 \leq \theta \leq 2\pi$
$y = 2 \sin \theta \quad 0 \leq z \leq 7$
$z = z$

16. $x = x \quad -5 \leq x \leq 5$
$y = x^2 \quad 0 \leq z \leq 7$
$z = z$

17. On a graphing calculator or a computer, plot $x = 2t/(t^2 + 1)$, $y = (t^2 - 1)/(t^2 + 1)$, first for $-50 \leq t \leq 50$ then for $-5 \leq t \leq 5$. Explain what you see. Is the curve really a circle?

18. Let $f(x, y) = \dfrac{x^2 - y^2}{x^2 + y^2}$.

    (a) In which direction should you move from the point $(1, 1)$ to obtain the maximum rate of increase of f?
    (b) Find a direction in which the directional derivative at the point $(1, 1)$ is equal to zero.
    (c) Suppose you move along the curve $x = e^{2t}$, $y = 2t^3 + 6t + 1$. What is df/dt at $t = 0$?

19. Find the parametric equation for the line of intersection of the planes $z = 4 + 2x + 5y$ and $z = 3 + x + 3y$.

20. Find parametric equations of the line passing through the points $(1, 2, 3)$, $(3, 5, 7)$ and calculate the shortest distance from the line to the origin.

21. Are the lines $x = 3 + 2t$, $y = 5 - t$, $z = 7 + 3t$ and $x = 3 + t$, $y = 5 + 2t$, $z = 7 + 2t$ parallel?

22. Are the lines $x = 3 + 2t$, $y = 5 - t$, $z = 7 + 3t$ and $x = 5 + 4t$, $y = 3 - 2t$, $z = 1 + 6t$ parallel?

23. Plot the Lissajous figure given by $x = \cos 2t$, $y = \sin t$ using a graphing calculator or computer. Explain why it looks like part of a parabola. [Hint: Use a double angle identity from trigonometry.]

24. Suppose that a planet P in the xy-plane orbits the star S counterclockwise in a circle of radius 10 units, completing one orbit in 2π units of time. Suppose in addition a moon M orbits the planet P counterclockwise in a circle of radius 3 units, completing one orbit in $2\pi/8$ units of time. The star S is fixed at the origin $x = 0$, $y = 0$, and at time $t = 0$ the planet P is at the point $(10, 0)$ and the moon M is at the point $(13, 0)$.

    (a) Find parametric equations for the x and y coordinates of the planet at time t.
    (b) Find parametric equations for the x and y coordinates of the moon at time t. [Hint: For the moon's position at time t, take a vector from the sun to the planet at time t and add a vector from the planet to the moon].
    (c) Plot the path of the planet using a graphing calculator or computer.
    (d) Experiment with different radii and speeds for the moon's orbit around the planet.

25. A particle travels along a line, with position at time t is given by

$$\vec{r}(t) = (2 + 5t)\vec{i} + (3 + t)\vec{j} + 2t\vec{k}.$$

    (a) Where is the particle when $t = 0$?
    (b) At what time does the particle reach the point $(12, 5, 4)$?
    (c) Does the particle ever reach the point $(12, 4, 4)$? Why or why not?

26. An ant, starting at the origin, moves at 2 units/sec along the x-axis to the point $(1, 0)$. The ant then moves counterclockwise along the unit circle to $(0, 1)$ at a speed of $3\pi/2$ units/sec, then straight down to the origin at a speed of 2 units/sec along the y-axis.

   (a)   Express the ant's coordinates as a function of time, t, in secs.

   (b)   Express the reverse path as a function of time.

27. A basketball player shoots the ball from 6 feet above the ground towards a basket that is 10 feet above the ground and 15 feet away horizontally.

   (a)   Suppose she shoots the ball at an angle of A degrees above the horizontal $(0 < A < \pi/2)$ with an initial speed V. Give the x and y-coordinates of the position of the basketball at time t. Assume the x-coordinate of the basket is 0 and that the x-coordinate of the shooter is -15. [Hint: There is an acceleration of -32 ft/sec^2 in the y-direction; there is no acceleration in the x-direction. Ignore air resistance.]

   (b)   Using the parametric equations you obtained above, experiment with different values for V and A, plotting the path of the ball on a graphing calculator or computer to see how close the ball comes to the basket. (The tick marks on the y-axis can be used to locate the basket.)

   (c)   Find the angle A that minimizes the velocity needed for the ball to reach the basket. (This is a lengthy computation. First find an equation in V and A that holds if the path of the ball passes through the point 15 feet from the shooter and 10 feet above the ground. Then minimize V.)

28. A cheerleader has a 0.4 m long baton with a light on one end. She throws the baton in such a way that its center moves along a parabola, and the baton rotates counterclockwise around the center with a constant angular velocity. The baton is thrown from a position where its center is 1.5 m above the ground; its initial velocity is 8 m/sec horizontally and 10 m/sec vertically, and its angular velocity is 2 revolutions per second. Also, the baton is horizontal when released. Find parametric equations describing the following motions:

   (a)   The center of the baton relative to the ground.

   (b)   The end of the baton relative to its center.

   (c)   The path traced out by the end of the baton relative to the ground.

   (d)   Sketch a graph of motion of the end of the baton.

29. Explain the significance of the constants $\alpha > 0$ and $\beta > 0$ in the family of helixes given by $\vec{r} = \alpha \cos t\vec{i} + \alpha \sin t\vec{j} + \beta t\vec{k}$.

30. Find parametric equations for motion along the line $y = 3x + 7$ such that the x-coordinate decreases by 2 units for each unit of time.

31. Find parametric equations to describe the loop shown in Figure 16.46. There are many possible answers.

Figure 16.46: A loop in the plane

32. Imagine a wheel of radius 1 meter resting on the x-axis with its center on the y-axis. There is a spot marked on the rim at the point $(1, 1)$, as in Figure 16.47. At time $t = 0$ the wheel starts rolling on the x-axis at a rotational rate of 1 radian per second.

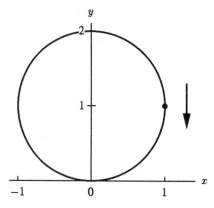

Figure 16.47

(a) Find parametric equations describing the motion of the center of the wheel.
(b) Find parametric equations describing the motion of the spot on the rim.

33. (a) The ellipse $2x^2 + 3y^2 = 8$ intersects the line of slope t through the point $P = (-2, 0)$ in two points, one of which is P. Compute the coordinate of the other point Q. See Figure 16.48.

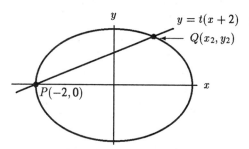

Figure 16.48: The ellipse $2x^2 + 3y^2 = 8$ and an intersecting line

(b) Give a parameterization of the ellipse $2x^2 + 3y^2 = 8$ by rational functions.

CHAPTER SEVENTEEN

VECTOR FIELDS

Some physical quantities (such as temperature) are best represented by scalars; others (such as velocity) are best represented by vectors. We have looked at functions of many variables whose values are scalars, for example, temperature as a function of position on a weather map. Such functions are called scalar-valued functions.

Some weather maps indicate wind velocity at various points by arrows. Wind velocity is an example of a vector-valued function, since its value at any point is the vector indicating the direction and strength of the wind. Such functions are also called *vector fields*. We have already seen one important example of a vector field, namely the gradient of a scalar-valued function. In this chapter will look at other examples, such as velocity vector fields describing a fluid flow. We will also look at the path followed by a particle moving with the flow, which is called a *flow line* of the vector field.

17.1 VECTOR FIELDS

Introduction to Vector Fields

A *vector field* is a function that assigns a vector to each point in the plane or in 3-space. One example of a vector field is the gradient of a function $f(x, y)$; at each point (x, y) the vector grad $f(x, y)$ points in the direction of maximum rate of increase of f. Gradients are not the only kinds of vector fields; in this section we will look at examples of vector fields representing velocities and forces.

Velocity Vector Fields

Figure 17.1 shows the flow of a part of the Gulf stream, a current in the Atlantic Ocean.[1] It is an example of a *velocity vector field*: each vector shows the velocity of the current at that point. The current is fastest where the velocity vectors are longest in the middle of the stream. Beside the stream are eddies where the water flows round and round in circles.

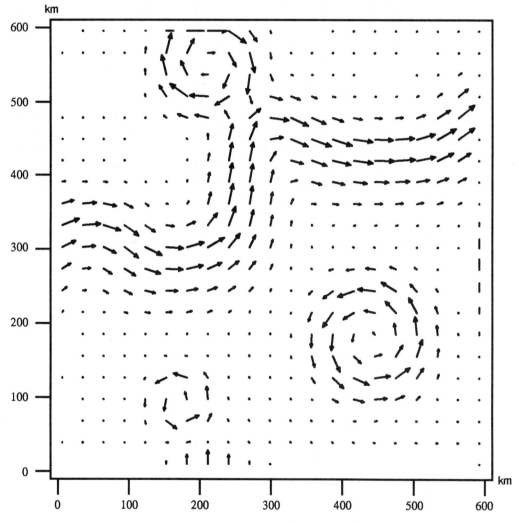

Figure 17.1: The velocity field of the Gulf stream

---

[1]Based on data supplied by Avijit Gangopadhyay of the Jet Propulsion laboratory

Force Fields

Another physical quantity represented by a vector is force. When we experience a force, sometimes it results from direct contact with the object that supplies the force (for example, a push). Many forces, however, can be felt at all points in space. For example, the earth exerts a gravitational pull on all other masses: the direction of this attractive force is toward the center of the earth and the magnitude decreases with increasing distance from the earth. Such forces can be modeled by vector fields.

Figure 17.2 shows the gravitational force exerted by the earth on a mass of one kilogram at different points in space. This is a sketch of the vector field in three-space. You can see that the vectors all point towards the earth (which is left out of the diagram for clarity) and that the vectors further from the earth are smaller in magnitude. Since the perspective in the drawing distorts the lengths and directions of the arrows, most of our sketches of vector fields will be in two-space.

Figure 17.2: The gravitational field of
the earth

Definition of a Vector Field

Now that you have seen some examples of vector fields we give a more formal definition.

> A **vector field** in 2-space is a function $\vec{F}(x, y)$ whose value at a point (x, y) is a 2-dimensional vector. Similarly, a vector field in 3-space is a function $\vec{F}(x, y, z)$ whose values are 3-dimensional vectors.

Notice that the function, $\vec{F}$, has an arrow over it to indicate that its value is a vector, not a scalar. Because it is easier to write, we shall often represent the point (x, y) or (x, y, z) by its position vector $\vec{r}$, and write the vector field as $\vec{F}(\vec{r})$.

Vector Fields Given By Formulas

Since a vector field is a function that assigns a vector to each point, we can often represent a vector field using a formula.

Visualizing a Vector Field Given by a Formula

Example 1   Draw a picture of the vector field in 2-space given by $\vec{F}(x, y) = -y\vec{i} + x\vec{j}$.

Solution   Table 17.1 shows the value of the vector field at a few points. Notice that each value is a vector. To plot the vector field, we plot the value $\vec{F}(x, y)$ with its tail at (x, y). See Figure 17.3.

TABLE 17.1: *A few values of*
$$\vec{F}(x, y) = -y\vec{i} + x\vec{j}$$

		y		
		-1	0	1
x	-1	$\vec{i} - \vec{j}$	$-\vec{j}$	$-\vec{i} - \vec{j}$
	0	$\vec{i}$	$\vec{0}$	$-\vec{i}$
	1	$\vec{i} + \vec{j}$	$\vec{j}$	$-\vec{i} + \vec{j}$

Now we look at the formula to get a better sketch. The magnitude of the vector at (x, y) is $\|\vec{F}(x, y)\| = \| -y\vec{i} + x\vec{j}\| = \sqrt{x^2 + y^2}$, which is the distance from (x, y) to the origin. Therefore, all the vectors at a fixed distance from the origin (that is, on a circle centered at the origin) have the same magnitude. The magnitude gets larger as we move further from the origin. What about the direction? Figure 17.3 suggests that at each point (x, y) the vector $\vec{F}(x, y)$ is perpendicular to the position vector $\vec{r} = x\vec{i} + y\vec{j}$. We verify this using the dot product: $\vec{r} \cdot \vec{F}(x, y) = (x\vec{i} + y\vec{j}) \cdot (-y\vec{i} + x\vec{j}) = 0$. This means that the vectors in this field are tangent to circles centered at the origin and get longer as we go out. Figure 17.4 is a plot of the vector field which has been scaled so that the vectors don't obscure each other.

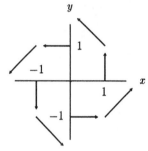

Figure 17.3: The value
$\vec{F}(x, y)$ is placed at the point
(x, y).

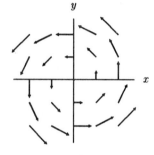

Figure 17.4: The vector field
$\vec{F}(x, y) = -y\vec{i} + x\vec{j}$, vectors
scaled smaller to fit in diagram.

Example 2   Draw pictures of the vector fields in 2-space given by (a) $\vec{F}(x, y) = x\vec{j}$ (b) $\vec{G}(x, y) = x\vec{i}$.

Solution   (a)   The vector $x\vec{j}$ is parallel to the y-direction, pointing up when x is positive and down when x is negative. Also, the larger $|x|$ is, the longer the vector. The vectors in the field are constant along vertical lines since the vector field does not depend on y. See Figure 17.5.

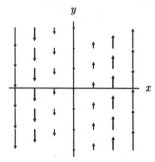

Figure 17.5: The vector field
$\vec{F}(x, y) = x\vec{j}$

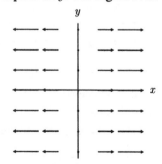

Figure 17.6: The vector field
$\vec{F}(x, y) = x\vec{i}$

(b)   This is similar to the previous example, except that the vector $x\vec{i}$ is parallel to the x-direction, pointing to the right when x is positive and to the left when x is negative. Again, the larger $|x|$ is the longer the vector, and the vectors are constant along vertical lines, since the vector field does not depend on y. See Figure 17.6.

Example 3   Describe the vector field in 3-space given by $\vec{F}(\vec{r}) = \vec{r}$, where $\vec{r} = x\vec{i} + y\vec{j} + z\vec{k}$.

Solution   The notation $\vec{F}(\vec{r}) = \vec{r}$ means that the value of $\vec{F}$ at the point (x, y, z) with position vector $\vec{r}$ is $\vec{r}$ itself, but positioned at (x, y, z). Thus the vector field points outward. See Figure 17.7. Note that the lengths of the vectors have been scaled down so as to fit into the diagram. This vector field can also be written without using the position vector $\vec{r}$ as $\vec{F}(x, y, z) = x\vec{i} + y\vec{j} + z\vec{k}$. You can see that the position vector notation is more concise.

Figure 17.7: The vector field
$\vec{F}(\vec{r}) = \vec{r}$

Finding a Formula for a Vector Field

Example 4   Newton's Law of Gravitation states that the magnitude of the gravitational force exerted by an object of mass M on an object of mass m is proportional to m and M and inversely proportional to the square of the distance between them. The direction of the force is from m to M along the line connecting them. (See Figure 17.8.) Find a formula for the vector field $\vec{F}(\vec{r})$ that represents the gravitational force, assuming M is located at the origin and m is located at the point with position vector $\vec{r}$.

Figure 17.8: Force
exerted on mass m by
mass M

Solution   If the mass m is located at $\vec{r}$, then Newton's law says that the magnitude of the force is

$$\frac{GMm}{\|\vec{r}\|^2},$$

where G is a constant of proportionality. A unit vector in the direction of the force is $-\vec{r}/\|\vec{r}\|$, where the negative sign indicates that the direction of force is towards the origin (gravity is attractive).

By taking the product of the magnitude of the forces and a unit vector in the direction of the force we obtain an expression for the force vector field:

$$\vec{F}(\vec{r}) = \frac{GMm}{\|\vec{r}\|^2}\left(-\frac{\vec{r}}{\|\vec{r}\|}\right) = \frac{-GMm\vec{r}}{\|\vec{r}\|^3}.$$

We have already seen a picture of this vector field in Figure 17.2.

Gradient Vector Fields

The gradient of a scalar function f is a function that assigns a vector to each point, and is therefore a vector field. It is called the *gradient field* of f. Many vector fields in physics are gradient fields.

Example 5   Sketch the gradient field of the functions in Figures 17.9–17.11.

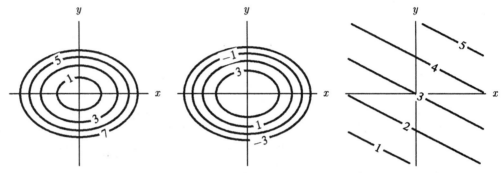

Figure 17.9: The contour map of $f(x, y) = x^2 + 2y^2$ **Figure 17.10**: The contour map of $g(x, y) = 5 - x^2 - 2y^2$ **Figure 17.11**: The contour map of $h(x, y) = x + 2y + 3$

Solution   See Figures 17.12–17.14. For a function $f(x, y)$, the gradient vector of f at a point is perpendicular to the contours in the direction of increasing f, and its magnitude is the rate of change in that direction. The rate of change is large when the contours are close together and small when they are far apart. Notice that in Figure 17.12 the vectors all point outward, away from the local minimum of f, and in Figure 17.13 the vectors of grad g all point inward, toward the local maximum of g. Since h is a linear function, its gradient is constant, so grad h in Figure 17.14 is a constant vector field.

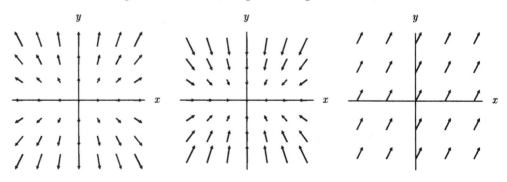

Figure 17.12: grad f **Figure 17.13**: grad g **Figure 17.14**: grad h

Problems for Section 17.1

1. Each vector field in Figures (I)–(IV) represents the force on a particle at different points in space as a result of another particle at the origin. Match up the vector fields with the descriptions below.

   (a) A repulsive force whose magnitude decreases as distance increases, such as between electric charges of the same sign.
   (b) A repulsive force whose magnitude increases as distance increases.
   (c) An attractive force whose magnitude decreases as distance increases, such as gravity.
   (d) An attractive force whose magnitude increases as distance increases.

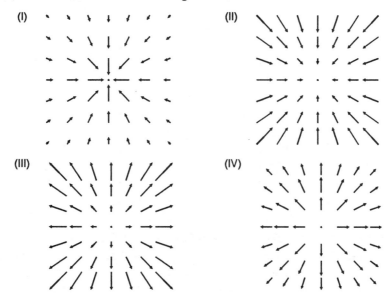

Sketch graphs of the vector fields in Problems 2–7.

2. $\vec{F}(x, y) = 2\vec{i} + 3\vec{j}$ 3. $\vec{F}(x, y) = y\vec{i}$ 4. $\vec{F}(x, y) = 2x\vec{i} + x\vec{j}$

5. $\vec{F}(\vec{r}) = 2\vec{r}$ 6. $\vec{F}(\vec{r}) = \dfrac{\vec{r}}{\|\vec{r}\|}$ 7. $\vec{F}(x, y) = (x + y)\vec{i}$
$\qquad\qquad\qquad\qquad\qquad\qquad\qquad\qquad\qquad\quad + (x - y)\vec{j}$

For Problems 8–13, find formulas for the vector fields. (There are many possible answers.)

8.

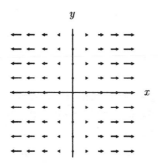

9.

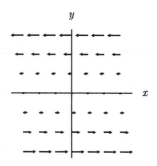

10.   11.

12.   13.

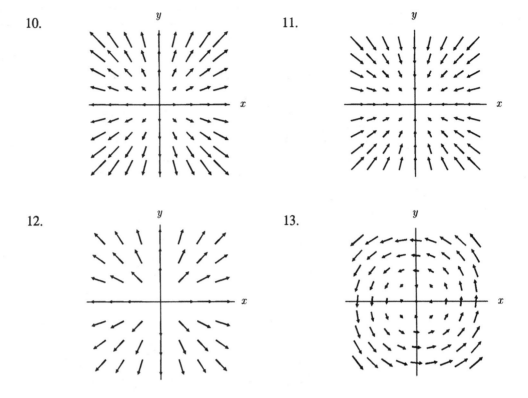

14.   Figures 17.15 and 17.16 show the gradient of the functions $z = f(x, y)$ and $z = g(x, y)$.
   (a)   For each function, draw a rough sketch of the level curves, showing possible z values.
   (b)   The xz-plane cuts each of the surfaces $z = f(x, y)$ and $z = g(x, y)$ in a curve. Sketch each of these curves, making clear how they are similar and how they are different from one another.

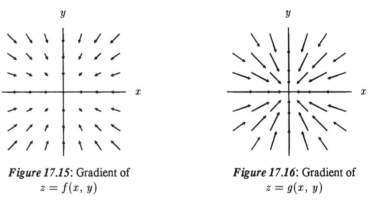

Figure 17.15: Gradient of
$z = f(x, y)$

Figure 17.16: Gradient of
$z = g(x, y)$

In Problems 15–17, use a computer to print out vector fields with the given properties. Show on your printout the formula used to generate it. (There are many possible answers.)

15.   All vectors are parallel to the x-axis, and all those on a vertical line have the same magnitude.

16.   All vectors point towards the origin and have constant length.

17.   All vectors are of unit length and perpendicular to the position vector at that point.

18. Imagine a wide, steadily flowing river in the middle of which there is a fountain that spouts water horizontally in all directions.

  (a) Suppose that the river flows in the $\vec{i}$-direction in the xy-plane and that the fountain is at the origin. Explain why the expression

$$\vec{v} = A\vec{i} + K(x^2 + y^2)^{-1/2}(x\vec{i} + y\vec{j}), \quad A > 0, K > 0$$

   could represent the velocity field for the combined flow of the river and the fountain.

  (b) What is the significance of the constants A and K?
  (c) Using a computer, sketch the vector field $\vec{v}$ for $K = 1$ and $A = 1$ and $A = 2$.

17.2 THE FLOW OF A VECTOR FIELD

When a dangerous iceberg is spotted in the North Atlantic, it is important to be able to predict where the iceberg is likely to be a day or a week later. To do this, one needs to know the velocity field of the ocean currents; how fast and in what direction the water is moving at each point.

In this section we will use differential equations to find the path of an object in a fluid flow. We call this path a *flow line*. Notice that it is important to know not only the shape of the path, but also the speed and direction of the object's motion along it. Figure 17.17 shows several flow lines for the Gulf stream velocity vector field in Figure 17.1 on page 324. The arrows on each flow line indicate the direction of flow along it.

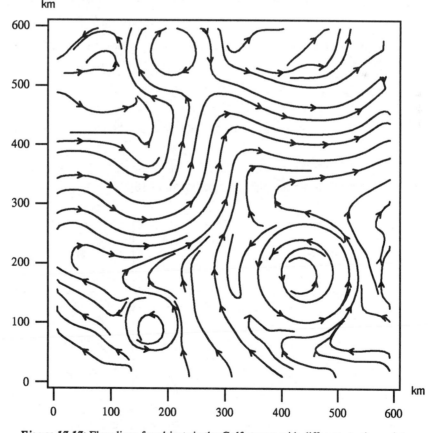

Figure 17.17: Flow lines for objects in the Gulf stream with different starting points

How Do We Find a Flow Line?

Suppose that $\vec{F}$ is the velocity field of water on the surface of a creek, and imagine a seed being carried along by the current. We want to know the position vector $\vec{r}(t)$ of the seed at time t. We know

$$
\begin{array}{ccc}
\text{Velocity of seed} & & \text{Velocity of current} \\
\text{at time } t & = & \text{at seed's position} \\
 & & \text{at time } t
\end{array}
$$

$$\vec{r}'(t) = \vec{F}(\vec{r}(t)).$$

Resolving into components $\vec{F} = F_1\vec{i} + F_2\vec{j}$ and $\vec{r}(t) = x(t)\vec{i} + y(t)\vec{j}$, this says

$$x'(t) = F_1 \quad \text{and} \quad y'(t) = F_2.$$

A solution to these differential equations will give a parameterization of the flow line.

Example 1   Find the flow line of the constant velocity field $\vec{v} = 3\vec{i} + 4\vec{j}$ ft/sec that passes through the point $(1, 2)$ at time $t = 0$.

Solution   Let $\vec{r}(t) = x(t)\vec{i} + y(t)\vec{j}$ be the position of a particle at time t, where t is measured in seconds. We must have $x'(t)\vec{i} + y'(t)\vec{j} = 3\vec{i} + 4\vec{j}$, so $x'(t) = 3$ or $x(t) = 3t + x_0$. Similarly, $y(t) = 4t + y_0$, where x_0 and y_0 are constants of integration. Therefore, the flow line is given by

$$\vec{r}(t) = (3t + x_0)\vec{i} + (4t + y_0)\vec{j}.$$

Since the path passes the point $(1, 2)$ at $t = 0$, we have $x_0 = 1$ and $y_0 = 2$ and so

$$\vec{r}(t) = (3t + 1)\vec{i} + (4t + 2)\vec{j}.$$

Equivalently,

$$x(t) = 3t + 1 \quad \text{and} \quad y(t) = 4t + 2.$$

To find an explicit equation for the path, eliminate t between these expressions to get

$$\frac{x - 1}{3} = \frac{y - 2}{4} \quad \text{or} \quad y = \frac{4}{3}x + \frac{2}{3}.$$

Thus the path is a straight line. See Figure 17.18.

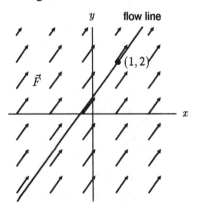

Figure 17.18: Vector field $\vec{F} = 3\vec{i} + 4\vec{j}$
with the flow line through $(1, 2)$

Example 2   The velocity of a flow at (x, y) is $\vec{F}(x, y) = \vec{i} + x\vec{j}$. Find the path of motion of an object in the flow that is at the point $(-2, 2)$ at time $t = 0$.

Solution   Figure 17.19 shows a sketch of this field. We are looking for the flow line $\vec{r}(t) = x(t)\vec{i} + y(t)\vec{j}$ of the vector field such that $\vec{r}(0) = -2\vec{i} + 2\vec{j}$. Since $\vec{r}'(t) = \vec{F}(\vec{r}(t))$, the flow line satisfies the system of differential equations

$$x'(t) = 1, \quad y'(t) = x(t).$$

Solving for $x(t)$ first, we get $x(t) = t + x_0$, where x_0 is a constant of integration. Thus, $y'(t) = t + x_0$, so $y(t) = (1/2)t^2 + x_0 t + y_0$, where y_0 is also a constant of integration. Since $x(0) = -2$ and $y(0) = 2$, we must have $x_0 = -2$ and $y_0 = 2$. The path of motion is given by

$$x(t) = t - 2, \quad y(t) = (1/2)t^2 - 2t + 2,$$

or equivalently

$$\vec{r}(t) = (t - 2)\vec{i} + ((1/2)t^2 - 2t + 2)\vec{j}.$$

The graph of this flow line in Figure 17.20 looks like a parabola. We can check this by finding an explicit equation for the path, $y = (1/2)x^2$.

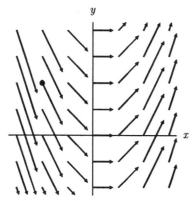

Figure 17.19: The velocity field $\vec{v} = \vec{i} + x\vec{j}$

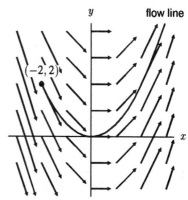

Figure 17.20: A flow line of the velocity field

Definition of Flow Lines for Arbitrary Vector Fields

> A **flow line** of a vector field $\vec{v} = \vec{F}(\vec{r})$ is a path $\vec{r}(t)$ whose velocity vector equals $\vec{v}$. Thus
>
> $$\vec{r}'(t) = \vec{v} = \vec{F}(\vec{r}(t)).$$
>
> The **flow** of a vector field is the family of all of its flow lines.

A flow line is also called an *integral curve* or a *streamline*.   The definition of flow line allows us to treat any vector field as if it were a velocity field and to think of how a particle would travel if it moved with velocity given by the field. It turns out to be profitable to study the flow of fields (for example, electric and magnetic) that are not velocity fields.

Example 3   Determine the flow of the vector field $\vec{v} = -y\vec{i} + x\vec{j}$.

Solution   Figure 17.21 suggests that the flow consists of concentric counterclockwise circles, centered at the origin. The system of differential equations for the flow is

$$x'(t) = -y(t)$$
$$y'(t) = x(t).$$

It is not hard to verify that $(x(t), y(t)) = (a \cos t, \, a \sin t)$ is a parameterized family that satisfies the system. These curves are the counterclockwise circles of radius a, centered at the origin.

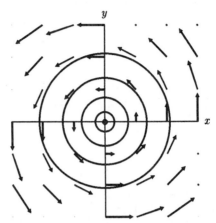

Figure 17.21: The flow of a vector field
$\vec{v} = -y\vec{i} + x\vec{j}$

Finding Flow Lines Numerically

Often it is not possible to find formulas for the flow lines of a vector field. However, we can approximate them numerically by Euler's method for solving differential equations. Since the flow lines $\vec{r}(t) = x(t)\vec{i} + y(t)\vec{j}$ of a vector field $\vec{v} = \vec{F}(x, y)$ satisfy the differential equation $\vec{r}'(t) = \vec{F}(\vec{r}(t))$, we have

$$\vec{r}(t + \Delta t) \approx \vec{r}(t) + (\Delta t)\vec{r}'(t)$$
$$= \vec{r}(t) + (\Delta t)\vec{F}(\vec{r}(t)) \quad \text{for } \Delta t \text{ near } 0.$$

To approximate a flow line, we start at a point $\vec{r}_0 = \vec{r}(0)$ and estimate the position $\vec{r}_1$ of a particle at time Δt later:

$$\vec{r}_1 = \vec{r}(\Delta t) \approx \vec{r}(0) + (\Delta t)\vec{F}(\vec{r}(0))$$
$$= \vec{r}_0 + (\Delta t)\vec{F}(\vec{r}_0).$$

We then repeat the same procedure starting at $\vec{r}_1$, and so on. The general formula for getting from one point to the next is

$$\vec{r}_{n+1} = \vec{r}_n + (\Delta t)\vec{F}(\vec{r}_n).$$

The points with position vectors $\vec{r}_0, \vec{r}_1, \ldots$ trace out the path. The next example shows a numerical approximation to a flow line.

Example 4   Use Euler's method to determine the flow line through $(1, 2)$ for the vector field $\vec{v} = y^2\vec{i} + 2x^2\vec{j}$.

Solution   The flow is determined by the differential equations $\vec{r}'(t) = \vec{v}$, or equivalently

$$x'(t) = y^2, \qquad y'(t) = 2x^2.$$

We use Euler's method with $\Delta t = 0.02$, giving

$$\vec{r}_{n+1} = \vec{r}_n + 0.02\, \vec{v}\,(x_n, y_n)$$
$$= x_n\vec{i} + y_n\vec{j} + 0.02(y_n^2\vec{i} + 2x_n^2\vec{j}),$$

or equivalently,

$$x_{n+1} = x_n + 0.02y_n{}^2,$$
$$y_{n+1} = y_n + 0.02 \cdot 2x_n{}^2.$$

When $t = 0$, $(x_0, y_0) = (1, 2)$. Then

$$x_1 = x_0 + 0.02 \cdot y_0{}^2 = 1 + 0.02 \cdot 2^2 = 1.08,$$
$$y_1 = y_0 + 0.02 \cdot 2x_0^2 = 2 + 0.02 \cdot 2 \cdot 1^2 = 2.04.$$

So after one step $x(0.02) \approx 1.08$ and $y(0.02) \approx 2.04$. Similarly, $x(0.04) = x(2\Delta t) \approx 1.16$, $y(0.04) = y(2\Delta t) \approx 2.09$ and so on. Further values along the flow line are given in Table 17.2 and plotted in Figure 17.22.

TABLE 17.2: *The computed path for the vector field $\vec{v} = y^2\vec{i} + 2x^2\vec{j}$ starting at the point $(1, 2)$*

t	0	.02	.04	.06	.08	.1	.12	.14	.16	.18
x	1	1.08	1.16	1.25	1.34	1.44	1.54	1.65	1.77	1.90
y	2	2.04	2.09	2.14	2.20	2.28	2.36	2.45	2.56	2.69

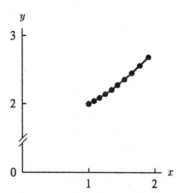

Figure 17.22: Euler's method
solution to $x' = y^2$, $y' = 2x^2$

Flows Arising From Differential Equations

We have seen how to find the flow of a vector field by treating it as a velocity vector field and solving the resulting system of differential equations. This process is reversible: any system of differential equations of the form

$$x'(t) = F_1(x(t), y(t)) \quad \text{and} \quad y'(t) = F_2(x(t), y(t))$$

can be thought of as equations for the flow of the velocity field $\vec{v} = F_1\vec{i} + F_2\vec{j}$.

Example 5 The Lanchester model for a battle between two armies is

$$\frac{dx}{dt} = -ay \qquad \frac{dy}{dt} = -bx,$$

where x and y represent the number of soldiers (in thousands) in each army at time t (in weeks), and a and b are positive constants.

(a) Find a vector field whose flow is the set of solutions of this system.

(b) Assume $a = b = 1$. Show that

$$x = e^t + 9e^{-t},$$
$$y = -e^t + 9e^{-t}$$

is a solution of the system of differential equations (equivalently, is a flow line of the vector field) satisfying the initial condition $x(0) = 10$, $y(0) = 8$. Sketch the solution curve. When is the battle over and what is the final result?

Solution (a) The system comes from the vector field $\vec{F} = -ay\vec{i} - bx\vec{j}$.

(b) To show that the given expressions for x and y are solutions to the differential equation, we check that

$$\frac{dx}{dt} = \frac{d}{dt}(e^t + 9e^{-t}) = e^t - 9e^{-t} = -y,$$

$$\frac{dy}{dt} = \frac{d}{dt}(-e^t + 9e^{-t}) = -e^t - 9e^{-t} = -x.$$

To check that this solution satisfies the initial conditions, we observe that

$$x(0) = 1 + 9 = 10, \qquad y(0) = -1 + 9 = 8.$$

To sketch the curve, we use a graphing calculator or computer. Alternatively, we can observe that $x + y = 18e^{-t}$ and $x - y = 2e^t$ so $(x + y)(x - y) = (18e^{-t})(2e^t) = 36$. Thus, the underlying curve is the hyperbola $x^2 - y^2 = 36$. Since the battle ends when one army has no soldiers left, we look for the point where one coordinate is 0. The curve intersects the x-axis at $x = 6$, $y = 0$, so the final result is a victory for the x-army, with six thousand of its soldiers surviving. The battle ends at the time when $y = 0$, that is when $-e^t + 9e^{-t} = 0$. Solving for t gives $e^t = 9e^{-t}$ so $e^{2t} = 9$ and $t = \frac{1}{2}\ln 9 = 1.09$ weeks.

If we were not given the solution, we could still have obtained the underlying curve (but not how it is traced out with time) using the chain rule

$$\frac{dy}{dx} = \frac{dy/dt}{dx/dt} = \frac{x}{y}$$

and separating variables

$$\int y \, dy = \int x \, dx$$
$$y^2/2 = x^2/2 + C.$$

The initial conditions $x = 10$, $y = 8$ give $64/2 = 100/2 + C$ so $C = -36/2$. Therefore, $y^2/2 = x^2/2 - 36/2$ so $x^2 - y^2 = 36$, as before.

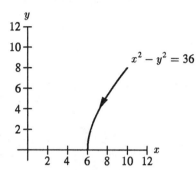

Figure 17.23: A solution to the Lanchester model for a battle.

Problems for Section 17.2

Sketch the vector field and flow for the vector fields in Problems 1 – 3.

1. $\vec{v} = 3\vec{i}$ 2. $\vec{v} = 2\vec{j}$ 3. $\vec{v} = 3\vec{i} - 2\vec{j}$

For each of the vector fields in Problems 4 –7, sketch the field and flow. Then find the system of differential equations associated with the field and verify that the flow satisfies the system.

4. $\vec{v} = y\vec{i} + x\vec{j}$; $x(t) = a(e^t + e^{-t})$, $y(t) = a(e^t - e^{-t})$.

5. $\vec{v} = y\vec{i} - x\vec{j}$; $x(t) = a\sin t$, $y(t) = a\cos t$.

6. $\vec{v} = x\vec{i} + y\vec{j}$; $x(t) = ae^t$, $y(t) = be^t$.

7. $\vec{v} = x\vec{i} - y\vec{j}$; $x(t) = ae^t$, $y(t) = be^{-t}$.

8. Use a computer or calculator with Euler's method to approximate the flow line through $(1, 2)$ for the vector field $\vec{v} = y^2\vec{i} + 2x^2\vec{j}$ using 5 steps with a time interval $\Delta t = 0.1$.

9. Consider a model for two interacting populations of robins and worms. Let $r(t)$ and $w(t)$ represent the robin population and the worm population as functions of time. Their interaction is governed by the system

$$w'(t) = aw - cwr, \quad r'(t) = -br + kwr$$

(where a, b, c, and k are positive constants). Find a vector field whose flow is the set of solutions to this system.

10. (a) Check that $x = e^{-t}\cos 7t$ is a solution of the second-order differential equation for a damped harmonic oscillator (that is, a spring with friction):

$$\frac{d^2 x}{dt^2} + 2\frac{dx}{dt} + 50x = 0.$$

(b) With a graphing calculator or a computer, sketch the phase portrait of the solution $x = e^{-t}\cos t$ for the differential equation given in part (a). That is, in the xv-plane sketch the curve given by the parametric equations:

$$x = e^{-t}\cos 7t$$
$$v = \frac{dx}{dt} = -e^{-t}\cos 7t + e^{-t}(-7\sin 7t).$$

(c) Explain in words what the phase portrait shows about the position x and velocity v of a mass attached to a spring with friction satisfying the differential equation in part (a).

REVIEW PROBLEMS FOR CHAPTER SEVENTEEN

1. (a) What is meant by a vector field?
   (b) Which of the following are vector fields? Why?

   (i) $\vec{r} + \vec{a}$ (ii) $\vec{r} \cdot \vec{a}$
   (iii) $x^2\vec{i} + y^2\vec{j} + z^2\vec{k}$ (iv) $x^2 + y^2 + z^2$

   where

   $$\vec{r} = x\vec{i} + y\vec{j} + z\vec{k}$$
   $$\vec{a} = a_1\vec{i} + a_2\vec{j} + a_3\vec{k}$$

   a_1, a_2, a_3 all constant.

Sketch the vector fields in Problems 2–4

2. $\vec{F} = \left(\dfrac{y}{\sqrt{x^2 + y^2}}\right)\vec{i} - \left(\dfrac{x}{\sqrt{x^2 + y^2}}\right)\vec{j}$ 3. $\vec{F} = \left(\dfrac{y}{x^2 + y^2}\right)\vec{i} - \left(\dfrac{x}{x^2 + y^2}\right)\vec{j}$

4. $\vec{F} = y\vec{i} - x\vec{j}$

5. The following problem concerns the vector field $\vec{F} = \vec{r}/\|\vec{r}\|^3$, where $\vec{r} = x\vec{i} + y\vec{j} + z\vec{k}$. In each case find the given quantity in terms of x, y, z, or t.

   (a) $\|\vec{F}\|$
   (b) $\vec{F} \cdot \vec{r}$
   (c) A unit vector parallel to $\vec{F}$ and pointing in the same direction.
   (d) A unit vector parallel to $\vec{F}$ and pointing in the opposite direction.
   (e) $\vec{F}$ if $\vec{r} = \cos t\vec{i} + \sin t\vec{j} + \vec{k}$
   (f) $\vec{F} \cdot \vec{r}$ if $\vec{r} = \cos t\vec{i} + \sin t\vec{j} + \vec{k}$

For Problems 6-9, find the region of the Gulf stream velocity field in figure 17.24 represented by the given table of velocity vectors (in cm/sec).

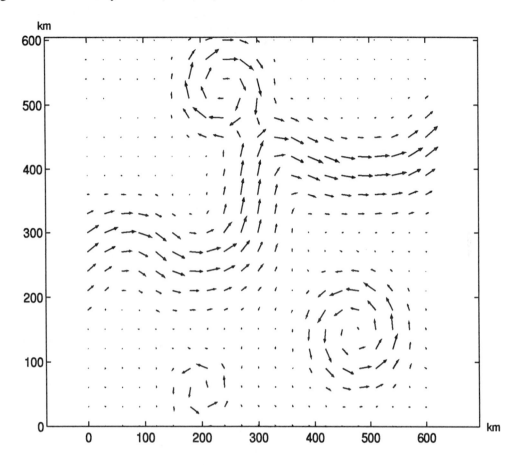

Figure 17.24: The velocity field of the Gulf stream

6.

$35\vec{i} + 131\vec{j}$	$48\vec{i} + 92\vec{j}$	$47\vec{i} + \vec{j}$
$-32\vec{i} + 132\vec{j}$	$-44\vec{i} + 92\vec{j}$	$-42\vec{i} + \vec{j}$
$-51\vec{i} + 73\vec{j}$	$-119\vec{i} + 84\vec{j}$	$-128\vec{i} + 6\vec{j}$

7.

$10\vec{i} - 3\vec{j}$	$11\vec{i} + 16\vec{j}$	$20\vec{i} + 75\vec{j}$
$53\vec{i} - 7\vec{j}$	$58\vec{i} + 23\vec{j}$	$64\vec{i} + 80\vec{j}$
$119\vec{i} - 8\vec{j}$	$121\vec{i} + 31\vec{j}$	$114\vec{i} + 66\vec{j}$

8.

$97\vec{i} - 41\vec{j}$	$72\vec{i} - 24\vec{j}$	$54\vec{i} - 10\vec{j}$
$134\vec{i} - 49\vec{j}$	$131\vec{i} - 44\vec{j}$	$129\vec{i} - 18\vec{j}$
$103\vec{i} - 36\vec{j}$	$122\vec{i} - 30\vec{j}$	$131\vec{i} - 17\vec{j}$

9.

$-95\vec{i} - 60\vec{j}$	$18\vec{i} - 48\vec{j}$	$82\vec{i} - 22\vec{j}$
$-29\vec{i} + 48\vec{j}$	$76\vec{i} + 63\vec{j}$	$128\vec{i} - 16\vec{j}$
$26\vec{i} + 105\vec{j}$	$49\vec{i} + 119\vec{j}$	$88\vec{i} + 13\vec{j}$

10. Each of the following vector fields represents an ocean current. Sketch the vector field, and sketch the path of an iceberg in this current. Determine the location of an iceberg at time $t = 7$ if it is at the point $(1, 3)$ at time $t = 0$.

    (a) The current everywhere is $\vec{i}$.

    (b) The current at (x, y) is $2x\vec{i} + y\vec{j}$.

    (c) The current at (x, y) is $-y\vec{i} + x\vec{j}$.

11. Suppose $q_1, \ldots, q_n$ are electric charges at points with position vectors $\vec{r}_1, \ldots, \vec{r}_n$ in 3-space. Coulomb's Law states that, at the point with position vector $\vec{r}$, the resulting electric field $\vec{E}$ is given by

$$\vec{E}(\vec{r}) = \sum_{i=1}^{n} q_i \frac{(\vec{r} - \vec{r}_i)}{|\vec{r} - \vec{r}_i|^3}.$$

    A charge configuration with just two charges q_1 and q_2 in 3-space is called an *electric dipole*. Suppose $\vec{r}_1 = \vec{i}$ and $\vec{r}_2 = -\vec{i}$.

    (a) If $q_1 = q$ and $q_2 = -q$, use a computer to sketch the vector field $\vec{E}$ in the xy-plane produced by these two opposite charges.

    (b) If $q_1 = q_2 = q$, sketch the vector field $\vec{E}$ in the xy-plane produced by these two like charges.

12. An *ideal electric dipole* in electrostatics is characterized by its position in 3-space and its dipole moment vector $\vec{p}$. The electric field $\vec{D}$, at the point with position vector $\vec{r}$, of an ideal electric dipole located at the origin with dipole moment $\vec{p}$ is given by

$$\vec{D}(\vec{r}) = 3\frac{(\vec{r} \cdot \vec{p})\vec{r}}{|\vec{r}|^5} - \frac{\vec{p}}{|\vec{r}|^3}.$$

    Assume $\vec{p} = p\vec{i}$, so that the dipole points in the $\vec{i}$ direction and has magnitude p.

    (a) Use a computer to plot the vector field $\vec{D}$ in the xy-plane for three different values of p.

    (b) The field $\vec{D}$ is a convenient approximation to the electric field $\vec{E}$ produced by two opposite charges, q at $\vec{r}_2$ and $-q$ at $\vec{r}_1$, where the distance $|\vec{r}_2 - \vec{r}_1|$ is very small. According to Coulomb's Law, the electric field $\vec{E}$, at the point with position vector $\vec{r}$, is given by

$$\vec{E}(\vec{r}) = q\frac{(\vec{r} - \vec{r}_2)}{|\vec{r} - \vec{r}_2|^3} - q\frac{(\vec{r} - \vec{r}_1)}{|\vec{r} - \vec{r}_1|^3}.$$

    The dipole moment of this configuration of charges is defined to be $q(\vec{r}_2 - \vec{r}_1)$. Suppose $\vec{r}_2 = (\ell/2)\vec{i}$, and $\vec{r}_1 = -(\ell/2)\vec{i}$, so that the dipole moment is $q\ell\vec{i}$.

    (i) Plot the vector field $\vec{E}$ in the xy-plane using the same values of $p = q\ell$ you used to plot $\vec{D}$.

    (ii) Where is the vector field $\vec{D}$ a good approximation to $\vec{E}$? Where is it a poor approximation?

    (iii) The magnitudes of each term in the expression for $\vec{E}$ decays like $1/|\vec{r}|^2$, while the magnitude of $\vec{D}$ decays like $1/|\vec{r}|^3$. If the vector field $\vec{D}$ is supposed to be a good approximation to $\vec{E}$ when the distance $|\vec{r}|$ from the origin is large, can you suggest a reason for this apparent discrepancy?

CHAPTER EIGHTEEN

LINE INTEGRALS

If a force $\vec{F}$ acts on an object as it moves along a displacement vector $\vec{d}$, the work done by the force is the dot product $\vec{F} \cdot \vec{d}$. What if the object moves along a curved path through a variable force field? In this chapter we will define a way of integrating a vector field along a curve, called a *line integral*, that calculates the work in this situation. It can also be used to measure the strength of eddies in the velocity field of a fluid flow.

Most importantly, it provides us with an analogue of the Fundamental Theorem of Calculus, which tells us how to recover a function from its derivative by means of the definite integral. The line integral analogue of the Fundamental Theorem shows how a line integral can be used to recover a multi-variable function from its gradient field. In contrast with the one-variable situation, not all vector fields are gradient fields. The line integral can be used to distinguish those that are, using the concept of a conservative field. We will study conservative fields, which have applications to physics, and finish with another analogue of the Fundamental Theorem, called Green's Theorem.

18.1 THE IDEA OF A LINE INTEGRAL

Imagine that you are rowing in a river with a noticeable current. At times you may be working against the current and at other times you may be moving with it. At the end you have a sense of whether you were helped or hindered by the current. In this section we find a way to measure the extent to which a curve in a vector field is, on balance, going with the vector field or against it.

Orientation of a Curve

There are two directions in which a curve can be traced out. See Figure 18.1. We need to choose one.

A curve is said to be **oriented** if we have chosen a direction of travel on it.

The orientation of a curve is shown graphically by an arrowhead on the curve.

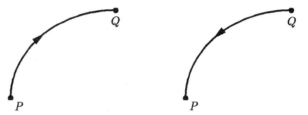

Figure 18.1: Two different orientations of the same curve

Definition of the Line Integral

Now consider a vector field $\vec{F}$ and an oriented curve C. We begin by dividing C into small pieces along which it is approximately straight and $\vec{F}$ is approximately constant. Then each piece can be represented approximately by the displacement vector $\Delta \vec{r}_i = \vec{r}_{i+1} - \vec{r}_i$. See Figure 18.2. At the point with position vector $\vec{r}_i$ we use the dot product $\vec{F}(\vec{r}_i) \cdot \Delta \vec{r}_i$ to compare $\Delta \vec{r}_i$ with the value of the vector field, $\vec{F}(\vec{r}_i)$. Summing over all such pieces, we get a Riemann sum:

$$\sum_i \vec{F}(\vec{r}_i) \cdot \Delta \vec{r}_i.$$

Taking the limit as $\|\Delta \vec{r}_i\| \to 0$ and the number of pieces goes to infinity, we have:

The **line integral** of a vector field $\vec{F}$ along an oriented curve C is

$$\int_C \vec{F} \cdot d\vec{r} = \lim_{\|\Delta \vec{r}_i\| \to 0} \sum_i \vec{F}(\vec{r}_i) \cdot \Delta \vec{r}_i.$$

How Does the Limit Defining a Line Integral Work?

The limit in the definition of a line integral exists if $\vec{F}$ is continuous on an open set containing C and if C is made by joining end to end a finite number of smooth curves, that is, curves which can be parameterized by smooth functions. We can use the parameterization to subdivide a smooth curve, by subdividing the parameter interval in the same way as for conventional one-variable integrals.

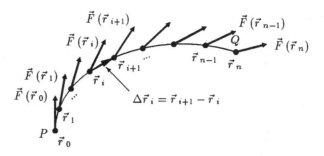

Figure 18.2

The parametrization must go from one end of the curve to the other, in the forward direction, without ever tracing the same portion of the curve twice. In this case, the line integral is independent of the way in which the subdivisions are made. All the curves we consider in this book are *piece-wise smooth* in the sense described above. We will see how to use a parameterization to compute a line integral in Section 18.2.

Example 1   Find the line integral of the constant vector field $\vec{F} = \vec{i} + 2\vec{j}$ along the path from $(1, 1)$ to $(10, 10)$ shown in Figure 18.3.

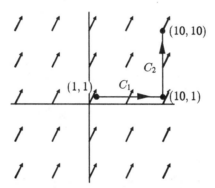

Figure 18.3: The constant vector field $\vec{i} + 2\vec{j}$

Solution   Let C_1 be the horizontal segment of the path going from $(1, 1)$ to $(10, 1)$. When we break this path into pieces, each piece $\Delta\vec{r}$ is horizontal, so $\Delta\vec{r} = \Delta x\vec{i}$, and $\vec{F} \cdot \Delta\vec{r} = (\vec{i} + 2\vec{j}) \cdot \Delta x\vec{i} = \Delta x$. Hence,

$$\int_{C_1} \vec{F} \cdot d\vec{r} = \int_{x=1}^{x=10} dx = 9.$$

Similarly, along the vertical segment C_2, we have $\Delta\vec{r} = \Delta y\vec{j}$ and $\vec{F} \cdot \Delta\vec{r} = (\vec{i} + 2\vec{j}) \cdot \Delta y\vec{j} = 2\Delta y$, so

$$\int_{C_2} \vec{F} \cdot d\vec{r} = \int_{y=1}^{y=10} 2 \, dy = 18.$$

Thus,

$$\int_C \vec{F} \cdot d\vec{r} = \int_{C_1} \vec{F} \cdot d\vec{r} + \int_{C_2} \vec{F} \cdot d\vec{r} = 9 + 18 = 27.$$

What Does the Line Integral Tell Us?

Remember that for any two vectors $\vec{u}$ and $\vec{v}$, $\vec{u} \cdot \vec{v}$ is positive if $\vec{u}$ and $\vec{v}$ point generally in the same direction (that is, if the angle between them is less than $\pi/2$), is 0 if $\vec{u}$ is perpendicular to $\vec{v}$, and is negative if they point generally in opposite directions (that is, if the angle between them is greater than $\pi/2$).

The line integral of $\vec{F}$ adds up the dot products of $\vec{F}$ with small portions of the path; thus, it gives you a positive number if $\vec{F}$ is mostly pointing in the same direction as the path at all points along the path, a negative number if $\vec{F}$ is mostly pointing in the opposite direction, and zero if $\vec{F}$ is perpendicular to the path at all points. In general, the line integral of a vector field $\vec{F}$ along a curve C tells you the extent to which C is going with $\vec{F}$ or against it. Notice that the value of a line integral depends on the values of the vector field along the curve, C, as well as on the orientation of C.

Example 2　The vector field $\vec{F}$ and the oriented curves $C_1, \ldots, C_4$ are shown in Figure 18.4. Which of the line integrals $\int_{C_i} \vec{F} \cdot d\vec{r}$ for $i = 1, \ldots, 4$ are positive? Which are negative? Arrange these line integrals in ascending order.

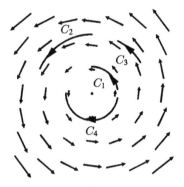

Figure 18.4: Vector field and paths

Solution　The vector field $\vec{F}$ and the line segments $\Delta\vec{r}$ are parallel and in the same direction for the curves C_1, C_2, and C_3. So the contributions of each term $\vec{F} \cdot \Delta\vec{r}$ are positive for these curves. Thus, $\int_{C_1} \vec{F} \cdot d\vec{r}$, $\int_{C_2} \vec{F} \cdot d\vec{r}$, and $\int_{C_3} \vec{F} \cdot d\vec{r}$ are each positive. For the curve C_4, the vector field and the line segments are in opposite directions, so each term $\vec{F} \cdot \Delta\vec{r}$ is negative, and therefore the integral $\int_{C_4} \vec{F} \cdot d\vec{r}$ is negative.

Since the magnitude of the vector field is smaller along C_1 than along C_3, and these two curves are the same length, we must have that:

$$\int_{C_1} \vec{F} \cdot d\vec{r} < \int_{C_3} \vec{F} \cdot d\vec{r}.$$

In addition, the magnitude of the vector field is the same along C_2 and C_3, but the curve C_2 is longer than the curve C_3. Thus,

$$\int_{C_3} \vec{F} \cdot d\vec{r} < \int_{C_2} \vec{F} \cdot d\vec{r}.$$

Putting this together, we have

$$\int_{C_4} \vec{F} \cdot d\vec{r} < \int_{C_1} \vec{F} \cdot d\vec{r} < \int_{C_3} \vec{F} \cdot d\vec{r} < \int_{C_2} \vec{F} \cdot d\vec{r}.$$

Interpretations of the Line Integral

Work

If a constant force $\vec{F}$ acts on an object while it moves along a straight path through a displacement $\vec{d}$, the work done by the force on the object is

$$\text{Work done} = \vec{F} \cdot \vec{d} = \|\vec{F}\|\|\vec{d}\|\cos\theta.$$

See Figure 18.5. Now suppose we want to find the work done by gravity on an object moving far above the surface of the earth. Since the force of gravity varies with distance from the earth, and the path may not be straight, we can't use the formula $\vec{F} \cdot \vec{d}$. We approximate the path by line segments which are small enough that the force is approximately constant on each one.

Suppose that the force at a point with position vector $\vec{r}$ is $\vec{F}(\vec{r})$ and that the object moves along the oriented curve C in Figure 18.6. Then

$$\begin{array}{c}\text{Work done by force } \vec{F}(\vec{r}) \\ \text{over small displacement } \Delta\vec{r}\end{array} \approx \vec{F}(\vec{r}) \cdot \Delta\vec{r}.$$

Thus,

$$\begin{array}{c}\text{Total work done by force} \\ \text{along oriented curve } C\end{array} \approx \sum \vec{F}(\vec{r}) \cdot \Delta\vec{r}.$$

Taking the limit, we get:

$$\begin{array}{c}\text{Work done by force } \vec{F}(\vec{r}) \\ \text{along curve } C\end{array} = \lim_{\|\Delta\vec{r}\|\to 0} \sum \vec{F}(\vec{r}) \cdot \Delta\vec{r} = \int_C \vec{F} \cdot d\vec{r}.$$

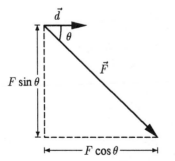

Figure 18.5: For a force $\vec{F}$ and displacement $\vec{d}$, work done $= \vec{F} \cdot \vec{d}$

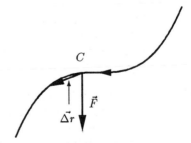

Figure 18.6: Defining the work done by force $\vec{F}(\vec{r})$

Example 3   A mass lying on a flat table is attached to a spring whose other end is fastened to the wall. (See Figure 18.7.) The spring is extended 20 cm beyond its rest position and released. If the axes are as shown in Figure 18.7, when the spring is extended by a distance of x, the force exerted by the spring on the mass is given by

$$\vec{F}(x) = -kx\vec{i},$$

where k is a positive constant that depends on the strength of the spring.

Suppose the mass moves back to the rest position. How much work is done by the force exerted by the spring?

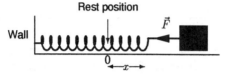

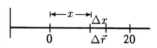

Figure 18.7: Force on mass due to an extended   *Figure 18.8*: Calculating the
spring   work done

Solution   The path from $x = 20$ to $x = 0$ is divided as shown in Figure 18.8, with a typical segment represented by

$$\Delta\vec{r} = \Delta x\vec{i}.$$

Since we are moving from $x = 20$ to $x = 0$, the quantity Δx will be negative. The work done by the force as the mass moves through this segment is approximated by

$$\text{Work done} \approx \vec{F} \cdot \Delta\vec{r} = (-kx\vec{i}) \cdot (\Delta x\vec{i}) = -kx\,\Delta x.$$

Thus, we have

$$\text{Total work done} \approx \sum -kx\,\Delta x.$$

In the limit, as $\Delta x \to 0$, this sum becomes an ordinary definite integral. Since the path starts at $x = 20$, this is the lower limit of integration; $x = 0$ is the upper limit. Thus, we get

$$\text{Total work done} = \int_{x=20}^{x=0} -kx\,dx = -\frac{kx^2}{2}\Big|_{20}^{0} = \frac{k(20)^2}{2} = 200k.$$

The previous example shows that line integrals over paths parallel to a coordinate axis reduce to one-variable integrals. Section 18.2 shows how to convert *any* line integral into a one-variable integral.

Example 4   Suppose a particle has position vector $\vec{r}$. The particle is subject to a force $\vec{F}$ due to gravity. What is the *sign* of the work done by $\vec{F}$ as the particle moves along the path C_1, a radial line through the center of the earth, starting 8000 km from the center and ending 10,000 km from the center (see Figure 18.9).

Solution   We divide the path into small radial segments $\Delta\vec{r}$, pointing away from the center of the earth, and parallel to the gravitational force. The vectors $\vec{F}$ and $\Delta\vec{r}$ point in opposite directions, so each term $\vec{F} \cdot \Delta\vec{r}$ is negative. Adding up all of these negative quantities and taking the limit results in a negative value for the total work. The negative sign indicates that we would have to do work *against* gravity to move the particle along the path C_1.

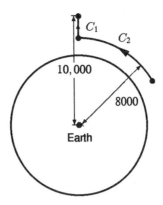

Figure 18.9: The earth

Example 5   Find the work done by gravity as a particle moves along C_2, an arc of a circle 8000 km long at a distance of 8000 km from the center of the earth (see Figure 18.9).

Solution   Since C_2 is everywhere perpendicular to the gravitational force, $\vec{F} \cdot \Delta \vec{r} = 0$ for all $\Delta \vec{r}$ along C_2. Thus,

$$\text{Work done} = \int_{C_2} \vec{F} \cdot d\vec{r} = 0,$$

so the work done is zero. This is why satellites can remain in orbit without expending any fuel, once they have attained orbital velocity.

Example 6   Find the sign of the work done by gravity along the curve C_1 in Example 4, but with the opposite orientation.

Solution   Tracing a curve in the opposite direction changes the sign of the line integral because all the segments $\Delta \vec{r}$ change direction, and so every term $\vec{F} \cdot \Delta \vec{r}$ changes sign. Thus, the result will be the negative of the answer found in Example 4. Therefore, the work done by gravity as a particle moves along C_1 toward the center of the earth is positive.

Circulation

The velocity vector field for the Gulf stream on page 324 shows distinct eddies, areas where the water circulates. We can measure this circulation using a line integral around a closed curve.

> If C is an oriented closed curve (that is, one that starts and ends at the same point), the line integral of $\vec{F}$ around C is called the **circulation** of $\vec{F}$ around C.

Circulation is a measure of the net tendency of the vector field to point around the curve C. To emphasize that C is closed, the circulation is sometimes denoted $\oint_C \vec{F} \cdot d\vec{r}$, with a small circle on the integral sign.

Example 7   Describe the rotation of the vector fields in Figures 18.10 and Figure 18.11.   Find the sign of the circulation of the vector fields around the indicated paths.

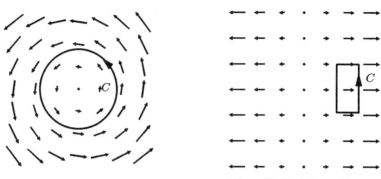

Figure 18.10: A circulating flow   *Figure 18.11*: A flow with zero circulation

Solution   First, consider Figure 18.10. If you think of this as the velocity vector field of water flowing in a pond, you can see that the water is circulating. The line integral around C, measuring the circulation around C, is positive, because the vectors of the field are all pointing in the same direction as the direction of the path. By way of contrast, look at the vector field shown in Figure 18.11. Here the line integral around C is zero because the vertical portions of the path are perpendicular to the field and the contributions from the two horizontal portions cancel out. This means that there is no net tendency for the water to circulate around C.

It turns out that the vector field in Figure 18.11 has the property that its circulation around *any* closed path is zero. Water moving according to this vector field has no tendency to circulate around any point and a leaf dropped anywhere in the flow will not spin around. We'll look at such special fields again later when we introduce the notion of the *curl* of a vector field.

Properties of Line Integrals

Line integrals share some basic properties with conventional one-variable integrals:

For a scalar constant λ and vector fields $\vec{F}$ and $\vec{G}$:

$$\int_C \lambda \vec{F} \cdot d\vec{r} = \lambda \int_C \vec{F} \cdot d\vec{r}$$

and

$$\int_C (\vec{F} + \vec{G}) \cdot d\vec{r} = \int_C \vec{F} \cdot d\vec{r} + \int_C \vec{G} \cdot d\vec{r}.$$

In other words, the line integral of a constant times a vector field is the constant times the line integral, and the line integral of a sum is the sum of the line integrals.

The next two properties are about the curve C over which the line integral is taken. If C_1 and C_2 are oriented curves that meet (so that C_1 ends where C_2 begins) we can construct a new oriented curve by joining them together. This new curve is called $C_1 + C_2$. See Figure 18.12.

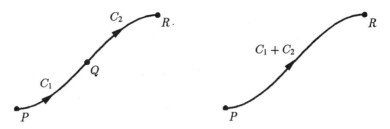

Figure 18.12: Joining two curves to make a new one

If C_1 and C_2 are two oriented curves, then

$$\int_{C_1+C_2} \vec{F} \cdot d\vec{r} = \int_{C_1} \vec{F} \cdot d\vec{r} + \int_{C_2} \vec{F} \cdot d\vec{r}.$$

In words:
   The line integral of $\vec{F}$ along the oriented curve obtained by joining C_1 and C_2 is
   the sum of the line integrals along the curves separately.

This property is the analogue for line integrals of the property for definite integrals which says that

$$\int_a^b f(x)\, dx = \int_a^c f(x)\, dx + \int_c^b f(x)\, dx.$$

Finally, if C is an oriented curve, we let $-C$ be the same curve traversed in the opposite direction, that is, with the opposite orientation. See Figure 18.13.

If C is an oriented curve, then

$$\int_{-C} \vec{F} \cdot d\vec{r} = -\int_{C} \vec{F} \cdot d\vec{r}.$$

In words:
   The line integral of $\vec{F}$ along C with the opposite orientation is the negative of
   the line integral along C.

This property holds because if we integrate in the opposite direction, the vectors $\Delta\vec{r}$ are facing in the opposite direction from before, so the dot products $\vec{F} \cdot \Delta\vec{r}$ are the negatives of what they were before.

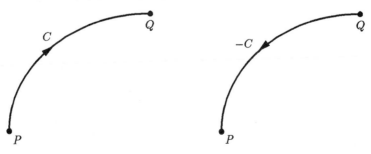

Figure 18.13: A curve and its opposite

Problems for Section 18.1

1. Consider the vector field $\vec{F}$ shown in Figure 18.14, together with the paths C_1, C_2, and C_3. Arrange the line integrals $\int_{C_1} \vec{F} \cdot d\vec{r}$, $\int_{C_2} \vec{F} \cdot d\vec{r}$ and $\int_{C_3} \vec{F} \cdot d\vec{r}$ in ascending order.

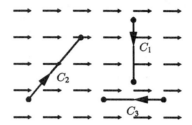

Figure 18.14

In Problems 2–5, say whether the line integral of the pictured vector field over the given curve is positive, negative, or zero.

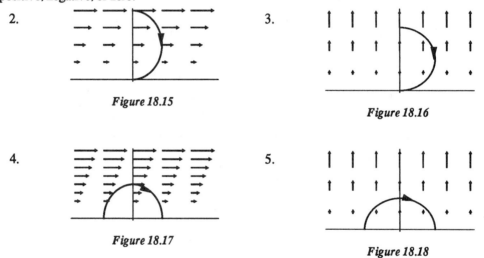

2.

Figure 18.15

3.

Figure 18.16

4.

Figure 18.17

5.

Figure 18.18

For Problems 6–10, say whether the given vector field has positive, negative, or zero circulation around the curve C shown in Figure 18.19. The segments, C_1 and C_3 are circular arcs centered at the origin. You might find it helpful to sketch the vector field first.

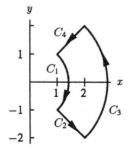

Figure 18.19: The closed curve C

6. $\vec{F}(x, y) = x\vec{i} + y\vec{j}$

7. $\vec{F}(x, y) = -y\vec{i} + x\vec{j}$

8. $\vec{F}(x, y) = y\vec{i} - x\vec{j}$

9. $\vec{F}(x, y) = -\frac{y}{x^2+y^2}\vec{i} + \frac{x}{x^2+y^2}\vec{j}$

10. $\vec{F}(x, y) = x^2\vec{i}$

11. Draw an oriented curve C and a vector field $\vec{F}$ along C that is not always perpendicular to C, but for which $\int_C \vec{F} \cdot d\vec{r} = 0$.

12. Given the force field $\vec{F}(x, y) = y\vec{i} + x^2\vec{j}$ and the right-angle curve, C, from the points $(0, -1)$ to $(4, -1)$ to $(4, 3)$, shown in Figure 18.20:

    (a) Evaluate $\vec{F}$ at the points $(0, -1)$, $(1, -1)$, $(2, -1)$, $(3, -1)$, $(4, -1)$, $(4, 0)$, $(4, 1)$, $(4, 2)$, $(4, 3)$.
    (b) Make a sketch showing the force field along C.
    (c) Estimate the work done by the indicated force field on an object traversing the curve C.

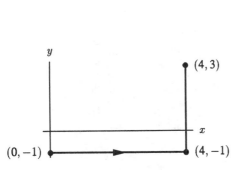

Figure 18.20

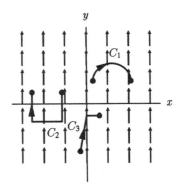

Figure 18.21

13. If $\vec{F}$ is the constant force field $\vec{j}$, consider the work done by the field on particles traveling on paths C_1, C_2 and C_3 of Figure 18.21. On which of these paths will zero work be done? Explain.

For Problems 14–19, use a computer to calculate the line integrals.

14. For $\vec{F} = x\vec{i} + y\vec{j}$,

    (a) Find the line integral of $\vec{F}$ around various rectangles, ellipses, and polygons. What do you get?
    (b) Find the line integral of $\vec{F}$ along three curves, each of which starts at the origin and ends at $(\frac{1}{2}, \frac{1}{2})$. What do you notice?

For Problems 15–17, answer the same questions as in Problems 14 for the given vector field.

15. $\vec{F} = -y\vec{i} + x\vec{j}$.

16. $\vec{F} = \vec{i} + y\vec{j}$.

17. $\vec{F} = \vec{i} + x\vec{j}$.

18. As a result of your answers to Problems 14–17, you should have noticed that the following statement is true: Whenever the line integral around any closed curve is zero, the line integral along a curve with fixed endpoints has a constant value (that is, the line integral is independent of the path the curve takes between the endpoints). Can you explain why this is so?

19. As a result of your answers to Problems 14–17, you should have noticed that the converse to the statement in Problem 18 is also true: Whenever the line integral depends only on endpoints and not on paths, the circulation is always zero. Can you explain why this is so?

20. It is a physical fact that an electric current gives rise to a magnetic field — this is the basis for some electric motors. Ampere's Law relates the magnetic field $\vec{B}$ to a steady current I. It says:

$$\int_C \vec{B} \cdot d\vec{r} = kI$$

where I is the current[1] flowing through a closed curve C, and k is a constant. Figure 18.22 shows a rod carrying current and the magnetic field induced around the rod. If the rod is very long and thin, experiments show that the magnetic field $\vec{B}$ is tangent to every circle that is perpendicular to the rod and has center on the axis of the rod. The magnitude of $\vec{B}$ is constant along every such circle (like C in Figure 18.22). Use Ampere's Law to show that along a circle of radius r, the magnetic field due to a current I has magnitude $kI/2\pi r$. (That is, the strength of the field is inversely proportional to the radial distance from the rod.)

Figure 18.22

18.2 COMPUTING LINE INTEGRALS OVER PARAMETERIZED CURVES

The goal of this section is to show how we can use a smooth parameterization of a curve to convert a line integral into a single-variable definite integral.

How Does a Parameterization of C Help In Evaluating $\int_C \vec{F} \cdot d\vec{r}$?

Recall the definition of the line integral:

$$\int_C \vec{F} \cdot d\vec{r} = \lim_{\|\Delta \vec{r}_i\| \to 0} \sum \vec{F}(\vec{r}_i) \cdot \Delta \vec{r}_i,$$

where the $\vec{r}_i$ are the position vectors of points subdividing the curve into short pieces, and the limit is taken as the length of the pieces goes to zero and the number of them goes to infinity. Now

---

[1]More precisely, I is the net current through any surface that has C as its boundary.

suppose we have a smooth parameterization of C, $\vec{r}(t) = x(t)\vec{i} + y(t)\vec{j}$ for $a \le t \le b$, so that $\vec{r}(a)$ is the beginning of the curve and $\vec{r}(b)$ is the end. Then we can divide C into n pieces by dividing the interval $a \le t \le b$ into n pieces, each of size $\Delta t = (b-a)/n$ (see Figures 18.23 and 18.24).

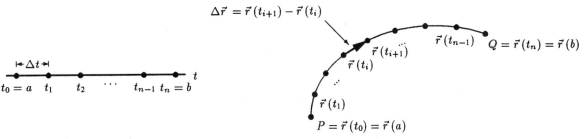

Figure 18.23: Subdivision of the interval $a \le t \le b$

Figure 18.24: Subdivision of a parameterized path

Now at each point $\vec{r}_i = \vec{r}(t_i)$ we want to compute

$$\vec{F}(\vec{r}_i) \cdot \Delta\vec{r}_i.$$

We use the velocity vector to approximate $\Delta\vec{r}_i$. Over a short interval, the velocity vector $\vec{v}(t) = x'(t)\vec{i} + y'(t)\vec{j}$ is approximately constant. Thus, the displacement vector $\Delta\vec{r}_i$ between $\vec{r}(t_i)$ and $\vec{r}(t_{i+1})$ is approximately $\vec{v}(t_i)\Delta t = (x'(t_i)\vec{i} + y'(t_i)\vec{j})\Delta t$. Then

$$\int_C \vec{F} \cdot d\vec{r} \approx \sum \vec{F}(\vec{r}_i) \cdot \Delta\vec{r}_i$$

$$\approx \sum \vec{F}(x(t_i), y(t_i)) \cdot (x'(t_i)\vec{i} + y'(t_i)\vec{j})\Delta t.$$

Notice that $\vec{F}(x(t_i), y(t_i)) \cdot (x'(t_i)\vec{i} + y'(t_i)\vec{j})$ is the value at t_i of a one-variable function of t. So this last sum is really a one-variable Riemann sum. Thus, in the limit as $\Delta t \to 0$, we get a definite integral:

$$\lim_{\Delta t \to 0} \sum \vec{F}(x(t_i), y(t_i)) \cdot (x'(t_i)\vec{i} + y'(t_i)\vec{j})\, \Delta t = \int_a^b \vec{F}(x(t), y(t)) \cdot (x'(t)\vec{i} + y'(t)\vec{j})\, dt.$$

Thus we have the following result:

If $\vec{r}(t) = x(t)\vec{i} + y(t)\vec{j}$, $a \le t \le b$, is a smooth parameterization of an oriented curve C, so that $\vec{r}(a)$ is the beginning of the curve and $\vec{r}(b)$ is the end, then

$$\int_C \vec{F} \cdot d\vec{r} = \int_a^b \vec{F}(x(t), y(t)) \cdot (x'(t)\vec{i} + y'(t)\vec{j})\, dt$$

In words:
   To compute the line integral of $\vec{F}$ over C, take the definite integral of $\vec{F}$ evaluated on C dotted with the velocity vector of the parameterization of C.

Example 1   Compute $\int_C \vec{F} \cdot d\vec{r}$ where $\vec{F} = (x + y)\vec{i} + y\vec{j}$ and C is the quarter unit circle, oriented counter-clockwise as shown in Figure 18.25.

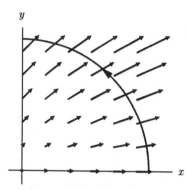

Figure 18.25: The vector field
$\vec{F} = (x + y)\vec{i} + y\vec{j}$ and the quarter circle C

Solution   Since all of the vectors in $\vec{F}$ along C point generally in a direction opposite to the orientation of C, we expect our answer to be negative. The first step is to parameterize C by

$$\vec{r}(t) = x(t)\vec{i} + y(t)\vec{j} = \cos t\vec{i} + \sin t\vec{j}, \quad 0 \le t \le \frac{\pi}{2}.$$

Substituting the parameterization into $\vec{F}$ we get $\vec{F}(x(t), y(t)) = (\cos t + \sin t)\vec{i} + \sin t\vec{j}$. The velocity vector is $\vec{v}(t) = x'(t)\vec{i} + y'(t)\vec{j} = -\sin t\vec{i} + \cos t\vec{j}$. Then

$$\int_C \vec{F} \cdot d\vec{r} = \int_0^{\pi/2} ((\cos t + \sin t)\vec{i} + \sin t\vec{j}) \cdot (-\sin t\vec{i} + \cos t\vec{j})dt$$

$$= \int_0^{\pi/2} (-\cos t \sin t - \sin^2 t + \sin t \cos t)dt$$

$$= \int_0^{\pi/2} -\sin^2 t\, dt = -\frac{\pi}{4} \approx -0.7854.$$

So the answer is negative, as expected. Also, the line integral is relatively small in absolute value because the vector field is nearly perpendicular to the curve C for much of its length. That means that the contributions made by the vectors $\vec{F}$ in the direction parallel to C are small.

Example 2   Consider the vector field $\vec{F} = x\vec{i} + y\vec{j}$.
(a)   Suppose C_1 is the line segment joining $(1, 0)$ to $(0, 2)$ and C_2 is a part of a parabola with its vertex at $(0, 2)$, joining the same points in the same order. (See Figure 18.26.) Verify that

$$\int_{C_1} \vec{F} \cdot d\vec{r} = \int_{C_2} \vec{F} \cdot d\vec{r}.$$

(b)   If C is the triangle shown in Figure 18.27, show that $\int_C \vec{F} \cdot d\vec{r} = 0$.

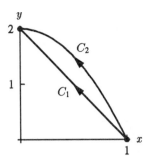

Figure 18.26   **Figure 18.27**

Solution   (a) We parameterize C_1 by $(x(t), y(t)) = (1 - t, 2t)$ with $0 \leq t \leq 1$. Then $\vec{v}(t) = -\vec{i} + 2\vec{j}$, so

$$\int_{C_1} \vec{F} \cdot d\vec{r} = \int_0^1 \vec{F}(1 - t, 2t) \cdot (-\vec{i} + 2\vec{j}) \, dt$$

$$= \int_0^1 ((1 - t)\vec{i} + 2t\vec{j}) \cdot (-\vec{i} + 2\vec{j}) \, dt$$

$$= \int_0^1 (5t - 1) \, dt = \frac{3}{2}.$$

To parameterize C_2, we use the fact that it is a parabola with vertex at $(0, 2)$, so its equation is of the form $y = -kx^2 + 2$ for some k. Since the parabola crosses the x-axis at $(1, 0)$, we find that $k = 2$ and $y = -2x^2 + 2$. Therefore, we use the parameterization $(x(t), y(t)) = (t, -2t^2 + 2)$ with $0 \leq t \leq 1$, which has velocity vector $\vec{i} - 4t\vec{j}$. This traces out C_2 in reverse, since it begins at the point $(0, 2)$ and ends at $(1, 0)$. Thus, if we use this parameterization we must must multiply by -1 to get what we want:

$$\int_{C_2} \vec{F} \cdot d\vec{r} = -\int_0^1 \vec{F}(t, -2(t^2 - 1)) \cdot (\vec{i} - 4t\vec{j}) \, dt$$

$$= -\int_0^1 (t\vec{i} - 2(t^2 - 1)\vec{j}) \cdot (\vec{i} - 4t\vec{j}) \, dt$$

$$= -\int_0^1 (8t^3 - 7t) \, dt = \frac{3}{2}.$$

So the line integrals along C_1 and C_2 are the same.

(b) We break $\int_C \vec{F} \cdot d\vec{r}$ up into three pieces, one of which we have already computed (namely, the piece connecting $(1, 0)$ to $(0, 2)$, where the line integral has value $3/2$). The piece running from $(0, 2)$ to $(0, 0)$ can be parameterized by $(x(t), y(t)) = (0, 2 - t)$ where $0 \leq t \leq 2$. The piece running from $(0, 0)$ to $(1, 0)$ can be parameterized by $(x(t), y(t)) = (t, 0)$ where $0 \leq t \leq 1$. Then

$$\int_C \vec{F} \cdot d\vec{r} = \frac{3}{2} + \int_0^2 \vec{F}(0, (2 - t)) \cdot (-\vec{j}) \, dt + \int_0^1 \vec{F}(t, 0) \cdot \vec{i} \, dt$$

$$= \frac{3}{2} + \int_0^2 (2 - t)\vec{j} \cdot (-\vec{j}) \, dt + \int_0^1 t\vec{i} \cdot \vec{i} \, dt$$

$$= \frac{3}{2} + \int_0^2 (t - 2) \, dt + \int_0^1 t \, dt = \frac{3}{2} + (-2) + \frac{1}{2} = 0.$$

Example 3   Let C be the closed curve consisting of the upper half-circle of radius 1 and the line forming its diameter along the x-axis, oriented counterclockwise. (See Figure 18.28.) Find $\int_C \vec{F} \cdot d\vec{r}$ where $\vec{F}(x, y) = -y\vec{i} + x\vec{j}$.

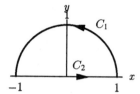

Figure 18.28: The curve
$C = C_1 + C_2$ for Example 3

Solution   We write $C = C_1 + C_2$ where C_1 is the half-circle and C_2 is the line, and compute $\int_{C_1} \vec{F} \cdot d\vec{r}$ and $\int_{C_2} \vec{F} \cdot d\vec{r}$ separately. We parameterize C_1 by $(x(t), y(t)) = (\cos t, \sin t), 0 \le t \le \pi$. Then

$$\int_{C_1} \vec{F} \cdot d\vec{r} = \int_0^\pi (-\sin t\vec{i} + \cos t\vec{j}) \cdot (-\sin t\vec{i} + \cos t\vec{j}) \, dt$$

$$= \int_0^\pi (\sin^2 t + \cos^2 t) \, dt = \int_0^\pi 1 \, dt = \pi.$$

For C_2, we have $\int_{C_2} \vec{F} \cdot d\vec{r} = 0$, since the vector field $\vec{F}$ has no $\vec{i}$ component along the x-axis (where $y = 0$) and is therefore perpendicular to C_2 at all points.

Finally, we can write

$$\int_C \vec{F} \cdot d\vec{r} = \int_{C_1} \vec{F} \cdot d\vec{r} + \int_{C_2} \vec{F} \cdot d\vec{r} = \pi + 0 = \pi.$$

It is no accident that the result for $\int_{C_1} \vec{F} \cdot d\vec{r}$ is the same as the length of the curve C_1. See Problems 10–11 on Page 359.

The next example illustrates the computation of a line integral over a path in 3-space.

Example 4   A particle is traveling along the circular helix C given by $\vec{r}(t) = \cos t\vec{i} + \sin t\vec{j} + 2t\vec{k}$ and is subject to a force $\vec{F} = x\vec{i} + z\vec{j} - yx\vec{k}$. Find the total work done on the particle by the force for $0 \le t \le 3\pi$.

Solution   Since the path is given in parametric form, we evaluate the line integral

$$\int_C \vec{F} \cdot d\vec{r} = \int_0^{3\pi} \vec{F}(\vec{r}(t)) \cdot \vec{r}'(t) \, dt$$

$$= \int_0^{3\pi} (\cos t\vec{i} + 2t\vec{j} - \cos t \sin t\vec{k}) \cdot (-\sin t\vec{i} + \cos t\vec{j} + 2\vec{k}) \, dt$$

$$= \int_0^{3\pi} (-\cos t \sin t + 2t \cos t - 2\cos t \sin t) \, dt$$

$$= \int_0^{3\pi} (-3\cos t \sin t + 2t \cos t) \, dt = -4.$$

The Notation $\int_C P\,dx + Q\,dy$

There is an alternative notation for line integrals that is quite common. Given functions $P(x, y)$ and $Q(x, y)$ and an oriented curve C, we can write

$$\int_C P(x, y)dx + Q(x, y)dy = \int_C \vec{F} \cdot d\vec{r}$$

where $\vec{F} = P\vec{i} + Q\vec{j}$. The relation between the two notations can be remembered using $d\vec{r} = dx\vec{i} + dy\vec{j}$.

Example 5   Evaluate $\int_C xy\,dx - y^2\,dy$ where C is the line segment from $(0, 0)$ to $(2, 6)$.

Solution   We parameterize C by $\vec{r}(t) = x(t)\vec{i} + y(t)\vec{j} = t\vec{i} + 3t\vec{j}, 0 \leq t \leq 2$. By definition,

$$\int_C xy\,dx - y^2\,dy = \int_C (xy\vec{i} - y^2\vec{j}) \cdot d\vec{r} = \int_0^2 (3t^2\vec{i} - 9t^2\vec{j}) \cdot (\vec{i} + 3\vec{j})\,dt$$

$$= \int_0^2 (-24t^2)\,dt = -64.$$

Independence of Parameterization

Since there are many different ways of parameterizing a given oriented curve, you may be wondering what happens to the value of a given line integral if you choose another parameterization. The answer is that the choice of parameterization makes no difference. Since we initially defined the line integral without reference to any particular parameterization, this is exactly as we would expect.

Example 6   Consider the oriented path which is a straight line segment L running from $(0, 0)$ to $(1, 1)$. Calculate the line integral of the vector field $\vec{F} = (3x - y)\vec{i} + y\vec{j}$ along L using each of the parameterizations:

(a) $A(t) = (t, t)$, $0 \leq t \leq 1$,   (b) $B(t) = (2t, 2t)$, $0 \leq t \leq 1/2$,

(c) $C(t) = \left(\dfrac{t^2 - 1}{3}, \dfrac{t^2 - 1}{3}\right)$, $1 \leq t \leq 2$,   (d) $D(t) = (e^t - 1, e^t - 1)$, $0 \leq t \leq \ln 2$.

Solution   First, check that each of these gives a parameterization of L: each has both coordinates equal (as do all points on L) and each begins at $(0,0)$ and ends at $(1,1)$. Now let's calculate the line integral of the vector field $\vec{F} = (3x - y)\vec{i} + y\vec{j}$ using each parameterization.
    (a) Using $A(t)$, we get

$$\int_L \vec{F} \cdot d\vec{r} = \int_0^1 ((3t - t)\vec{i} + t\vec{j}) \cdot (\vec{i} + \vec{j})\,dt = \int_0^1 3t\,dt = \frac{3t^2}{2}\bigg|_0^1 = \frac{3}{2}.$$

(b) Using $B(t)$ gives

$$\int_L \vec{F} \cdot d\vec{r} = \int_0^{1/2} ((6t - 2t)\vec{i} + 2t\vec{j}) \cdot (2\vec{i} + 2\vec{j}) \, dt = \int_0^{1/2} 12t \, dt = 6t^2 \Big|_0^{1/2} = \frac{3}{2}.$$

Both $A(t)$ and $B(t)$ are parameterizations that traverse L at a constant speed: $A(t)$ has speed 1 and $B(t)$ has speed 2. You can see how the speed factor of 2 cancels when evaluating the definite integral.

(c) Now we use $C(t)$:

$$\int_L \vec{F} \cdot d\vec{r} = \int_1^2 \left(\left(\frac{3(t^2 - 1)}{3} - \frac{(t^2 - 1)}{3} \right) \vec{i} + \frac{t^2 - 1}{3} \vec{j} \right) \cdot \left(\frac{2t}{3}\vec{i} + \frac{2t}{3}\vec{j} \right) dt$$

$$= \int_1^2 \frac{2t}{3}(t^2 - 1) \, dt = \frac{2}{3} \int_1^2 (t^3 - t) \, dt$$

$$= \frac{2}{3} \left(\frac{t^4}{4} - \frac{t^2}{2} \right) \Big|_1^2 = \frac{3}{2}.$$

(d) Finally, using $D(t)$, we get

$$\int_L \vec{F} \cdot d\vec{r} = \int_0^{\ln 2} \left((3(e^t - 1) - (e^t - 1)) \vec{i} + (e^t - 1)\vec{j} \right) \cdot (e^t\vec{i} + e^t\vec{j}) \, dt$$

$$= \int_0^{\ln 2} 3e^t(e^t - 1) \, dt = 3 \int_0^{\ln 2} (e^{2t} - e^t) \, dt$$

$$= 3 \left(\frac{e^{2t}}{2} - e^t \right) \Big|_0^{\ln 2} = \frac{3}{2}.$$

The fact that the four answers are the same illustrates that the value of a line integral is independent of the parameterization of the path. Problems 12–14 at the end of this section give another way of seeing this.

Problems for Section 18.2

In Problems 1–6, compute the line integral of the given vector field along the given path.

1. $\vec{F}(x, y) = x^2\vec{i} + y^2\vec{j}$ and C is the line from the point $(1, 2)$ to the point $(3, 4)$.

2. $\vec{F}(x, y) = \ln y\vec{i} + \ln x\vec{j}$ and C is the curve given parametrically by $(2t, t^3)$ for $2 \le t \le 4$.

3. $\vec{F}(x, y) = e^x\vec{i} + e^y\vec{j}$ and C is the part of the ellipse $x^2 + 4y^2 = 4$ joining the point $(0, 1)$ to the point $(2, 0)$ in the clockwise direction.

4. $\vec{F}(x, y) = xy\vec{i} + (x - y)\vec{j}$ and C is the triangle joining the points $(1, 0), (0, 1)$ and $(-1, 0)$ in the clockwise direction.

5. $\vec{F} = x\vec{i} + 2zy\vec{j} + x\vec{k}$ and C is given by $\vec{r} = t\vec{i} + t^2\vec{j} + t^3\vec{k}$ for $1 \le t \le 2$.

6. $\vec{F} = e^y\vec{i} + \ln(x^2 + 1)\vec{j} + \vec{k}$ and C is the circle of radius 2 in the yz-plane centered at the origin and traversed as shown in Figure 18.29.

7. Find parameterizations for the oriented curves shown in Figure 18.30. C_1 is a semicircle of radius 1, centered at the point $(1, 0)$. C_2 is a portion of a parabola, with vertex at the point $(1, 0)$ and y-intercept -2. C_3 is one arc of the sine curve.

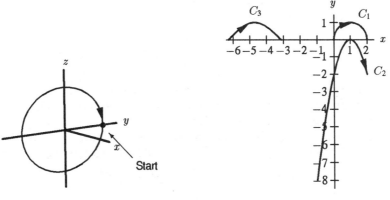

Figure 18.29   **Figure 18.30**

8. Let C be the oriented line segment from the point $(0, 0)$ to the point $(4, 12)$, and suppose that $\vec{F} = xy\vec{i} + x\vec{j}$.

   (a) Is $\int_C \vec{F} \cdot d\vec{r}$ greater than, less than, or equal to zero? Give a geometric explanation.

   (b) A simple parameterization of C is $(x(t), y(t)) = (t, 3t)$ for $0 \le t \le 4$. Use this to compute $\int_C \vec{F} \cdot d\vec{r}$.

   (c) Suppose a particle leaves the point $(0, 0)$, moves along the line towards the point $(4, 12)$, stops before reaching it and backs up, stops again and reverses direction, then completes its journey to the endpoint. All travel takes place along the line segment joining the point $(0, 0)$ to the point $(4, 12)$. If we call this path C', explain why $\int_{C'} \vec{F} \cdot d\vec{r} = \int_C \vec{F} \cdot d\vec{r}$.

   (d) A parameterization for a path like C' is given by

   $$(x(t), y(t)) = \left(\frac{1}{3}(t^3 - 6t^2 + 11t), (t^3 - 6t^2 + 11t)\right), \qquad 0 \le t \le 4.$$

   Check that this begins at the point $(0, 0)$ and ends at the point $(4, 12)$. Check also that all points of C' lie on the line segment connecting the point $(0, 0)$ to the point $(4, 12)$. What are the values of t where the particle changes direction?

   (e) Find $\int_{C'} \vec{F} \cdot d\vec{r}$ using the vector field $xy\vec{i} + x\vec{j}$ and the parameterization in part (d). Do you get the same answer as in part (b)?

9. In Example 6 on page 357 we integrated $\vec{F} = (3x - y)\vec{i} + x\vec{j}$ over two parameterizations of the line from $(0, 0)$ to $(1, 1)$, getting $3/2$ each time. Now we compute the line integral along the following two different paths with the same endpoints, and show that the answers are different.

   (a) The path (t, t^2), with $0 \le t \le 1$

   (b) The path (t^2, t), with $0 \le t \le 1$

10. Consider the vector field $\vec{F} = -y\vec{i} + x\vec{j}$. Let C be the unit circle oriented counterclockwise.

    (a) Show that $\vec{F}$ has a constant magnitude of 1 on the circle C.

(b) Show that $\vec{F}$ is always tangent to the circle C.

(c) Show that $\int_C \vec{F} \cdot d\vec{r} = $ Length of C.

11. Suppose that along a curve C, a vector field $\vec{F}$ has direction always tangent to C in direction of orientation and has constant magnitude $\|\vec{F}\| = m$. Use the definition of the line integral to explain why

$$\int_C \vec{F} \cdot d\vec{r} = m \cdot \text{Length of } C.$$

Problems 12–14 concern the line integral in Example 6 on page 357. In the example, several different parameterizations are used to convert the line integral into a definite integral. In each of the following cases, show that the two definite integrals corresponding to the two given parameterizations are equal by finding a substitution which converts one integral to the other. This gives us another way of seeing why changing the parameterization of the curve does not change the value of the line integral.

12. $A(t)$ and $B(t)$ 13. $A(t)$ and $C(t)$ 14. $A(t)$ and $D(t)$

18.3 GRADIENT FIELDS AND CONSERVATIVE FIELDS

Finding the Total Change in f from grad f: The Fundamental Theorem

In one-variable calculus, we saw how the definite integral of a rate of change gives the total change. This is the Fundamental Theorem of Calculus:

$$\int_a^b f'(t)\, dt = f(b) - f(a).$$

What about functions of two variables? The quantity that describes the rate of change is now a vector field, the gradient. If we know the gradient of a function f, can we compute the total change in f between two points? The answer is yes, and we will see how trying to reconstruct the total change leads to a line integral. We begin with the case where the gradient is constant.

If grad f is Constant

If grad f is constant, then the contours of f are equally spaced with the same slope everywhere, so f is a linear function. (See Figure 18.31). Going along the displacement vector $\Delta \vec{r}$ takes us to

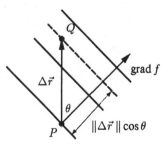

Figure 18.31: $f(Q) - f(P) =$ grad $f \cdot \Delta \vec{r}$

the same contour as if we had gone a distance of $\|\Delta \vec{r}\| \cos \theta$ perpendicular to the contours. Since the rate of change perpendicular to the contours is $\| \operatorname{grad} f \|$, the change in f is

$$\text{Distance} \times \text{Rate of change} = \|\Delta \vec{r}\| \cos \theta \, \| \operatorname{grad} f \| = \operatorname{grad} f \cdot \Delta \vec{r}.$$

Thus, we have the following result:

If grad f is constant, then Δf, the change in f, is given by

$$\Delta f = f(Q) - f(P) = \operatorname{grad} f \cdot \Delta \vec{r},$$

where $\Delta \vec{r}$ is the displacement vector from P to Q.

If grad f is Not Constant

To find the change in f between two points P and Q when grad f is not constant, we choose a smooth path C from P to Q, then divide the path into many small pieces. See Figure 18.32. Writing $\Delta \vec{r}_i$ for the displacement vector from $\vec{r}_i$ to $\vec{r}_{i+1}$, and assuming grad f is approximately constant along this vector, we have

$$\Delta f = f(\vec{r}_{i+1}) - f(\vec{r}_i) \approx \operatorname{grad} f(\vec{r}_i) \cdot \Delta \vec{r}_i,$$

so

$$f(Q) - f(P) \approx \sum \operatorname{grad} f(\vec{r}_i) \cdot \Delta \vec{r}_i.$$

In the limit as $\|\Delta \vec{r}\|$ approaches zero and the number n of subintervals gets larger, we obtain the following result:

The Fundamental Theorem of Calculus for Line Integrals

Suppose C is a smooth oriented path with P as its starting point and Q as its endpoint. If f is a function whose gradient is continuous in an open set containing the path C, then

$$\int_C \operatorname{grad} f \cdot d\vec{r} = f(Q) - f(P).$$

Notice that there are many different paths from P to Q. (See Figure 18.33.) However, the value of the line integral $\int_C \operatorname{grad} f \cdot d\vec{r}$ depends only on the endpoints of C; it does not depend on where C goes in between. Exercise 15 shows how the Fundamental Theorem for Line Integrals can be derived from the one-variable Fundamental Theorem of Calculus.

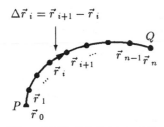

Figure 18.32: Subdivision of the path from P to Q

Figure 18.33: Different paths from P to Q

Example 1   Suppose that grad f is shown in Figure 18.34. Explain why $f(P) = f(Q)$:
(a)   Using a line integral.   (b)   By looking at contours.

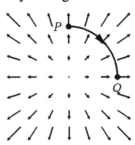

Figure 18.34: The gradient
vector field of the function f

Solution   (a)   The change in f from P to Q, namely $f(Q) - f(P)$, is given by the line integral of grad f
along any path joining P to Q. If we choose the path shown in Figure 18.34, we see that the
direction of grad f along the path is perpendicular to the path. This means that the component
of grad f in the direction of the path is 0 at each point on the path, so the line integral equals 0.

(b)   Recall that the gradient vector is perpendicular to the contours of f. Since P and Q lie on a
path which is everywhere perpendicular to the gradient vector, this path is part of a contour.
Thus, P and Q lie on the same contour; hence, $f(P) = f(Q)$.

Example 2   Consider the vector field $\vec{F} = x\vec{i} + y\vec{j}$. In Example 2 on page 354 we calculated $\int_{C_1} \vec{F} \cdot d\vec{r}$ and
$\int_{C_2} \vec{F} \cdot d\vec{r}$ over the oriented curves shown in Figure 18.35 and found they were the same. Find
a scalar function f with grad $f = \vec{F}$. Hence, find an easy way to calculate the line integrals, and
explain how you could have expected them to be the same.

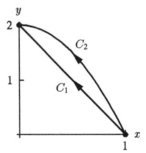

Figure 18.35

Solution   One possibility for f is $f(x, y) = x^2/2 + y^2/2$. You can check that grad $f = x\vec{i} + y\vec{j}$. Now we
can use the Fundamental Theorem to compute the line integral. The Fundamental Theorem says
that the line integral of grad f is the total change in the function f between the endpoints. Thus, if
$\vec{F} = \text{grad } f$ we have

$$\int_{C_1} \vec{F} \cdot d\vec{r} = \int_{C_2} \vec{F} \cdot d\vec{r} = f(0, 2) - f(1, 0) = 2 - \frac{1}{2} = \frac{3}{2}.$$

Since the integral depends only on the value of f at the endpoints, it should be the same no matter
what path you choose.

Conservative Vector Fields

If $\vec{F} = \text{grad } f$, then the Fundamental Theorem for Line Integrals tells us that if C is a path from P to Q, then

$$\int_C \vec{F} \cdot d\vec{r} = f(Q) - f(P).$$

Notice that the right hand side of this equation does not depend on the path, but only on the endpoints of the path. We give vector fields with this property a special name.

> A vector field $\vec{F}$ is said to be **conservative** if for any two points P and Q, the line integral $\int_C \vec{F} \cdot d\vec{r}$ has the same value along any path C from P to Q.

Thus,

> Every gradient field is conservative.

Being conservative means that the line integral of $\vec{F}$ is the same along *any* two paths with the same endpoints. We say that the line integral is *path-independent*, meaning that the value of the line integral depends on the endpoints of the path, but not on where the path goes in between.

Why do We Care about Conservative Vector Fields?

It turns out that some of the fundamental vector fields of nature are conservative — two important examples are the gravitational field and the electric field of particles at rest. The fact that the gravitational field is conservative means that the work done in moving an object subject to gravity depends only on the starting and ending points, and not on the path taken. For example the work done by gravity (computed by the line integral) on a bicycle being carried to a sixth floor apartment is the same whether it is carried up the stairs in a zig-zag path or taken straight up in an elevator.

When a force field is conservative we can define a quantity called the *potential energy* of a body that depends only on its position. When the body moves to another position, the potential energy changes by an amount equal to the work done by the force, which then depends only on the starting and ending positions. If the work done had not been path-independent, the potential energy would depend both on the body's current position *and* on how it got there, and so a useful potential energy could not have been defined. It turns out that the potential energy function is related to the force field by means of a gradient, as we see below.

Conservative Fields and Gradient Fields

We have seen that every gradient field is conservative. What about the converse? That is, given a conservative vector field $\vec{F}$, can we always find a function f such that $\vec{F} = \text{grad } f$? The answer is yes, and we will show how to construct f from $\vec{F}$.

First, notice that there are many different choices for f, since you can always add a constant to f without changing grad f. If we pick a fixed point P arbitrarily, then by adding or subtracting a constant to f we can make sure that $f(P) = 0$. Now we define $f(Q)$ for any other point Q by the formula:

$$f(Q) = \int_C \vec{F} \cdot d\vec{r}, \quad \text{where } C \text{ is a path from } P \text{ to } Q.$$

Since $\vec{F}$ is conservative, it doesn't matter which path we choose from P to Q. On the other hand, if $\vec{F}$ is not conservative, then different choices might give different values for $f(Q)$, so f would not be a function (a function has to have a single value at each point).

We will show at the end of this section that the gradient of the function f is indeed $\vec{F}$. Thus, by constructing a function f in this manner, we have the following:

If $\vec{F}$ is conservative, then $\vec{F} = \text{grad } f$ for some f.

The function f is sufficiently important that it is given a special name:

If a vector field $\vec{F}$ is of the form $\vec{F} = \text{grad } f$ for some scalar function f, then f is called a **potential function** for the vector field $\vec{F}$.

Warning

Physicists call a function f a potential function for a vector field $\vec{F}$ if $\vec{F} = -\text{grad } f$. Problem 22 on page 369 explains the minus sign.

Example 3   Show that the vector field $\vec{F}(x, y) = y \cos x\vec{i} + \sin x\vec{j}$ is conservative.

Solution   If we can find a potential function f, then $\vec{F}$ must be conservative. We want $\partial f/\partial x = y \cos x$, so f must be of the form $y \sin x + g(y)$ where $g(y)$ is a function of y only. Now, $\partial f/\partial y = \sin x + g'(y)$. For this to be the same as the second component of $\vec{F}$, that is, $\sin x$, we must have $g'(y) = 0$, so $g(y) = C$ where C is some constant. Thus,

$$f(x, y) = y \sin x + C$$

is a potential function for $\vec{F}$ and therefore $\vec{F}$ is conservative.

Example 4   Show that the gravitational field

$$\vec{F} = -\frac{GM}{r^3}\vec{r}$$

of an object of mass M is a gradient field, so is therefore conservative, and find a potential function for $\vec{F}$.

Solution   All the force vectors point in towards the origin, and if they are going to be gradient vectors of some function f they must be perpendicular to the level surfaces of f, so the level surfaces of f must be spheres. Also, the magnitude of the vector is GM/r^2, and this is the rate of change of the function f in the inward direction. Since

$$\frac{d}{dr}\left(\frac{1}{r}\right) = \frac{-1}{r^2},$$

we might guess that $\vec{F}$ is the gradient of the function

$$f(x, y, z) = \frac{GM}{r} = \frac{GM}{\sqrt{x^2 + y^2 + z^2}}.$$

We will try this:

$$f_x = \frac{\partial}{\partial x} \frac{GM}{\sqrt{x^2 + y^2 + z^2}} = \frac{-GMx}{(x^2 + y^2 + z^2)^{3/2}},$$

$$f_y = \frac{\partial}{\partial y} \frac{GM}{\sqrt{x^2 + y^2 + z^2}} = \frac{-GMy}{(x^2 + y^2 + z^2)^{3/2}},$$

$$f_z = \frac{\partial}{\partial z} \frac{GM}{\sqrt{x^2 + y^2 + z^2}} = \frac{-GMz}{(x^2 + y^2 + z^2)^{3/2}}.$$

So

$$\operatorname{grad} f = f_x \vec{i} + f_y \vec{j} + f_z \vec{k} = \frac{-GM}{(x^2 + y^2 + z^2)^{3/2}}(x\vec{i} + y\vec{j} + z\vec{k})$$

$$= \frac{-GM}{r^3}\vec{r} = \vec{F}.$$

Our computations show that $\vec{F}$ is a gradient field and that $f = GM/r$ is a potential function for $\vec{F}$.

Why Conservative Vector Fields are Gradient Fields

Suppose $\vec{F}$ is a conservative vector field. On page 364 we defined the function $f(Q)$ by the formula

$$f(Q) = \int_C \vec{F} \cdot d\vec{r},$$

where C is a path from some fixed starting place P to a variable point Q. This works because $\vec{F}$ is conservative, so this integral has the same value for any path from P to Q.

But why is $\operatorname{grad} f = \vec{F}$? First, we write out the line integral in terms of the vector field components $\vec{F} = F_1\vec{i} + F_2\vec{j}$ and the components $d\vec{r} = dx\vec{i} + dy\vec{j}$ of the infinitesimal plane vector:

$$f(Q) = \int_C F_1 dx + F_2 dy.$$

Now we consider small increments Δx and Δy in x and y and imagine adding to the end of the path C either a small line L_x from Q to $Q + \Delta x \cdot \vec{i}$ or a small line L_y from Q to $Q + \Delta y \cdot \vec{j}$ (see Figure 18.36). If $Q = (x, y)$, then we can calculate $f(x + \Delta x, y)$ by the path $C(P, Q) + L_x$ which starts at P, goes to Q and then on to $Q + \Delta x \cdot \vec{i} = (x + \Delta x, y)$. Therefore

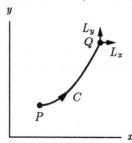

Figure 18.36: Path C with small increments Δx and Δy

$$f(x + \Delta x, y) = \int_{C+L_x} (F_1 \, dx + F_2 \, dy)$$

$$= \int_C (F_1 dx + F_2 dy) + \int_{L_x} (F_1 \, dx + F_2 \, dy)$$

$$\approx f(x, y) + F_1(x, y) \cdot \Delta x$$

Here the line integral $\int_{L_x} F_2 \, dy$ is zero because y is constant on the path L_x and $\int_{L_x} F_1 \, dx \approx F_1(x, y) \Delta x$ because F_1 is approximately constant on the small path L_x whose length is Δx. Putting $f(x, y)$ on the left hand side and dividing by Δx, we get in the limit:

$$\frac{\partial f}{\partial x}(x, y) = F_1(x, y).$$

The same argument, now using the line segment L_y, shows that

$$\frac{\partial f}{\partial y}(x, y) = F_2(x, y).$$

Therefore,

$$\operatorname{grad} f = \frac{\partial f}{\partial x}\vec{i} + \frac{\partial f}{\partial y}\vec{j} = F_1\vec{i} + F_2\vec{j} = \vec{F}.$$

Summary

We have studied two apparently different sorts of vector fields: conservative vector fields and gradient vector fields. It turns out that these are the same. It is still useful, however, to have the different terminology, since depending on the context we may want to emphasize one or another property of these types of fields. Here is a summary of the properties and the connections between them:

- **Conservative vector fields** are fields with the property that for any two points P and Q, the integral along a path from P to Q is the same no matter what path you choose.
- **Gradient vector fields** are fields of the form grad f for some scalar function f, which is called the potential function of the vector field.
- **Gradient vector fields are conservative** by the Fundamental Theorem of Line Integrals, which says that integral of a gradient field is the change in its potential between the beginning and ending points, and so must be independent of the path.
- **Conservative vector fields are gradient fields** because you can use a line integral and path independence to construct the potential function.

Problems for Section 18.3

For Problems 1–4, decide whether or not the given vector fields could be gradient vector fields. Give a justification for your answer.

1. $\vec{F}(x, y) = x\vec{i}$

2. $\vec{F}(x, y, z) = \dfrac{-z}{\sqrt{x^2 + z^2}}\vec{i} + \dfrac{y}{\sqrt{x^2 + z^2}}\vec{j} + \dfrac{x}{\sqrt{x^2 + z^2}}\vec{k}$

3. $\vec{G}(x, y) = (x^2 - y^2)\vec{i} - 2xy\vec{j}$

4. $\vec{F}(\vec{r}) = \vec{r}/r^3$, where $\vec{r} = x\vec{i} + y\vec{j} + z\vec{k}$

5. Consider the vector field $\vec{F}$ graphed in Figure 18.37.

   (a) Is the line integral $\int_C \vec{F} \cdot d\vec{r}$ positive, negative, or zero?

   (b) From your answer to part (a), can you determine whether or not $\vec{F} = \text{grad } f$ for some function f?

   (c) Which of the following formulas best fits this vector field?

   $$\vec{F_1} = \frac{x}{x^2 + y^2}\vec{i} + \frac{y}{x^2 + y^2}\vec{j}, \quad \vec{F_2} = -y\vec{i} + x\vec{j}, \quad \vec{F_3} = \frac{-y}{(x^2 + y^2)^2}\vec{i} + \frac{x}{(x^2 + y^2)^2}\vec{j}.$$

6. Consider the vector field $\vec{F}(x, y) = x\vec{j}$ shown in Figure 18.38.

   (a) Find paths C_1, C_2 and C_3 from P to Q such that

   $$\int_{C_1} \vec{F} \cdot d\vec{r} = 0, \qquad \int_{C_2} \vec{F} \cdot d\vec{r} > 0, \qquad \text{and} \int_{C_3} \vec{F} \cdot d\vec{r} < 0.$$

   (b) Is $\vec{F}$ a gradient field?

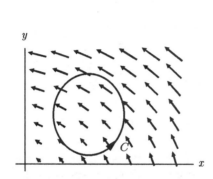

Figure 18.37: Vector field and path for Problem 5

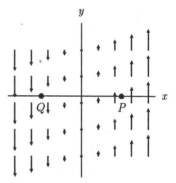

Figure 18.38: Vector field for Problem 6

7. Let $\vec{F} = x\vec{i} + y\vec{j}$, and let C_1 be the line joining the point $(1, 0)$ to the point $(0, 2)$ and let C_2 be the line joining the point $(0, 2)$ to the point $(-1, 0)$. Is $\int_{C_1} \vec{F} \cdot d\vec{r} = -\int_{C_2} \vec{F} \cdot d\vec{r}$? Explain.

8. What is the approximate value of $\int_C \vec{F} \cdot d\vec{r}$ if C is an oriented curve that runs from the point $(2, -6)$ to the point $(4, 4)$ and if $\vec{F} \approx 6\vec{i} - 7\vec{j}$ on C?

In Problems 9–13, each of the statements is *false*. Explain why or give a counterexample.

9. If $\int_C \vec{F} \cdot d\vec{r} = 0$ for one particular closed path C, then $\vec{F}$ is conservative.

10. $\int_C \vec{F} \cdot d\vec{r}$ is the total change in $\vec{F}$ along C.

11. If the vector fields $\vec{F}$ and $\vec{G}$ have $\int_C \vec{F} \cdot d\vec{r} = \int_C \vec{G} \cdot d\vec{r}$ for a particular path C, then $\vec{F} = \vec{G}$.

12. If the vector fields $\vec{F}$ and $\vec{G}$ have $\int_C \vec{F} \cdot d\vec{r} = \int_C \vec{G} \cdot d\vec{r}$ for a particular path C, then for each point (x, y) on the curve C we must have $\vec{F}(x, y) = \vec{G}(x, y)$.

13. If the total change of a function f along a curve C is zero, then C must be a contour of f.

14. Let $\vec{F}(x, y)$ be the conservative vector field in Figure 18.39. The vector field $\vec{F}$ associates with each point a unit vector pointing radially outward. The curves $C_1, C_2, \ldots, C_7$ have the directions shown. Consider the line integrals $\int_{C_i} \vec{F} \cdot d\vec{x}$, $i = 1, \ldots, 7$. Without computing any integrals:

     (a) List all the line integrals which are zero.
     (b) List all the negative line integrals.
     (c) List all the positive line integrals in ascending order.

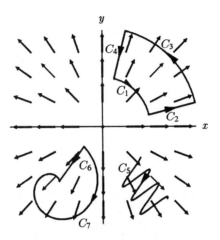

Figure 18.39

15. In this problem, we see how the Fundamental Theorem for Line Integrals can be derived from the Fundamental Theorem for ordinary definite integrals. Suppose that $(x(t), y(t))$, $a \le t \le b$, is a parameterization of C, which has endpoints $P = (x(a), y(a))$ and $Q = (x(b), y(b))$. The values of f along C are given by the single variable function $h(t) = f(x(t), y(t))$.

     (a) Show that
     $$h'(t) = f_x(x(t), y(t))\frac{dx}{dt} + f_y(x(t), y(t))\frac{dy}{dt}.$$

     (b) Use the Fundamental Theorem of Calculus applied to $h(t)$ to show $\int_C \operatorname{grad} f \cdot d\vec{r} = f(Q) - f(P)$.

16. Suppose that $\operatorname{grad} f = 2xe^{x^2} \sin y\vec{i} + e^{x^2} \cos y\vec{j}$. Find the change in f between $(0, 0)$ and $(1, \pi/2)$:   (a) By computing a line integral   (b) By computing f.

17. Suppose a particle subject to a force $\vec{F}(x, y) = y\vec{i} - x\vec{j}$ moves along a the arc of the unit circle, centered at the origin, that begins at $(-1, 0)$ and ends at $(0, 1)$.

     (a) Find the work done by $\vec{F}$. Explain the sign of your answer.
     (b) Is $\vec{F}$ conservative? Explain.

18. The vector field $\vec{F}(x, y) = x\vec{i} + y\vec{j}$ is conservative. Compute geometrically the line integrals over the three paths A, B and C shown in Figure 18.40 from $(1, 0)$ to $(0, 1)$ and verify that they are equal. Here A is a portion of a circle, B is a line, and C consists of two line segments meeting at a right angle.

19. The vector field $\vec{F}(x,y) = x\vec{i} + y\vec{j}$ is conservative. Compute algebraically the line integrals over the three paths A, B and C shown in Figure 18.41 from $(0,0)$ to $(1,1)$ and verify that they are equal. Here A is a line segment, B is the graph of $f(x) = x^2$ and C consists of two line segments meeting at a right angle.

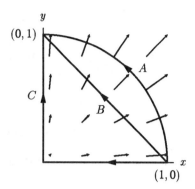

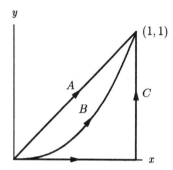

Figure 18.40: Paths for Problem 18   *Figure 18.41*: Paths for Problem 19

20. The line integral of $\vec{F} = (x+y)\vec{i} + x\vec{j}$ along each of the paths below is $3/2$:
    (i)   The path (t, t^2), with $0 \leq t \leq 1$
    (ii) The path (t^2, t), with $0 \leq t \leq 1$
    (iii) The path (t, t^n), with $n > 0$ and $0 \leq t \leq 1$

    Verify this
    (a) Using a parameterization to compute the line integral.
    (b) Using the Fundamental Theorem of Calculus for Line Integrals.

21. A particle moves with position vector $\vec{r}(t) = x(t)\vec{i} + y(t)\vec{j} + z(t)\vec{k}$. Let $\vec{v}(t)$ and $\vec{a}(t)$ be its velocity and acceleration vectors. Show that

$$\frac{1}{2}\frac{d}{dt}\|\vec{v}(t)\|^2 = \vec{a}(t) \cdot \vec{v}(t).$$

22. Let $\vec{F}$ be a conservative vector field. It is customary in physics to write $\vec{F} = -\nabla f$ for some potential function f. This exercise gives an example illustrating the significance of the minus sign. [2]
    (a) Let the xy-plane represent part of the Earth's surface with the z-axis pointing away from the Earth (we assume the scale is small enough so that a flat plane is a good approximation to the Earth's surface). Let $\vec{r} = x\vec{i} + y\vec{j} + z\vec{k}$, with $z \geq 0$, x, y, z measured in meters, be the position vector of a rock of unit mass. The gravitational potential energy function for the rock is $f(x, y, z) = gz$, where $g \approx 9.8$ m/s^2. Draw three distinct level surfaces of f. Does the potential energy increase or decrease with height above the Earth?
    (b) On a level surface of f sketch vectors ∇f and force vector, $\vec{F}$, of gravity. Explain the significance of the minus sign in the equation $\vec{F} = -\nabla f$ for the gravitational potential, f.

---

[2]Adapted from V.I. Arnold, *Mathematical Methods of Classical Mechanics*, 2nd Edition, Graduate Texts in Mathematics, Springer

23. In this problem we derive the principle of Conservation of Energy. The kinetic energy of a particle moving with speed v is $(1/2)mv^2$. For a conservative vector field $\vec{F}$, with potential function $f(x, y, z)$ (in the physicist's sense) so that $\vec{F} = -\nabla f$, the potential energy of a particle at position $\vec{r}$ is $f(\vec{r})$. The Conservation of Energy principle says that the expression

$$\text{total energy} = \text{kinetic energy} + \text{potential energy} = \frac{1}{2}m\|\vec{v}(t)\|^2 + f(\vec{r}(t))$$

is constant for a particle moving in the field with position vector $\vec{r}(t)$ and velocity vector $\vec{v}(t)$. Let P and Q be two points in space and let C be a path from P to Q parameterized by $\vec{r}(t)$, $t_0 \leq t \leq t_1$, where $\vec{r}(t_0) = P, \vec{r}(t_1) = Q$.

(a) Using Problem 21 and Newton's law $\vec{F} = m\vec{a}$, show

Work done by $\vec{F}$ as particle moves along $C =$ kinetic energy at Q – kinetic energy at P.

(b) If a vector field is conservative, $\vec{F} = -\nabla f$ for some f. Use the Fundamental Theorem of Calculus for Line Integrals to show that

Work = potential energy at P – potential energy at Q.

(c) Use (a) and (b) to show the total energy at P is the same as at Q.

24. A *central field* is a vector field whose direction is always toward (or away from) a fixed point C (the center) and whose magnitude at a point P is a function only of the distance from P to C. In two dimensions this means that the field has constant magnitude on circles centered at C. The gravitational and electrical fields of spherically symmetric sources are both central fields.

(a) Sketch an example of a central field $\vec{F}$.

(b) Suppose that the central field $\vec{F}$ is a gradient field, i.e. $\vec{F} = \text{grad } f$. What must be the shape of the contours of f? Sketch in some contours for this case.

(c) Is every gradient field a central field? Explain.

(d) In Figure 18.42, two paths are shown between the points Q and P. Assuming that there is a central force $\vec{F}$ with center C, explain why the work done by $\vec{F}$ is the same for either path.

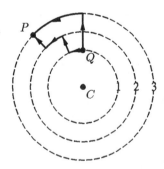

Figure 18.42

(e) It is in fact true that every central field is a gradient field. Use an argument suggested by Figure 18.42 to explain why the line integral of any central field must be path independent.

18.4   NONCONSERVATIVE FIELDS AND GREEN'S THEOREM

What if we are given a vector field but are not told whether it is conservative or not? How can we tell if it has a potential function, that is, if it is a gradient field?

How to Tell if a Vector Field is Nonconservative Using Line Integrals

Example 1   Is the vector field $\vec{F}$ shown in Figure 18.43 conservative? At any point $\vec{F}$ has magnitude equal to the distance from the origin and direction perpendicular to the line joining the point to the origin.

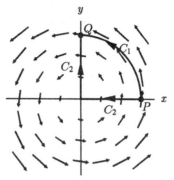

Figure 18.43: Is this field
conservative?

Solution   We choose $P = (1, 0)$ and $Q = (0, 1)$, and two paths between them: C_1, a quarter circle of radius 1, and C_2, formed by parts of the x and y-axes. (See Figure 18.43.) Along C_1, the line integral $\int_{C_1} \vec{F} \cdot d\vec{r} > 0$, since $\vec{F}$ points along the direction of the curve. Along C_2, however, we have $\int_{C_2} \vec{F} \cdot d\vec{r} = 0$, since $\vec{F}$ is perpendicular to C_2 everywhere. So $\vec{F}$ is not conservative.

Notice that the vector field in the previous example has non-zero circulation around the origin.

Non-Conservative Fields and Circulation

If $\vec{F}$ is a conservative vector field, what can we say about its circulation around a closed curve C? If P and Q are any two points on the path, then we can think of C as made up of the path C_1 followed by $-C_2$ (see Figure 18.44). Since $\vec{F}$ is conservative

$$\int_{C_1} \vec{F} \cdot d\vec{r} = \int_{C_2} \vec{F} \cdot d\vec{r},$$

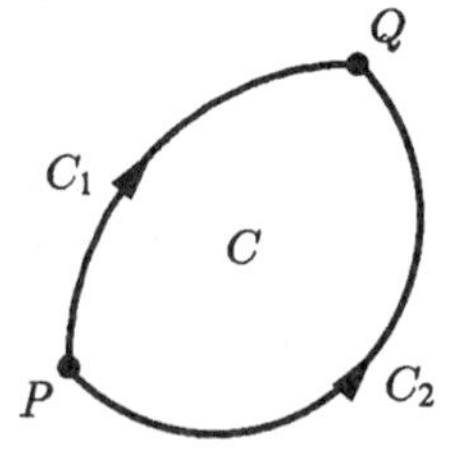

Figure 18.44: The path
C broken into two
pieces, C_1 and C_2

so

$$\int_C \vec{F} \cdot d\vec{r} = \int_{C_1} \vec{F} \cdot d\vec{r} - \int_{C_2} \vec{F} \cdot d\vec{r} = 0.$$

Thus, the circulation around C is zero. The reverse argument shows that if the circulation around any closed curve is zero, the line integral must be path-independent and so the vector field is conservative. We define a vector field to be *circulation free* if it has zero circulation around any closed path. Thus, we have the following result:

A vector field is conservative if and only if it is circulation free.

Hence, to tell that a field is non-conservative, we look for closed paths with non-zero circulation. For instance, the vector field in Example 1 has non-zero circulation along any circle around the origin, so it is not conservative.

How to Tell If a Vector Field is Nonconservative Algebraically

Example 2   Does the vector field $\vec{F} = 2xy\vec{i} + xy\vec{j}$ have a potential function? If so, find it.

Solution   Let's suppose $\vec{F}$ does have a potential function, f, so $\vec{F} = \text{grad } f$. This means that

$$\frac{\partial f}{\partial x} = 2xy \quad \text{and} \quad \frac{\partial f}{\partial y} = xy.$$

Integrating the expression for $\partial f/\partial x$ shows that we must have $f(x,y) = x^2 y + C(y)$ where $C(y)$ is a function of y. Differentiating this expression for $f(x,y)$ with respect to y and using the fact that $\partial f/\partial y = xy$, we get

$$\frac{\partial f}{\partial y} = x^2 + C'(y) = xy.$$

Thus, we must have

$$C'(y) = xy - x^2.$$

But this expression for $C'(y)$ is impossible because $C'(y)$ is a function of y alone. This argument shows that there is no potential function for the vector field $\vec{F}$.

Is there an easier way to see that a vector field has no potential function, other than by trying to find the potential function and failing? The answer is yes.

First we look at a 2-dimensional vector field $\vec{F} = F_1\vec{i} + F_2\vec{j}$. If $\vec{F}$ is a gradient field then there is a potential function f such that

$$\vec{F} = F_1\vec{i} + F_2\vec{j} = \frac{\partial f}{\partial x}\vec{i} + \frac{\partial f}{\partial y}\vec{j}.$$

Thus,

$$F_1 = \frac{\partial f}{\partial x} \quad \text{and} \quad F_2 = \frac{\partial f}{\partial y}.$$

Let us assume that f has continuous second partial derivatives. Then, by the equality of mixed partials:

$$\frac{\partial F_1}{\partial y} = \frac{\partial^2 f}{\partial y \partial x} = \frac{\partial^2 f}{\partial x \partial y} = \frac{\partial F_2}{\partial x}$$

Thus we have the following result:

> If $\vec{F} = F_1\vec{i} + F_2\vec{j}$ is a gradient vector field with continuous partial derivatives, then
>
> $$\frac{\partial F_1}{\partial y} = \frac{\partial F_2}{\partial x}.$$

Example 3   Show that $\vec{F} = 2xy\vec{i} + xy\vec{j}$ cannot be a gradient vector field.

Solution   We have $F_1 = 2xy$ and $F_2 = xy$. Since $\frac{\partial F_1}{\partial y} = 2x$ and $\frac{\partial F_2}{\partial x} = y$, in this case $\frac{\partial F_1}{\partial y} \neq \frac{\partial F_2}{\partial x}$ so $\vec{F}$ cannot be a gradient field.

Green's Theorem

We now have two ways of seeing that a vector field $\vec{F}$ is nonconservative. We can evaluate $\int_C \vec{F} \cdot d\vec{r}$ for some closed curve and find it is not zero, or we can show that $\frac{\partial F_1}{\partial y} \neq \frac{\partial F_2}{\partial x}$, that is, that $\frac{\partial F_1}{\partial y} - \frac{\partial F_2}{\partial x} \neq 0$. It's natural to think that

$$\int_C \vec{F} \cdot d\vec{r} \quad \text{and} \quad \frac{\partial F_1}{\partial y} - \frac{\partial F_2}{\partial x}$$

might be related. The relation is a very important result combining calculus and geometry, Green's theorem.

We restrict ourselves to 2 dimensions and consider $\int_C \vec{F} \cdot d\vec{r}$, where C is the closed curve, that does not cross itself, oriented as shown in Figure 18.45 and $\vec{F} = F_1\vec{i} + F_2\vec{j}$. We divide up the region R bounded by C, into small pieces, each one bounded by a closed curve with the orientation shown. Figure 18.46 shows that if we add the circulation around each of these small closed curves, each common edge of a pair of adjacent closed curves is counted twice, once in each direction. Hence the integrals along these edges cancel. Thus, the line integrals along all the edges inside the region R cancel, giving

$$\begin{array}{c} \text{Circulation of } \vec{F} \\ \text{around } C \end{array} = \sum_{\Delta C} \begin{array}{c} \text{Circulation of } \vec{F} \\ \text{around small curve, } \Delta C \end{array}$$

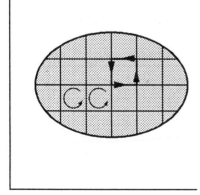

Figure 18.45: Region R bounded by a closed curve C and split into many small regions, ΔR

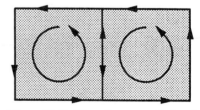

Figure 18.46: Two adjacent small closed curves

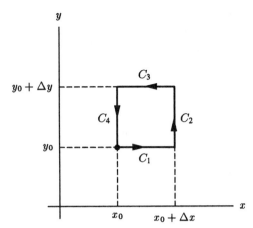

Figure 18.47: A small closed curve ΔC broken into
C_1, C_2, C_3, C_4

Now we estimate the line integral around one of these small closed curves, shown in Figure 18.47. Assuming that ΔC is small, we can estimate the line integral by replacing $\vec{F}$ by its linear approximation:

$$F_1(x, y) \approx F_1(x_0, y_0) + \frac{\partial F_1}{\partial x}(x_0, y_0)(x - x_0) + \frac{\partial F_1}{\partial y}(x_0, y_0)(y - y_0)$$

$$F_2(x, y) \approx F_2(x_0, y_0) + \frac{\partial F_2}{\partial x}(x_0, y_0)(x - x_0) + \frac{\partial F_2}{\partial y}(x_0, y_0)(y - y_0).$$

Thus for (x, y) near (x_0, y_0), we can approximate F_1 and F_2 as follows:

$$F_1(x, y) \approx a + b(x - x_0) + c(y - y_0)$$
$$F_2(x, y) \approx d + e(x - x_0) + f(y - y_0),$$

where a, b, c, d, e, and f, are constants, with $c = \partial F_1/\partial y$, and so on. When we integrate around ΔC, we have

$$\int_{\Delta C} \vec{F} \cdot d\vec{r} = \int_{\Delta C} (F_1\, dx + F_2\, dy)$$

$$\approx \int_{\Delta C} \left(a + b(x - x_0) + c(y - y_0)\right) dx + \left(d + e(x - x_0) + f(y - y_0)\right)\, dy.$$

A constant vector field and a vector field whose components are $bx\vec{i} + fy\vec{j}$ are both conservative (because they are gradient fields), and so have line integrals of 0 around ΔC. Thus, this line integral reduces to:

$$\int_{\Delta C} \vec{F} \cdot d\vec{r} \approx \int_{\Delta C} cy\, dx + \int_{\Delta C} ex\, dy$$

$$\approx \int_{C_1} cy\, dx + \int_{C_3} cy\, dx + \int_{C_2} ex\, dy + \int_{C_4} ex\, dy$$

$$\approx cy_0\, \Delta x - c(y_0 + \Delta y)\, \Delta x + e(x_0 + \Delta x)\, \Delta y - ex_0\, \Delta y$$

$$= -c\, \Delta y\, \Delta x + e\, \Delta x\, \Delta y$$

$$= \left(-\frac{\partial F_1}{\partial y} + \frac{\partial F_2}{\partial x}\right) \Delta x \Delta y.$$

Summing over all small regions gives

$$\int_C \vec{F} \cdot d\vec{r} \approx \sum_{\Delta C} \int_{\Delta C} \vec{F} \cdot d\vec{r} \approx \sum_{\Delta R} \left(-\frac{\partial F_1}{\partial y} + \frac{\partial F_2}{\partial x} \right) \Delta x \, \Delta y.$$

The last sum is a Riemann sum approximating a two-variable integral. Thus, in the limit as the size of each subdivision shrinks to zero, we have the following result:

Green's Theorem

Suppose C is a closed curve, that does not cross itself, surrounding a region R in the plane and oriented so that the region is on the left as you move around the curve in the forward direction. Suppose $\vec{F} = F_1\vec{i} + F_2\vec{j}$ is a vector field with continuous partial derivatives at every point in the region R. Then

$$\int_C \vec{F} \cdot d\vec{r} = \int_R \left(-\frac{\partial F_1}{\partial y} + \frac{\partial F_2}{\partial x} \right) dx \, dy$$

What Does Green's Theorem Tell Us About Conservative Vector Fields?

Let us suppose we have a 2-dimensional vector field $\vec{F} = F_1\vec{i} + F_2\vec{j}$ defined in a region T in the plane with no holes in it. Suppose also that

$$\frac{\partial F_1}{\partial y} = \frac{\partial F_2}{\partial x}$$

throughout T. If C is any closed curve[3] contained in T and oriented for Green's Theorem, let R be the region inside C, then

$$\int_C \vec{F} \cdot d\vec{r} = \int_R \left(-\frac{\partial F_1}{\partial y} + \frac{\partial F_2}{\partial x} \right) dx \, dy = 0$$

since the integrand in the two-variable integral is identically 0. Therefore,

$$\int_C \vec{F} \cdot d\vec{r} = 0,$$

so $\vec{F}$ is conservative. Thus, with the boxed result on page 373, we have the following result:

If $\vec{F}$ has continuous second partials at all points within a region T in the plane with no holes in it, then the following statements are equivalent.

- $\dfrac{\partial F_1}{\partial y} = \dfrac{\partial F_2}{\partial x}$ at all points in T

- $\vec{F}$ is a gradient vector field.

Are Holes in T Really Important?

The reason for assuming T has no holes in it is that we need to be sure that every point enclosed by C is really inside T so that we can apply Green's Theorem to C. Let's see what can go wrong if T has holes in it.

---

[3]We assume the curve doesn't cross itself. It it does, we break the curve up and apply Green's Theorem to each piece.

Example 4   Let $\vec{F} = \dfrac{-y\vec{i} + x\vec{j}}{x^2 + y^2}$. Calculate:

(a) $\displaystyle\int_C \vec{F} \cdot d\vec{r}$ where C is the unit circle centered at the origin and traversed counterclockwise.

(b) $-\dfrac{\partial F_1}{\partial y} + \dfrac{\partial F_2}{\partial x}$

(c)   Explain why the answers to parts (a) and (b) do not contradict Green's Theorem.

Solution   (a)   On the unit circle, $\vec{F}$ is tangent to the circle and $||\vec{F}|| = 1$. Thus,

$$\int_C \vec{F} \cdot d\vec{r} = ||\vec{F}|| \cdot (\text{Length of curve}) = 1 \cdot 2\pi 1 = 2\pi.$$

(b)

$$-\frac{\partial F_1}{\partial y} = -\frac{\partial}{\partial y}\left(\frac{-y}{x^2 + y^2}\right) = \frac{1}{x^2 + y^2} - \frac{y \cdot 2y}{(x^2 + y^2)^2} = \frac{x^2 - y^2}{(x^2 + y^2)^2}.$$

Similarly,

$$\frac{\partial F_2}{\partial x} = \frac{\partial}{\partial x}\left(\frac{x}{x^2 + y^2}\right) = \frac{1}{x^2 + y^2} - \frac{x \cdot 2x}{(x^2 + y^2)^2} = \frac{y^2 - x^2}{(x^2 + y^2)^2}$$

Thus,

$$-\frac{\partial F_1}{\partial y} + \frac{\partial F_2}{\partial x} = 0.$$

(c)   Since the partial derivatives of F_1 and F_2 do not exist at the origin, Green's Theorem does not hold for any region containing the origin.

Problems for Section 18.4

1.   Example 1 on page 371 showed that the vector field in Figure 18.48 could not be a gradient field by showing that it is not conservative. Here is another way to see the same thing. Suppose that the vector field were the gradient of a function f. Draw and label a diagram showing what the contours of f would have to look like, and explain why it would not be possible for f to have a single value at any given point.

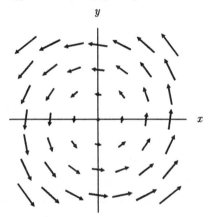

Figure 18.48

2. Repeat Problem 1 for the vector field in Problem 6 on page 367.

3. Find f if grad $f = 2xy\vec{i} + (x^2 + 8y^3)\vec{j}$

4. Find f if grad $f = (yze^{xyz} + z^2 \cos(xz^2))\vec{i} + (xze^{xyz})\vec{j} + (xye^{xyz} + 2xz \cos(xz^2))\vec{k}$

For Problems 5–6, decide whether the given vector field is the gradient of a function f. If so, find such an f. If not, explain why not.

5. $y\vec{i} + y\vec{j}$ 6. $(x^2 + y^2)\vec{i} + 2xy\vec{j}$

Do the vector fields in Problems 7–10 have potential functions? If so, compute them.

7. $\vec{F} = (2xy^3 + y)\vec{i} + (3x^2y^2 + x)\vec{j}$

8. $\vec{F} = \dfrac{\vec{i}}{x} + \dfrac{\vec{j}}{y} + \dfrac{\vec{k}}{xy}$

9. $\vec{F} = \dfrac{\vec{i}}{x} + \dfrac{\vec{j}}{y} + \dfrac{\vec{k}}{z}$

10. $\vec{F} = 2x \cos(x^2 + z^2)\vec{i} + \sin(x^2 + z^2)\vec{j} + 2z \cos(x^2 + z^2)\vec{k}$

11. Suppose $\vec{F} = x\vec{j}$. Show that the line integral of $\vec{F}$ around a closed curve in the xy-plane measures the area of the region enclosed by the curve.

12. Consider the vector field $\vec{F} = y\vec{i}$.

 (a) Sketch $\vec{F}$ and hence decide the sign of the circulation of $\vec{F}$ around the unit circle centered at the origin and traversed counterclockwise.

 (b) Use Green's Theorem to compute the circulation in part (a) exactly.

REVIEW PROBLEMS FOR CHAPTER EIGHTEEN

For Problems 1–2, consider the vector field $\vec{F}$ shown. Say whether the line integral $\int_C \vec{F} \cdot d\vec{r}$ is positive, negative, or zero along (a) A (b) C_1, C_2, C_3, C_4 (c) C, the closed curve consisting of all the C's together.

1.
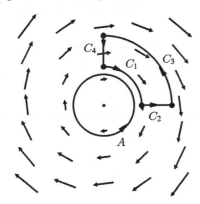

2.
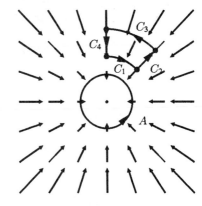

For Problems 3–5, compute $\int_C \vec{F} \cdot d\vec{r}$ for the given $\vec{F}$ and C.

3. $\vec{F} = (x^2 - y)\vec{i} + (y^2 + x)\vec{j}$, C is the parabola $y = x^2 + 1$ traversed from $(0, 1)$ to $(1, 2)$.

4. $\vec{F} = (2x - y + 4)\vec{i} + (5y + 3x - 6)\vec{j}$, C is the triangle with vertices $(0,0), (3,0), (3,2)$ traversed counterclockwise.

5. $\vec{F} = (3x - 2y)\vec{i} + (y + 2z)\vec{j} - x^2\vec{k}$, C is the path consisting of the straight line joining the points $(0, 0, 0)$ to $(1, 1, 1)$.

Are the statements in Problems 6–9 true or false? Explain why or give a counterexample.

6. $\int_C \vec{F} \cdot d\vec{r}$ is a vector.

7. $\int_C \vec{F} \cdot d\vec{r} = \vec{F}(Q) - \vec{F}(P)$ when P and Q are the endpoints of C.

8. The fact that the line integral of a vector field $\vec{F}$ is zero around the unit circle $x^2 + y^2 = 1$ means that $\vec{F}$ must be a gradient vector field.

9. Suppose C_1 is the unit square joining the points $(0, 0), (1, 0), (1, 1), (0, 1)$ oriented clockwise and C_2 is the same square but traversed twice in the opposite direction. If $\int_{C_1} \vec{F} \cdot d\vec{r} = 3$, then $\int_{C_2} \vec{F} \cdot d\vec{r} = -6$.

10. Suppose P and Q both lie on the same contour of f. What can you say about the total change in f from P to Q? Explain your answer in terms of $\int_C \operatorname{grad} f \cdot d\vec{r}$ where C is a portion of the contour that goes from P to Q.

11. A *free vortex* circulating about the origin in the xy-plane (or about the z-axis in 3-space) has vector field $\vec{v} = K(x^2 + y^2)^{-1}(-y\vec{i} + x\vec{j})$ where K is a constant. The Rankine model of a tornado hypothesizes an inner core that rotates at constant angular velocity, surrounded by a free vortex. Suppose that the inner core has radius 100 m and that $\|\vec{v}\| = 3 \cdot 10^5$ m/hr at a distance of 100 m from the center.

    (a) Assuming that the tornado rotates counterclockwise (viewed from above the xy-plane) and that $\vec{v}$ is continuous, determine ω and K such that

    $$\vec{v} = \begin{cases} \omega(-y\vec{i} + x\vec{j}) & \text{if } \sqrt{x^2 + y^2} < 100 \\ K(x^2 + y^2)^{-1}(-y\vec{i} + x\vec{j}) & \text{if } \sqrt{x^2 + y^2} \geq 100 \end{cases}$$

    (b) Sketch the vector field $\vec{v}$.

    (c) Compute the circulation of $\vec{v}$ about the circle of radius r centered at the origin, traversed counterclockwise.

12. Figure 18.49 shows the tangential velocity as a function of radius for the tornado that hit Dallas on April 2, 1957. Use it to estimate K and ω for the tornado.[4]

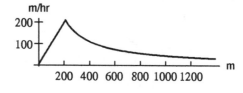

Figure 18.49

---

[4] Adapted from *Encyclopedia Britannica, Macropedia*, Vol. 16, page 477, "Climate and the Weather", Tornadoes and Waterspouts, 1991.

CHAPTER NINETEEN

FLUX INTEGRALS

In the previous chapter we saw how to integrate vector fields along curves. In this chapter we shall define a new sort of integral, the flux integral, which goes over a surface rather than along a curve. If we view a vector field as representing the velocity of a fluid flow, the flux integral tells us about the rate at which fluid is flowing through the surface. In addition, the flux integral appears in the theory of electricity and magnetism.

19.1   THE IDEA OF A FLUX INTEGRAL

Flow Through a Surface

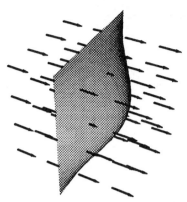

Figure 19.1: Flux measures rate of
flow through a surface

Suppose we want to measure the flow rate of a fluid through a porous surface, that is, the volume of fluid that passes through the surface per unit time. For example, imagine water flowing through a fishing net stretched across a stream. (See Figure 19.1.) This flow rate is called the *flux* of the fluid through the surface.

Orientation of a Surface

Flux can be positive or negative depending on the direction of flow. Before computing the flux, we need to decide which direction is positive.

> A surface is said to be *oriented* if one direction of flow through the surface has been chosen as the positive direction. This choice is called a choice of *orientation*.

Figure 19.2 shows two different orientations for the same surface. Often the orientation of a surface is indicated by putting on the surface a normal vector that points in the direction of positive flow. A normal vector pointing in the direction of positive flow is called a *positive normal*.

The Area Vector

The flux through a flat surface depends both on the area of the surface and the direction in which the surface faces. Thus, in computing flux, it is useful to represent area by a vector quantity.

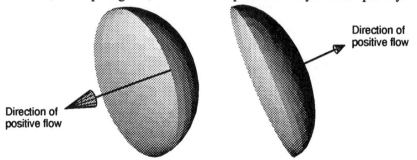

Direction of
positive flow

Direction of
positive flow

Figure 19.2: Two different orientations of the same surface

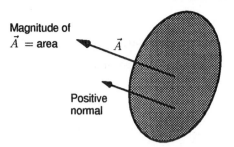

Figure 19.3: The area vector $\vec{A}$ of an oriented
flat piece of surface

The *area vector* of a flat, oriented surface is a vector $\vec{A}$ such that:
- The magnitude of $\vec{A}$ is the area of the surface.
- The direction of $\vec{A}$ is the direction of a positive normal.

See Figure 19.3.

The Flux of a Constant Flow Through a Flat Surface

We first consider the simple situation where the velocity vector field of the fluid is constant and the surface is flat, with area vector $\vec{A}$. The flux through the area $\vec{A}$ is the volume of fluid that flows past in one unit of time—the volume of the skewed cylinder in Figure 19.4. It has cross-sectional area $\|\vec{A}\|$ and height $\|\vec{v}\|\cos\theta$, so its volume is $(\|\vec{v}\|\cos\theta)\|\vec{A}\| = \vec{v}\cdot\vec{A}$. Thus,

$$\text{Flux through } \vec{A} = \vec{v}\cdot\vec{A}.$$

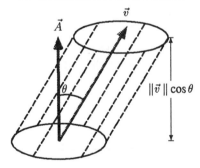

Figure 19.4: Flux of $\vec{v}$ through area $\vec{A}$

Example 1 Water is flowing down a cylindrical pipe 2 cm in radius with a velocity of 3 cm/sec. Find the flux of the vector field through the following regions:

(a) A 2 cm radius circular region perpendicular to the pipe.

(b) An ellipse-shaped region whose normal makes an angle of θ with the direction of flow, and which cuts all the way across the pipe. The area of this region is $4\pi/\cos\theta$ cm^2.

Both surfaces are oriented so that the flow is positive.

Solution   The answer to both parts of this question will be the same, because in both cases the flux tells us the rate at which water is flowing down the pipe. Therefore,

$$\text{Flux through circle} = \text{Flux through ellipse}$$
$$= \frac{\text{Rate of flow}}{\text{of water}} \times \frac{\text{Area of}}{\text{circle}}$$
$$= \left(3\,\frac{\text{cm}}{\text{sec}}\right)(\pi 2^2\,\text{cm}^2) = 12\pi\,\text{cm}^3/\text{sec}.$$

(a)   Let's calculate the flux using the formula above to confirm that water is flowing down the pipe at $12\pi\,\text{cm}^3/\text{sec}$. Let $\vec{A}$ be the area vector of the circle, and let $\vec{v}$ be the velocity vector of the fluid. Since $\vec{v}$ and $\vec{A}$ are parallel and in the same direction (see Figure 19.5), the flux through the circular region is given by

$$\vec{v} \cdot \vec{A} = \|\vec{v}\|\|\vec{A}\| = 3(\text{Area of circle}) = 3(\pi 2^2) = 12\pi\,\text{cm}^3/\text{sec}.$$

(b)   For the ellipse-shaped region in Figure 19.6,

$$\vec{v} \cdot \vec{A} = \|\vec{v}\| \cdot \|\vec{A}\|\cos\theta = 3(\text{Area of ellipse})\cos\theta$$
$$= 3\left(\frac{4\pi}{\cos\theta}\right)\cos\theta = 12\pi\,\text{cm}^3/\text{sec}.$$

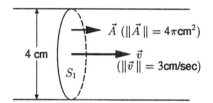

Figure 19.5: Flux through circular region

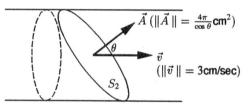

Figure 19.6: Flux through ellipse-shaped region

What if the Flow is not Constant?

When the flow is not constant, we follow the usual strategy of dividing the surface up into pieces where the flow is approximately constant. In the next example, we give a more realistic model for water flowing down a pipe.

Example 2   Suppose water is flowing down a cylindrical pipe of radius 2 cm, and that the speed is $(3 - (3/4)r^2)$ cm/sec at a distance r cm from the center of the pipe. Find the flux through the circular cross-section of the pipe, oriented so that the flow is positive.

Solution   Notice that the speed is 3 cm/sec at the center of the pipe and 0 cm/sec at the sides. Suppose $\vec{i}$ is the unit vector parallel to the direction of flow. Then, at a distance r from the center of the pipe, the velocity is given by

$$\vec{v} = \left(3 - \frac{3}{4}r^2\right)\vec{i}\,\text{cm/sec}.$$

Divide the circular cross-section into concentric rings of width Δr, so that the velocity is approximately constant on each one. The area of a typical ring is $\Delta A \approx 2\pi r \Delta r$. Then since $\vec{v}$ and $\Delta \vec{A}$ are parallel (see Figure 19.7), we have

$$\vec{v} \cdot \Delta \vec{A} = \|\vec{v}\| \cdot \|\Delta \vec{A}\| \approx \left(3 - \frac{3}{4}r^2\right)\frac{\text{cm}}{\text{sec}} \cdot (2\pi r \Delta r)\,\text{cm}^2.$$

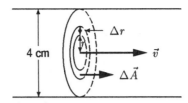

Figure 19.7: Flux through pipe when
velocity varies with distance from the center

Thus, the flux through the circular cross-section is given by

$$\lim_{\|\Delta \vec{A}\| \to 0} \sum \vec{v} \cdot \Delta \vec{A} = \lim_{\Delta r \to 0} \sum \left(3 - \frac{3}{4}r^2\right)(2\pi r \Delta r) = \int_{r=0}^{r=2} \left(3 - \frac{3}{4}r^2\right) 2\pi r \, dr$$

$$= 6\pi \int_0^2 \left(r - \frac{r^3}{4}\right) dr$$

$$= 6\pi \text{ cm}^3/\text{sec}.$$

We evaluated the flux integral by converting it into a Riemann sum over a single variable r. This, in turn, we recognized as a definite integral, which we evaluated by antidifferentiation.

The Flux Integral

The idea of the flux of a fluid through a surface can be applied to any vector field, whether or not it represents a fluid flow. To calculate the flux of a vector field $\vec{F}$ through a curved surface, we divide the surface into a patchwork of small, approximately flat pieces (like a wire-frame representation of the surface). If the patches are small enough, we can assume that $\vec{F}$ is approximately constant on each piece. We let $\Delta \vec{A}$ be an approximate area vector of a small piece. See Figure 19.8. Thus,

$$\text{Flux through } \Delta \vec{A} \approx \vec{F} \cdot \Delta \vec{A}$$

and

$$\text{Flux through whole surface} \approx \sum \vec{F} \cdot \Delta \vec{A},$$

where the sum adds up the fluxes through all the small pieces. As each of the patches becomes smaller and smaller, and $\|\Delta \vec{A}\| \to 0$, the approximation gets better and better, and we get

$$\text{Flux through } S = \lim_{\|\Delta \vec{A}\| \to 0} \sum \vec{F} \cdot \Delta \vec{A}.$$

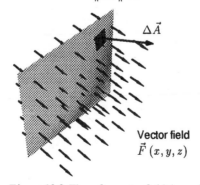

Figure 19.8: Flux of a vector field through
a curved surface with a patch of area $\Delta \vec{A}$

The **flux integral** of the vector field $\vec{F}$ through the oriented surface S is

$$\int_S \vec{F} \cdot d\vec{A} = \lim_{\|\Delta \vec{A}\| \to 0} \sum \vec{F} \cdot \Delta \vec{A}.$$

In the case that the vector field is the velocity field $\vec{v}$ of a fluid, we have

$$\begin{array}{ccc} \text{Rate fluid flows through} \\ \text{whole surface} \end{array} = \begin{array}{c} \text{Flux of } \vec{v} \\ \text{through } S \end{array} = \int_S \vec{v} \cdot d\vec{A}$$

How Does the Limit Defining a Flux Integral Work?

In computing a flux integral, you have to divide the surface up in a reasonable way, or the limit won't exist. The limit exists if $\vec{F}$ is continuous on an open region containing the surface, and if the surface is made by piecing together a finite number of smooth surfaces, that is, surfaces which are parameterized by smooth functions. We can use the parameterization to subdivide the surface by subdividing the parameter region in the same way as for a conventional two-variable integral. This leads to a subdivision of the surface into parameter rectangles. In this case, not only does the area $\|\Delta \vec{A}\|$ of each patch tend to zero, but the diameter tends to zero also. This ensures that each patch becomes approximately flat, so that $\vec{F}$ is approximately constant on each patch. In Sections 19.2 and 19.3 we will see how to use a parameterization to compute flux integrals over graphs of functions and parameterized surfaces.

Example 3 The vector field $\vec{v}$ in Figure 19.9 is circling around the z-axis. The magnitude of the vector at a distance of r units from the z-axis is $2r$, in the direction shown. Find the flux of $\vec{v}$ through the square S of side length 2 shown in Figure 19.10. The square is oriented so that the positive x-direction is positive.

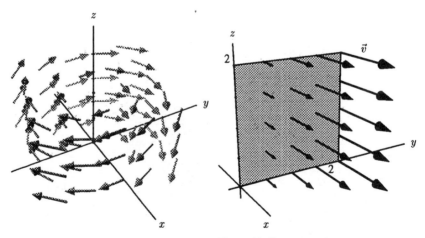

Figure 19.9: The vector field $\vec{v}$

Figure 19.10: Flux of $\vec{v}$ through the square S of side 2

Solution   Consider a small patch with area vector $\Delta \vec{A}$ in S, with sides Δy and Δz so that $\|\Delta \vec{A}\| = \Delta y \, \Delta z$. See Figure 19.11. At the point $(0, y, z)$ in S, the vector $\vec{v}$ is perpendicular to the plane, in the same direction as $\Delta \vec{A}$, and $\|\vec{v}\| = 2y$ since y is the distance from the z-axis. Hence,

$$\vec{v} \cdot \Delta \vec{A} = \|\vec{v}\| \|\Delta \vec{A}\| = 2y \, \Delta y \, \Delta z.$$

So

$$\int_S \vec{v} \cdot d\vec{A} = \lim_{\|\Delta \vec{A}\| \to 0} \sum \vec{v} \cdot \Delta \vec{A}$$

$$= \lim_{\substack{\Delta y \to 0 \\ \Delta z \to 0}} \sum 2y \, \Delta y \, \Delta z.$$

This last expression is a Riemann sum for the double integral $\int_R (2y) dA$, where R is the square $0 \leq y \leq 2$, $0 \leq z \leq 2$. Thus

$$\int_S \vec{v} \cdot d\vec{A} = \int_R (2y) \, dA$$

$$= \int_0^2 \int_0^2 (2y) \, dy \, dz = 8.$$

The answer is positive, as we would expect, since the vector field is passing through the surface in the positive direction.

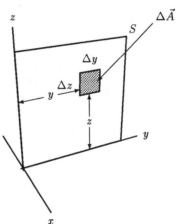

Figure 19.11: A small patch of area with $\|\Delta \vec{A}\| = \Delta y \Delta z$

Example 4   Find the flux of the vector field

$$\vec{F}(x, y, z) = (x^2 + y^2)\vec{i} + xy\vec{j}$$

through the square region in the xy-plane with corners at $(1, 1, 0)$, $(-1, 1, 0)$, $(1, -1, 0)$, and $(-1, -1, 0)$, oriented with upward pointing normal.

Solution   All the vectors in the vector field point horizontally (because their z-component is zero), and the surface is horizontal, so there is no flow through the surface and the flux is zero.

Flux through a Closed Surface

A closed surface is a surface with no boundary, such as a sphere. See Figure 19.12. It is conventional to orient a closed surface so that the positive direction of flow is from inside to outside. Thus, in computing a flux integral, the area vectors $\Delta \vec{A}$ all point outwards.

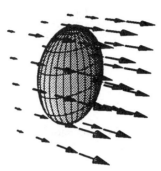

Figure 19.12: Flux through a closed surface

For a closed surface S and a vector field $\vec{F}$

$$\text{Flux through } S = \int_S \vec{F} \cdot d\vec{A} = \text{flux out of the region enclosed by the surface.}$$

Example 5   Each of the vector fields in Figure 19.13 consists entirely of vectors parallel to the xy-plane, and is constant in the z direction (that is, the vector field looks the same in any plane parallel to the xy-plane). For each one, decide whether the flux through a closed surface surrounding the origin is positive, negative, or zero. In (a) the surface is a cube with faces parallel to the axes; in (b) and (c) the surface is a cylinder with end caps. Cross sections of each surface are shown in Figure 19.13.

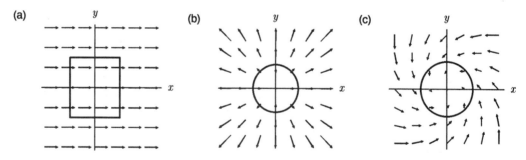

Figure 19.13: Flux of a vector field through a closed surface

Solution   (a)   Since the vector field is parallel to the faces of the cube which are perpendicular to the y- and z-axes, the flux through these faces is zero. The fluxes through the two faces which are perpendicular to the x-axis are equal in magnitude and opposite in sign, so the net flux is zero.

(b)   Since the end caps are parallel to the flow, the flux through them is zero. Along the cylinder, using the outward normal, $\vec{v}$ and $\Delta \vec{A}$ are everywhere parallel and in the same direction, so each term $\vec{v} \cdot \Delta \vec{A}$ is positive, and therefore the flux integral $\int_S \vec{v} \cdot d\vec{A}$ is positive.

(c) As in (b), the flux through the endcaps is zero. In this case $\vec{v}$ and $\Delta\vec{A}$ are not parallel along the cylinder, but since the fluid is flowing inwards as well as swirling, each term $\vec{v} \cdot \Delta\vec{A}$ is negative, and so the flux integral is negative.

Example 6 Find the flux of the vector field $\vec{v} = \vec{r}$ out of the sphere of radius R.

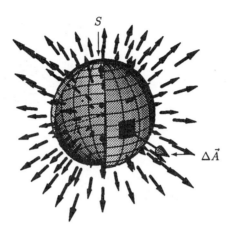

Figure 19.14: Flux of $\vec{v}$ through surface of a sphere

Solution Since this vector field points radially outward from the origin, it always points in the same direction as the surface area vector, $\Delta\vec{A}$. In addition, $\|\vec{v}\| = R$ on the surface, S, so we have

$$\vec{v} \cdot \Delta\vec{A} = \|\vec{v}\|\|\Delta\vec{A}\| = R\|\Delta\vec{A}\|.$$

Thus,

$$\int_S \vec{v} \cdot d\vec{A} = \lim_{\|\Delta\vec{A}\|\to 0}\sum \vec{v} \cdot \Delta\vec{A} = \lim_{\|\Delta\vec{A}\|\to 0}\sum R\|\Delta\vec{A}\| = R\lim_{\|\Delta\vec{A}\|\to 0}\sum \|\Delta\vec{A}\|.$$

The last sum adds up all the little areas $\|\Delta\vec{A}\|$, so it approximates the surface area of the sphere. The approximation gets better as the subdivisions get finer, so that in the limit we have

$$\lim_{\|\Delta\vec{A}\|\to 0}\sum \|\Delta\vec{A}\| = \text{Surface area of sphere}.$$

Thus, the flux is given by

$$\int_S \vec{v} \cdot d\vec{A} = R\lim_{\|\Delta\vec{A}\|\to 0}\sum \|\Delta\vec{A}\| = R(\text{Surface area of sphere}) = R(4\pi R^2) = 4\pi R^3.$$

Problem 17 on page 391 generalizes this result.

Problems for Section 19.1

1. Let $\vec{F}(x, y, z) = z\vec{i}$. For each of the surfaces in (a)–(e), say whether the flux of $\vec{F}$ through the surface is positive, negative, or zero. In each case, the orientation of the surface is indicated by the given normal vector.

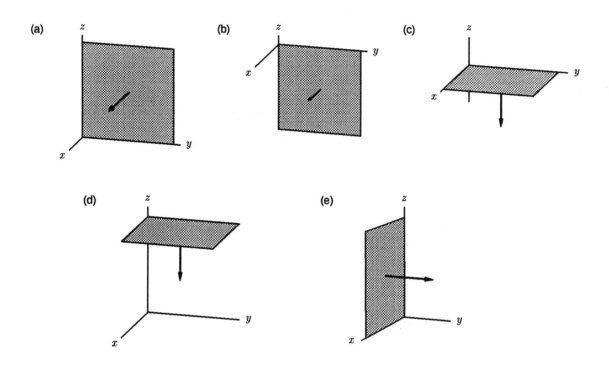

2. Repeat Problem 1 with $F(x, y, z) = -z\vec{i} + x\vec{k}$.
3. Repeat Problem 1 with the vector field $\vec{F}(\vec{r}) = \vec{r}$.
4. Arrange the following flux integrals,

$$\int_{S_i} \vec{F} \cdot d\vec{A},$$

with $i = 1, 2, 3, 4$, in ascending order if $\vec{F} = -\vec{i} - \vec{j} + \vec{k}$ and S_i are the following surfaces:
- S_1 is a horizontal square of side 1, oriented upward with one corner at $(0, 0, 2)$ and above the first quadrant of the xy-plane.
- S_2 is a horizontal square of side 1, oriented upward with one corner at $(0, 0, 3)$ and above the third quadrant of the xy-plane.
- S_3 is a square of side $\sqrt{2}$ in the xz-plane with one corner at the origin, one edge along the positive x-axis, one along the negative z-axis, and oriented in the negative y- direction.
- S_4 is a square of side $\sqrt{2}$, one corner at the origin, one edge along the positive y-axis, one corner at $(1, 0, 1)$, and oriented upwards.

5. Compute the flux of the vector field $\vec{v} = 2\vec{i} + 3\vec{j} + 5\vec{k}$ through each of the rectangular regions in (a)–(d), assuming each is oriented as shown.

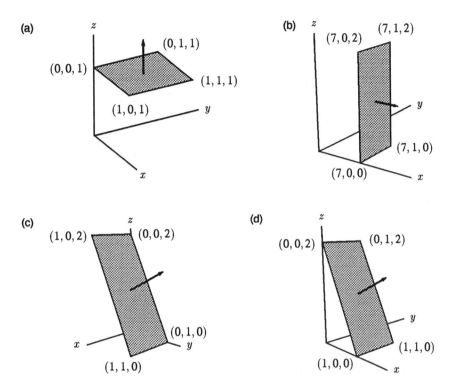

6. Figure 19.15 shows a schematic cross-section of the earth's magnetic field. Say whether the magnetic flux through a horizontal plate, oriented skyward, is positive, negative, or zero if the plate is (a) At the north pole.   (b) At the south pole.   (c) On the equator.

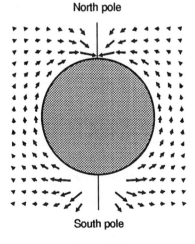

Figure 19.15

7. (a) What do you think will be the electric flux through the cylindrical surface that is placed as shown in the constant electric field in Figure 19.16? Why?

   (b) What if the cylinder is placed upright, as shown in Figure 19.17? Explain.

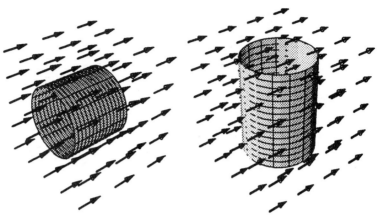

<table>
<tr><td align="center">*Figure 19.16*</td><td align="center">*Figure 19.17*</td></tr>
</table>

For Problems 8–12, compute the flux integral of the given vector field through the given surface S.

8. $\vec{F} = 2\vec{i}$ and S is a disk of radius 2 on the plane $x + y + z = 2$, oriented upward.

9. $\vec{F} = -y\vec{i} + x\vec{j}$ and S is the square plate in the yz-plane with corners at $(0, 1, 1), (0, -1, 1)$, $(0, 1, -1)$, and $(0, -1, -1)$, oriented in the positive x-direction.

10. $\vec{F} = -y\vec{i} + x\vec{j}$ and S is the disk in the xy-plane with radius 2, oriented upwards and centered at the origin.

11. $\vec{F} = \vec{r}$ and S is the disk of radius 2 parallel to the xy-plane oriented upwards and centered at $(0, 0, 2)$.

12. $\vec{F}(x, y, z) = (2 - x)\vec{i}$ and S is the cube whose vertices include the points $(0, 0, 0), (3, 0, 0)$, $(0, 3, 0), (0, 0, 3)$, and oriented outward.

13. Find the flux of $\vec{F}(\vec{r}) = \vec{r}/r^3$ through the sphere of radius R centered at the origin.

14. Find the flux of $\vec{F}(\vec{r}) = \vec{r}/r^2$ through the sphere of radius R centered at the origin.

15. Let S be the cube with side 2, faces parallel to the coordinate planes, and centered at the origin.

    (a) Calculate the total flux of the constant vector field $\vec{v} = -\vec{i} + 2\vec{j} + \vec{k}$ out of S by computing the flux through each face separately.

    (b) Calculate the flux out of S for any constant vector field $\vec{v} = a\vec{i} + b\vec{j} + c\vec{k}$.

    (c) Do your answers in parts (a) and (b) make sense? Explain.

16. Let S be the tetrahedron with vertices at the origin and at $(1, 0, 0), (0, 1, 0)$ and $(0, 0, 1)$.

    (a) Calculate the total flux of the constant vector field $\vec{v} = -\vec{i} + 2\vec{j} + \vec{k}$ out of S by computing the flux through each face separately.

    (b) Calculate the flux out of S in part (a) for any constant vector field $\vec{v}$.

    (c) Do your answers in parts (a) and (b) make sense? Explain.

17. Explain why if $\vec{F}$ has constant magnitude on S and is everywhere normal to S and in the direction of orientation, then

$$\int_S \vec{F} \cdot d\vec{A} = \|\vec{F}\| \text{Area of } S.$$

For Problems 18–19 let $\vec{F}(\vec{r}) = \vec{r}$, and let S be a square plate perpendicular to the z-axis and centered on the z-axis. Sketch as a function of time the flux of $\vec{F}$ through S as S moves in the given manner.

18. S moves from far up the positive z-axis to far down the negative z-axis. Assume S is oriented upward.

19. S rotates about an axis parallel to the x-axis, through the center of S. Assume S is far up the z-axis so that $\vec{r}$ is approximately constant on S as S rotates, and S is initially oriented upward.

20. Repeat Problems 18–19 with $\vec{F}(\vec{r}) = \vec{r}/r^3$.

21. Consider a body of fluid with an xyz-coordinate system to locate points in it. Let $P(x, y, z)$ be the pressure at the point (x, y, z). Let $\vec{F}(x, y, z) = P(x, y, z)\vec{k}$. Let S be the surface of a body submerged in the fluid. Show that $\int_S \vec{F} \cdot d\vec{A}$ is the buoyant force on the body, that is, the force upwards on the body due to the pressure of the fluid surrounding it. [Hint: $\vec{F} \cdot d\vec{A} = P(x, y, z)\vec{k} \cdot d\vec{A} = (P(x, y, z) \, d\vec{A}) \cdot \vec{k}$.]

22. Consider the function $\rho(x, y, z)$ which gives the electrical charge density at all points in space. The vector field $\vec{J}(x, y, z)$ gives the electric current density at any point in space and is defined so that the current through a small area $d\vec{A}$ is given by

$$\text{Current through small area} \approx \vec{J} \cdot d\vec{A}.$$

Suppose S is a closed surface enclosing a volume W.

(a) What does the integral

$$\int_W \rho \, dV$$

represent, in terms of electricity?

(b) What does the integral

$$\int_S \vec{J} \cdot d\vec{A}$$

represent, in terms of electricity?

(c) Using the fact that an electric current is the rate of change of charge with time, explain why

$$\int_S \vec{J} \cdot d\vec{A} = -\frac{\partial}{\partial t}\left(\int_W \rho \, dV\right).$$

23. A fluid is flowing along in a cylindrical pipe of radius a running in the $\vec{i}$ direction. The velocity of the fluid at a distance r from the center of the pipe is $\vec{v} = u(1 - r^2/a^2)\vec{i}$.

(a) What is the significance of the constant u?

(b) What is the velocity of the fluid at the wall of the pipe?

(c) Find the flux through a circular cross-section of the pipe.

24. Suppose a region of 3-space has a non-uniform temperature. Let $T(x, y, z)$ be the temperature at a point (x, y, z). One form of Newton's law of cooling says that grad T is proportional to the vector field $\vec{F}$ that gives the heat flow. (Here $\vec{F}$ points in the direction that heat is flowing and has magnitude equal to the rate of flow of heat.)

    (a)   Suppose $\vec{F} = k \operatorname{grad} T$ for some constant k. What is the sign of k?
    (b)   Explain why this form of Newton's law of cooling makes sense.
    (c)   Let W be a region of space bounded by the surface S. Explain why

$$\begin{array}{l} \text{Rate of heat} \\ \text{loss from } W \end{array} = k \int_S (\operatorname{grad} T) \cdot d\vec{A}.$$

25. The purpose of this problem is to investigate the behavior of the electric field produced by an infinitely long, straight, uniformly charged wire. (There is no current running through the wire — all charges are fixed.) Making the wire infinitely long, rather than very long but finite, is simpler because in that case it is reasonable to assume that the electric field is normal to any cylinder that has the wire as an axis, and that the magnitude of the field is constant on any such cylinder. Denote by E_r the magnitude of the electric field due to the wire on a cylinder of radius r (see Figure 19.18).

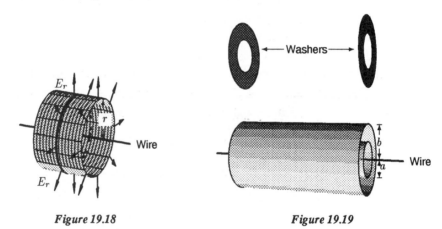

Figure 19.18   **Figure 19.19**

    Imagine a closed surface S made up of two cylinders, one of radius a and one of larger radius b, both coaxial with the wire, and the two washers that cap the ends (see Figure 19.19). Note that the outward orientation of S means that a normal on the outer cylinder points away from the wire and a normal on the inner cylinder points towards the wire.

    (a)   Explain why the flux of $\vec{E}$ through the washers is 0.
    (b)   Gauss' Law states that the flux of an electric field through a closed surface S is proportional to the amount of electric charge inside S. Explain why Gauss' Law implies that the flux through the inner cylinder is the same as the flux through the outer cylinder. (Note that the charge on the wire is *not* inside the surface S).
    (c)   Use part (b) to show that $E_b/E_a = a/b$.
    (d)   Explain why part (c) shows that the strength of the field due to an infinitely long uniformly charged wire is proportional to $1/r$.

26. Consider an infinite flat sheet uniformly covered with charge. As in the case of the charged wire from Problem 25, symmetry considerations force the electric field $\vec{E}$ to be perpendicular to the sheet, and to have the same magnitude for all points that are at the same distance from

the sheet. By considering the flux through a box with sides parallel to the sheet, shown in Figure 19.20, use Gauss' Law (described in Problem 25) to explain why the field due to the charged sheet is the same at all points in space off the sheet, on any given side of the sheet.

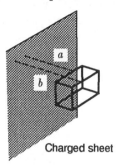

Charged sheet

Figure 19.20

19.2 CALCULATING FLUX INTEGRALS

So far we have calculated flux integrals only in a few special cases. In this section and the next we will see how to compute them in more general situations. Recall the definition:

$$\int_S \vec{F} \cdot d\vec{A} = \lim_{\|\Delta\vec{A}\| \to 0} \sum \vec{F} \cdot \Delta\vec{A},$$

where the sum is over a subdivision of the surface into small pieces with area vector $\Delta\vec{A}$.

How do we divide a general surface into small pieces? If S is the graph of a function $f(x, y)$, we can use the sections of f with x or y constant, and take the pieces to be the patches in a wire frame representation of the surface. This will enable us to express the flux integral as an ordinary double integral. Before seeing how this works, we need to know how to calculate the area vector of one of these patches.

The Area Vector of a Parallelogram

On page 100 we saw that the area of the parallelogram determined by two vectors $\vec{v}$ and $\vec{w}$ is $\|\vec{v} \times \vec{w}\|$. Since $\vec{v} \times \vec{w}$ is also perpendicular to this parallelogram, we have

---

The Area Vector of a Parallelogram

The area vector of the parallelogram determined by $\vec{v}$ and $\vec{w}$, with positive normal determined by the right hand rule, is the cross product $\vec{v} \times \vec{w}$.

See Figure 19.21.

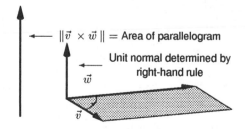

$\|\vec{v} \times \vec{w}\| = $ Area of parallelogram

Unit normal determined by right-hand rule

$\vec{w}$

$\vec{v}$

Figure 19.21: Area vector of a parallelogram $= \vec{v} \times \vec{w}$

When the Surface Is of the Form $z = f(x, y)$

Example 1   Compute $\int_S \vec{F} \cdot d\vec{A}$ where $\vec{F}(x, y, z) = -y\vec{i} + z\vec{k}$ and S is the portion of the curved surface $z = f(x, y) = x^2$ above the square region R given by $0 \leq x \leq 1, 0 \leq y \leq 1$. Let the orientation of S be given by an upward pointing normal. See Figure 19.22 and Figure 19.23.

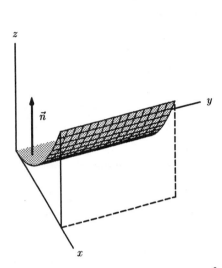

Figure 19.22: The curved surface $S : z = x^2$ with orientation

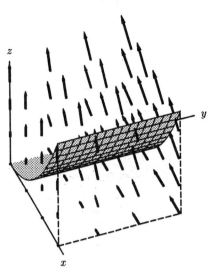

Figure 19.23: The vector field $\vec{F}$ on the surface S

Solution   Figures 19.22 and 19.23 shows that both $\vec{F}$ and the normal vectors point to the same side of the surface S. This leads us to expect that the flux integral will be a positive number. To compute the value exactly we divide the surface S into small pieces by cutting it first into thin strips parallel to the x-axis, then cutting across the strips in the y-direction. See Figure 19.24 and Figure 19.25.

From Figure 19.25 you can see that the cuts on the surface S correspond to a division of the square $0 \leq x \leq 1, 0 \leq y \leq 1$ into rectangles which we will take to be of dimension $\Delta x \times \Delta y$,

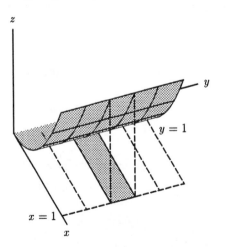

Figure 19.24: After the first cut

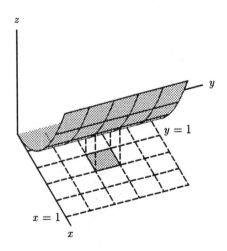

Figure 19.25: After the second cut

with edges parallel to the x and y-axes. The little pieces of the surface S above these rectangles are called parameter rectangles (for the parameters x and y). If Δx and Δy are both small enough, then the parameter rectangles will be nearly flat, and the vector field $\vec{F}$ will be nearly constant on each parameter rectangle, which is what we want.

To calculate the flux, we need to find the approximate area vector of the parameter rectangle S_{mn} of S in Figure 19.26, lying above the rectangle with corner (x_m, y_n). Figure 19.27 shows that the parameter rectangle is approximately a parallelogram bounded by the vectors

$$\vec{a}_{mn} = (\vec{i} + 2x_m\vec{k})\,\Delta x$$

and

$$\vec{b}_{mn} = \vec{j}\,\Delta y.$$

As we saw on page 393, the cross product of two vectors with the same initial point is the area vector of the parallelogram between them. Thus the area vector of the parameter rectangle is given by the cross product:

$$\Delta\vec{A}_{mn} = \vec{a}_{mn} \times \vec{b}_{mn} = (\vec{i} + 2x_m\vec{k})\,\Delta x \times \vec{j}\,\Delta y = (-2x_m\vec{i} + \vec{k})\,\Delta x\,\Delta y.$$

Notice that $\Delta\vec{A}_{mn}$ has a positive $\vec{k}$-component, as we need for an upward orientation.

Observe next that when both Δx and Δy are sufficiently small, then the vector field $\vec{F}$ will be approximated well at all points of S_{mn} by the constant vector field $\vec{F}(x_m, y_n, f(x_m, y_n))$. It follows that the flux of $\vec{F}$ through the parameter rectangle S_{mn} will be approximated by

$$\int_{S_{mn}} \vec{F}\cdot d\vec{A} \approx \vec{F}(x_m, y_n, f(x_m, y_n))\cdot\Delta\vec{A}_{mn}$$

$$= (-y_n\vec{i} + x_m^2\vec{k})\cdot(-2x_m\vec{i} + \vec{k})\,\Delta x\,\Delta y$$

$$= (2x_m y_n + x_m^2)\,\Delta x\,\Delta y.$$

Thus, we have the approximation

$$\int_S \vec{F}\cdot d\vec{A} = \sum_{m,n}\int_{S_{mn}} \vec{F}\cdot d\vec{A} \approx \sum_{m,n}(2x_m y_n + x_m^2)\,\Delta x\,\Delta y.$$

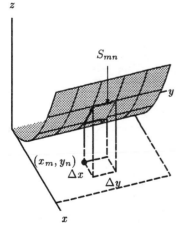

Figure 19.26: The parameter rectangle S_{mn}

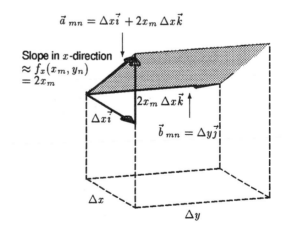

Figure 19.27: A closer look at the parameter rectangle S_{mn}

Finally, taking the limit as Δx and Δy go to zero, the double sum becomes a double integral which we can evaluate over the square region R in the xy-plane:

$$\int_S \vec{F} \cdot d\vec{A} = \lim_{\Delta y \to 0, \Delta x \to 0} \sum_{m,n} (2x_m y_n + x_m^2) \, \Delta x \, \Delta y$$

$$= \int_0^1 \int_0^1 (2xy + x^2) dx \, dy = \frac{5}{6}.$$

In the last example we needed the area vector of a parameter rectangle on the surface $z = f(x, y)$. Now we see how to find this in general.

The Area Vector of a Parameter Rectangle on the Graph of $z = f(x, y)$

Suppose we have a surface given by $z = f(x, y)$. Imagine a parameter rectangle with one corner at the point P on the surface where $x = a, y = b$. See Figure 19.28. Then since a point on the surface has position vector $\vec{r} = x\vec{i} + y\vec{j} + z\vec{k} = x\vec{i} + y\vec{j} + f(x, y)\vec{k}$, the parameter curve $y = b$ on the surface has a tangent vector at P:

$$\vec{v}_x(P) = \frac{\partial \vec{r}}{\partial x} = \vec{i} + f_x(a, b)\vec{k}.$$

Similarly, the parameter curve $x = a$ on the surface has tangent vector at P

$$\vec{v}_y(P) = \frac{\partial \vec{r}}{\partial y} = \vec{j} + f_y(a, b)\vec{k}.$$

We will now consider the vector $\Delta \vec{r}$ along the $y = b$ side of the parameter rectangle. Figure 19.29 shows that $\Delta \vec{r}$ has components $\Delta x\, \vec{i}$ in the x-direction and $\Delta z\, \vec{k}$ in the z-direction.

By local linearity,

$$\Delta z \approx f_x(a, b) \, \Delta x$$

so

$$\Delta \vec{r} \approx \Delta x \, \vec{i} + f_x(a, b)\Delta x \, \vec{k} = \vec{v}_x \, \Delta x.$$

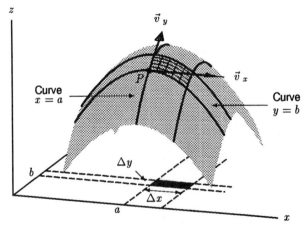

Figure 19.28: Surface showing parameter rectangle and tangent vectors $\vec{v}_x$ and $\vec{v}_y$ at P.

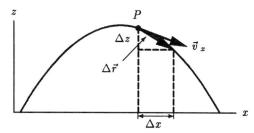

Figure 19.29: Cross-section of surface showing
vector $\Delta\vec{r}$ along the edge of the parameter rectangle

Similarly, the vector along the other side of the parameter rectangle is approximately

$$\Delta y\,\vec{j} + f_y(a,b)\Delta y\,\vec{k} = \vec{v}_y\,\Delta y.$$

Thus, the area vector of the parameter rectangle $\Delta\vec{A}(P)$ is approximated by the cross-product:

$$
\begin{aligned}
\Delta\vec{A}(P) &\approx (\vec{v}_x\,\Delta x) \times (\vec{v}_y\,\Delta y) \\
&= \begin{vmatrix} \vec{i} & \vec{j} & \vec{k} \\ 1 & 0 & f_x(a,b) \\ 0 & 1 & f_y(a,b) \end{vmatrix} \Delta x\,\Delta y \\
&= (-f_x(a,b)\vec{i} - f_y(a,b)\vec{j} + \vec{k}\,)\,\Delta x\,\Delta y.
\end{aligned}
$$

We will often write this relationship in its infinitesimal form:

Area vector of a parameter rectangle at the point (a,b) **on the surface** $z = f(x,y)$**:**

$$d\vec{A}(a,b) = (-f_x(a,b)\vec{i} - f_y(a,b)\vec{j} + \vec{k}\,)dx\,dy$$

Computing Flux Integrals over the Graph of $z = f(x,y)$

In general, we have the following method of computing flux integrals when S is part of the surface $z = f(x,y)$, oriented in the positive z-direction. On S, as we just saw,

$$d\vec{A} = (-f_x\vec{i} - f_y\vec{j} + \vec{k}\,)\,dx\,dy.$$

This means that

If $\vec{F}(x,y,z)$ is a vector field, and S is the surface $z = f(x,y)$ oriented upwards, then the flux integral may be evaluated using

$$\int_S \vec{F} \cdot d\vec{A} = \int_R \vec{F}(x,y,f(x,y)) \cdot (-f_x\vec{i} - f_y\vec{j} + \vec{k}\,)\,dx\,dy$$

where R is the "shadow" region obtained by projecting the surface S onto the xy-plane.

Example 2   Compute $\int_S \vec{F} \cdot d\vec{A}$ where $\vec{F}(x, y, z) = z\vec{k}$ and S is the rectangular plate with corners $(0, 0, 0)$, $(1, 0, 0)$, $(0, 1, 3)$, and $(1, 1, 3)$, oriented upwards (that is, in the positive z-direction).

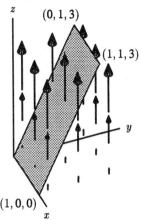

Figure 19.30: The vector field
$\vec{F} = z\vec{k}$ on the surface S

Solution   Since S is part of a plane, we can begin by finding the equation for the plane in the form $z = f(x, y)$. Since f is linear, with x-slope equal to 0 and y-slope equal to 3, and $f(0, 0) = 0$, we have

$$z = f(x, y) = 0 + 0x + 3y = 3y.$$

Thus, in xy-coordinates,

$$d\vec{A} = (-f_x\vec{i} - f_y\vec{j} + \vec{k})dx\,dy = (0\vec{i} - 3\vec{j} + \vec{k})dx\,dy = (-3\vec{j} + \vec{k})dx\,dy.$$

The flux integral is therefore

$$\int_S \vec{F} \cdot d\vec{A} = \int_0^1 \int_0^1 3y\vec{k} \cdot (-3\vec{j} + \vec{k})dx\,dy$$

$$= \int_0^1 \int_0^1 3y\,dx\,dy = 1.5$$

Example 3   Find the flux of $\vec{F}(x, y, z) = x\vec{i} + y\vec{j} + z\vec{k}$ through the paraboloid $z = x^2 + y^2 - 1$, $-1 \le z \le 0$ oriented upward. (See Figure 19.31.)

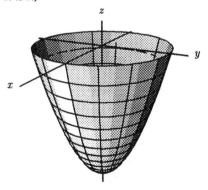

Figure 19.31

Solution   We have

$$d\vec{A} = -f_x\vec{i} - f_y\vec{j} + \vec{k} = -2x\vec{i} - 2y\vec{j} + \vec{k},$$

so

$$
\begin{aligned}
\int_S \vec{F} \cdot d\vec{A} &= \int_R (x\vec{i} + y\vec{j} + (x^2 + y^2 - 1)\vec{k}) \cdot (-2x\vec{i} - 2y\vec{j} + \vec{k})dA \\
&= \int_R (-2x^2 - 2y^2 + x^2 + y^2 - 1)dA \\
&= \int_R (-x^2 - y^2 - 1)dA,
\end{aligned}
$$

where R is the unit circle in the xy-plane $x^2 + y^2 \leq 1$. This last integral can be easily evaluated using polar coordinates:

$$\int_R (-x^2 - y^2 - 1)dA = \int_0^{2\pi} \int_0^1 (-r^2 - 1)r \, dr \, d\theta = -3\pi/2.$$

Problems for Section 19.2

In Problems 1–8 compute the flux of the given vector field, $\vec{F}$, through the given surface, S.

1. $\vec{F} = x\vec{i} + y\vec{j} + (z^2 + 3)\vec{k}$; S is the rectangle $z = 4, 0 \leq x \leq 2, 0 \leq y \leq 3$, oriented in the positive z-direction.

2. $\vec{F} = z\vec{i} + y\vec{j} + 2x\vec{k}$; S is the rectangle $z = 4, 0 \leq x \leq 2, 0 \leq y \leq 3$, oriented in the positive z-direction.

3. $\vec{F} = (x + \cos z)\vec{i} + y\vec{j} + 2x\vec{k}$; S is the rectangle $x = 2, 0 \leq y \leq 3, 0 \leq z \leq 4$, oriented in the positive x-direction.

4. $\vec{F} = x^2\vec{i} + (x + e^y)\vec{j} - \vec{k}$, S is the rectangle $y = -1, 0 \leq x \leq 2, 0 \leq z \leq 4$, oriented in the negative y-direction.

5. $\vec{F} = (x - y)\vec{i} + z\vec{j} + 3x\vec{k}$; S is the region in the plane $z = x + y$ above the rectangle $0 \leq x \leq 2, 0 \leq y \leq 3$, oriented upward.

6. $\vec{F} = \vec{r}$; S is the region in the plane $x + y + z = 1$ above the rectangle $0 \leq x \leq 2, 0 \leq y \leq 3$, oriented downward.

7. $\vec{F} = \vec{r}$; S is the surface $z = x^2 + y^2$, oriented downward, above the disc $x^2 + y^2 \leq 1$.

8. $\vec{F} = y\vec{i} + \vec{j} - xz\vec{k}$; S is the surface $y = x^2 + z^2$, with $x^2 + z^2 \leq 1$, oriented in the positive y-direction.

In Problems 9–15, compute $\int_S \vec{F} \cdot d\vec{A}$ for the given vector field $\vec{F}$ and the given surface S.

9. $\vec{F}(x, y, z) = 2x\vec{j} + y\vec{k}$, S is the part of the surface $z = -y + 1$ above the square region R given by $0 \leq x \leq 1, 0 \leq y \leq 1$.

10. $\vec{F}(x, y, z) = \ln(x^2)\vec{i} + e^x\vec{j} + \cos y\vec{k}$, S is as in Problem 9.

11. $\vec{F}(x,y,z) = x\vec{i} + y\vec{j}$, S is the part of the surface $z = 25 - (x^2 + y^2)$ above R, where R is the disc of radius 5 centered at the origin.

12. $\vec{F}(x,y,z) = \cos(x^2 + y^2)\vec{k}$, S is as in Problem 11.

13. $\vec{F}(x,y,z) = -y\vec{j} + z\vec{k}$, S is the part of the surface $z = y^2 + 5$ over the rectangle R given by $-2 \leq x \leq 1, 0 \leq y \leq 1$.

14. $\vec{F}(x,y,z) = 3x\vec{i} + y\vec{j} + z\vec{k}$, S is the part of the surface $z = -2x - 4y + 1$ above the triangle R in the xy-plane with vertices $(0,0), (0,2), (1,0)$.

15. $\vec{F}(x,y,z) = -xz\vec{i} - yz\vec{j} + z^2\vec{k}$, S is the cone $z = f(x,y) = \sqrt{x^2 + y^2}$ for $0 \leq z \leq 6$.

19.3 NOTES ON FLUX INTEGRALS OVER PARAMETERIZED SURFACES

As we saw in Chapter 16, the most convenient parameters for a surface may not be x and y. Fortunately, we can compute flux integrals over a surface given any parameters p and q for the surface. Let us consider the general case of the flux of a vector field $\vec{F}$ through a surface, S, given parametrically by the equations

$$x = x(p,q), \quad y = y(p,q), \quad z = z(p,q).$$

Following the approach of Example 1 on page 394, we divide the surface S into parameter rectangles as in Figure 19.32, where the rectangles in the parameter space R are of dimension $\Delta p \times \Delta q$. If Δp and Δq are small enough, the parameter rectangles will be nearly flat, and the vector field $\vec{F}$ will be nearly constant on each one, which is what we want.

The parameter rectangle S_{mn} in Figure 19.32, comes from a rectangle of dimension $\Delta p \times \Delta q$ in the parameter space with corner (p_m, q_n). As before, the parameter rectangle S_{mn} is approximately a parallelogram spanned by the vectors

$$\vec{a}_{mn} = \vec{v}_p(p_m, q_n)\Delta p \quad \text{and} \quad \vec{b}_{mn} = \vec{v}_q(p_m, q_n)\Delta q$$

where $\vec{v}_p$ and $\vec{v}_q$ are the velocity vectors to the parameter curves. The area vector of this parallelogram is

$$\Delta \vec{A}_{mn} = \vec{a}_{mn} \times \vec{b}_{mn} = (\vec{v}_p(p_m, q_n) \times \vec{v}_q(p_m, q_n))\Delta p \Delta q.$$

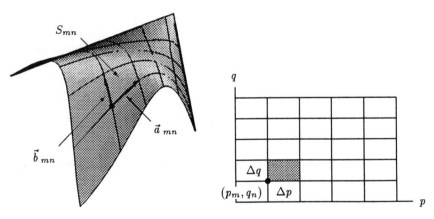

Figure 19.32: The surface S and the parameter space R

Summing fluxes over all the parameter rectangles and making a Riemann sum shows that the flux integral $\int_S \vec{F} \cdot d\vec{A}$ can be computed in pq coordinates by making the substitution

$$d\vec{A} = (\vec{v}_p \times \vec{v}_q)dpdq.$$

In other words,

The flux integral over a parameterized oriented surface is given by

$$\int_S \vec{F} \cdot d\vec{A} = \int_R \vec{F}\left(x(p,q), y(p,q), z(p,q)\right) \cdot (\vec{v}_p \times \vec{v}_q)\, dp\, dq$$

where R is the region in the pq-parameter space and $\vec{v}_p$ and $\vec{v}_q$ are the velocity vectors to the parameter curves. The direction of $\vec{v}_p \times \vec{v}_q$ must agree with the orientation of S.

If necessary, we can reverse the direction of $\vec{v}_p \times \vec{v}_q$ by reversing the order of the parameters.

Example 1   Compute $\int_S \vec{F} \cdot d\vec{A}$ where $\vec{F}(x,y,z) = y\vec{j}$ and S is the portion of the circular cylinder of radius 2 centered on the z-axis such that $x \geq 0$, $y \geq 0$, and $0 \leq z \leq 3$, and the surface is oriented towards the z-axis.

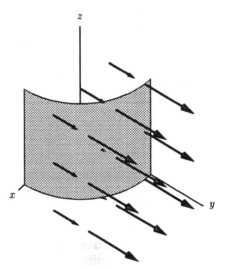

Figure 19.33: The vector field $\vec{F} = y\vec{j}$ on the surface S

Solution   Since the orientation of S is towards the z-axis, Figure 19.33 shows that the flux across S will be negative. To compute the exact value we parameterize the surface S. The shape of S suggests that we parameterize with cylindrical coordinates θ and z. (The coordinate r is unnecessary, because $r = 2$ at all points on the cylinder.) We have the parameterization

$$\begin{aligned} x(\theta, z) &= 2\cos\theta \\ y(\theta, z) &= 2\sin\theta \qquad \text{for } 0 \leq \theta \leq \pi/2, 0 \leq z \leq 3. \\ z(\theta, z) &= z \end{aligned}$$

The tangents to the parameter curves for this parameterization are

$$\vec{v}_\theta = -2\sin\theta\vec{i} + 2\cos\theta\vec{j} \quad \text{and} \quad \vec{v}_z = \vec{k}$$

and so

$$\vec{v}_\theta \times \vec{v}_z = 2\cos\theta\vec{i} + 2\sin\theta\vec{j}.$$

Since the components of $\vec{v}_\theta \times \vec{v}_z$ in the $\vec{i}$ and $\vec{j}$ directions are positive in the range $0 \leq \theta \leq \pi/2$, the vector $\vec{v}_\theta \times \vec{v}_z$ points away from the z-axis, and so our parameterization doesn't match the orientation of S. To re-orient it, we reverse the order of θ and z, taking z first. Finally,

$$
\begin{aligned}
\int_S \vec{F} \cdot d\vec{A} &= \int_{\theta=0}^{\pi/2} \int_{z=0}^{3} 2\sin\theta\vec{j} \cdot (\vec{v}_z \times \vec{v}_\theta)\, dz\, d\theta \\
&= \int_{\theta=0}^{\pi/2} \int_{z=0}^{3} 2\sin\theta\vec{j} \cdot (-2\cos\theta\vec{i} - 2\sin\theta\vec{j})\, dz\, d\theta \\
&= \int_{\theta=0}^{\pi/2} \int_{z=0}^{3} -4\sin^2\theta\, dz\, d\theta = -3\pi.
\end{aligned}
$$

Note on Surface Area

Although it does not involve a flux integral, we can use parameter rectangles to find surface area. Let us see how to compute the surface area of a parameterized surface $S : x = x(p,q), y = y(p,q), z = z(p,q)$. We will set up an integral for the surface area by thinking of the Riemann sums that will approximate it. First we divide S into parameter rectangles S_{mn}, as in Figure 19.32. Since the piece S_{mn} in Figure 19.32 is approximated by a parallelogram with area vector

$$\Delta\vec{A}_{mn} \approx (\vec{v}_p(p_m, q_n) \times \vec{v}_q(p_m, q_n))\Delta p\Delta q,$$

we have

$$\text{Area of } S_{mn} = \|\Delta\vec{A}_{mn}\| \approx \|\vec{v}_p \times \vec{v}_q\|\Delta p\Delta q.$$

Thus,

$$\text{Area of } S = \sum_{m,n} \text{Area of } S_{mn} \approx \sum_{m,n} \|(\vec{v}_p(p_m, q_n) \times \vec{v}_q(p_m, q_n))\|\Delta p\Delta q.$$

In the limit as Δp and Δq go to 0, we have

$$\text{Surface area of } S = \int_R \|\vec{v_p} \times \vec{v_q}\|\,dpdq,$$

where R is the region in the pq-parameter space and $\vec{v}_p$ and $\vec{v}_q$ are the velocity vectors of the parameter curves. Thus:

The surface area of a parameterized surface S is given by

$$\text{Area of } S = \int_R \|d\vec{A}\|$$

where, in pq coordinates, $\|d\vec{A}\| = \|\vec{v}_p \times \vec{v}_q\|\, dp\, dq$, and R is the parameter region.

Example 2   Compute the surface area of a sphere of radius a.

Solution   We will take the sphere S of radius a centered at the origin, and parameterize it with the spherical coordinates ϕ and θ. The parameterization is

$$x = a \sin \phi \cos \theta, \quad y = a \sin \phi \sin \theta, \quad z = a \cos \phi,$$

for $0 \le \theta \le 2\pi, 0 \le \phi \le \pi$. We compute

$$\vec{v}_\phi \times \vec{v}_\theta = (a \cos \theta \cos \phi \vec{i} + a \sin \theta \cos \phi \vec{j} - a \sin \phi \vec{k}) \times (-a \sin \theta \sin \phi \vec{i} + a \cos \theta \sin \phi \vec{j})$$
$$= a^2 (\cos \theta \sin^2 \phi \vec{i} + \sin \theta \sin^2 \phi \vec{j} + \sin \phi \cos \phi \vec{k})$$

and so

$$\|\vec{v}_\phi \times \vec{v}_\theta\| = a^2 \sin \phi.$$

Finally

$$\text{Surface area of } S = \int_{\phi=0}^{\pi} \int_{\theta=0}^{2\pi} a^2 \sin \phi \, d\theta \, d\phi = 4\pi a^2.$$

Problems for Section 19.3

1. Integrate the vector field $\vec{F} = \vec{i} + y\vec{j}$ over the surface S given by $0 \le p \le 1, 0 \le q \le 1$, and

$$x = 2p,$$
$$y = p + q,$$
$$z = 1 + p - q,$$

oriented towards the origin.

2. Integrate the vector field $\vec{F} = z\vec{k}$ over the surface S given by $0 \le p \le 1, 0 \le q \le 1$, and

$$x = p + q$$
$$y = p - q,$$
$$z = p^2 + q^2,$$

oriented towards the z-axis.

3. Integrate the vector field $\vec{F} = y\vec{i} + x\vec{j}$ over the surface S given by $0 \le p \le \pi, 0 \le q \le 1$,

$$x = 3 \sin p$$
$$y = 3 \cos p,$$
$$z = q + 1,$$

oriented away from the z-axis.

4. Integrate the vector field $\vec{F} = x\vec{i} + y\vec{j}$ over the surface S given by $p^2 + q^2 \le 1, p, q \ge 0$, and

$$x = p$$
$$y = q,$$
$$z = \sqrt{1 - p^2 - q^2},$$

oriented away from the origin.

5. Compute the flux of the vector field $\vec{F} = (x+y)\vec{i} + y\vec{j} + \vec{k}$ through the surface S given by $z = x^2 + y^2$, $1/2 \le x^2 + y^2 \le 1$ oriented outward.

6. Compute the flux of the vector field $\vec{F} = \vec{i} + \vec{j} + z\vec{k}$ through the surface S given by $z = \sin x \sin y$, $0 \le x \le \pi/2$, $0 \le y \le \pi/2$ oriented outwards.

7. Evaluate $\int_S \vec{F} \cdot d\vec{A}$ where $\vec{F} = x\vec{i} + y\vec{j}$ and S is the lateral surface of a cylinder as shown in Figure 19.34.

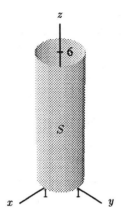

Figure 19.34

8. Compute $\int_S \vec{F} \cdot d\vec{A}$ for $\vec{F} = xz\vec{i} + yz\vec{j} + z^3\vec{k}$ and S is the surface in Problem 7.

9. For $\vec{F} = z^2\vec{k}$, evaluate $\int_S \vec{F} \cdot d\vec{A}$ where S is the upper hemisphere of the sphere $x^2 + y^2 + z^2 = 25$.

10. Compute the flux of $\vec{F}(x, y, z) = x\vec{i}$ over the surface S with parametric equations

$$x = e^p$$
$$y = \cos(3q)$$
$$z = 6p$$

where $0 \le p \le 4$, $0 \le q \le \frac{\pi}{6}$.

11. Compute the flux of $\vec{F} = z\vec{i} + x\vec{j}$ over the surface S with parametric equations:

$$x = p^2$$
$$y = 2p + q^2$$
$$z = 5q$$

where $0 \le p \le 1$, $1 \le q \le 3$.

12. Compute $\int_S \vec{F} \cdot d\vec{A}$ for $\vec{F} = -\frac{2}{x}\vec{i} + \frac{2}{y}\vec{j}$ and S the surface with parametric equations:

$$x = r\cos\theta$$
$$y = r\sin\theta$$
$$z = \sin r^2$$

where $1 \le r \le 3$, $0 \le \theta \le \pi$.

13. Consider a sphere of radius 2 parameterized by the usual spherical coordinates θ and ϕ.

    (a) Compute $d\vec{A}$, the area vector of a parameter rectangle, oriented outward.

    (b) Use part (a) to compute the flux integral

    $$\int_S \vec{F} \cdot d\vec{A}$$

    where $\vec{F}(x, y, z) = z\vec{k}$, and S is the upper hemisphere of radius 2 centered at the origin, oriented outward.

14. If $\vec{F} = x\vec{i} + y\vec{j} + z\vec{k}$, evaluate $\int_S \vec{F} \cdot d\vec{A}$, where S is the surface of the sphere $x^2 + y^2 + z^2 = a^2$, oriented outward.

15. Evaluate $\int_S \vec{F} \cdot d\vec{A}$, where $\vec{F} = x^2\vec{i} + y^2\vec{j} + z^2\vec{k}$ and S is the oriented surface of the triangle $\triangle ABC$ as shown in Figure 19.35.

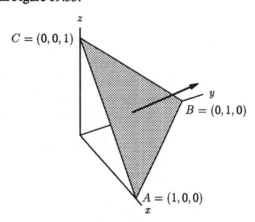

Figure 19.35

16. Evaluate $\int_S \vec{F} \cdot d\vec{A}$, where $\vec{F} = x^2y^2z\vec{k}$ and S is the surface of the cone $\sqrt{x^2 + y^2} = z$ $(0 \leq z \leq R)$, oriented downward. See Figure 19.36.

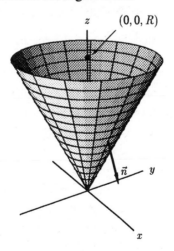

Figure 19.36

17. If $\vec{F} = x^2\vec{i} + y^2\vec{j} + z^2\vec{k}$ Evaluate $I = \int_S \vec{F} \cdot d\vec{A}$ where S is the surface of the sphere $(x-a)^2 + (y-b)^2 + (z-c)^2 = d^2$, oriented outward.

18. Evaluate $\int_S \vec{F} \cdot d\vec{A}$ where $\vec{F} = (bx/a)\vec{i} + (ay/b)\vec{j}$ and S is the curved surface of the elliptic cylinder $x^2/a^2 + y^2/b^2 = 1$, and $|z| \le c$, oriented outward.

19. Evaluate $\int_S \vec{F} \cdot d\vec{A}$ where $\vec{F} = x^3\vec{i} + y^3\vec{j} + z^3\vec{k}$ and S is the surface of the ellipsoid $x^2/a^2 + y^2/b^2 + z^2/c^2 = 1$, oriented outward. [Hint: parameterize the ellipsoid by adapting a parameterization of the sphere.]

20. Compute the flux of the radius vector $\vec{r} = x\vec{i} + y\vec{j} + z\vec{k}$ through the lateral surface of the circular cylinder $x^2 + y^2 = 1$ bounded below by the plane $x + y + z = 1$ and from above by the plane $x + y + z = 2$, oriented outward.

21. Find the area of the ellipse S cut on the plane $2x + y + z = 2$ by the circular cylinder $x^2 + y^2 = 2x$. (See Figure 19.37)

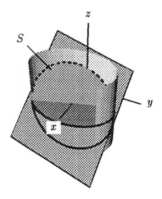

Figure 19.37

REVIEW PROBLEMS FOR CHAPTER NINETEEN

For Problems 1–4 find the flux of the constant vector field $\vec{v} = \vec{i} - \vec{j} + 3\vec{k}$ through the given surfaces.

1. A disk of radius 2 in the xy-plane oriented upward.

2. A triangular plate of area 4 in the yz-plane oriented in the positive x-direction.

3. A square plate of area 4 in the yz-plane oriented in the positive x-direction.

4. The triangular plate with vertices $(1, 0, 0)$, $(0, 1, 0)$, and $(0, 0, 1)$ oriented away from the origin.

5. Suppose that $\vec{E}$ is a *uniform* electric field on 3-space, in the sense that $\vec{E}(x, y, z) = a\vec{i} + b\vec{j} + c\vec{k}$, for all points (x, y, z), where a, b, c are constants. Show, with the aid of symmetry, that the flux of $\vec{E}$ through each of the following closed surfaces S is zero:
    (a) S is the cube bounded by the planes $x = \pm 1$, $y = \pm 1$, and $z = \pm 1$
    (b) S is the sphere $x^2 + y^2 + z^2 = 1$
    (c) S is the cylinder bounded by $x^2 + y^2 = 1$, $z = 0$, and $z = 2$

6. According to Coulomb's Law the electrostatic field $\vec{E}$, at the point with position vector $\vec{r}$ in 3-space, due to a charge q at the origin is given by

    $$\vec{E}(\vec{r}) = q\frac{\vec{r}}{|\vec{r}|^3}.$$

    Let S_a be the outward oriented sphere in 3-space of radius $a > 0$ and center at the origin. Show that the flux of the resulting electric field $\vec{E}$ through the surface S_a is equal to $4\pi q$, for any radius a. This is Gauss's Law for a single point charge.

7. An infinitely long straight wire lying along the z-axis carries an electric current I flowing in the $\vec{k}$ direction. Ampère's Law in magnetostatics says that the current gives rise to a magnetic field $\vec{B}$ given by

    $$\vec{B}(x, y, z) = \frac{I}{2\pi}\frac{-y\vec{i} + x\vec{j}}{x^2 + y^2}.$$

    (a) Sketch the field $\vec{B}$ in the xy plane.
    (b) Suppose S_1 is a disc with center at $(0, 0, h)$, radius a, and parallel to the xy-plane, oriented in the $\vec{k}$ direction. What is the flux of $\vec{B}$ through S_1? Does your answer seem reasonable?
    (c) Suppose S_2 is the rectangle given by $x = 0$, $a \leq y \leq b$, $0 \leq z \leq h$, and oriented in the $-\vec{i}$ direction. What is the flux of $\vec{B}$ through S_2? Does your answer seem reasonable?

8. An *ideal electric dipole* in electrostatics is characterized by its position in 3-space and its dipole moment vector $\vec{p}$. The electric field $\vec{D}$, at the point with position vector $\vec{r}$, of an ideal electric dipole located at the origin with dipole moment $\vec{p}$ is given by

    $$\vec{D}(\vec{r}) = 3\frac{(\vec{r} \cdot \vec{p})\vec{r}}{|\vec{r}|^5} - \frac{\vec{p}}{|\vec{r}|^3}.$$

    Assume $\vec{p} = p\vec{k}$, so the dipole points in the $\vec{k}$ direction and has magnitude p.

    (a) What is the flux of $\vec{D}$ through a sphere S with center at the origin and radius $a > 0$?
    (b) The field $\vec{D}$ is a useful approximation to the electric field $\vec{E}$ produced by two "equal and opposite" charges, q at $\vec{r}_2$ and $-q$ at $\vec{r}_1$, where the distance $|\vec{r}_2 - \vec{r}_1|$ is small. The dipole moment of this configuration of charges is defined to be $q(\vec{r}_2 - \vec{r}_1)$. Gauss's Law in electrostatics says that the flux of $\vec{E}$ through S is equal to 4π times the total charge enclosed by S. What is the flux of $\vec{E}$ through S if the charges at $\vec{r}_1$ and $\vec{r}_2$ are enclosed by S? How does this compare with your answer for the flux of $\vec{D}$ through S if $\vec{p} = q(\vec{r}_2 - \vec{r}_1)$?

9. Evaluate $\int_S \vec{F} \cdot d\vec{A}$, where S is the 2×2 square plate in the yz-plane centered at the origin, oriented in the positive x-direction, and $\vec{F} = (5 + xy)\vec{i} + z\vec{j} + yz\vec{k}$.

10. Find the flux of the vector field $\vec{F} = x\vec{i} + y\vec{j}$ through the surface of a closed cylinder of radius 2 and height 3 centered on the z-axis with its base in the xy-plane.

11. Find the flux of the vector field $\vec{F} = -y\vec{i} + x\vec{j} + z\vec{k}$ through the surface of a closed cylinder of radius 1 centered on the z-axis with base in the plane $z = -1$ and top in the plane $z = 1$.

CHAPTER TWENTY

CALCULUS OF VECTOR FIELDS

We have seen two ways of integrating vector fields in three dimensions: along curves and over surfaces. Now we will look at two ways of differentiating them. If we view the vector field as the velocity field of a fluid flow, then one method of differentiation (the divergence) tells us about the strength of outflow from a point, and the other method (the curl) tells us about the strength of rotation around a point. Each method fits together with one of the ways of integrating to form a vector analogue of the Fundamental Theorem of Calculus: the Divergence Theorem relating the divergence to flux, and Stokes' theorem relating the curl to circulation around a closed path.

20.1   THE DIVERGENCE OF A VECTOR FIELD

Look at the vector fields pictured in Figure 20.1 and 20.2. Imagine that they are velocity vector fields describing the flow of a fluid. Figure 20.1 suggests an outflow from the origin; for example, it could be water flowing from a fountain or spring, or it could be an expanding cloud of gas. We say that the origin is a *source*. Figure 20.2 suggests the flow of water into a drain at the origin, or the compression of a gas. In this case we say that the origin is a *sink*.

Note that a source could be a genuine source, as in the case of water flowing out of a spring, or it could be the result of expansion in a gas, in which case there may not be any new material being introduced; however, the picture of the flow is the same, and we will use the word in either case. The same remarks apply to the word sink.

In this section we will use the flux out of a closed surface surrounding a point to measure the outflow per unit volume there. This is called the *divergence* of the vector field at the point.

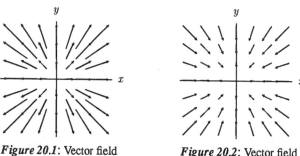

Figure 20.1: Vector field
showing a source

Figure 20.2: Vector field
showing a sink

Example 1   A fluid has velocity vector field $\vec{F}(\vec{r}) = \vec{r}$. Find the outflow per unit volume at the origin.

Solution   In Example 6 on page 387, we calculated the flux across the sphere of radius R, centered at the origin, to be $4\pi R^3$. Since this is the combined outflow from all points in the enclosed ball of volume $\frac{4}{3}\pi R^3$, we say that the average outflow per unit volume at a point in the sphere is

$$\frac{\text{Flux}}{\text{Volume}} = \frac{4\pi R^3}{\frac{4}{3}\pi R^3} = 3 \text{ cubic units of fluid per unit time per cubic unit of space.}$$

To get the outflow per unit volume *at the origin*, we consider the limit of this flux to volume ratio as R tends to zero, so that the sphere contracts around the origin. In this case the limit is 3, so we say that the outflow per unit volume is 3 cubic units of fluid per unit of time per cubic unit of space, or simply

$$\text{Divergence of the vector field } \vec{F} \text{ at the origin} = 3.$$

Definition of Divergence

The definition of divergence of a general vector field at a point follows the same idea.

The **divergence** of a vector field $\vec{F}$ at a point P is defined by the limit

$$\text{div}\,\vec{F}\,(P) = \lim_{\text{volume}\to 0} \frac{\int_S \vec{F} \cdot d\vec{A}}{\text{Volume inclosed by } S}.$$

Here S is a surface surrounding P that contracts down to P in the limit, in such a way that the diameter of S tends to zero. We orient S with the outward pointing normal.

How Does the Limit Defining the Divergence Work?

The limit defining the divergence exists if $\vec{F} = F_1\vec{i} + F_2\vec{j} + F_3\vec{k}$ has continuous partial derivatives at every point of an open region containing P (that is, if F_1, F_2, and F_3 do), and if S is a closed surface, patched together out of finitely many smooth pieces (such as a sphere or a cube). In that case the limit does not depend on the shape of S as it contracts; all that is required is that the diameter of S tend to zero.

The divergence of a vector field is a scalar-valued function. A vector field $\vec{F}$ is said to be *divergence free* or *solenoidal* if $\text{div}\,\vec{F}\,(P) = 0$ at every point P.

How to Calculate the Divergence in Cartesian Coordinates

When we know the components of a vector field $\vec{F}$, we can use the following formula:

If $\vec{F} = F_1\vec{i} + F_2\vec{j} + F_3\vec{k}$ has continuous partial derivatives, then

$$\text{div}\,\vec{F} = \frac{\partial F_1}{\partial x} + \frac{\partial F_2}{\partial y} + \frac{\partial F_3}{\partial z}.$$

To get an idea of why this is true, imagine a small box at the point (x, y, z) in the vector field $\vec{F}$, as in Figure 20.3. This figure shows a cross-section parallel to the xy-plane. Suppose the box is small enough that $\vec{F}$ is approximately constant on each side. Since the left face of the box is perpendicular to the x-direction, the flux in through that face is approximately the x-component of the vector field times the area of the face, that is, $F_1(x, y, z)\Delta y \Delta z$. Similarly, the flux out of the right face is approximately $F_1(x + \Delta x, y, z)\Delta y \Delta z$. Thus the net out flow through these faces is approximately

$$F_1(x + \Delta x, y, z)\Delta y \Delta z - F_1(x, y, z)\Delta y \Delta z = \frac{F_1(x + \Delta x, y, z) - F_1(x, y, z)}{\Delta x}\Delta x \Delta y \Delta z$$

$$\approx \frac{\partial F_1}{\partial x}\Delta x \Delta y \Delta z.$$

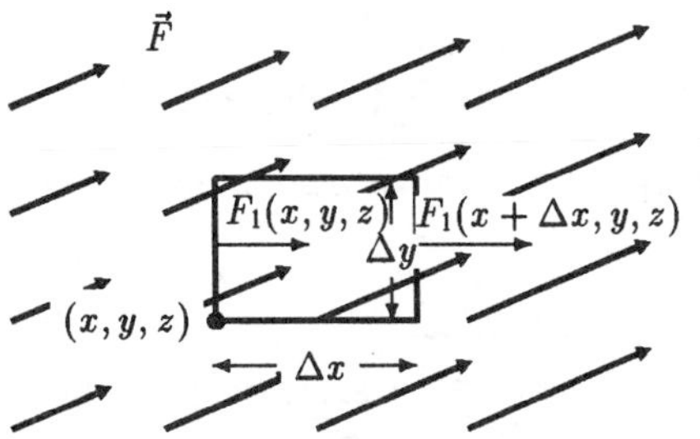

Figure 20.3: Small box at (x, y, z) in vector field $\vec{F}$.

Similarly, the net flow through the faces perpendicular to the y-direction is approximately $\frac{\partial F_2}{\partial y}\Delta x\Delta y\Delta z$, and the net flow through the faces perpendicular to the z-direction is $\frac{\partial F_3}{\partial z}\Delta x\Delta y\Delta z$. The outflow per unit volume is

$$\frac{\text{Net outflow from box}}{\text{Volume of box}} \approx \frac{\frac{\partial F_1}{\partial x}\Delta x\Delta y\Delta z + \frac{\partial F_2}{\partial y}\Delta x\Delta y\Delta z + \frac{\partial F_3}{\partial z}\Delta x\Delta y\Delta z}{\Delta x\Delta y\Delta z}$$

$$= \frac{\partial F_1}{\partial x} + \frac{\partial F_2}{\partial y} + \frac{\partial F_3}{\partial z}.$$

We will give a more detailed explanation of the formula for divergence in Section 20.5.

What Does the Divergence Tell Us?

If you think of a vector field as the velocity flow of a fluid, then the divergence gives the outflow per unit volume at a point. If the vector field represents flow in toward a point, the divergence is negative (or zero) there; if the vector field represents a flow away from a point, the divergence is positive (or zero) there.

Example 2   Figure 20.4 shows the part of a vector field $\vec{E}$ that lies in the xy-plane. Assume that the vector field is independent of z, so that any horizontal cross section looks the same. What can you say about the divergence of the vector field at the points marked P and Q, assuming it is defined?

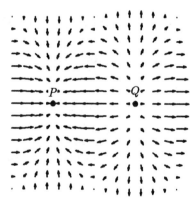

Figure 20.4: Divergence of the vector
field $\vec{E}$

Solution   The flux of $\vec{E}$ through a small sphere of radius R around the point marked P is negative, because all the arrows are pointing into the sphere. The divergence at P is

$$\text{div}\,\vec{E}\,(P) = \lim_{\text{vol}\to 0}\left(\frac{\int_S \vec{E}\cdot d\vec{A}}{\text{Volume of sphere}}\right) = \lim_{R\to 0}\left(\frac{\text{Negative number}}{\frac{4}{3}\pi R^3}\right) \leq 0.$$

By a similar argument, the divergence at Q must be positive or zero.

Example 3   Use the definition to find div $\vec{v}$ at the origin, when $\vec{v}$ is the vector field in 3-space given by
(a) $\vec{v} = -2\vec{r}$ (b) $\vec{v} = x\vec{i}$.
Check your answer using the formula in Cartesian coordinates.

Solution   (a)   Figure 20.5 shows a two dimensional cross-section of the vector field $\vec{v} = -2\vec{r}$. The vector field points radially inwards, so if we take S to be a sphere of radius R centered at the origin, oriented outward, we have

$$\vec{v} \cdot \Delta \vec{A} = -2R \|\Delta \vec{A}\|,$$

for a small area vector $\Delta \vec{A}$ on the sphere. Therefore,

$$\int_S \vec{v} \cdot d\vec{A} = \int_S -2R \|d\vec{A}\| = -2R(\text{Surface area of sphere}) = -2R(4\pi R^2) = -8\pi R^3.$$

Thus, we find that

$$\text{div } \vec{v}\,(0,0,0) = \lim_{\text{vol} \to 0} \left(\frac{\int_S \vec{v} \cdot d\vec{A}}{\text{Volume of sphere}} \right) = \lim_{R \to 0} \left(\frac{-8\pi R^3}{\frac{4}{3}\pi R^3} \right) = -6.$$

Notice that the divergence is negative. This is what you would expect, since the vector field represents an inward flow at the origin.

Since $\vec{v} = -2\vec{r} = -2x\vec{i} - 2y\vec{j} - 2z\vec{k}$, the formula gives

$$\text{div } \vec{v} = \frac{\partial}{\partial x}(-2x) + \frac{\partial}{\partial y}(-2y) + \frac{\partial}{\partial z}(-2z) = -2 - 2 - 2 = -6.$$

(b)   The vector field $\vec{v} = x\vec{i}$ is parallel to the x axis, as shown in Figure 20.6. Let's take S to be a cube centered at the origin with edges parallel to the axes, of length $2R$. Then the flux through the faces perpendicular to the y and z-axes is zero (because the vector field is parallel to these faces). On the faces perpendicular to the x-axis, the vector field and the outward normal are parallel. On the face at $x = R$, we have

$$\vec{v} \cdot \Delta \vec{A} = R \|\Delta \vec{A}\|.$$

On the face at $x = -R$, the dot product is still positive, and

$$\vec{v} \cdot \Delta \vec{A} = R \|\Delta \vec{A}\|.$$

Therefore, the flux through the box is given by

$$\int_S \vec{v} \cdot d\vec{A} = \int_{x=-R} \vec{v} \cdot d\vec{A} + \int_{x=R} \vec{v} \cdot d\vec{A}$$

$$= 2 \int_{x=R} R \|d\vec{A}\| = 2R(\text{Area of one face}) = 2R(2R)^2 = 8R^3.$$

Thus,

$$\text{div } \vec{v}\,(0,0,0) = \lim_{\text{vol} \to 0} \left(\frac{\int_S \vec{v} \cdot d\vec{A}}{\text{Volume of box}} \right) = \lim_{R \to 0} \left(\frac{8R^3}{(2R)^3} \right) = 1.$$

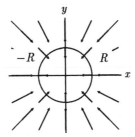

Figure 20.5: The vector field $\vec{v} = -2\vec{r}$

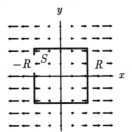

Figure 20.6: Vector field $\vec{v} = x\vec{i}$ and cross section of the surface S

Since the vector field points outward from the yz-plane, it makes sense that the divergence is positive at the origin. Since $\vec{v} = x\vec{i} + 0\vec{j} + 0\vec{k}$, the formula gives

$$\text{div}\,\vec{v} = \frac{\partial}{\partial x}(x) + \frac{\partial}{\partial y}(0) + \frac{\partial}{\partial z}(0) = 1 + 0 + 0 = 1.$$

Notice that the divergence is positive if the *net* outflow from small regions around the point is positive. This could happen even if there is some inflow, as long as there is a greater outflow to counteract it.

Example 4   According to the formula, the divergence of the vector field $\vec{v} = x\vec{i}$ at the point $(2, 2, 0)$ is $\frac{\partial}{\partial x}(x) = 1$. Explain this in terms of flow.

Solution   In Example 3 we used the definition to compute the divergence at the origin. This time, let's take S to be a cube with edges parallel to the axes, of length $2R$, centered at the point $(2, 2, 0)$. See Figure 20.7. As before, the flux through the faces perpendicular to the y and z-axes is zero (because the vector field is parallel to these faces).

On the face at $x = 2 + R$,

$$\vec{v} \cdot \Delta \vec{A} = (2 + R)\,\|\Delta \vec{A}\|.$$

On the face at $x = 2 - R$ with outward normal, the dot product is negative, and

$$\vec{v} \cdot \Delta \vec{A} = -(2 - R)\,\|\Delta \vec{A}\|.$$

Therefore, the flux through the box is given by

$$\int_S \vec{v} \cdot d\vec{A} = \int_{x=2-R} \vec{v} \cdot d\vec{A} + \int_{x=2+R} \vec{v} \cdot d\vec{A}$$
$$= ((2 + R) - (2 - R))(\text{Area of one face}) = 2R(2R)^2 = 8R^3.$$

Then, as before,

$$\text{div}\,\vec{v}\,(2, 2, 0) = \lim_{\text{vol}\to 0} \left(\frac{\int_S \vec{v} \cdot d\vec{A}}{\text{Volume of box}} \right) = \lim_{R\to 0} \left(\frac{8R^3}{(2R)^3} \right) = 1.$$

Note that the vector field does not appear to flow away from the point $(2, 2, 0)$; however, because the inflow on the left of this point is less than the outflow on the right, the divergence is positive.

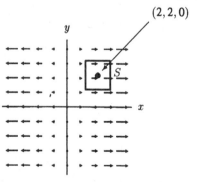

Figure 20.7: Vector field $\vec{v} = x\vec{i}$ and cross section of the surface S

Alternative Notation

Using $\nabla = \dfrac{\partial}{\partial x}\vec{i} + \dfrac{\partial}{\partial y}\vec{j} + \dfrac{\partial}{\partial z}\vec{k}$, we can write

$$\operatorname{div}\vec{F} = \nabla \cdot \vec{F} = \left(\frac{\partial}{\partial x}\vec{i} + \frac{\partial}{\partial y}\vec{j} + \frac{\partial}{\partial z}\vec{k}\right) \cdot (F_1\vec{i} + F_2\vec{j} + F_3\vec{k}) = \frac{\partial F_1}{\partial x} + \frac{\partial F_2}{\partial y} + \frac{\partial F_3}{\partial z}.$$

Problems for Section 20.1

1. Which of the two vector fields in Figure 20.8 has the greater divergence at the origin? Assume the scales are the same on each.

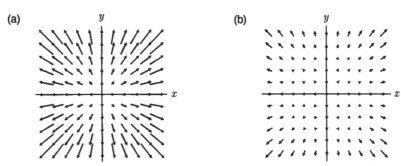

Figure 20.8

2. For each of the following vector fields, say whether the divergence is positive, zero, or negative at the indicated point.

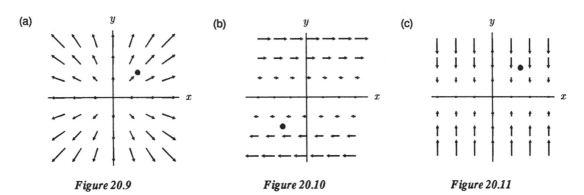

Figure 20.9   Figure 20.10   Figure 20.11

3. Draw two vector fields that have positive divergence everywhere.
4. Draw two vector fields that have negative divergence everywhere.
5. Draw two vector fields that have zero divergence everywhere.
6. Let $\vec{F}(\vec{r}) = \dfrac{\vec{r}}{r^3}$ (in 3-space), $\vec{r} \neq \vec{0}$.

   (a) Calculate $\operatorname{div}\vec{F}$.
   (b) Sketch $\vec{F}$. Does it appear to be diverging? Does this agree with your answer to part (a)?

7. Let $\vec{F}(x, y, z) = z\vec{k}$.

    (a) Calculate div $\vec{F}$.
    (b) Sketch $\vec{F}$. Does it appear to be diverging? Does this agree with your answer to part (a)?

In Problems 8–12, find the divergence of the given vector field.

8. $\vec{F}(x, y) = (x^2 - y^2)\vec{i} + 2xy\vec{j}$

9. $\vec{F}(\vec{r}) = \dfrac{\vec{r}}{r}$ (in 3-space), $\vec{r} \neq \vec{0}$

10. $\vec{F}(x, y) = -x\vec{i} + y\vec{j}$

11. $\vec{F}(x, y) = -y\vec{i} + x\vec{j}$

12. $\vec{F}(x, y, z) = (-x + y)\vec{i} + (y + z)\vec{j} + (-z + x)\vec{k}$

13. (a) Find the flux of the vector field $\vec{F} = x\vec{i}$ through a box with edge length c in the first octant with one corner at the origin and sides along the axes.
     (b) Use your answer to part (a) to find div $\vec{F}$ at the origin.

14. (a) Find the flux of $\vec{F} = 2\vec{i} + y\vec{j} + 3\vec{k}$ through a box with edge length c in the first octant, with one corner at the origin and edges parallel to the axes.
     (b) Use your answer to part (a) to find div $\vec{F}$ at the origin.

15. (a) Find the flux of the vector field $\vec{F} = x\vec{i} + y\vec{j} + 0\vec{k}$ through a box of side c in the first octant with one corner at the origin, edges parallel to the axes.
     (b) Use your answer to part (a) to find div $\vec{F}$ at the origin.

16. (a) Find the flux of the vector field $\vec{F} = x\vec{i} + y\vec{j}$ through the surface of the closed cylinder of radius c and height c, centered on the z-axis with base in the xy-plane. (See Figure 20.12.)
     (b) Use your answer to part (a) to find div $\vec{F}$ at the origin.
     (c) Compute div $\vec{F}$ at the origin using partial derivatives.

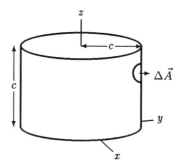

Figure 20.12

Problems 17–18 involve electric fields. Electric charge produces a vector field $\vec{E}$, called the electric field, which represents the force on a unit positive charge placed at the point. Two positive or two negative charges are observed to repel one another, whereas two charges of opposite sign attract one another. In addition it can be shown that the divergence of $\vec{E}$ is proportional to the density of the electric charge (that is, the charge per unit area or charge per unit volume), with positive constant of proportionality.

17. Suppose a certain distribution of electric charge produces the electric field shown in Figure 20.13. Where are the charges that produced this electric field concentrated? Which concentrations are positive and which are negative?

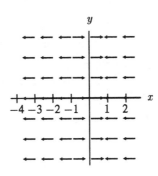

Figure 20.13

18. The electric field produced by a charge located at the origin is given by

$$\vec{E}(\vec{r}) = \frac{k\vec{r}}{r^3}.$$

   (a) Calculate div $\vec{E}$.
   (b) What can you say about div $\vec{E}$ at the point $(0, 0, 0)$?
   (c) Explain what your answers mean in terms of charge density.

19. The divergence of a magnetic vector field $\vec{B}$ must be zero everywhere. Which of the following vector fields cannot be a magnetic vector field?

   (a) $\vec{B}(x, y, z) = -y\vec{i} + x\vec{j} + (x + y)\vec{k}$
   (b) $\vec{B}(x, y, z) = -z\vec{i} + y\vec{j} + x\vec{k}$
   (c) $\vec{B}(x, y, z) = (x^2 - y^2 - x)\vec{i} + (y - 2xy)\vec{j}$

20. Show that if $\vec{a}$ is a constant vector and $f(x, y, z)$ is a scalar valued function, and if $\vec{F} = f\vec{a}$, then div $\vec{F} = (\text{grad } f) \cdot \vec{a}$.

21. If $f(x, y, z)$ and $g(x, y, z)$ are scalar-valued functions with continuous second partial derivatives, show that
$$\text{div}(\text{grad } f \times \text{grad } g) = 0.$$

22. In Problem 24 on page 392 it was shown that the rate of heat loss from a volume V in a region of non-uniform temperature is $k \int_S (\text{grad } T) \cdot d\vec{A}$, where k is a constant, S is the surface bounding V, and $T(x, y, z)$ is the temperature at the point (x, y, z) in space. By taking the limit as V contracts to a point, show that $\partial T/\partial t = A \text{ div grad } T$ at that point, where A is a constant with respect to x, y, z, but may depend on t.

23. A vector field in the plane is a *point source* at the origin if its direction is away from the origin at every point, its magnitude depends only on the distance from the origin, and its divergence is zero away from the origin.

   (a) Explain why a source must be of the form $\vec{v} = [f(x^2 + y^2)](x\vec{i} + y\vec{j})$ for some positive function f.

(b) Show that $\vec{v} = K(x^2 + y^2)^{-1}(x\vec{i} + y\vec{j})$ is a point source at the origin (that is, div $\vec{v} = 0$) if $K > 0$.

(c) Determine the magnitude $\|\vec{v}\|$ of the source in part (b) as a function of the distance from its center.

(d) Sketch the vector field $\vec{v} = (x^2 + y^2)^{-1}(x\vec{i} + y\vec{j})$.

(e) Show that $\phi = \frac{K}{2} \log(x^2 + y^2)$ is a potential function for the source in part (b).

24. A vector field in the plane is a *point sink* at the origin if its direction is toward the origin at every point, its magnitude depends only on the distance from the origin, and its divergence is zero away from the origin.

(a) Explain why a sink must be of the form $\vec{v} = \left[f(x^2 + y^2) \right] (x\vec{i} + y\vec{j})$ for some negative function f.

(b) Show that $\vec{v} = K(x^2 + y^2)^{-1}(x\vec{i} + y\vec{j})$ is a point sink at the origin (that is, div $\vec{v} = 0$) if $K < 0$.

(c) Determine the magnitude $\|\vec{v}\|$ of the sink in part (a) as a function of the distance from its center.

(d) Sketch the vector field $\vec{v} = -(x^2 + y^2)^{-1}(x\vec{i} + y\vec{j})$.

(e) Show that $\phi = \frac{K}{2} \log(x^2 + y^2)$ is a potential function for the sink in part (b).

25. A vector field is a *point source* at the origin in 3-space if its direction is away from the origin at every point, its magnitude depends only on the distance from the origin, and its divergence is zero except at the origin. (Such a vector field might be used to model the photon flow out of a star or the neutrino flow out of a supernova.)

(a) Show that $\vec{v} = K(x^2 + y^2 + z^2)^{-3/2}(x\vec{i} + y\vec{j} + z\vec{k})$ is a point source at the origin if $K > 0$.

(b) Determine the magnitude $\|\vec{v}\|$ of the source in (a) as a function of the distance from its center.

(c) Compute the flux of $\vec{v}$ through a sphere of radius r centered at the origin.

(d) Compute the flux of $\vec{v}$ through a closed surface that does not contain the origin.

20.2 THE DIVERGENCE THEOREM

If $\vec{v}$ is the velocity field of an incompressible fluid flow, the flux integral of $\vec{v}$ through a closed surface S tells you the net outflow of fluid from the region enclosed by the surface. On the other hand, the divergence of $\vec{v}$ tells you the outflow per unit volume at a point. Thus, we should be able to compute the flux from the divergence.

Example 1   Calculate the flux of the vector field $\vec{F}(\vec{r}) = \vec{r}$ through the sphere of radius R, using the fact that div $\vec{F} = 3$.

Solution   In Example 6 on page 387 we computed the flux integral directly:

$$\text{Flux through sphere} = \int_{\text{Sphere}} \vec{r} \cdot d\vec{A} = 4\pi R^3.$$

Now we will compute it another way. We calculate the divergence

$$\text{div } \vec{F} = \text{div}(x\vec{i} + y\vec{j} + z\vec{k}) = \frac{\partial x}{\partial x} + \frac{\partial y}{\partial y} + \frac{\partial z}{\partial z} = 1 + 1 + 1 = 3.$$

Thus, if $\vec{F}$ is the velocity field of an incompressible fluid, fluid is being created at the rate of 3 units of fluid per unit volume at every point. Hence, the total production of fluid in the ball of radius R enclosed by the sphere is

$$3(\text{Volume of ball}) = 3(\frac{4}{3}\pi R^3) = 4\pi R^3.$$

This is the same answer we found before.

What if the Divergence is Not Constant?

In the previous example it was easy to obtain the total flux from the divergence because the divergence was constant. If the divergence is not constant, we can still compute the flux out of a volume W by means of a volume integral over W. Break W up into many small pieces, as shown in Figure 20.14, and for each piece multiply the divergence $\operatorname{div} \vec{F}(x, y, z)$ at a point (x, y, z) in the piece by the volume ΔV of the piece. This gives an approximation

$$\text{Flux out of small piece} \approx \operatorname{div} \vec{F}(x, y, z)\, \Delta V.$$

What happens if we add together the fluxes for two adjacent pieces? The adjacent wall has the flux through it counted twice, once out of the cube on one side and once out of the cube on the other. These two contributions cancel. (See Figure 20.15.) Thus, if we add up the fluxes for all the pieces, we get the flux out of the entire volume. So

$$\text{Flux out of the entire volume} = \sum \text{Flux out of small pieces} \approx \sum \operatorname{div} \vec{F}\, \Delta V.$$

We have approximated the flux by a Riemann sum. As the subdivision gets finer, the sum approaches an integral, so we can say

$$\text{Flux out of entire volume} = \int_W \operatorname{div} \vec{F}\ dV.$$

We now have two ways of calculating the flux of a vector field $\vec{F}$ through a closed surface S: we can calculate the flux integral, or we can calculate the integral of $\operatorname{div} \vec{F}$ over the volume W enclosed by S. Thus we have the following result:

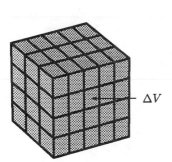

Figure 20.14: Subdivision of the cube

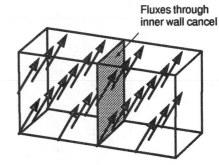

Figure 20.15: Adding the flux out of adjacent pieces

> ### The Divergence Theorem
>
> If $\vec{F}$ is a vector field, and if W is a volume whose boundary S is a smooth closed surface, then
>
> $$\int_S \vec{F} \cdot d\vec{A} = \int_W \operatorname{div} \vec{F} \, dV,$$
>
> assuming S is oriented outward and $\operatorname{div} \vec{F}$ is defined at every point of W.

The Divergence Theorem is like the Fundamental Theorem of Calculus. The Fundamental Theorem of Calculus says that the integral of the rate of change of a function of one variable gives us the total change. The Divergence Theorem says that the integral over a volume of the outflow per unit volume of a vector field gives the total flux out of the volume.

The Divergence Theorem sometimes provides an easy way of computing a flux integral for a closed surface.

Example 2   Calculate the flux of the vector field

$$\vec{F}(x, y, z) = (x^2 + y^2)\vec{i} + (y^2 + z^2)\vec{j} + (x^2 + z^2)\vec{k}$$

through the cube in Figure 20.16.

Solution   The divergence of $\vec{F}$ is $\operatorname{div} \vec{F} = 2x + 2y + 2z$. Since $\operatorname{div} \vec{F}$ is positive everywhere in the first quadrant, the flux through S is positive. By the Divergence Theorem,

$$\int_S \vec{F} \cdot d\vec{A} = \int_0^1 \int_0^1 \int_0^1 2(x + y + z)\, dx\, dy\, dz = \int_0^1 \int_0^1 x^2 + 2x(y + z)\Big|_0^1 dy\, dz$$

$$= \int_0^1 \int_0^1 1 + 2(y + z)\, dy\, dz = \int_0^1 y + y^2 + 2yz\Big|_0^1 dz$$

$$= \int_0^1 (2 + 2z)\, dz = 2z + z^2\Big|_0^1 = 3.$$

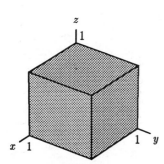

Figure 20.16: Find the flux out of this cube

Figure 20.17: Find the flux out of the sphere of radius R

Example 3   Use the Divergence Theorem to calculate the flux of the vector field $\vec{F}(x, y, z) = -z\vec{i} + x\vec{k}$ through the sphere of radius R centered at the origin.

Solution   The divergence of the field is

$$\operatorname{div} \vec{F} = \frac{\partial(-z)}{\partial x} + \frac{\partial(0)}{\partial y} + \frac{\partial x}{\partial z} = 0.$$

Hence,

$$\int_{\text{Sphere}} \vec{F} \cdot d\vec{A} = \int_{\text{Ball}} \operatorname{div} \vec{F} \, dV = \int_{\text{Ball}} 0 \, dV = 0,$$

and so the flux through the sphere is zero. This makes sense, because, as Figure 20.17 shows, the vector field is flowing around the y-axis and is always tangent to the sphere.

The Divergence Theorem applies to any volume W and its boundary S, even in cases where the boundary consists of two or more surfaces. For example, if W is the solid region between the sphere S_1 of radius 1 and the sphere S_2 of radius 2, both centered at the same point, then the boundary of W consists of both S_1 and S_2. The orientation of these surfaces that is required to apply the Divergence Theorem is an outward normal, which on S_2 points away from the center and on S_1 points towards the center (see Figure 20.18).

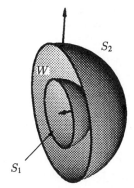

Figure 20.18: Cut away view of the region W between two spheres

Harmonic Functions

A function ϕ of three variables x, y, and z is said to be *harmonic* in a region if $\operatorname{div}(\operatorname{grad} \phi) = 0$ at every point in the region. (This equation is also written $\nabla^2 \phi = 0$, because $\operatorname{div}(\operatorname{grad} \phi) = \nabla \cdot (\nabla \phi)$.) In Problem 12 on page 424 you will show that

$$\nabla^2 \phi(x, y, z) = \frac{\partial^2 \phi}{\partial x^2} + \frac{\partial^2 \phi}{\partial y^2} + \frac{\partial^2 \phi}{\partial z^2}.$$

Harmonic functions are of great importance in pure and applied mathematics. For example, the temperature in a region of space that is not undergoing temperature change is harmonic, as is the electric potential in a charge-free region of space. Several of the basic properties of harmonic functions can be deduced from the Divergence Theorem.

Example 4   Show that a nonconstant harmonic function ϕ cannot have a local maximum.

Solution   Suppose that a nonconstant function ϕ has a local maximum at a point P. Then the vector field grad ϕ will point approximately towards P at all points near P, because it points in the direction of increasing ϕ. Taking a small sphere S centered at P, oriented outwards, we will therefore have

$$\int_S \text{grad}\,\phi \cdot d\vec{A} < 0.$$

On the other hand, if ϕ is harmonic, then by the Divergence Theorem

$$\int_S \text{grad}\,\phi \cdot d\vec{A} = \int_W \text{div}(\text{grad}\,\phi)dV = 0,$$

where W is the ball enclosed by S. Clearly if ϕ has a local maximum, then it cannot be harmonic.

The most famous fact about harmonic functions is their mean value property, discovered by Gauss. If ϕ is a harmonic function in the region enclosed by a sphere, then the value of ϕ at the center of the sphere equals the average value of ϕ on the sphere. For example, the temperature at a point in space equals the average value of the temperature on any sphere centered at the point.

Problems for Section 20.2

For Problems 1–3, compute the flux integral $\int_S \vec{F} \cdot d\vec{A}$ in two ways, if possible, directly and using the Divergence Theorem. In all cases, S is closed and oriented outwards.

1. $\vec{F}(\vec{r}) = \vec{r}$, where S is the cube enclosing the volume $0 \leq x \leq 2, 0 \leq y \leq 2$, and $0 \leq z \leq 2$.

2. $\vec{F}(x, y, z) = y\vec{j}$, where S a vertical cylinder of height 2, with its base a circle of radius 1 on the xy-plane, centered at the origin. S includes the disks that close it off at top and bottom.

3. $\vec{F}(x, y, z) = -z\vec{i} + x\vec{k}$, where S is a square pyramid with the base on the xy-plane of side length 1 and height 3.

4. Suppose V_1 and V_2 are the two cube-shaped volumes in the first quadrant shown in Figure 20.19. Both have sides of length 1 parallel to the axes; V_1 has one corner at the origin, while V_2 has the corresponding corner at the point $(1, 0, 0)$. Suppose S_1 and S_2 are the six-faced surfaces of V_1 and V_2, respectively. Suppose V is the box-shaped volume consisting of V_1 and V_2 together, and having outside surface S.

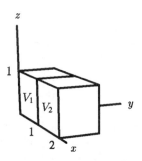

Figure 20.19

Are the following true or false? Give reasons.

(a) If $\vec{F}$ is a constant vector field, $\int_S \vec{F} \cdot d\vec{A} = 0$

(b) If S_1, S_2 and S are all oriented outward and $\vec{F}$ is any vector field:

$$\int_S \vec{F} \cdot d\vec{A} = \int_{S_1} \vec{F} \cdot d\vec{A} + \int_{S_2} \vec{F} \cdot d\vec{A}.$$

5. Calculate $\int_{S_2} \vec{F} \cdot d\vec{A}$ where $\vec{F} = x^2\vec{i} + 2y^2\vec{j} + 3z^2\vec{k}$ and S_2 is as in Problem 4.

    Do this directly and using the Divergence Theorem.

6. Use the Divergence Theorem to evaluate the flux integral $\int_S (x^2\vec{i} + (y - 2xy)\vec{j} + 10z\vec{k}) \cdot d\vec{A}$, where S is the sphere of radius 5 centered at the origin, oriented outward.

7. Suppose that a vector field $\vec{F}$ satisfies $\operatorname{div}\vec{F} = 0$ everywhere. Show that $\int_S \vec{F} \cdot d\vec{A} = 0$ for every closed surface S.

8. The gravitational field, $\vec{F}$, of a planet of mass m at the origin is given by

$$\vec{F} = -G\frac{m\vec{r}}{r^3}.$$

    Use the Divergence Theorem to show that the flux of the gravitational field through the sphere of radius R is independent of R. [Hint: Consider a volume bounded by two concentric spheres.]

9. A basic property of the electric field $\vec{E}$ is that its divergence is zero at points where there is no charge. Suppose that the only charge is along the z-axis, and that the electric field $\vec{E}$ points radially out from the z-axis and its magnitude depends only on the distance r from the z-axis. Use the Divergence Theorem to show that the magnitude of the field is proportional to $1/r$.

10. Some people have suggested that there may be "magnetic charges" which give rise to magnetic fields in the same way that electric charges give rise to electric fields. Suppose the vector field $\vec{B}(x, y, z)$ represents the magnetic field at any point in space. Then one of Maxwell's equations tells us that

$$\operatorname{div}\vec{B} = 0.$$

    This equation contradicts the hypothesis that magnetic charges exists. Explain why.

11. If a surface S is submerged in an incompressible fluid, a force $\vec{F}$ is exerted on one side of the surface by the pressure in the fluid. The component of force in the direction of a unit vector $\vec{u}$ is given by the following

$$\vec{F} \cdot \vec{u} = -\int_S \rho gz\vec{u} \cdot d\vec{A}$$

    where ρ is the density of the fluid (mass/volume), g is the acceleration due to gravity, and the surface is oriented away from the side on which the force is exerted. Suppose the z-axis is vertical, with the positive direction upward and the fluid level at $z = 0$. In this problem we will consider a totally submerged closed surface enclosing a volume V. We are interested in the force of the liquid on the external surface, so S is oriented inward.

    (a) Use the Divergence Theorem to show that the force in the $\vec{i}$ and $\vec{j}$ directions is zero.

    (b) Use the Divergence Theorem to show that the force in the $\vec{k}$ direction is ρgV, the weight of the volume of fluid with the same volume as V. This is *Archimedes' Principle*.

12. Show that $\nabla^2\phi(x, y, z) = \partial^2\phi/\partial x^2 + \partial^2\phi/\partial y^2 + \partial^2\phi/\partial z^2$

13. Show that linear functions are harmonic.

14. What is the condition on the constant coefficients a, b, c, d, e, f such that $ax^2 + by^2 + cz^2 + dxy + exz + fyz$ is harmonic?

15. Use the Divergence Theorem to show that if ϕ is harmonic in a region W, then $\int_S \nabla\phi \cdot d\vec{A} = 0$ for every closed surface S in W such that the volume enclosed by S lies completely within W.

16. Show that a nonconstant harmonic function cannot have a local minimum, and that it can achieve a minimum value in a closed region only on the boundary.

17. Let $\phi = 1/(x^2 + y^2 + z^2)^{1/2}$.
     (a) Show that ϕ is harmonic everywhere except the origin.
     (b) Give a geometric explanation for the fact that ϕ has no local maximum or minimum. [Hint: $\phi = 1/r$.]
     (c) Compute $\int_S \nabla\phi \cdot d\vec{A}$ where S is the sphere of radius 1 centered at the origin. Does your answer contradict the assertion of Problem 15?

18. Show that if ϕ is a harmonic function, then $\mathrm{div}(\phi \operatorname{grad}\phi) = \|\operatorname{grad}\phi\|^2$.

19. Suppose that ϕ is a harmonic function in the region enclosed by a closed surface S, and suppose that $\phi = 0$ at all points of S. Show that $\phi = 0$ at all points of the region enclosed by S. [Hint: Apply the Divergence Theorem to $\int_S \phi \operatorname{grad}\phi \cdot d\vec{A}$ and use Problem 18.]

20. Suppose that ϕ_1 and ϕ_2 are harmonic functions in the region enclosed by a closed surface S, and suppose that $\phi_1 = \phi_2$ at all points of S. Show that $\phi_1 = \phi_2$ at all points of the region enclosed by S. [Hint: Use Problem 19.]

21. Show that if u and v are harmonic functions in a region W, then

$$\int_S u \operatorname{grad} v \cdot d\vec{A} = \int_S v \operatorname{grad} u \cdot d\vec{A}$$

for every closed surface S in W such that the volume enclosed by S lies completely within W. [Hint: Use the Divergence Theorem.]

Problems 22 and 23 outline two different proofs of the same result.

22. Let ϕ be a harmonic function, and let $M(R)$ equal the average value of ϕ on the sphere of radius R centered at the origin. This exercise outlines a proof that $M(R) = \phi(0)$ for all R.
     (a) Explain why $M(R) = 1/(4\pi) \int_S \phi(Rx, Ry, Rz)\|d\vec{A}\|$ where S is the sphere of radius 1 centered at the origin.
     (b) Assuming that you can differentiate with respect to R under the integral sign in the formula for $M(R)$ given in part (a), show that $dM/dR = 1/(4\pi) \int_S \nabla\phi(Rx, Ry, Rz) \cdot d\vec{A}$.
     (c) Applying the Divergence Theorem to the integral in part (b), show that $dM/dR = 0$, and hence conclude that the average value of ϕ on spheres centered at the origin is the same for all such spheres.
     (d) By considering smaller and smaller values of R and thus contracting the sphere of radius R to the origin, show that $M(R) = \phi(0, 0, 0)$ for all R.

23. Let ϕ be a harmonic function, and let $M(R)$ equal the average value of ϕ on the sphere of radius R centered at the origin. This exercise outlines a proof that $M(R) = \phi(0, 0, 0)$ for all R.

(a)   Explain why $M(R) = -\frac{1}{4\pi}\int_{S(R)}\phi\nabla(1/r)\cdot d\vec{A}$ where $S(R)$ is the sphere of radius R centered at the origin, oriented outwards.

(b)   Show that $\int_S (1/r)\nabla\phi\cdot d\vec{A} = 0$ where S is the surface consisting of two spheres $S(a)$ and $S(b)$ centered at the origin of radii a and b, with $a < b$, where $S(a)$ is oriented towards the origin, and $S(b)$ is oriented away from the origin.

(c)   Show that $M(a) = M(b)$, so that the average value of ϕ on spheres centered at the origin is the same for all such spheres. [Hint: Use Problems 21 and 23(a) and (b).]

(d)   By considering smaller and smaller values of R and thus contracting the sphere of radius R to the origin, show that $M(R) = \phi(0,0,0)$ for all R.

20.3   THE CURL OF A VECTOR FIELD

In Section 20.1 we studied the divergence of a vector field, which measures the tendency of its flow to be away from or toward a point. In this section we will study a way of measuring the tendency of the flow of a vector field to go around a point. Let's start with a vector field in 3 dimensions, each of whose vectors lies parallel to some fixed plane.

For example, consider the vector field which looks like Figure 20.20. Suppose that it represents the velocity field of a fluid, and imagine holding the paddle-wheel illustrated in Figure 20.21 in the flow. Then the speed at which the paddle-wheel spins (its angular velocity) will be an indication of the strength of circulation. Note that the angular velocity will depend on the direction in which the stick is pointing. For example, in Figure 20.20, the paddle-wheel will spin one way if the stick is pointing up and the opposite way if it is pointing down. If the stick is pointing horizontally it won't spin at all, because the velocity field will strike opposite vanes of the paddle with equal force.

This suggests that the circulation of the field can be described by a vector quantity. On page 427 we will define the curl of a vector field at a point to be the vector pointing in the direction where the circulation has maximum strength, and whose magnitude will be the strength of circulation. But first we have to make precise the idea of measuring the strength of circulation in a given direction.

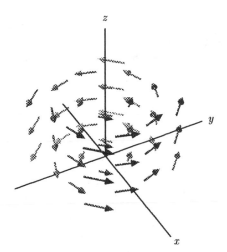

Figure 20.20: A vector field with circulation about the z-axis

Figure 20.21: A device for measuring circulation

The Circulation Density in a Given Direction

Consider the vector field $\vec{F}$ shown in Figure 20.20. Suppose that at a distance r from the z-axis it has magnitude $2r$ in the direction shown. We use the idea of circulation around a closed curve, introduced on page 347, to measure the strength of circulation at the origin. We will calculate the circulation per unit area around the z-axis. Take a circle C of radius R in the xy-plane and centered at the origin, traversed in a direction determined by the right hand rule as in Figure 20.23. Then, since $\vec{F}$ is tangent to C everywhere, and points in the forward direction around C, we have

$$\text{Circulation around } C = \int_C \vec{F} \cdot d\vec{r} = \|\vec{F}\| \cdot (\text{Circumference of circle } C) = 2R(2\pi R) = 4\pi R^2.$$

Thus the circulation per unit area is:

$$\frac{\text{Circulation around } C}{\text{Area inside } C} = \frac{\int_C \vec{F} \cdot d\vec{r}}{\pi R^2} = \frac{2R \cdot 2\pi R}{\pi R^2} = 4.$$

If the vector field had magnitude $3r$ in the same direction, the ratio would be

$$\frac{\text{Circulation around } C}{\text{Area inside } C} = \frac{\int_C \vec{F} \cdot d\vec{r}}{\pi R^2} = \frac{3R \cdot 2\pi R}{\pi R^2} = 6.$$

If it had magnitude $5r$ in the opposite direction, then the ratio would be

$$\frac{\text{Circulation around } C}{\text{Area inside } C} = \frac{\int_C \vec{F} \cdot d\vec{r}}{\pi R^2} = \frac{-5R \cdot 2\pi R}{\pi R^2} = -10.$$

The negative sign indicates that the field is circulating around the z-axis in the opposite direction to that determined by the right hand rule.

This circulation per unit area determines the angular velocity of the paddle-wheel in Figure 20.21[1] (assuming you could make one sufficiently small and light, and insert it without disturbing the flow).

Of course, in each case we are using a very special vector field whose magnitude is a constant multiple of the distance r. Now we define the circulation density of an arbitrary vector field $\vec{F}$ at a point P about an arbitrary direction, by taking the limit using closed curves C in a plane perpendicular to the direction. Suppose that $\vec{n}$ is the unit vector pointing in the direction about which we want to measure the circulation density. The curve C around which we compute the circulation is in a plane perpendicular to $\vec{n}$ and oriented with respect to $\vec{n}$ by the right hand rule (see Figures 20.22 and 20.23).

Then we make the following definition:

The **circulation density** of $\vec{F}$ at the point P around the direction of the unit vector $\vec{n}$ is defined by the limit

$$\text{Circulation density} = \lim_{\text{area} \to 0} \left(\frac{\text{Circulation around } C}{\text{Area inside } C} \right) = \lim_{\text{area} \to 0} \left(\frac{\int_C \vec{F} \cdot d\vec{r}}{\text{Area inside } C} \right).$$

Here C is a closed curve about P in the plane perpendicular to $\vec{n}$, traversed in the direction determined by the right-hand rule. The diameter of C should go to zero as well as the area inside it.

---

[1] In fact it is twice the angular velocity. See Example 4 on page 430

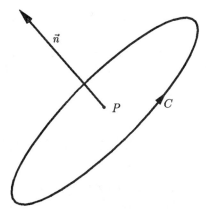

Figure 20.22: Direction of C relates to direction of $\vec{n}$ by the right-hand rule

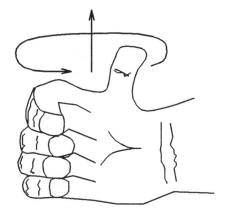

Figure 20.23: When the thumb points up, the fingers curl in the forward direction around C

How Does the Limit Defining the Circulation Density Work?

The limit exists if $\vec{F}$ has continuous partial derivatives at every point of an open set containing P, and if C is a closed, piece-wise smooth curve, which does not cross itself, in the plane through P perpendicular to $\vec{n}$. In that case, the limit does not depend on the shape of C as it contracts around P; all that is required is that the diameter of C tend to zero.

In Section 20.5 we will derive the following formula for the circulation density in terms of the circulation densities around the coordinate directions:

$$
\begin{pmatrix} \text{Circulation} \\ \text{density} \\ \text{around } \vec{n} \end{pmatrix} = \begin{pmatrix} \text{Circulation} \\ \text{density} \\ \text{around } \vec{i} \end{pmatrix} n_1 + \begin{pmatrix} \text{Circulation} \\ \text{density} \\ \text{around } \vec{j} \end{pmatrix} n_2 + \begin{pmatrix} \text{Circulation} \\ \text{density} \\ \text{around } \vec{k} \end{pmatrix} n_3,
$$

where $\vec{n} = n_1\vec{i} + n_2\vec{j} + n_3\vec{k}$ is a unit vector.

This formula is analogous to the formula on page 140 expressing the directional derivative in an arbitrary direction in terms of the partial derivatives.

Definition of the Curl Vector

The Curl of a Vector Field

The curl of a vector field $\vec{F}(x, y, z)$ at a point P, written curl $\vec{F}(P)$, is the vector with the following properties

- The direction of curl $\vec{F}(P)$ is the direction around which the circulation density of $\vec{F}$ is the greatest, if it exists.

- The magnitude of curl $\vec{F}$ is the circulation density around that direction.

If the circulation density is zero around every direction, then we define the curl to be $\vec{0}$.

The curl of a vector field is itself a vector field. A vector field is said to be *curl free* or *irrotational* if curl $\vec{F}(P) = \vec{0}$ at every point P.

Using the formula on page 427 for the circulation density in terms of circulation densities in the coordinate directions, we can write

$$\text{Circulation density around } \vec{n} = \vec{v} \cdot \vec{n},$$

where

$$\vec{v} = \left(\begin{array}{c} \text{Circulation} \\ \text{density} \\ \text{around } \vec{i} \end{array} \right) \vec{i} + \left(\begin{array}{c} \text{Circulation} \\ \text{density} \\ \text{around } \vec{j} \end{array} \right) \vec{j} + \left(\begin{array}{c} \text{Circulation} \\ \text{density} \\ \text{around } \vec{k} \end{array} \right) \vec{k}.$$

The maximum value of $\vec{v} \cdot \vec{n}$, and therefore of the circulation density, occurs when $\vec{n}$ points in the direction of $\vec{v}$. By definition, this is also the direction of curl $\vec{F}$. Thus, curl $\vec{F}$ and $\vec{v}$ point in the same direction. Also, the maximum value of $\vec{v} \cdot \vec{n}$ is $\|\vec{v}\|\|\vec{n}\| = \|\vec{v}\|$, so $\vec{v}$ and curl $\vec{F}$ also have the same magnitude. Hence they are the same vector. Thus we have the following result:

Components of the Curl

$$\text{curl } \vec{F} = \left(\begin{array}{c} \text{Circulation} \\ \text{density} \\ \text{around } \vec{i} \end{array} \right) \vec{i} + \left(\begin{array}{c} \text{Circulation} \\ \text{density} \\ \text{around } \vec{j} \end{array} \right) \vec{j} + \left(\begin{array}{c} \text{Circulation} \\ \text{density} \\ \text{around } \vec{k} \end{array} \right) \vec{k}.$$

Example 1   For each field in Figure 20.24, say whether the z-component of the curl points up, down, or is $\vec{0}$. Each of the vector fields has z-component $\vec{0}$, and is independent of z.

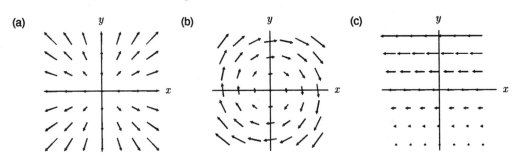

Figure 20.24: Three vector fields

Solution   (a)   This vector field shows no rotation, and the circulation around any closed curve looks like it will be zero, so we suspect a zero curl here.

(b)   This vector field definitely looks like it is swirling, so we expect a nonzero curl here. By the right-hand rule, the circulation density around $\vec{k}$ is negative, so the z-component of the curl will point down.

(c)   At first glance, you might expect this field to have zero curl, as all the vectors are parallel to the x-axis. However, if you find the circulation around the curve C in Figure 20.25, the sides contribute nothing (they are perpendicular to the vector field), the bottom contributes a negative quantity (the curve is in the opposite direction to the vector field), and the top contributes a larger positive quantity (the curve is in the same direction as the vector field and the magnitude

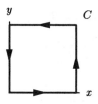

Figure 20.25

of the vector field is larger at the top than at the bottom). Thus, the circulation around C is positive and hence the curl must be nonzero and have upward pointing z-component.

Another way to see that the curl is nonzero in this case is to imagine the vector field representing the velocity of moving water. A boat sitting in the water will tend to rotate, as the water will be moving faster on one side than the other.

How to Calculate the Curl in Cartesian Coordinates

We have the following formula for curl $\vec{F}$ in Cartesian coordinates.

If $\vec{F} = F_1\vec{i} + F_2\vec{j} + F_3\vec{k}$ and $\vec{F}$ has continuous partial derivatives, then

$$\text{curl}\,\vec{F} = \left(\frac{\partial F_3}{\partial y} - \frac{\partial F_2}{\partial z}\right)\vec{i} + \left(\frac{\partial F_1}{\partial z} - \frac{\partial F_3}{\partial x}\right)\vec{j} + \left(\frac{\partial F_2}{\partial x} - \frac{\partial F_1}{\partial y}\right)\vec{k}$$

Comparing this with the formula for the components of the curl vector on page 428, we see that it is saying, for example, that the circulation density around $\vec{k}$ at a point P is $\frac{\partial F_2}{\partial x}(P) - \frac{\partial F_1}{\partial y}(P)$.

To see why this is true, recall that the circulation density around $\vec{k}$ at a point (x_0, y_0, z_0) is calculated using curves C around P that lie perpendicular to $\vec{k}$; hence the $\vec{k}$-component of $\vec{F}$ has no effect, that is, the circulation density of $\vec{F}$ around $\vec{k}$ is the same as the circulation density of $F_1\vec{i} + F_2\vec{j}$ around $\vec{k}$. But in any fixed plane perpendicular to $\vec{k}$, z is constant, so F_1 and F_2 are functions of x and y alone. Thus $F_1\vec{i} + F_2\vec{j}$ is a two-dimensional vector field, and Green's Theorem says that

$$\int_C (F_1\vec{i} + F_2\vec{j}) \cdot d\vec{r} = \int_R \left(\frac{\partial F_2}{\partial x} - \frac{\partial F_1}{\partial y}\right) dA \approx \left(\frac{\partial F_2}{\partial x}(P) - \frac{\partial F_1}{\partial y}(P)\right) \cdot (\text{Area of } R).$$

Here R is the region enclosed by C, and the approximation holds if R is sufficiently small, so that the partial derivatives are approximately constant on R. Hence

$$\begin{array}{c}\text{Circulation density} \\ \text{around } \vec{k}\end{array} \approx \frac{\int_C (F_1\vec{i} + F_2\vec{j}) \cdot d\vec{r}}{\text{Area of } R} \approx \frac{\partial F_2}{\partial x}(P) - \frac{\partial F_1}{\partial y}(P).$$

In the limit, the approximation becomes exact. We will explain this formula in more detail in Section 20.5.

Shorthand Notation for Curl

Using $\nabla = \dfrac{\partial}{\partial x}\vec{i} + \dfrac{\partial}{\partial y}\vec{j} + \dfrac{\partial}{\partial z}\vec{k}$, we can write

$$\text{curl } \vec{F} = \nabla \times \vec{F} = \begin{vmatrix} \vec{i} & \vec{j} & \vec{k} \\ \dfrac{\partial}{\partial x} & \dfrac{\partial}{\partial y} & \dfrac{\partial}{\partial z} \\ F_1 & F_2 & F_3 \end{vmatrix}.$$

Example 2   Use the definition to find the curl of the vector field $\vec{F}(\vec{r}) = \vec{r}$, then check your answer using the formula.

Solution   The part of this vector field in the xy-plane looks like Figure 20.1 on page 410, and shows no rotational tendency. Thus we expect the curl to be $\vec{0}$. In fact it is, because the circulation around *every* closed curve C is zero, since

$$\vec{F} = x\vec{i} + y\vec{j} + z\vec{k} = \text{grad}(x^2/2 + y^2/2 + z^2/2),$$

so $\vec{F}$ is a gradient field. Thus the circulation density is zero in any direction, and hence curl $\vec{F}(P) = \vec{0}$ for every point P. Using the formula:

$$\text{curl } \vec{F} = \nabla \times \vec{F} = \begin{vmatrix} \vec{i} & \vec{j} & \vec{k} \\ \dfrac{\partial}{\partial x} & \dfrac{\partial}{\partial y} & \dfrac{\partial}{\partial z} \\ x & y & z \end{vmatrix}$$

$$= \left(\dfrac{\partial z}{\partial y} - \dfrac{\partial y}{\partial z}\right)\vec{i} + \left(\dfrac{\partial x}{\partial z} - \dfrac{\partial z}{\partial x}\right)\vec{j} + \left(\dfrac{\partial y}{\partial x} - \dfrac{\partial x}{\partial y}\right)\vec{k} = 0\vec{i} + 0\vec{j} + 0\vec{k} = \vec{0}.$$

Example 3   Let $\vec{F}$ be the vector field in Figure 20.20 on page 425. Find the curl using the formula and relate your answer to circulation density.

Solution   The vector field always points perpendicularly to both $\vec{k}$ and $\vec{r}$, in the direction determined by the right hand rule, and its magnitude is twice the magnitude of $\vec{r}$. Thus

$$\vec{F}(\vec{r}) = 2\vec{k} \times \vec{r} = -2y\vec{i} + 2x\vec{j}.$$

Thus

$$\text{curl } \vec{v} = \begin{vmatrix} \vec{i} & \vec{j} & \vec{k} \\ \dfrac{\partial}{\partial x} & \dfrac{\partial}{\partial y} & \dfrac{\partial}{\partial z} \\ -2y & 2x & 0 \end{vmatrix}$$

$$= (\dfrac{\partial}{\partial y}(0) - \dfrac{\partial}{\partial z}(2x))\vec{i} + (\dfrac{\partial}{\partial z}(-2y) - \dfrac{\partial}{\partial x}(0))\vec{j} + (\dfrac{\partial}{\partial x}(2x) - \dfrac{\partial}{\partial y}(-2y))\vec{k} = 4\vec{k}.$$

This makes sense, because we computed the circulation density of this vector field in the z-direction and found it was 4, and we would expect the z-direction to give the maximum circulation density from the symmetry of the vector field.

Example 4   A flywheel is rotating with angular velocity $\vec{\omega}$, and the velocity of a point P is given by $\vec{v} = \vec{\omega} \times \vec{r}$. See Figure 20.26. Find curl $\vec{v}$.

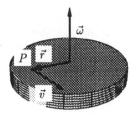

Figure 20.26: Rotating
flywheel

Solution   If $\vec{\omega} = \omega_1\vec{i} + \omega_2\vec{j} + \omega_3\vec{k}$, we have

$$\vec{v} = \vec{\omega} \times \vec{r} = \begin{vmatrix} \vec{i} & \vec{j} & \vec{k} \\ \omega_1 & \omega_2 & \omega_3 \\ x & y & z \end{vmatrix} = (\omega_2 z - \omega_3 y)\vec{i} + (\omega_3 x - \omega_1 z)\vec{j} + (\omega_1 y - \omega_2 x)\vec{k}.$$

Thus,

$$\text{curl } \vec{v} = \begin{vmatrix} \vec{i} & \vec{j} & \vec{k} \\ \frac{\partial}{\partial x} & \frac{\partial}{\partial y} & \frac{\partial}{\partial z} \\ \omega_2 z - \omega_3 y & \omega_3 x - \omega_1 z & \omega_1 y - \omega_2 x \end{vmatrix}$$

$$= \left(\frac{\partial}{\partial y}(\omega_1 y - \omega_2 x) - \frac{\partial}{\partial z}(\omega_3 x - \omega_1 z) \right)\vec{i} + \left(\frac{\partial}{\partial z}(\omega_2 z - \omega_3 y) - \frac{\partial}{\partial x}(\omega_1 y - \omega_2 x) \right)\vec{j}$$

$$+ \left(\frac{\partial}{\partial x}(\omega_3 x - \omega_1 z) - \frac{\partial}{\partial y}(\omega_2 z - \omega_3 y) \right)\vec{k}$$

$$= 2\omega_1\vec{i} + 2\omega_2\vec{j} + 2\omega_3\vec{k} = 2\vec{\omega}.$$

Thus, as we would expect, curl $\vec{v}$ is parallel to the axis of rotation of the flywheel (namely, the direction of $\vec{\omega}$) and the magnitude of curl $\vec{v}$ is larger the faster the flywheel is rotating (that is, the larger the magnitude of $\vec{\omega}$).

Problems for Section 20.3

Compute the curl of the vector fields in Problems 1–7.

1. $\vec{F}(x, y, z) = (x^2 - y^2)\vec{i} + 2xy\vec{j}$

2. $\vec{F}(\vec{r}) = \vec{r}/r$

3. $\vec{F} = x^2\vec{i} + y^3\vec{j} + z^4\vec{k}.$

4. $\vec{F} = e^x\vec{i} + \cos y\vec{j} + e^{z^2}\vec{k}.$

5. $\vec{F} = 2yz\vec{i} + 3xz\vec{j} + 7xy\vec{k}.$

6. $\vec{F}(x, y, z) = (-x + y)\vec{i} + (y + z)\vec{j} + (-z + x)\vec{k}$

7. $\vec{F} = (x + yz)\vec{i} + (y^2 + xzy)\vec{j} + (zx^3y^2 + x^7y^6)\vec{k}.$

8. Decide whether each of the following vector fields has a nonzero curl. In each case, a cross-section of the vector field is shown; assume that the cross-section is the same in all other planes parallel to the given cross-section.

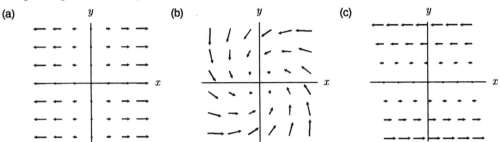

Figure 20.27

9. Using your answers to Problems 3–4, make a conjecture about the value of curl $\vec{F}$ when the vector field $\vec{F}$ has a certain form. (What form?) Show why your conjecture is true.

10. A large fire becomes a fire-storm when the nearby air acquires a circulatory motion. (This motion has the effect of bringing more air to the fire, causing it to burn faster.) Records show that a fire-storm developed during the Chicago Fire of 1871 and during the Second World War bombing of Hamburg, Germany, but there was no fire-storm during the Great Fire of London in 1666. Explain how a fire-storm could be identified using the curl of a vector field.

11. Assume f is a scalar function with continuous second partial derivatives. Use the Fundamental Theorem of Calculus for Line Integrals to show that $\int_C \operatorname{grad} f \cdot d\vec{r} = 0$ for any smooth closed path C. Deduce that curl grad $f = \vec{0}$.

12. If $\vec{F}$ is any vector field whose components have continuous second partial derivatives, show that div curl $\vec{F} = 0$.

13. For any constant vector field $\vec{c}$, and any vector field, $\vec{F}$, show that $\operatorname{div}(\vec{F} \times \vec{c}) = \vec{c} \cdot \operatorname{curl} \vec{F}$.

14. Show that curl $(\vec{F} + \vec{C}) = \operatorname{curl} \vec{F}$ for a constant vector field $\vec{C}$.

15. Show that curl $(\phi \vec{F}) = \phi \operatorname{curl} \vec{F} + (\operatorname{grad} \phi) \times \vec{F}$ where ϕ is a function and $\vec{F}$ is a vector field.

16. A vortex that rotates at constant angular velocity ω about the z-axis has velocity vector field $\vec{v} = \omega(-y\vec{i} + x\vec{j})$.
    (a) Sketch the vector field if $\omega = 1$.
    (b) Sketch the vector field if $\omega = -1$.
    (c) Determine the speed $\|\vec{v}\|$ of the vortex as a function of the distance from its center.
    (d) Compute div $\vec{v}$.
    (e) Compute curl $\vec{v}$.
    (f) Compute the circulation of $\vec{v}$ counterclockwise about the circle of radius R in the xy-plane, centered at the origin.

Problems 17–19 concern the vector fields in Figure 20.28.

17. Three of the vector fields have zero curl. Which are they? How do you know?

18. Three of the vector fields have zero divergence. Which are they? How do you know?

19. Four of the line integrals $\int_{C_i} \vec{F} \cdot d\vec{r}$ are zero. Which are they? How do you know?

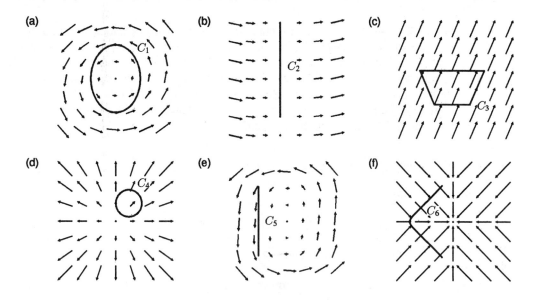

Figure 20.28

20. Show that if ϕ is a harmonic function, then grad ϕ is both curl free and divergence free.

21. It is a theorem of Helmholtz that every vector field $\vec{F}$ equals the sum of a curl free vector field and a divergence free vector field. Show how to do this, assuming that there is a function ϕ such that $\nabla^2 \phi = \text{div}\vec{F}$.

22. Express $(3x + 2y)\vec{i} + (4x + 9y)\vec{j}$ as the sum of a curl free vector field and a divergence free vector field.

23. Find a vector field $\vec{F}$ such that curl $\vec{F} = 2\vec{i} - 3\vec{j} + 4\vec{k}$. [Hint: Try $\vec{F} = \vec{v} \times \vec{r}$ for a suitable vector $\vec{v}$.]

24. Figure 20.29 gives a sketch of a velocity vector field $\vec{F} = y\vec{i} + x\vec{j}$ in the xy-plane.
    (a) What is the direction of rotation of a thin twig placed at the origin along the x-axis?
    (b) What is the direction of rotation of a thin twig placed at the origin along the y-axis?
    (c) Compute curl $\vec{F}$.

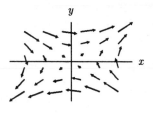

Figure 20.29

25. A central force field is a vector field of the form $\vec{F} = f(r)\vec{r}$ where f is any function of $r = \|\vec{r}\|$. Show that any central force field is irrotational.

20.4 STOKES' THEOREM

In Section 20.2, we discussed the Divergence Theorem, which says that the flux of a vector field out of a closed surface equals the integral of the divergence of the vector field over the volume enclosed by the surface. This is because the flux integral can be interpreted as a total outflow from the volume, and the divergence at a point is the outflow per unit volume at that point.

In a similar way, we now find the circulation of a vector field $\vec{F}$ around a closed curve C by integrating the circulation density over a surface S bounded by the curve. First we see how to calculate the circulation density from the curl.

Relation Between the Curl and the Circulation Density

We saw on page 427 that

$$\begin{pmatrix} \text{Circulation} \\ \text{density} \\ \text{around } \vec{n} \end{pmatrix} = \begin{pmatrix} \text{Circulation} \\ \text{density} \\ \text{around } \vec{i} \end{pmatrix} n_1 + \begin{pmatrix} \text{Circulation} \\ \text{density} \\ \text{around } \vec{j} \end{pmatrix} n_2 + \begin{pmatrix} \text{Circulation} \\ \text{density} \\ \text{around } \vec{k} \end{pmatrix} n_3$$

where $\vec{n} = n_1 \vec{i} + n_2 \vec{j} + n_3 \vec{k}$ is a unit vector, and also on page 428 that

$$\text{curl } \vec{F} = \begin{pmatrix} \text{Circulation} \\ \text{density} \\ \text{around } \vec{i} \end{pmatrix} \vec{i} + \begin{pmatrix} \text{Circulation} \\ \text{density} \\ \text{around } \vec{j} \end{pmatrix} \vec{j} + \begin{pmatrix} \text{Circulation} \\ \text{density} \\ \text{around } \vec{k} \end{pmatrix} \vec{k}.$$

Putting these together, we get

$$\boxed{\text{Circulation density of } \vec{F} \text{ around } \vec{n} = \text{curl } \vec{F} \cdot \vec{n}.}$$

This is analogous to the formula relating the directional derivative to the gradient vector.

Calculating the Circulation from the Circulation Density

Break the surface S up into many small pieces, as in Figure 20.30. We make certain that the pieces are small enough so that each is approximately flat. Then we compute the circulation of the vector field $\vec{F}$ around the boundary of each piece. If $\vec{n}$ is a positive unit normal vector to a piece of surface with area ΔA, then the vector $\vec{n} \Delta A$ is approximately the area vector $\Delta \vec{A}$ of the piece. If the piece is small enough, then

$$\begin{matrix} \text{Circulation of } \vec{F} \text{ around} \\ \text{boundary of the piece} \end{matrix} \approx \begin{pmatrix} \text{Circulation} \\ \text{density} \\ \text{around } \vec{n} \end{pmatrix} \Delta A = ((\text{curl } \vec{F}) \cdot \vec{n}) \Delta A = (\text{curl} \vec{F}) \cdot \Delta \vec{A}.$$

The next step is to add up the circulations around all the small pieces. When we do this, the line integral along the common edge of a pair of adjacent pieces gets counted twice, once each way. See Figure 20.31. Hence, these integrals cancel. All that is left are the line integrals around the outside edges. As we continue to add up all the circulations around all the small pieces, the line

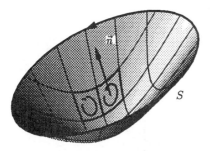

Figure 20.30: The circulations of the pieces add up to the circulation of the whole

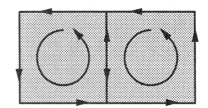

Figure 20.31: Two adjacent pieces of the surface

integrals along inside edges cancel, and we are left with the circulation around the boundary of the entire surface. Thus,

$$\begin{array}{c} \text{Circulation around} \\ \text{boundary of } S \end{array} = \sum \begin{array}{c} \text{Circulation around} \\ \text{boundary of pieces} \end{array} \approx \sum \operatorname{curl} \vec{F} \cdot \Delta \vec{A}.$$

Taking the limit as $\Delta A \to 0$, we get

$$\text{Circulation around boundary of } S = \int_S \operatorname{curl} \vec{F} \cdot d\vec{A}.$$

Since the circulation is also the line integral of $\vec{F}$ around C, the boundary of S, we have the following result:

Stokes' Theorem

If $\vec{F}$ is a vector field and S is a smooth oriented surface with oriented boundary C, then

$$\text{Circulation around boundary of } S = \int_C \vec{F} \cdot d\vec{r} = \int_S \operatorname{curl} \vec{F} \cdot d\vec{A}.$$

The orientation of C must be determined from the orientation of S by the right-hand rule, and curl $\vec{F}$ must be defined at every point of S.

How Do We Choose an Orientation for the Boundary of S?

An orientation for a surface S determines an orientation for its boundary C by the right hand rule, as follows. Pick a positive normal $\vec{n}$ on S, and use the right hand rule to determine a direction of travel around $\vec{n}$. This in turn determines a direction of travel around the boundary C. See Figure 20.30.

Example 1   Suppose that at a distance r from the z-axis the vector field $\vec{F}$ in Figure 20.32 has magnitude $2r$ in the direction shown. Use Stokes' Theorem to find the circulation of $\vec{F}$ around the circle C of radius R in the xy-plane, centered at the origin. We take the orientation of C to be counterclockwise as viewed from above the xy-plane.

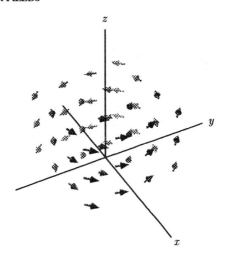

Figure 20.32: A vector field

Solution   As we saw on page 426

$$\int_C \vec{F} \cdot d\vec{r} = \|\vec{F}\| \times \text{Length of } C = 2R(2\pi R) = 4\pi R^2,$$

and as we saw in Example 3 on page 430

$$\text{curl } \vec{F}(x, y, z) = 4\vec{k}.$$

Let S be the region in the xy-plane enclosed by the circle. We orient S upward to be consistent with the orientation of C. The curl of $\vec{F}$ is in the same direction as the positive normal vector to the surface, so by Stokes' Theorem,

$$\text{Circulation around } C = \int_S \text{curl } \vec{F} \cdot d\vec{A} = \int_S 4\vec{k} \cdot d\vec{A} = 4(\text{Area of } S) = 4\pi R^2.$$

This is the same answer we obtained before.

It should be noted that Stokes' Theorem applies to any oriented surface S and its boundary C, even in cases where the boundary may consist of two or more curves. For example,

$$\int_C \vec{F} \cdot d\vec{r} = \int_S \text{curl } \vec{F} \cdot d\vec{A}$$

where S is a ring or a cylinder and C consists of two curves as in Figure 20.33. Notice how the right hand rule determines the orientation of each of the boundary curves.

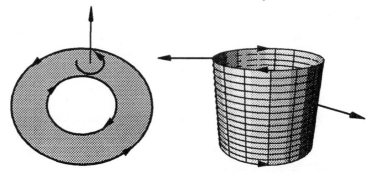

Figure 20.33: Stokes' Theorem applies to a ring and a cylinder

The Curl and the Gradient

In Problem 11 on page 432 we saw that

$$\text{curl grad } f = \vec{0}$$

for a function f with continuous second partial derivatives.

What about the converse? What can be said about vector fields $\vec{F}$ such that curl $\vec{F} = \vec{0}$? Suppose that curl $\vec{F} = \vec{0}$ and let us consider the line integral $\int_C \vec{F} \cdot d\vec{A}$ for a closed curve C in the region where $\vec{F}$ is defined. If C is the boundary curve of an orientable surface S that lies wholly in the domain of curl $\vec{F}$, then Stokes' Theorem asserts that

$$\int_C \vec{F} \cdot d\vec{r} = \int_S \text{curl } \vec{F} \cdot d\vec{A} = \int_S \vec{0} \cdot d\vec{A} = 0.$$

If $\int_C \vec{F} \cdot d\vec{r} = 0$ for every closed curve C, then we know that $\vec{F}$ is a gradient field. Thus, we have

The Curl Criterion for a Gradient Vector Field

If curl $\vec{F}$ exists and equals zero wherever $\vec{F}$ is defined, and if every closed curve in the domain of $\vec{F}$ bounds an orientable surface that also lies in the domain of $\vec{F}$, then $\vec{F}$ is a gradient field.

It can be shown that every closed curve is the boundary of some orientable surface. Thus, we get the following result from the curl criterion for a gradient field:

If a vector field $\vec{F}$ is defined everywhere and its curl is zero everywhere, then $\vec{F}$ is a gradient field.

Example 2 Which of the following vector fields is a gradient field?
(a) $\vec{F} = yz\vec{i} + (xz + z^2)\vec{j} + (xy + 2yz)\vec{k}$ (b) $\vec{G} = -y\vec{i} + x\vec{j}$.

Solution (a) Since curl $\vec{F} = \vec{0}$, we know that $\vec{F}$ is a gradient field. In fact, $\vec{F} = \text{grad} f$, where $f(x, y, z) = xyz + yz^2$, so f is a potential function for $\vec{F}$.
(b) Since curl $\vec{G} = 2\vec{k} \neq \vec{0}$, the vector field $\vec{G}$ is not a gradient field.

The Curl and The Divergence

In Problem 12 on page 432 we showed

$$\text{div curl } \vec{F} = 0,$$

if $\vec{F}$ is a vector field whose components have continuous second partial derivatives.

Conversely, it turns out (though we will not prove it), that if $\text{div}\vec{F} = 0$ everywhere then $\vec{F} = \text{curl }\vec{G}$ for some vector field $\vec{G}$. A field with this property is given a special name.

$\vec{F}$ is a *curl field* if $\vec{F} = \text{curl }\vec{G}$ for some vector field $\vec{G}$.

Recall that a function f such that $\vec{F} = \text{grad} f$ is called a potential function for the vector field $\vec{F}$. By analogy, if a vector field $\vec{F}$ is a curl field, then any vector field $\vec{G}$ such that $\vec{F} = \text{curl }\vec{G}$ is called a *vector potential* for $\vec{F}$. The vector potential of a given curl field is not unique.

Example 3   Which of the following vector fields is a curl field?
(a) $\vec{F} = xy\vec{i} - 2yz\vec{j} + (z^2 - yz)\vec{k}$ (b) $\vec{G} = xy\vec{i} + z^2\vec{j} - 3xz\vec{k}$.

Solution   (a)   Since $\text{div}\vec{F} = 0$, we know that $\vec{F}$ is a curl field. In fact, $\vec{F} = \text{curl }\vec{H}$, where $\vec{H} = -yz^2\vec{i} - xyz\vec{j}$, so $\vec{H}$ is a vector potential for $\vec{F}$.
(b)   Since $\text{div}\vec{G} = y - 3x \neq 0$, the vector field $\vec{G}$ is not a curl field.

Properties of Curl Fields

Recall that if a vector field $\vec{F}$ is a gradient field, then $\vec{F}$ has some particularly nice properties: The line integral $\int_C \vec{F} \cdot d\vec{r}$ depends only on the endpoints of the curve C, and if C is a closed curve then $\int_C \vec{F} \cdot d\vec{r} = 0$. We will now consider the properties curl fields.

Example 4   Suppose $\vec{F} = \text{curl }\vec{G}$.
(a)   Suppose that S_1 and S_2 are two oriented surfaces with the same oriented boundary curve C (that is, they have the same boundary and they determine the same orientation on it). Show that the flux integrals of $\vec{F}$ over the two surfaces are equal:

$$\int_{S_1} \vec{F} \cdot d\vec{A} = \int_{S_2} \vec{F} \cdot d\vec{A}.$$

Thus, the flux integral of a curl field over a surface depends only on the boundary of the surface. See Figure 20.34.
(b)   If S is a closed surface, show that $\int_S \vec{F} \cdot d\vec{A} = 0$. Thus, the flux integral of a curl field over a closed surface equals zero.

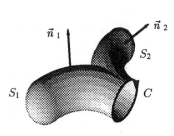

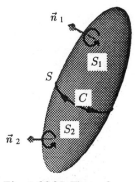

Figure 20.34: The flux of a curl is the same through these two surfaces S_1 and S_2 which have the same boundary, C

Figure 20.35: The surfaces S_1 and S_2 have the same boundary, C, but with different orientations.

Solution　(a)　By Stokes' Theorem,

$$\int_{S_1} \vec{F} \cdot d\vec{A} = \int_{S_1} \text{curl}\, \vec{G} \cdot d\vec{A} = \int_C \vec{G} \cdot d\vec{r}$$

and

$$\int_{S_2} \vec{F} \cdot d\vec{A} = \int_{S_2} \text{curl}\, \vec{G} \cdot d\vec{A} = \int_C \vec{G} \cdot d\vec{r}.$$

Thus,

$$\int_{S_1} \vec{F} \cdot d\vec{A} = \int_{S_2} \vec{F} \cdot d\vec{A}$$

(b)　Draw a closed curve C on the surface S, thus dividing S into two surfaces S_1 and S_2 as shown in Figure 20.35. The outward orientations on S_1 and S_2 determine opposite orientations on C, so by part (a) of this example,

$$\int_{S_1} \vec{F} \cdot d\vec{A} = -\int_{S_2} \vec{F} \cdot d\vec{A}.$$

Thus,

$$\int_S \vec{F} \cdot d\vec{A} = \int_{S_1} \vec{F} \cdot d\vec{A} + \int_{S_2} \vec{F} \cdot d\vec{A} = 0.$$

Example 5　Evaluate $\int_S \vec{F} \cdot d\vec{A}$ where $\vec{F} = (8yz - z)\vec{j} + (3 - 4z^2)\vec{k}$, and S is the hemisphere of radius 5 shown in Figure 20.36, oriented upwards. Use the fact that $\vec{F} = \text{curl}(4yz^2\vec{i} + 3x\vec{j} + xz\vec{k})$ to simplify the computation.

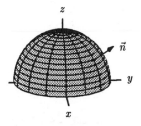

Figure 20.36: The hemispherical surface S

Solution   Because $\vec{F}$ is a curl field, its flux through S will be the same as its flux through any surface with the same boundary as S. Let us replace S by the circular disc S_1 of radius 5 in the xy-plane. On S_1, $z = 0$, and so the vector field reduces to $\vec{F} = 3\vec{k}$. Since the vector $3\vec{k}$ is normal to S_1 and in the direction of the orientation of S_1, we have

$$\int_S \vec{F} \cdot d\vec{A} = \int_{S_1} \vec{F} \cdot d\vec{A} = \int_{S_1} 3\vec{k} \cdot d\vec{A} = \|3\vec{k}\|(\text{area of } S_1) = 75\pi.$$

Three Differential Operators and Three Integral Theorems

In three-variable calculus we have studied three operators computed in terms of derivatives: grad, curl, and div. Here is a summary of their differences:

- The gradient of a scalar function is a vector field.
- The curl of a vector field is vector field.
- The divergence of a vector field is a scalar function.

There is a fundamental integral theorem for each of these three operators. In all three cases an integral over a region is expressed in terms of the boundary of the region.

Fundamental Theorem of Calculus for Line Integrals

$$\int_C \text{grad} f \cdot d\vec{r} = f(Q) - f(P).$$

This expresses the line integral of the gradient of a function f over a (1-dimensional) curve C in terms of the values of f at the boundary (endpoints P and Q) of the curve.

Stokes' Theorem

$$\int_S \text{curl}\,\vec{F} \cdot d\vec{A} = \int_C \vec{F} \cdot d\vec{r}.$$

This expresses the flux integral of the curl of a vector field $\vec{F}$ over a (2-dimensional) surface S in terms of the line integral of $\vec{F}$ over the boundary curve C of the surface.

Divergence Theorem

$$\int_W \text{div}\,\vec{F}\, dV = \int_S \vec{F} \cdot d\vec{A}.$$

This expresses the integral of the divergence of a vector field $\vec{F}$ over a (3-dimensional) volume W in terms of the flux integral of $\vec{F}$ over the boundary surface S of the volume.

Problems for Section 20.4

In Problems 1-2 compute the given line integral using Stokes' Theorem.

1. $\int_C (-z\vec{i} + y\vec{j} + x\vec{k}) \cdot d\vec{r}$, where C is a circle of radius 2 around the y-axis with orientation indicated in Figure 20.37.

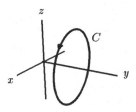

Figure 20.37

2. $\int_C \vec{F} \cdot d\vec{r}$, with $\vec{F} = \vec{r}/r^3$ and C is the path consisting of straight line segments from $(1, 0, 1)$ to $(1, 0, 0)$ to $(0, 0, 1)$ back to $(1, 0, 1)$.

3. Use Stokes' Theorem to evaluate $\int_C \vec{F} \cdot d\vec{r}$ where $\vec{F} = (2x - y)\vec{i} + (x + 4y)\vec{j}$ and C is as given.
   (a) C is the circle of radius 10 in the xy-plane centered at the origin, oriented clockwise as viewed from way up the positive z-axis.
   (b) C is the circle of radius 10 in the yz-plane centered at the origin, oriented clockwise as viewed from way out the positive x-axis.

4. Can you use Stokes' Theorem to compute the line integral $\int_C (2x\vec{i} + 2y\vec{j} + 2z\vec{k}) \cdot d\vec{r}$ where C is the straight line from the point $(1, 2, 3)$ to the point $(4, 5, 6)$? Why or why not?

5. For this exercise, $\vec{F} = -z\vec{j} + y\vec{k}$, C is the circle of radius R in the yz-plane oriented clockwise as viewed from the positive x-axis, and S is the disk in the yz-plane enclosed by C, oriented in the positive x-direction. See Figure 20.38.

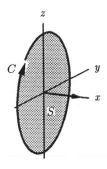

Figure 20.38

   (a) Evaluate directly $\int_C \vec{F} \cdot d\vec{r}$.
   (b) Evaluate directly $\int_S \text{curl}\vec{F} \cdot d\vec{A}$.
   (c) The answers in parts (a) and (b) are not equal. Explain why this does not contradict Stokes' Theorem.

6. Use Stokes' Theorem to show that $\int_S \text{curl } \vec{F} \cdot d\vec{A} = 0$ for any surface S which is the complete boundary surface of a volume V. Deduce that div curl $\vec{F} = 0$.

7. A basic property of the magnetic field $\vec{B}$ is that curl $\vec{B} = 0$ in a region where there are no currents flowing. Consider the magnetic field around a long thin wire carrying a constant current. You are given that the magnitude of the magnetic field depends only on the distance from the wire, and that its direction is always tangent to the circle around the wire traversed in a direction related to the direction of the current by the right hand rule. Use Stokes' Theorem to deduce that the magnitude of the magnetic field is proportional to the reciprocal of the distance from the wire. [Hint: Consider an annulus around the wire. Its boundary has two pieces: the inner circle and the outer circle.]

8. You might guess that the speed of a naturally occurring vortex (tornado, waterspout, whirlpool) is a decreasing function of the distance from its center, so the constant angular velocity model of Problem 16 on page 432 would be inappropriate. A *free vortex* circulating about the z-axis has vector field $\vec{v} = K(x^2 + y^2)^{-1}(-y\vec{i} + x\vec{j})$ where K is a constant.
   (a) Sketch the vector field if $K = 1$.
   (b) Sketch the vector field if $K = -1$.
   (c) Determine the speed $\|\vec{v}\|$ of the vortex as a function of the distance from its center.
   (d) Compute div $\vec{v}$.
   (e) Show that curl $\vec{v} = \vec{0}$.
   (f) Compute the circulation of $\vec{v}$ counterclockwise about the circle of radius R at the origin.
   (g) The computations in parts (e) and (f) show that $\vec{v}$ has curl $\vec{0}$, but has nonzero circulation around the closed curve in part (d). Explain why this does not contradict Stokes' Theorem.

9. Is there a vector field $\vec{G}$ such that curl $\vec{G} = y\vec{i} + x\vec{j}$? How do you know?

10. Determine whether vector potentials for $\vec{F}$ and $\vec{G}$ exist, and if so, find one.
   (a) $\vec{F} = 2x\vec{i} + (3y - z^2)\vec{j} + (x - 5z)\vec{k}$
   (b) $\vec{F} = x^2\vec{i} + y^2\vec{j} + z^2\vec{k}$

11. Suppose that C is a closed curve in the xy-plane, oriented counterclockwise as viewed from above. Show that $(1/2)\int_C(-y\vec{i} + x\vec{j}) \cdot d\vec{r}$ equals the area of the region R in the xy-plane enclosed by C.

12. Imagine the following vector fields in 3-space:

$$\vec{F} = F_1(x, y)\vec{i} + F_2(x, y)\vec{j} \quad \text{and} \quad \vec{G} = G_1(x, y)\vec{i} + G_2(x, y)\vec{j}$$

sketched in Figures 20.39 and 20.40.

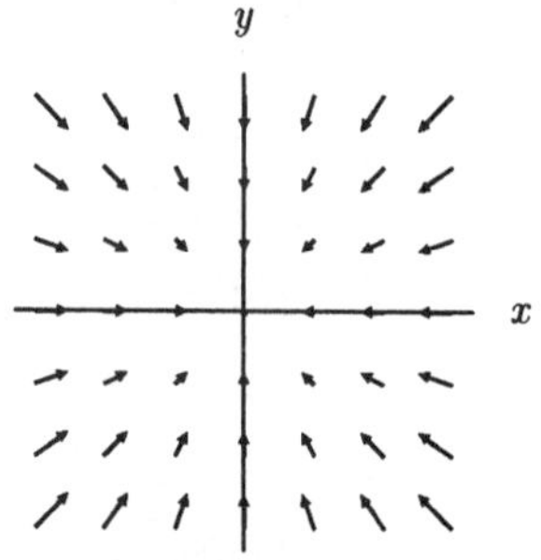

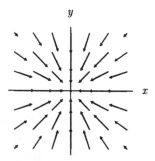

Figure 20.39: Cross-section of $\vec{F}$

Figure 20.40: Cross-section of $\vec{G}$

(a) What can you say about div $\vec{F}$ and div $\vec{G}$ at the origin?

(b) What can you say about curl $\vec{F}$ and curl $\vec{G}$ at the origin?

(c) Can you draw a closed surface around the origin such that $\vec{F}$ has a nonzero flux through it?

(d) Repeat part (c) for $\vec{G}$.

(e) Can you draw a closed curve around the origin such that $\vec{F}$ has a nonzero circulation around it?

(f) Repeat part (e) for $\vec{G}$.

13. A vector field $\vec{F}$ is defined everywhere except on the z-axis, and curl $\vec{F} = \vec{0}$ everywhere where $\vec{F}$ is defined. What can you say about $\int_C \vec{F} \cdot d\vec{r}$ if C is a circle of radius 1 in the xy-plane, and if the center of C is at   (a)   the origin,   (b)   the point $(2, 0)$?

14. Compute the line integral $\int_C ((yz^2 - y)\vec{i} + (xz^2 + x)\vec{j} + 2xyz\vec{k}) \cdot d\vec{r}$ where C is the circle of radius 3 in the xy-plane, centered at the origin, oriented counterclockwise as viewed from the positive z-axis. Do it two ways:

(a)   Directly   (b)   Using Stokes' Theorem

20.5 NOTES ON THE DIVERGENCE AND CURL

In this section we will give a more detailed explanation of the formulas for the divergence and curl in Cartesian coordinates. We start with the divergence.

The Divergence in Cartesian Coordinates

Suppose we want to find div $\vec{F}$, where $\vec{F} = F_1\vec{i} + F_2\vec{j} + F_3\vec{k}$. To find div $\vec{F}$ at the point (x_0, y_0, z_0) we use the definition

$$\text{div } \vec{F}(x_0, y_0, z_0) = \lim_{\text{vol} \to 0} \frac{\int_S \vec{F} \cdot d\vec{A}}{\text{Volume inside } S}.$$

Here we take for S a small cube-shaped surface with one corner at (x_0, y_0, z_0) and edges of length Δx, Δy, and Δz parallel to the axes; see Figure 20.41. (It turns out not to matter in this case that (x_0, y_0, z_0) is on S rather than surrounded by it.)

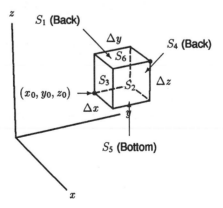

Figure 20.41: Cube used to find div $\vec{F}$ at (x_0, y_0, z_0)

First we'll calculate the flux through the surfaces S_1 and S_2 perpendicular to the x-axis. On S_1 (the back face of the cube shown in Figure 20.41), the outward normal is in the negative x direction, so $d\vec{A} = -dy\,dz\,\vec{i}$. Therefore, on S_1,

$$\vec{F} \cdot d\vec{A} = (F_1\vec{i} + F_2\vec{j} + F_3\vec{k}) \cdot (-dy\,dz\,\vec{i}) = -F_1\,dy\,dz.$$

On S_2 the outward normal points in the positive x direction, so $d\vec{A} = dy\,dz\,\vec{i}$. Therefore, on S_2

$$\vec{F} \cdot d\vec{A} = (F_1\vec{i} + F_2\vec{j} + F_3\vec{k}) \cdot (dy\,dz\,\vec{i}) = F_1\,dy\,dz.$$

Notice, however, that on S_1 the function F_1 is evaluated at the point (x_0, y, z) whereas on S_2, the function F_1 is evaluated at $(x_0 + \Delta x, y, z)$. But, by local linearity,

$$F_1(x_0 + \Delta x, y, z) \approx F_1(x_0, y, z) + \frac{\partial F_1}{\partial x}\,\Delta x.$$

Thus, the total contribution to the flux from S_1 and S_2 is

$$\int_{S_1} \vec{F} \cdot d\vec{A} + \int_{S_2} \vec{F} \cdot d\vec{A}$$
$$= \int_{z=z_0}^{z_0+\Delta z} \int_{y=y_0}^{y_0+\Delta y} -F_1(x_0, y, z)\,dy\,dz + \int_{z=z_0}^{z_0+\Delta z} \int_{y=y_0}^{y_0+\Delta y} F_1(x_0 + \Delta x, y, z)\,dy\,dz$$
$$\approx \int_{z=z_0}^{z_0+\Delta z} \int_{y=y_0}^{y_0+\Delta y} -F_1(x_0, y, z)\,dy\,dz + \int_{z=z_0}^{z_0+\Delta z} \int_{y=y_0}^{y_0+\Delta y} \left(F_1(x_0, y, z) + \frac{\partial F_1}{\partial x}\Delta x \right) dy\,dz$$
$$= \int_{z=z_0}^{z_0+\Delta z} \int_{y=y_0}^{y_0+\Delta y} \left(\frac{\partial F_1}{\partial x}\,\Delta x \right) dy\,dz.$$

Thus, we have to integrate the function $\left(\dfrac{\partial F_1}{\partial x}\,\Delta x \right)$ over the rectangular region $y_0 \le y \le y_0 + \Delta y$, $z_0 \le z \le z_0 + \Delta z$. Over this region, the function $\left(\dfrac{\partial F_1}{\partial x}\,\Delta x \right)$ is approximately constant, so

$$\int_{z=z_0}^{z_0+\Delta z} \int_{y=y_0}^{y_0+\Delta y} \left(\frac{\partial F_1}{\partial x}\,\Delta x \right) dy\,dz \approx \left(\frac{\partial F_1}{\partial x}\,\Delta x \right) \int_{z=z_0}^{z_0+\Delta z} \int_{y=y_0}^{y_0+\Delta y} dy\,dz = \frac{\partial F_1}{\partial x}\,\Delta x\,\Delta y\,\Delta z.$$

By an analogous argument, the contribution to the flux from S_3 and S_4 (the surfaces perpendicular to the y-axis) is approximately

$$\frac{\partial F_2}{\partial y}\,\Delta x\,\Delta y\,\Delta z,$$

and the contribution to the flux from S_5 and S_6 is approximately

$$\frac{\partial F_3}{\partial z}\,\Delta x\,\Delta y\,\Delta z.$$

Thus, adding these contributions we have

$$\text{Total flux through } S \approx \frac{\partial F_1}{\partial x}\,\Delta x\,\Delta y\,\Delta z + \frac{\partial F_2}{\partial y}\,\Delta x\,\Delta y\,\Delta z + \frac{\partial F_3}{\partial z}\,\Delta x\,\Delta y\,\Delta z.$$

Since the volume of the cube is $\Delta x \, \Delta y \, \Delta z$,

$$\text{div } \vec{F} \, (x_0, y_0, z_0) = \lim_{\text{vol} \to 0} \left(\frac{\text{Flux through } S}{\text{Volume of } S} \right)$$

$$= \lim_{\Delta x \, \Delta y \, \Delta z \to 0} \frac{\left(\frac{\partial F_1}{\partial x} + \frac{\partial F_2}{\partial y} + \frac{\partial F_3}{\partial z} \right) \Delta x \, \Delta y \, \Delta z}{\Delta x \, \Delta y \, \Delta z}$$

$$= \frac{\partial F_1}{\partial x} + \frac{\partial F_2}{\partial y} + \frac{\partial F_3}{\partial z}.$$

Thus, we have the following result:

If $\vec{F} = F_1 \vec{i} + F_2 \vec{j} + F_3 \vec{k}$, then

$$\text{div } \vec{F} = \frac{\partial F_1}{\partial x} + \frac{\partial F_2}{\partial y} + \frac{\partial F_3}{\partial z}.$$

The Curl in Cartesian Coordinates

Circulation Density

First we give a justification for the formula for the circulation density given on page 427. We will use the tetrahedron in Figure 20.42 to express the circulation density of a vector field $\vec{F}$ around the direction $\vec{n}$ in terms of the circulation density around each of the coordinate directions. The tetrahedron has three faces S_x, S_y, and S_z facing in the direction of the coordinate axes (toward the inside of the tetrahedron), and a face S in the direction of $\vec{n}$ (away from the tetrahedron). Notice that if $\vec{n}$ makes an angle θ_x with the x-axis, θ_y with the y-axis, and θ_z with the z-axis, then

$$\vec{n} = \cos \theta_x \vec{i} + \cos \theta_y \vec{j} + \cos \theta_z \vec{k}.$$

See Figure 20.43.

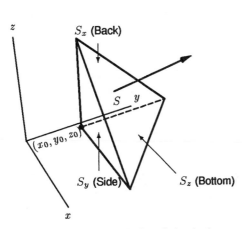

Figure 20.42: A tetrahedron facing in the direction of $\vec{n}$

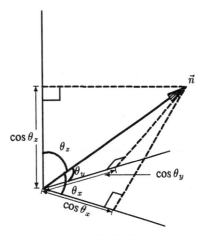

Figure 20.43: $\vec{n} = \cos \theta_x \vec{i} + \cos \theta_y \vec{j} + \cos \theta_z \vec{k}$

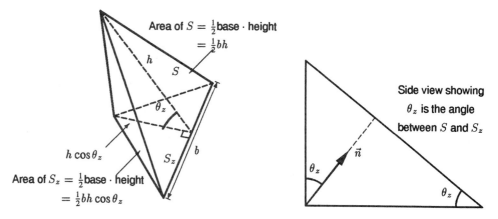

Figure 20.44: $A_z = (1/2)bh \cos \theta_z = A \cos \theta_z$

We approximate the circulation density around $\vec{n}$ by dividing the circulation around the boundary of S by the area, A, of S. Similarly, the circulation density around the z-direction is approximately the circulation around S_z divided by the area, A_z, of S_z, and so on. Figure 20.44 shows that

$$A_z = A \cos \theta_z.$$

Similarly,

$$A_x = A \cos \theta_x \quad \text{and} \quad A_y = A \cos \theta_y.$$

Figure 20.45 shows that if we add up the circulations around the boundaries of S_x, S_y, and S_z, the integrals along the axes cancel, and we get

Circulation around S = Circulation around S_x + Circulation around S_y + Circulation around S_z,

If we divide both sides of this equation by the A, we get

Circulation density around $\vec{n}$

$$\approx \frac{\text{Circulation around } S}{A}$$

$$= \frac{\text{Circulation around } S_x}{A} + \frac{\text{Circulation around } S_y}{A} + \frac{\text{Circulation around } S_z}{A}$$

$$= \frac{\text{Circulation around } S_x}{A_x} \cos \theta_x + \frac{\text{Circulation around } S_y}{A_y} \cos \theta_y + \frac{\text{Circulation around } S_z}{A_z} \cos \theta_y$$

$$\approx \left(\begin{array}{c} \text{Circulation} \\ \text{density} \\ \text{around } \vec{i} \end{array} \right) \cos \theta_x + \left(\begin{array}{c} \text{Circulation} \\ \text{density} \\ \text{around } \vec{j} \end{array} \right) \cos \theta_y + \left(\begin{array}{c} \text{Circulation} \\ \text{density} \\ \text{around } \vec{k} \end{array} \right) \cos \theta_z.$$

Taking the limit as the tetrahedron shrinks around the point P, we get:

$$\left(\begin{array}{c} \text{Circulation} \\ \text{density} \\ \text{around } \vec{n} \end{array} \right) = \left(\begin{array}{c} \text{Circulation} \\ \text{density} \\ \text{around } \vec{i} \end{array} \right) n_1 + \left(\begin{array}{c} \text{Circulation} \\ \text{density} \\ \text{around } \vec{j} \end{array} \right) n_2 + \left(\begin{array}{c} \text{Circulation} \\ \text{density} \\ \text{around } \vec{k} \end{array} \right) n_3$$

where $\vec{n} = n_1 \vec{i} + n_2 \vec{j} + n_3 \vec{k}$ is a unit vector.

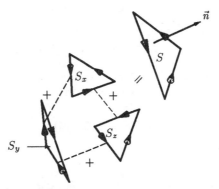

Figure 20.45: A blown apart view of the tetrahedron

Justification of the Formula for the Curl

Suppose we have a vector field $\vec{F}$ in 3-space and want to calculate the components of the curl in Cartesian coordinates at some point $\vec{r}_0 = x_0\vec{i} + y_0\vec{j} + z_0\vec{k}$. The formula

$$\text{curl } \vec{F} = \left(\begin{array}{c} \text{Circulation} \\ \text{density} \\ \text{around } \vec{i} \end{array} \right) \vec{i} + \left(\begin{array}{c} \text{Circulation} \\ \text{density} \\ \text{around } \vec{j} \end{array} \right) \vec{j} + \left(\begin{array}{c} \text{Circulation} \\ \text{density} \\ \text{around } \vec{k} \end{array} \right) \vec{k} .$$

shows that we must compute the circulation densities in the three coordinate directions. First we will calculate the circulation density in the z-direction:

$$\left(\begin{array}{c} \text{Circulation} \\ \text{density} \\ \text{around } \vec{k} \end{array} \right) = \lim_{\text{area} \to 0} \left(\frac{\int_C \vec{F} \cdot d\vec{r}}{\text{Area inside } C} \right) .$$

In this definition, C is a closed curve around the point with position vector $\vec{r}_0$, and in the plane perpendicular to the vector $\vec{k}$. We take C to be a rectangular curve parallel to the xy-plane, with one corner at the point (x_0, y_0, z_0) and edges of length Δx and Δy parallel to the axes. See Figure 20.46. (The fact that the point $\vec{r}_0$ is actually on C rather than enclosed by it turns out not to matter.)

We need to calculate the circulation $\int_C \vec{F} \cdot d\vec{r}$, where $\vec{F} = F_1\vec{i} + F_2\vec{j} + F_3\vec{k}$. Break C into C_1, C_2, C_3, C_4 (see Figure 20.47), and start by calculating the line integrals along C_1 and C_3, where $\Delta \vec{r}$ is parallel to the x-axis, so

$$\Delta \vec{r} = \Delta x\vec{i} .$$

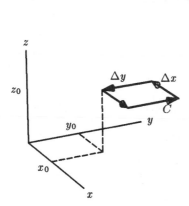

Figure 20.46: Curve C used to calculate $(\text{curl } \vec{F}) \cdot \vec{k}$

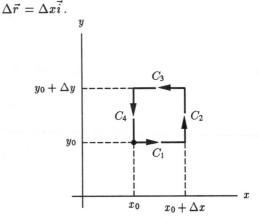

Figure 20.47: Breaking C into C_1, C_2, C_3, C_4

Thus, on C_1 and C_3,

$$\vec{F} \cdot \Delta\vec{r} = (F_1\vec{i} + F_2\vec{j} + F_3\vec{k}) \cdot \Delta x\vec{i} = F_1 \Delta x.$$

However, the function F_1 is evaluated at (x, y_0, z_0) on C_1 and at $(x, y_0 + \Delta y, z_0)$ on C_3. (In each case, $x_0 \leq x \leq x_0 + \Delta x$.) By local linearity

$$F_1(x, y_0 + \Delta y, z_0) \approx F_1(x, y_0, z_0) + \frac{\partial F_1}{\partial y} \Delta y,$$

so

$$
\begin{aligned}
\int_{C_1} \vec{F} \cdot d\vec{r} + \int_{C_3} \vec{F} \cdot d\vec{r} &= \int_{C_1} F_1(x, y_0, z_0)\, dx + \int_{C_3} F_1(x, y_0 + \Delta y, z_0)\, dx \\
&\approx \int_{x=x_0}^{x_0+\Delta x} F_1(x, y_0, z_0)\, dx + \int_{x_0+\Delta x}^{x=x_0} \left(F_1(x, y_0, z_0) + \frac{\partial F_1}{\partial y} \Delta y \right) dx \\
&= \int_{x=x_0}^{x_0+\Delta x} F_1(x, y_0, z_0)\, dx - \int_{x=x_0}^{x_0+\Delta x} \left(F_1(x, y_0, z_0) + \frac{\partial F_1}{\partial y} \Delta y \right) dx \\
&= -\int_{x=x_0}^{x_0+\Delta x} \left(\frac{\partial F_1}{\partial y} \Delta y \right) dx.
\end{aligned}
$$

Thus, we have to integrate the function $\left(\dfrac{\partial F_1}{\partial y} \right) \Delta y$ over the interval $x_0 \leq x \leq x_0 + \Delta x$. Over this interval, the function $\left(\dfrac{\partial F_1}{\partial y} \right) \Delta y$ is approximately constant, so

$$-\int_{x=x_0}^{x_0+\Delta x} \left(\frac{\partial F_1}{\partial y} \Delta y \right) dx \approx - \left(\frac{\partial F_1}{\partial y} \Delta y \right) \int_{x=x_0}^{x_0+\Delta x} dx = -\frac{\partial F_1}{\partial y} \Delta y\, \Delta x.$$

Similarly, on C_2 and C_4, we have $\Delta\vec{r} = \Delta y\vec{j}$ and so

$$\vec{F} \cdot \Delta\vec{r} = (F_1\vec{i} + F_2\vec{j} + F_3\vec{k}) \cdot \Delta y\vec{j} = F_2 \Delta y.$$

Now, on C_2 we evaluate F_2 at $(x_0 + \Delta x, y, z_0)$, while on C_4 we evaluate F_2 at (x_0, y, z_0), and

$$F_2(x_0 + \Delta x, y, z_0) \approx F_2(x_0, y, z_0) + \frac{\partial F_2}{\partial x} \Delta x.$$

Thus,

$$
\begin{aligned}
\int_{C_2} \vec{F} \cdot d\vec{r} + \int_{C_4} \vec{F} \cdot d\vec{r} &= \int_{y=y_0}^{y_0+\Delta y} F_2(x_0 + \Delta x, y, z_0)\, dy + \int_{y_0+\Delta y}^{y=y_0} F_2(x_0, y, z_0)\, dy \\
&\approx \int_{y=y_0}^{y_0+\Delta y} \left(F_2(x_0, y, z_0) + \frac{\partial F_2}{\partial x} \Delta x \right) dy - \int_{y=y_0}^{y_0+\Delta y} F_2(x_0, y, z_0)\, dy \\
&= \int_{y=y_0}^{y_0+\Delta y} \left(\frac{\partial F_2}{\partial x} \Delta x \right) dy \\
&\approx \left(\frac{\partial F_2}{\partial x} \Delta x \right) \int_{y=y_0}^{y_0+\Delta y} dy \\
&= \frac{\partial F_2}{\partial x} \Delta x\, \Delta y.
\end{aligned}
$$

Combining these results we get that the circulation is

$$\int_C \vec{F} \cdot d\vec{r} = \int_{C_1} \vec{F} \cdot d\vec{r} + \int_{C_2} \vec{F} \cdot d\vec{r} + \int_{C_3} \vec{F} \cdot d\vec{r} + \int_{C_4} \vec{F} \cdot d\vec{r}$$

$$\approx \frac{\partial F_2}{\partial x} \Delta x \, \Delta y - \frac{\partial F_1}{\partial y} \Delta y \, \Delta x.$$

Now the area inside C is $\Delta x \, \Delta y$, so the circulation density in the z-direction is given by

$$\begin{pmatrix} \text{Circulation} \\ \text{density} \\ \text{around } \vec{k} \end{pmatrix} = \lim_{\text{Area} \to 0} \left(\frac{\int_C \vec{F} \cdot d\vec{r}}{\text{Area enclosed by } C} \right)$$

$$= \lim_{\Delta x \, \Delta y \to 0} \frac{\frac{\partial F_2}{\partial x} \Delta x \, \Delta y - \frac{\partial F_1}{\partial y} \Delta y \, \Delta x}{\Delta x \, \Delta y}$$

$$= \frac{\partial F_2}{\partial x} - \frac{\partial F_1}{\partial y}.$$

The other two components of the curl can be computed in a similar manner, giving the following formula:

If $\vec{F} = F_1 \vec{i} + F_2 \vec{j} + F_3 \vec{k}$, then

$$\text{curl } \vec{F} = \left(\frac{\partial F_3}{\partial y} - \frac{\partial F_2}{\partial z} \right) \vec{i} + \left(\frac{\partial F_1}{\partial z} - \frac{\partial F_3}{\partial x} \right) \vec{j} + \left(\frac{\partial F_2}{\partial x} - \frac{\partial F_1}{\partial y} \right) \vec{k}$$

Problems for Section 20.5

1. Sketch the curves you would use to calculate the x-and y- components of curl $\vec{F}$ in Cartesian coordinates. (Note: Figure 20.47 on page 447 contains the curve used to find the z-component.)

2. Derive the expressions for the x- and y-components of curl $\vec{F}$, by a method similar to that used in this section.

REVIEW PROBLEMS FOR CHAPTER TWENTY

1. Can you evaluate the flux integral in Problem 9 on page 407 by application of the Divergence Theorem? Why or why not?

2. (a) Find the flux of the vector field $\vec{F} = 2x\vec{i} - 3y\vec{j} + 5z\vec{k}$ through a cubical box of edge length w with four of its eight corners at the points $(a, b, c), (a + w, b, c), (a, b + w, c), (a, b, c + w)$. See Figure 20.48.

(b) Use your answer to part (a) to find div $\vec{F}$ at the point (a, b, c).

(c) Find div $\vec{F}$ using partial derivatives.

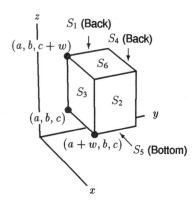

Figure 20.48

3. Suppose $\vec{F} = (3x + 2)\vec{i} + 4x\vec{j} + (5x + 1)\vec{k}$. Use the method of Problem 2 to find div $\vec{F}$ at the point (a, b, c) by two different methods.

4. If V is a volume surrounded by a closed surface S, show that

$$\frac{1}{3}\int_S \vec{r} \cdot d\vec{A} = V.$$

5. Use the result of Problem 4 to compute the volume of a sphere given that the surface area of a sphere of radius R is $4\pi R^2$.

6. Use the result of Problem 4 to compute the volume of a cone of base radius b and height h.[Hint: Stand the cone with its point downward and its axis along the positive z-axis.]

7. Compute the flux integral $\int_S (x^3\vec{i} + 2y\vec{j} + 3\vec{k}) \cdot d\vec{A}$, where S is the $2 \times 2 \times 2$ cubical surface centered at the origin, oriented outward. Do this in two ways:

(a) Directly

(b) By means of the Divergence Theorem

8. Show that if $g(x, y, z)$ is a scalar valued function and $\vec{F}(x, y, z)$ is a vector field, then

$$\text{div}(g\vec{F}) = (\text{grad } g) \cdot \vec{F} + g \, \text{div } \vec{F}.$$

9. True or false? div $\vec{F}$ is a scalar whose value can vary from point to point. Explain your answer.

10. True or false? If $\int_S \vec{F} \cdot d\vec{A} = 12$ and S is a flat disc of area 4π, then div $\vec{F} = 3/\pi$. Explain your answer.

11. True or false? curl $\vec{F}$ is a vector field. Explain your answer.

12. True or false? If $\vec{F}$ is as shown in Figure 20.49, curl $\vec{F} \cdot \vec{j} > 0$. Explain your answer.

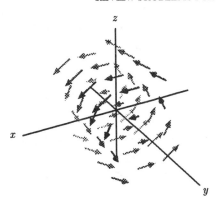

Figure 20.49

13. True or false? $\operatorname{div}(\vec{F} + \vec{G}) = \operatorname{div} \vec{F} + \operatorname{div} \vec{G}$, where $\vec{F} = F_1\vec{i} + F_2\vec{j} + F_3\vec{k}$ and $\vec{G} = G_1\vec{i} + G_2\vec{j} + G_3\vec{k}$ are vector fields in 3-space. Explain your answer.

14. True or false? $\operatorname{grad}(fg) = (\operatorname{grad} f) \cdot (\operatorname{grad} g)$. Explain your answer.

15. True or false? $\operatorname{grad}(\vec{F} \cdot \vec{G}) = \vec{F}(\operatorname{div} \vec{G}) + (\operatorname{div} \vec{F})\vec{G}$, where $\vec{F} = F_1\vec{i} + F_2\vec{j} + F_3\vec{k}$ and $\vec{G} = G_1\vec{i} + G_2\vec{j} + G_3\vec{k}$ are vector fields in 3-space. Explain your answer.

16. True or false? $\operatorname{curl}(f\vec{G}) = (\operatorname{grad} f) \times \vec{G} + f(\operatorname{curl} \vec{G})$, where f is a function of three variables and $\vec{G} = G_1\vec{i} + G_2\vec{j} + G_3\vec{k}$ is a vector field in 3-space. Explain your answer.

17. According to Coulomb's Law the electrostatic field $\vec{E}$, at the point with position vector $\vec{r}$ in 3-space, due to a charge q at the origin is given by

$$\vec{E}(\vec{r}) = q\frac{\vec{r}}{|\vec{r}|^3}.$$

   (a) Compute $\operatorname{div} \vec{E}$.
   (b) Let S_a be the sphere in 3-space of radius $a > 0$ and center at the origin and oriented outwards. Show that the flux of the resulting electric field $\vec{E}$ through the surface S_a is equal to $4\pi q$, for any radius a.
   (c) Could you have used the Divergence Theorem in part (b)? Explain why or why not.
   (d) Let S be an arbitrary, closed, outward oriented surface surrounding the origin. The origin does not lie on S. Show that the flux of $\vec{E}$ through S is again equal to $4\pi q$. [Hint: The Divergence Theorem applies to the solid region lying between a sufficiently small sphere S_a and the surface S.]

18. According to Coulomb's Law the electrostatic field $\vec{E}$, at the point with position vector $\vec{r}$ in 3-space, due to charge q at the point with position vector $\vec{r}_0$ is given by

$$\vec{E}(\vec{r}) = q\frac{(\vec{r} - \vec{r}_0)}{|\vec{r} - \vec{r}_0|^3}.$$

Suppose S is a closed outward oriented surface and that $\vec{r}_0$ does not lie on S. Show that

$$\int_S \vec{E} \cdot d\vec{A} = \begin{cases} 4\pi q & \text{if } q \text{ lies inside } S, \\ 0 & \text{if } q \text{ lies outside } S. \end{cases}$$

[Hint: See the preceding problem.]

19. The electric field $\vec{D}$, at the point with position vector $\vec{r}$, of an ideal electric dipole with dipole moment $\vec{p} = p_1\vec{i} + p_2\vec{j} + p_3\vec{k}$ located at the origin is given by

$$\vec{D}(\vec{r}) = 3\frac{(\vec{r} \cdot \vec{p})\vec{r}}{|\vec{r}|^5} - \frac{\vec{p}}{|\vec{r}|^3}.$$

   (a) What is div $\vec{D}$?
   (b) Suppose S is an outward oriented, closed smooth surface which surrounds the origin. The origin does not lie on S. Compute the flux of $\vec{D}$ through S. Can you use the Divergence Theorem directly to compute the flux? Explain why or why not. [Hint: First compute the flux of the dipole field $\vec{D}$ through an outward-oriented sphere S_a with center at the origin and radius $a > 0$. Then apply the Divergence Theorem to the region W lying between S and a sufficiently small sphere S_a.]

20. Due to roadwork ahead, the traffic on a highway slows linearly from 55 miles/hour to 15 miles/hour over a 2000 foot stretch of road, then crawls along at 15 miles/hour for 5000 feet, then speeds back up linearly to 55 miles/hour in the next 1000 feet, after which it moves steadily at 55 miles/hour.

   (a) Sketch a velocity vector field for the traffic flow.
   (b) Write a formula for the velocity vector field $\vec{v}$ (miles/hour) as a function of the distance x feet from the initial point of slowdown. (Take the direction of motion to be $\vec{i}$ and consider the various units separately.)
   (c) Compute div $\vec{v}$ at $x = 1000, 5000, 7500, 10000$. Be sure to include the proper units.

21. The velocity field $\vec{v}$ in Problem 20 on page 452 does not give a complete description of the traffic flow, for it takes no account of the spacing between vehicles. Let ρ be the density (cars/mile) of highway, where we assume that ρ depends only on x.

   (a) Using your highway experience, arrange in ascending order: $\rho(0), \rho(1000), \rho(5000)$.
   (b) What are the units and interpretation of the vector field $\rho\vec{v}$?
   (c) Why would you expect $\rho\vec{v}$ to be constant? What does this mean for div$(\rho\vec{v})$?
   (d) Determine $\rho(x)$ if $\rho(0) = 75$ cars/mile and $\rho\vec{v}$ is constant.
   (e) If the highway has two lanes, find the approximate number of feet between cars at $x = 0, 1000$, and 5000.

22. (a) An idealized model of a river flowing around a rock considers a circular rock of radius 1 in the river, which flows across the xy-plane in the positive x-direction. A simple model begins with the potential function $\phi = x + (x/(x^2 + y^2))$. Compute the velocity vector field, $\vec{v} = \text{grad}\,\phi$.
   (b) Show that div $\vec{v} = 0$.
   (c) Show that the flow of $\vec{v}$ is tangent to the circle $x^2 + y^2 = 1$. This means that no water crosses the circle. The water on the outside must therefore all flow around the circle.
   (d) Use a computer to sketch the vector field $\vec{v}$ in the region outside the unit circle.

23. The relations between the electric field, $\vec{E}$, the magnetic field, $\vec{B}$, the charge density, ρ, and the current density, $\vec{J}$, at a point in space are described by the equations

$$\text{div}\,\vec{E} = 4\pi\rho,$$

$$\text{curl}\,\vec{B} - \frac{1}{c}\frac{\partial\vec{E}}{\partial t} = \frac{4\pi}{c}\vec{J},$$

where c is a constant (the speed of light).

(a)   Using the results of Problem 12 on page 432, show that

$$\frac{\partial \rho}{\partial t} + \text{div}\, \vec{J} = 0.$$

(b)   What does the equation in part (a) say about charge and current density? Explain in intuitive terms why this is reasonable.

(c)   Why do you think the equation in part (a) is called the charge conservation equation?

24.   Evaluate

$$\vec{F} = \text{grad}\, \phi + \vec{v} \times \vec{r}$$

where

$$\phi(x, y, z) = \frac{1}{2}(a_1 x^2 + b_2 y^2 + c_3 z^2 + (a_2 + b_1)xy + (a_3 + c_1)xz + (b_3 + c_2)yz)$$

and

$$\vec{v} = \frac{1}{2}((c_2 - b_3)\vec{i} + (a_3 - c_1)\vec{j} + (b_1 - a_2)\vec{k})$$

and

$$\vec{r} = x\vec{i} + y\vec{j} + z\vec{k}.$$

Explain why every linear vector field can be written in the form $\text{grad}\, \phi + \vec{v} \times \vec{r}$.

APPENDICES

A. Review of Local Linearity for One Variable
B. Maxima and Minima of Functions of One Variable
C. Determinants
D. Review of One-Variable Integration
E. Table of Integrals
F. Review of Density Functions and Probability
G. Review of Polar Coordinates
H. Change of Variables
I. The Implicit Function Theorem
J. Notes on Kepler, Newton and Planetary Motion

A REVIEW OF LOCAL LINEARITY FOR ONE VARIABLE

If you zoom in on the graph of a smooth function of one variable, $y = f(x)$, around the point $x = a$, the graph looks more and more like a straight line, and thus becomes indistinguishable from its tangent line at that point. See Figure A.1.

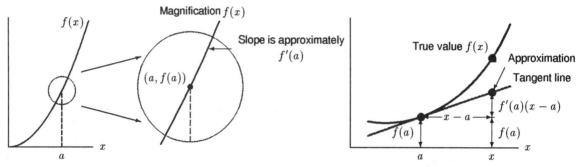

Figure A.1: Zooming in on a portion of a function of one variable until the graph is almost straight

Figure A.2: Local linearization: by the tangent line approximation

The slope of the tangent line is the derivative $f'(a)$, and the line passes through the point $(a, f(a))$, so its equation is

$$y = f(a) + f'(a)(x - a).$$

(See Figure A.2.) Now we approximate the values of f by the y values from the tangent line, giving the result in the following box.

The Tangent Line Approximation for values of x near a

$$f(x) \approx f(a) + f'(a)(x - a)$$

We are thinking of a as fixed, so that $f(a)$ and $f'(a)$ are constants, and the expression on the right-hand side is linear in x. The fact that f is approximately a linear function of x for x near a is expressed by saying f is *locally linear* near $x = a$.

Example 1   Find the local linearization at $x = 2$ of the one-variable function m in Example 1 on page 110.

Solution   Recall that $m(2) = 135$, and $m'(2) = 16$, so the tangent line approximation to $m(x)$ at $x = 2$ is

$$m(x) \approx m(2) + m'(2)(x - 2) = 135 + 16(x - 2) \quad \text{for } x \text{ near } 2.$$

Figure A.3 shows the function m and its tangent line at $x = 2$. Notice that $m(x)$ is not well approximated by values from the tangent line as x moves away from 2.

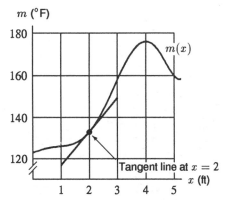

Figure A.3: Local linearization of $m(x)$ at
$x = 2$

B MAXIMA AND MINIMA OF FUNCTIONS OF ONE VARIABLE

If f is a function of one variable and x is a point in its domain, we say

- p is a *critical point* of f if $f'(p) = 0$ or $f'(p)$ is undefined
- f has a *local maximum* at a critical point, x_0, if $f(x) \leq f(x_0)$ for all x near x_0
- f has a *local minimum* at a critical point, x_0, if $f(x) \geq f(x_0)$ for all x near x_0
- f has a *global maximum* at x_0 if $f(x) \leq f(x_0)$ for all x
- f has a *global minimum* at x_0 if $f(x) \geq f(x_0)$ for all x

Notice that global extrema (that is, global maxima and minima) can only occur at local extrema. To find local extrema, first find the critical points. To find global extrema, evaluate the function at the critical points and the endpoints of the domain (if they are included). For example, if we are maximizing over a closed interval $a \leq x \leq b$, a function could have a global maximum at an endpoint. See Figure B.4.

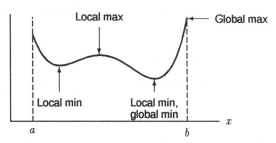

Figure B.4: Local and global extrema on a closed
interval $a \leq x \leq b$

Functions do not necessarily have local or global extrema — it depends on the function and on the domain under consideration. For example, $f(x) = x^2$ has a local minimum at $x = 0$, and this local minimum is also the global minimum, but it has no local or global maxima (see Figure B.5). If, on the other hand we look at the same function on the domain $1 \leq x \leq 2$ then f has a global minimum at $x = 1$ and a global maximum at $x = 2$ (see Figure B.6).

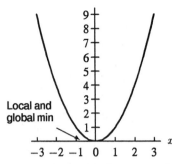

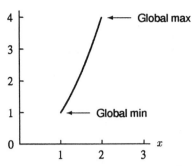

Figure B.5: Extrema of $f(x) = x^2$
on real line

Figure B.6: Extrema of $f(x) = x^2$
restricted to $1 \leq x \leq 2$

To find the critical points, we solve the equation

$$f' = 0 \qquad \text{(or } f' \text{ undefined)}.$$

To decide if a critical point is a local maximum, local minimum, or neither, use the Second
Derivative Test:
- If $f'(p) = 0$ and $f''(p) < 0$, then f has a local maximum at p.
- If $f'(p) = 0$ and $f''(p) > 0$, then f has a local minimum at p.

C DETERMINANTS

In this appendix we introduce the determinant of an array of numbers. Each 2 by 2 array of numbers
has another number associated with it, called its determinant, which is given by

$$\begin{vmatrix} a_1 & a_2 \\ b_1 & b_2 \end{vmatrix} = a_1 b_2 - a_2 b_1.$$

For example

$$\begin{vmatrix} 2 & 5 \\ -4 & -6 \end{vmatrix} = 2(-6) - 5(-4) = 8.$$

Each 3 by 3 array of numbers also has a number associated with it, also called a determinant,
which can be given in terms of 2 by 2 determinants as follows:

$$\begin{vmatrix} a_1 & a_2 & a_3 \\ b_1 & b_2 & b_3 \\ c_1 & c_2 & c_3 \end{vmatrix} = a_1 \begin{vmatrix} b_2 & b_3 \\ c_2 & c_3 \end{vmatrix} - a_2 \begin{vmatrix} b_1 & b_3 \\ c_1 & c_3 \end{vmatrix} + a_3 \begin{vmatrix} b_1 & b_2 \\ c_1 & c_2 \end{vmatrix}$$

Notice that the determinant of the 2 by 2 array multiplied by a_i is the determinant of the array found
by removing the row and column containing a_i. Also, note the minus sign in the second term. An
example is given by

$$\begin{vmatrix} 2 & 1 & -3 \\ 0 & 3 & -1 \\ 4 & 0 & 5 \end{vmatrix} = 2 \begin{vmatrix} 3 & -1 \\ 0 & 5 \end{vmatrix} - 1 \begin{vmatrix} 0 & -1 \\ 4 & 5 \end{vmatrix} + (-3) \begin{vmatrix} 0 & 3 \\ 4 & 0 \end{vmatrix}$$

$$= 2(15 + 0) - 1(0 - (-4)) + (-3)(0 - 12) = 32$$

We have shown that the cross product $\vec{a} \times \vec{b}$ can be computed using the expression:

$$\vec{a} \times \vec{b} = (a_2 b_3 - a_3 b_2)\vec{i} + (a_3 b_1 - a_1 b_3)\vec{j} + (a_1 b_2 - a_2 b_1)\vec{k} .$$

A useful way of remembering this formula is to write it as a determinant

$$\vec{a} \times \vec{b} = \begin{vmatrix} \vec{i} & \vec{j} & \vec{k} \\ a_1 & a_2 & a_3 \\ b_1 & b_2 & b_3 \end{vmatrix} .$$

Expanding the determinant as above gives

$$\begin{vmatrix} \vec{i} & \vec{j} & \vec{k} \\ a_1 & a_2 & a_3 \\ b_1 & b_2 & b_3 \end{vmatrix} = \vec{i}\,(a_2 b_3 - a_3 b_2) - \vec{j}\,(a_1 b_3 - a_3 b_1) + \vec{k}\,(a_1 b_2 - a_2 b_1) = \vec{a} \times \vec{b} .$$

D REVIEW OF ONE-VARIABLE INTEGRATION

Definition of the One-Variable Integral

The one-variable integral

$$\int_a^b f(x)\,dx$$

is defined to be a limit of *Riemann sums*, which may be constructed as follows. We divide the interval $a \le x \le b$ into n equal subdivisions, and we call the width of an individual subdivision Δx. Thus,

$$\Delta x = \frac{b - a}{n} .$$

We will let $x_0, x_1, x_2, \ldots, x_n$ be endpoints of the subdivisions, as in Figures D.7 and D.8. We construct two special Riemann sums:

$$\text{Left-hand sum} = f(x_0)\Delta x + f(x_1)\Delta x + \cdots + f(x_{n-1})\Delta x$$

and

$$\text{Right-hand sum} = f(x_1)\Delta x + f(x_2)\Delta x + \cdots + f(x_n)\Delta x.$$

To define the definite integral, we take the limit of these sums as n goes to infinity.

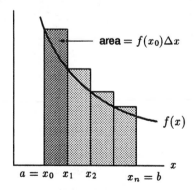

Figure D.7: Left-hand sum

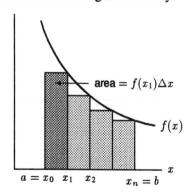

Figure D.8: Right-hand sum

The **definite integral** of f from a to b, written

$$\int_a^b f(x)\,dx,$$

is the limit of the left-hand or right-hand sums with n subdivisions as n gets arbitrarily large. In other words,

$$\int_a^b f(x)\,dx = \lim_{n\to\infty}\ (\text{left-hand sum}) = \lim_{n\to\infty}\left(\sum_{i=0}^{n-1} f(x_i)\Delta x\right)$$

and

$$\int_a^b f(x)\,dx = \lim_{n\to\infty}\ (\text{right-hand sum}) = \lim_{n\to\infty}\left(\sum_{i=1}^{n} f(x_i)\Delta x\right).$$

Each of these sums is called a *Riemann sum*, f is called the *integrand*, and a and b are called the *limits of integration*.

There are other sorts of Riemann sums in addition to left- and right-hand sums; for one thing, you can evaluate the function at any point in each subinterval, not just at the left- or right-hand endpoint. Graphically, this means that the graph of the function can intersect the top of each rectangle at any point, not just at the endpoints. Figure D.9 shows why you might want this to happen: if you want to get a lower estimate for the integral, you need to have all your rectangles sitting below the curve. A Riemann sum where the rectangles all sit just below the graph is called a *lower sum*. Figure D.10 shows an *upper sum* for the same integral.

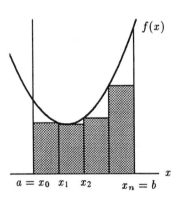

Figure D.9: A lower sum

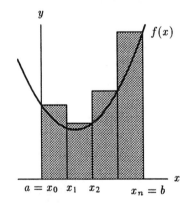

Figure D.10: An upper sum

Interpretations of the Definite Integral

As an Area

If $f(x)$ is positive we can interpret each term $f(x_0)\Delta x$, $f(x_1)\Delta x$, … in a left- or right-hand Riemann sum as the area of a rectangle, as we saw in the previous section. As the width Δx of the rectangles approaches zero, the rectangles fit the curve of the graph more exactly, and the sum of their areas gets closer and closer to the area under the curve, shaded in Figure D.11. Thus, we conclude that:

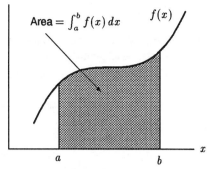

Figure D.11: The definite integral
$\int_a^b f(x)\, dx$

If $f(x) \geq 0$ and $a < b$:

$$\text{Area under graph of } f \text{ between } a \text{ and } b = \int_a^b f(x)\, dx.$$

As an Average Value

The definite integral can be used to compute the average value of a function:

$$\text{Average value of } f \text{ from } a \text{ to } b = \frac{1}{b-a}\int_a^b f(x)\, dx$$

If we interpret the integral as the area under the graph of f, then we can think of the average value of f as the height of the rectangle with the same area that is on the same base. (See Figure D.12.)

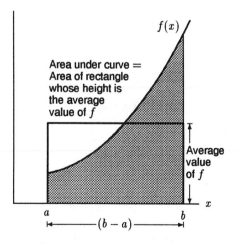

Figure D.12: Area and average value

When $f(x)$ Represents a Density

If $f(x)$ represents a density, say a population density or the density of a substance, then we can calculate the total mass or total population using a definite integral. To find the total quantity, we divide the region into small pieces and find the population or mass of each one by multiplying the density by the size of that piece.

If $f(x)$ represents the density of a substance along the interval $a \leq x \leq b$, the

$$\text{Total mass} = \int_a^b f(x)\, dx.$$

As the Total Change

The integral of the rate of change of any quantity will give the total change in that quantity; for example, the integral of velocity gives the total change in position (or distance moved).

If $r(t)$ is the rate of change of Q with respect to t, then

$$Q(b) - Q(a) = \begin{matrix} \text{Total Change in } Q \\ \text{between } t = a \text{ and } t = b \end{matrix} = \int_a^b r(t)\, dt.$$

The Fundamental Theorem of Calculus

Since

$$F'(t) = \text{rate of change of } F(t) \text{ with respect to } t,$$

and since the definite integral of a rate of change of some quantity is the total change in that quantity, we find that

$$\int_a^b F'(t)\, dt = \int_a^b \left[\text{rate of change of } F(t)\right]\, dt$$
$$= \text{Total change in } F(t) \text{ between } a \text{ and } b$$
$$= F(b) - F(a).$$

This is

The Fundamental Theorem of Calculus

If $f = F'$, then

$$\int_a^b f(t)\, dt = F(b) - F(a). - 1z$$

The Fundamental Theorem is the basic tool for evaluating definite integrals algebraically; it enables us to avoid using Riemann sums whenever we can find an antiderivative, or indefinite integral, for the integrand. Appendix E gives a brief table of indefinite integrals.

Problems for Section D

In Exercises 1—20, find an antiderivative for each of the functions given.

1. $\int (x^2 + 2x + \frac{1}{x}) \, dx.$

2. $\int \frac{t+1}{t^2} \, dt.$

3. $\int \frac{(t+2)^2}{t^3} \, dt.$

4. $\int \sin t \, dt.$

5. $\int \cos 2t \, dt.$

6. $\int \frac{x}{x^2+1} \, dx.$

7. $\int \tan \theta \, d\theta.$

8. $\int e^{5z} \, dz.$

9. $\int te^{t^2+1} \, dt.$

10. $\int \frac{dz}{1+z^2}.$

11. $\int \frac{dz}{1+4z^2}.$

12. $\int \sin^2 \theta \cos \theta \, d\theta.$

13. $\int \sin 5\theta \cos^3 5\theta \, d\theta.$

14. $\int \sin^3 z \cos^3 z \, dz.$

15. $\int \frac{(\ln x)^2}{x} \, dx.$

16. $\int \cos \theta \sqrt{1 + \sin \theta} \, d\theta.$

17. $\int xe^x \, dx.$

18. $\int t^3 e^t \, dt.$

19. $\int x \ln x \, dx.$

20. $\int \frac{1}{\cos^2 \theta} \, d\theta.$

In Exercises 21—25 find the definite integral by two methods (Fundamental Theorem and Numerically).

21. $\int_1^3 x(x^2+1)^{70} \, dx.$

22. $\int_0^1 \frac{dx}{x^2+1}.$

23. $\int_0^{10} ze^{-z} \, dz.$

24. $\int_{-\pi/3}^{\pi/4} \sin^3 \theta \cos \theta \, d\theta.$

25. $\int_1^4 \frac{e^{\sqrt{x}}}{\sqrt{x}} \, dx.$

26. Water is leaking out of a tank at a rate of $R(t)$ gallons/hour, where t is measured in hours.

   (a) Write a definite integral that expresses the total amount of water that leaks out in the first two hours.

   (b) Figure D.13 is a graph of $R(t)$. On a sketch, shade in the region whose area represents the total amount of water that leaks out in the first two hours.

   (c) Give an upper and lower estimate of the total amount of water that leaks out in the first two hours.

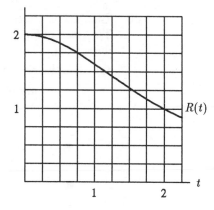

Figure D.13

27. The rate at which the world's oil is being consumed is continuously increasing. Suppose the rate (in billions of barrels per year) is given by the function $r = f(t)$, where t is measured in

years and $t = 0$ is the start of 1990.

(a) Write a definite integral which represents the total quantity of oil used between the start of 1990 and the start of 1995.

(b) Suppose $r = 32e^{0.05t}$. Using a left-hand sum with five subdivisions, find an approximate value for the total quantity of oil used between the start of 1990 and the start of 1995.

(c) Interpret each of the five terms in the sum from part (b) in terms of oil consumption.

28. Figure D.14 shows the graph of the derivative $g'(x)$ of a function $g(x)$. It is given that $g(0) = 50$. Sketch the graph of $g(x)$, showing all critical points and inflection points of g and giving their coordinates.

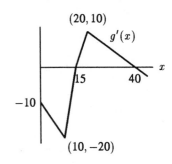

Figure D.14

29. The graph of dy/dt against t is in Figure D.15. Suppose the three shaded regions each have area 2. Given that $y = 0$ when $t = 0$, draw the graph of y against t, indicating all special features the graph might have (known heights, maxima and minima, inflection points, etc.). Pay particular attention to the relationship between the graphs. Mark $t_1, t_2, \ldots, t_5$ on the t axis.[1]

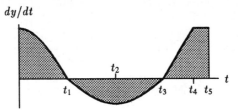

Figure D.15

30. The Quabbin Reservoir in the western part of Massachusetts provides most of Boston's water. The graph in Figure D.16 represents the flow of water in and out of the Quabbin Reservoir throughout 1993.

(a) Sketch a possible graph for the quantity of water in the reservoir, as a function of time.

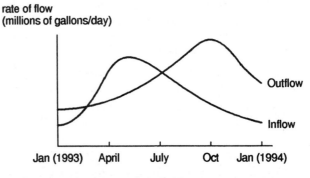

Figure D.16

[1]From *Calculus: The Analysis of Functions*, by Peter D. Taylor (Toronto: Wall & Emerson, Inc., 1992)

(b) When, in the course of 1993, was the quantity of water in the reservoir largest? Smallest? Mark and label these points on the graph you drew in part (a).

(c) When was the quantity of water decreasing most rapidly? Again, mark and label this time on both graphs.

(d) By July 1994 the quantity of water in the reservoir was about the same as in January 1993. Draw plausible graphs for the flow into and the flow out of the reservoir for the first half of 1994. Explain your graph.

31. A rod has length 2 meters. At a distance x meters from its left end, the density of the rod is given by

$$\rho(x) = 2 + 6x \text{ g/m.}$$

(a) Write a Riemann sum approximating the total mass of the rod.

(b) Find the exact mass by converting the sum into an integral.

32. The density of cars (in cars per mile) down a 20-mile stretch of the Pennsylvania Turnpike can be approximated by

$$\rho(x) = 300 \left(2 + \sin\left(4\sqrt{x + 0.15}\right)\right),$$

where x is the distance in miles from the Breezewood toll plaza.

(a) Sketch a graph of this function for $0 \le x \le 20$.

(b) Write a sum that approximates the total number of cars on this 20-mile stretch.

(c) Find the total number of cars on the 20-mile stretch.

33. Circle City, a typical metropolis, is very densely populated near its center, and its population gradually thins out toward the city limits. In fact, its population density is $10,000(3 - r)$ people/square mile at distance r miles from the center.

(a) Assuming that the population density at the city limits is zero, find the radius of the city.

(b) What is the total population of the city?

34. The density of oil in a circular oil slick on the surface of the ocean at a distance r meters from the center of the slick is given by $\rho(r) = 50/(1 + r)$ kg/m^2.

(a) If the slick extends from $r = 0$ to $r = 10,000$ m, find a Riemann sum approximating the total mass of oil in the slick.

(b) Find the exact value of the mass of oil in the slick by turning your sum into an integral and evaluating it.

(c) Within what distance r is half the oil of the slick contained?

35. An exponential model for the density of the earth's atmosphere says that if the temperature of the atmosphere were constant, then the density of the atmosphere as a function of height, h (in meters), above the surface of the earth would be given by

$$\rho(h) = 1.28e^{-0.000124h} \text{ kg/m}^3.$$

(a) Write (but do not evaluate) a sum that approximates the mass of the portion of the atmosphere from $h = 0$ to $h = 100$ m (i.e., the first 100 meters above sea level). Assume the radius of the earth is 6370 km.

(b) Find the exact answer by turning your sum in part (a) into an integral. Evaluate the integral.

36. Water is flowing in a cylindrical pipe of radius 1 inch. Because water is viscous and sticks to the pipe, the rate of flow varies with the distance from the center. The speed of the water at a distance r inches from the center is $10(1 - r^2)$ inches per second. What is the rate (in cubic inches per second) at which water is flowing through the pipe?

37. The reflector behind a car headlight is made in the shape of the parabola, $x = \frac{4}{9}y^2$, with a circular cross-section, as shown in Figure D.17.

    (a) Find a Riemann sum approximating the volume contained by this headlight.
    (b) Find the volume exactly.

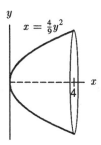

Figure D.17

38. Rotate the bell-shaped curve $y = e^{-x^2/2}$ shown in Figure D.18 around the y-axis, forming a hill-shaped solid of revolution. By slicing horizontally, find the volume of this hill.

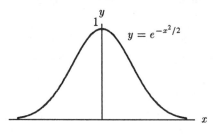

Figure D.18

39. The circumference of the trunk of a certain tree at different heights above the ground is given in the following table.

Height (feet)	0	20	40	60	80	100	120
Circumference (feet)	26	22	19	14	6	3	1

    Assume all horizontal cross-sections of the trunk are circles. Estimate the volume of the tree trunk using the trapezoid rule.

40. Most states expect to run out of space for their garbage soon. In New York, solid garbage is packed into pyramid-shaped dumps with square bases. (The largest such dump is on Staten Island.) A small community has a dump with a base length of 100 yards. One yard vertically above the base, the length of the side parallel to the base is 99 yards; the dump can be built up to a vertical height of 20 yards. (The top of the pyramid is never reached.) If 65 cubic yards of garbage arrive at the dump every day, how long will it be before the dump is full?

E TABLE OF INTEGRALS

A Short Table of Indefinite Integrals

I. Basic Functions

1. $\displaystyle\int x^n \, dx = \frac{1}{n+1}x^{n+1} + C, \quad n \neq -1$

5. $\displaystyle\int \sin x \, dx = -\cos x + C$

2. $\displaystyle\int \frac{1}{x} \, dx = \ln|x| + C$

6. $\displaystyle\int \cos x \, dx = \sin x + C$

3. $\displaystyle\int a^x \, dx = \frac{1}{\ln a}a^x + C$

7. $\displaystyle\int \tan x \, dx = -\ln|\cos x| + C$

4. $\displaystyle\int \ln x \, dx = x \ln x - x + C, \quad x > 0$

II. Products of e^x, $\cos x$, and $\sin x$

8. $\displaystyle\int e^{ax} \sin(bx) \, dx = \frac{1}{a^2+b^2}e^{ax}[a\sin(bx) - b\cos(bx)] + C$

9. $\displaystyle\int e^{ax} \cos(bx) \, dx = \frac{1}{a^2+b^2}e^{ax}[a\cos(bx) + b\sin(bx)] + C$

10. $\displaystyle\int \sin(ax) \sin(bx) \, dx = \frac{1}{b^2-a^2}[a\cos(ax)\sin(bx) - b\sin(ax)\cos(bx)] + C, \quad a \neq b$

11. $\displaystyle\int \cos(ax) \cos(bx) \, dx = \frac{1}{b^2-a^2}[b\cos(ax)\sin(bx) - a\sin(ax)\cos(bx)] + C, \quad a \neq b$

12. $\displaystyle\int \sin(ax) \cos(bx) \, dx = \frac{1}{b^2-a^2}[b\sin(ax)\sin(bx) + a\cos(ax)\cos(bx)] + C, \quad a \neq b$

III. Product of Polynomial $p(x)$ with $\ln x$, e^x, $\cos x$, $\sin x$

13. $\displaystyle\int x^n \ln x \, dx = \frac{1}{n+1}x^{n+1}\ln x - \frac{1}{(n+1)^2}x^{n+1} + C, \quad n \neq -1, \quad x > 0$

14. $\displaystyle\int p(x)e^{ax} \, dx = \frac{1}{a}p(x)e^{ax} - \frac{1}{a}\int p'(x)e^{ax} \, dx$

$$= \frac{1}{a}p(x)e^{ax} - \frac{1}{a^2}p'(x)e^{ax} + \frac{1}{a^3}p''(x)e^{ax} - \cdots$$
$$(+ - + - \ldots) \quad \text{(signs alternate)}$$

15. $\displaystyle\int p(x)\sin ax \, dx = -\frac{1}{a}p(x)\cos ax + \frac{1}{a}\int p'(x)\cos ax \, dx$

$$= -\frac{1}{a}p(x)\cos ax + \frac{1}{a^2}p'(x)\sin ax + \frac{1}{a^3}p''(x)\cos ax - \cdots$$
$$(- + + - - + + \ldots) \quad \text{(signs alternate in pairs after first term)}$$

16. $\displaystyle\int p(x)\cos ax \, dx = \frac{1}{a}p(x)\sin ax - \frac{1}{a}\int p'(x)\sin ax \, dx$

$$= \frac{1}{a}p(x)\sin ax + \frac{1}{a^2}p'(x)\cos ax - \frac{1}{a^3}p''(x)\sin ax - \cdots$$
$$(+ + - - + + - - \ldots) \quad \text{(signs alternate in pairs)}$$

IV. Integer Powers of $\sin x$ and $\cos x$

17. $\displaystyle\int \sin^n x\, dx = -\frac{1}{n}\sin^{n-1} x \cos x + \frac{n-1}{n}\int \sin^{n-2} x\, dx,\quad n \text{ positive}$

18. $\displaystyle\int \cos^n x\, dx = \frac{1}{n}\cos^{n-1} x \sin x + \frac{n-1}{n}\int \cos^{n-2} x\, dx,\quad n \text{ positive}$

19. $\displaystyle\int \frac{1}{\sin^m x}\, dx = \frac{-1}{m-1}\frac{\cos x}{\sin^{m-1} x} + \frac{m-2}{m-1}\int \frac{1}{\sin^{m-2} x}\, dx,\quad m \neq 1, m \text{ positive}$

20. $\displaystyle\int \frac{1}{\sin x}\, dx = \frac{1}{2}\ln\left|\frac{(\cos x)-1}{(\cos x)+1}\right| + C$

21. $\displaystyle\int \frac{1}{\cos^m x}\, dx = \frac{1}{m-1}\frac{\sin x}{\cos^{m-1} x} + \frac{m-2}{m-1}\int \frac{1}{\cos^{m-2} x}\, dx,\quad m \neq 1, m \text{ positive}$

22. $\displaystyle\int \frac{1}{\cos x}\, dx = \frac{1}{2}\ln\left|\frac{(\sin x)+1}{(\sin x)-1}\right| + C$

23. $\displaystyle\int \sin^m x \cos^n x\, dx$: If m is odd, let $w = \cos x$. If n is odd, let $w = \sin x$. If both m and n are even and non-negative, convert all to $\sin x$ or all to $\cos x$ (using $\sin^2 x + \cos^2 x = 1$), and use IV-17 or IV-18. If m and n are even and one of them is negative, convert to whichever function is in the denominator and use IV-19 or IV-21. The case in which both m and n are even and negative is omitted.

V. Quadratic in the Denominator

24. $\displaystyle\int \frac{1}{x^2 + a^2}\, dx = \frac{1}{a}\arctan \frac{x}{a} + C,\quad a \neq 0$

25. $\displaystyle\int \frac{bx + c}{x^2 + a^2}\, dx = \frac{b}{2}\ln|x^2 + a^2| + \frac{c}{a}\arctan \frac{x}{a} + C,\quad a \neq 0$

26. $\displaystyle\int \frac{1}{(x-a)(x-b)}\, dx = \frac{1}{a-b}(\ln|x-a| - \ln|x-b|) + C,\quad a \neq b$

27. $\displaystyle\int \frac{cx + d}{(x-a)(x-b)}\, dx = \frac{1}{a-b}\left[(ac+d)\ln|x-a| - (bc+d)\ln|x-b|\right] + C,\quad a \neq b$

VI. Integrands Involving $\sqrt{a^2 + x^2}$, $\sqrt{a^2 - x^2}$, $\sqrt{x^2 - a^2}$, $a > 0$

28. $\displaystyle\int \frac{1}{\sqrt{a^2 - x^2}}\, dx = \arcsin \frac{x}{a} + C$

29. $\displaystyle\int \frac{1}{\sqrt{x^2 \pm a^2}}\, dx = \ln\left|x + \sqrt{x^2 \pm a^2}\right| + C$

30. $\displaystyle\int \sqrt{a^2 \pm x^2}\, dx = \frac{1}{2}\left(x\sqrt{a^2 \pm x^2} + a^2\int \frac{1}{\sqrt{a^2 \pm x^2}}\, dx\right) + C$

31. $\displaystyle\int \sqrt{x^2 - a^2}\, dx = \frac{1}{2}\left(x\sqrt{x^2 - a^2} - a^2\int \frac{1}{\sqrt{x^2 - a^2}}\, dx\right) + C$

F REVIEW OF DENSITY FUNCTIONS AND PROBABILITY

Understanding the distribution of various quantities through the population can be important to decision makers. For example, the income distribution gives useful information about the economic structure of a society. In this section we will look at the distribution of ages in the US. To allocate funding for education, health care, and social security, the government needs to know how many people are in each age group. We will see how to represent such information by a density function.

US Age Distribution

TABLE F.1: *Distribution of ages in the US in 1990*

Age group	Percentage of total population
0–20	30%
20–40	31%
40–60	24%
60–80	14%
Over 80	1%

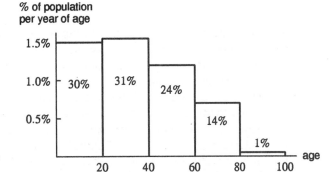

Figure F.19: How ages were distributed in the US in 1990

Suppose we have the data in Table F.1 showing how the ages of the US population were distributed in 1990. To represent this information graphically we use a *histogram*, putting a vertical bar above each age group in such a way that the *area* of each bar represents the percentage in that age group. The total area of all the rectangles is 100% = 1. We will assume that there is nobody over 100 years old, so that the last age group is 80–100. For the 0–20 age group, the base of the rectangle is 20, and we want the area to be 30%, so the height must be 30%/20 = 1.5%. Notice that the vertical axis is measured in percent/year. (See Figure F.19.)

Example 1 In 1990, what percentage of the US population was:
(a) Between 20 and 60 years old?
(b) Less than 10 years old?
(c) Between 75 and 80 or between 80 and 85 years old?

Solution (a) We add the percentages, so 31% + 24% = 55%.
(b) To find the percentage less than 10 years old, we could assume, for example, that the population was distributed evenly over the 0–20 group. (This means we are assuming that babies were born at a fairly constant rate over the last 20 years, which is probably reasonable.) If we make this assumption, then we can say that the population less than 10 years old was about half that in that 0–20 group, that is, 15%. Notice that we get the same result by computing the area of the rectangle from 0 to 10. (See Figure F.20.)
(c) To find the population between 75 and 80 years old, since 14% of Americans in 1990 were in the 60-80 group, we might apply the same reasoning and say that $\frac{1}{4}(14\%) = 3.5\%$ of the

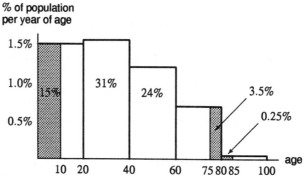

% of population
per year of age

Figure F.20: Ages in the US in 1990 — various subgroups (for Example 1)

population was in this age group. This result is represented as an area in Figure F.20. The assumption that the population was evenly distributed is not a good one here; certainly there were more people between the ages of 60 and 65 than between 75 and 80. Thus, the estimate of 3.5% is certainly too high.

Again using the (faulty) assumption that ages in each group were distributed uniformly, we would find that the percentage between 80 and 85 was $\frac{1}{4}(1\%) = 0.25\%$. (See Figure F.20.) This estimate is also poor — there were certainly more people in the 80–85 group than, say, the 95–100 group, and so the 0.25% is too low. In addition, although the percentage of 80–85-year-olds was certainly smaller than the percentage of 75–80-year-olds, the difference between 0.25% and 3.5% (a factor of 14) is unreasonably large. We can expect the transition from one age group to the next to be smoother and more gradual.

Smoothing Out the Histogram

We could get better estimates if we had smaller age groups (each age group in Figure F.19 is 20 years, which is quite large) or if the histogram were smoother. Suppose we have the more detailed data in Table F.2, which leads to the new histogram in Figure F.21.

As we get more detailed information, the upper silhouette of the histogram becomes smoother, but the area of any of the bars still represents the percentage of the population in that age group. Imagine, in the limit, replacing the upper silhouette of the histogram by a smooth curve in such a way that area under the curve above one age group is the same as the area in the corresponding rectangle. The total area under the whole curve is again 100% = 1. (See Figure F.21.)

The Age Density Function

If t is age in years, we define $p(t)$, the age *density function*, to be a function which "smoothes out" the age histogram. This function has the property that

$$\left(\begin{array}{c} \text{Fraction of population} \\ \text{between ages } a \text{ and } b \end{array} \right) = \left(\begin{array}{c} \text{Area under} \\ \text{graph of } p \\ \text{between } a \text{ and } b \end{array} \right) = \int_a^b p(t)dt.$$

If a and b are the smallest and largest possible ages (say, $a = 0$ and $b = 100$), so that the ages of all of the population are between a and b, then

$$\int_a^b p(t)dt = \int_0^{100} p(t)dt = 1.$$

TABLE F.2 *Ages in the US in 1990 (more detailed)*

Age group	Percentage of total population
0–10	15%
10–20	15%
20–30	16%
30–40	15%
40–50	13%
50–60	11%
60–70	9%
70–80	5%
80–90	1%

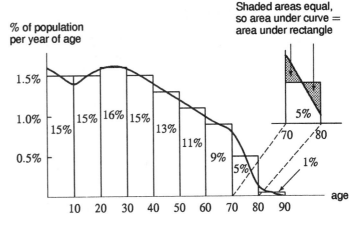

Figure F.21: Smoothing out the age histogram

What does the age density function p tell us? Notice that we have not talked about the meaning of $p(t)$ itself, but *only* of the integral $\int_a^b p(t)\,dt$. Let's look at this in a bit more detail. Suppose, for example, that $p(10) = 0.015 = 1.5\%$ per year. This is *not* telling us that 1.5% of the population is precisely 10 years old (where 10 years old means exactly 10, not $10\frac{1}{2}$, not $10\frac{1}{4}$, not 10.1). However, $p(10) = 0.015$ does tell us that for some small interval Δt around 10, the fraction of the population with ages in this interval is approximately $p(10)\,\Delta t = 0.015\,\Delta t$. Notice also that the units of $p(t)$ are *% per year*, so $p(t)$ must be multiplied by years to give a percentage of the population.

The Density Function

In order to generalize the idea of the age distribution, let us look at a general density function. Suppose we are interested in how a certain characteristic, x, is distributed through a population. For example, x might be height, age, or wattage, and the population might be people, or any set of objects such as light bulbs. Then we define a general density function with the following properties:

> The function, $p(x)$, is a **density function** if
> $$\begin{pmatrix} \text{Fraction of population} \\ \text{for which } x \text{ is} \\ \text{between } a \text{ and } b \end{pmatrix} = \begin{pmatrix} \text{Area under} \\ \text{graph of } p \\ \text{between } a \text{ and } b \end{pmatrix} = \int_a^b p(x)\,dx.$$
> $$\int_{-\infty}^{\infty} p(x)\,dx = 1 \qquad \text{and} \qquad p(x) \geq 0 \quad \text{for all } x.$$

The density function must be nonnegative if its integral always gives a fraction of the population. Also, the fraction of the population with x between $-\infty$ and ∞ is 1 because the entire population has the characteristic x between $-\infty$ and ∞. The function $p(t)$ used to smooth out the age histogram satisfies this definition of a density function. We do not assign a meaning to the value of $p(x)$ alone, but rather interpret $p(x)\,\Delta x$ as the fraction of the population with the characteristic in a short interval of length Δx around x.

Example 2   The graph in Figure F.22 shows the distribution of the number of years of education completed by adults in a population. What does the graph tell us?

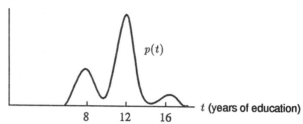

Figure F.22: Distribution of years of education

Solution   The fact that most of the area under the graph of the density function is concentrated in two humps, centered at 8 and 12 years, indicates that most of the population belong to one of two groups, those who leave school after finishing approximately 8 years and those who finish about 12 years. There is a smaller group of people who finish approximately 16 years of school.

The density function is often approximated by formulas, as in the next example.

Example 3   Find reasonable formulas representing the density function for the US age distribution, assuming the function is constant at 1.5% up to age 40 and then drops linearly.

Solution   We need to construct a linear function sloping downward from age 40 in such a way that $p(40) = 1.5\%$ per year $= 0.015$ and that $\int_0^{100} p(t)\,dt = 1$. Suppose b is as in Figure F.23. Since

$$\int_0^{100} p(t)\,dt = \int_0^{40} p(t)\,dt + \int_{40}^{100} p(t)\,dt = 40(0.015) + \frac{1}{2}(0.015)b = 1,$$

we have

$$\frac{0.015}{2}b = 0.4, \quad \text{giving} \quad b \approx 53.3.$$

Thus the slope of the line is $-0.015/53.3 \approx -0.00028$, so for $40 \le t \le 93.3$,

$$p(t) - 0.015 = -0.00028(t - 40),$$
$$p(t) = 0.0263 - 0.00028t.$$

According to this way of smoothing the data, there is no one over 93.3 years old.

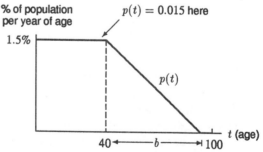

Figure F.23: Age density function

Probability

Suppose we pick a member of the US population at random and ask what is the probability that the person is between, say, the ages of 60 and 65. The probability, or chance, that the person is in a certain age group is equal to the fraction of the population in that age group. Consider the density function $p(t)$ defined on page 470 to describe the distribution of ages in the US. We can use the density function to calculate probabilities as follows:

$$\begin{pmatrix} \text{Probability that} \\ \text{a person is between} \\ \text{ages } a \text{ and } b \end{pmatrix} = \begin{pmatrix} \text{Fraction of population} \\ \text{between ages } a \text{ and } b \end{pmatrix} = \int_a^b p(t)\, dt.$$

The Median and Mean

It is often useful to be able to give an "average" value for a distribution. Two measures that are in common use are the *median* and the *mean*.

The Median

A **median** is a value T such that half the population has values of x less than (or equal to) T, and half the population has values of x greater than (or equal to) T. Thus, a median T satisfies:

$$\int_{-\infty}^{T} p(x)\, dx = 0.5,$$

where p is the density function. In other words, half the area under the graph of p lies to the left of T.

Example 4   Find the median age in the US in 1990, using the age density function given by

$$p(t) = \begin{cases} 0.015 & \text{for } 0 \le t \le 40 \\ 0.0263 - 0.00028t & \text{for } 40 < t \le 93.3. \end{cases}$$

Solution   We want to find the value of T such that

$$\int_{-\infty}^{T} p(t)\, dt = \int_0^T p(t)\, dt = 0.5.$$

Since $p(t) = 1.5\%$ up to age 40, we have

$$\text{Median} = T = \frac{50\%}{1.5\%} \approx 33 \text{ years.}$$

(See Figure F.24.)

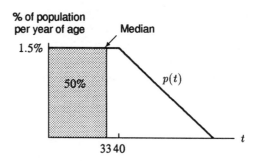

Figure F.24: Median of age distribution

The Mean

Another commonly used average value is the *mean*. To find the mean of N numbers, you add the numbers and divide the sum by N. For example, the mean of the numbers 1, 2, 7, and 10 is $(1 + 2 + 7 + 10)/4 = 5$. The mean age of the entire US population is therefore defined as

$$\frac{\sum \text{ Ages of all people in the US}}{\text{Total number of people in the US}}.$$

Calculating the sum of all the ages directly would be an enormous task; we will approximate the sum by an integral. The idea is to "slice up" the age axis and consider the people whose age is between t and $t + \Delta t$. How many are there?

The percentage of the population between t and $t + \Delta t$ is the area under the graph of p between these points, which is well approximated by the area of the rectangle, $p(t)\Delta t$. (See Figure F.25.) If the total number of people in the population is N, then

$$\begin{matrix}\text{Number of people with age} \\ \text{between } t \text{ and } t + \Delta t\end{matrix} \approx p(t)\Delta t N.$$

The age of all of these people is approximately t:

$$\begin{matrix}\text{Sum of ages of people} \\ \text{between age } t \text{ and } t + \Delta t\end{matrix} \approx t p(t)\Delta t N.$$

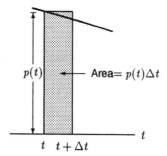

Figure F.25: Shaded area is percentage of population with age between t and $t + \Delta t$

Therefore, adding and factoring out an N gives us

$$\text{Sum of ages of all people} \approx \left(\sum tp(t)\Delta t \right) N.$$

In the limit, as we allow Δt to shrink to 0, the sum becomes an integral, so as an approximation:

$$\text{Sum of ages of all people} = \left(\int_0^{100} tp(t)dt \right) N.$$

Therefore, with N equal to the total number of people in the US, and assuming no person is over 100 years old,

$$\text{Mean age} = \frac{\text{Sum of ages of all people in US}}{N} = \int_0^{100} tp(t)dt.$$

We can give the same argument for any[2] density function $p(x)$.

If a quantity has density function $p(x)$,

$$\text{\textbf{Mean value} of the quantity} = \int_{-\infty}^{\infty} xp(x)\, dx.$$

It can be shown that the mean is the point on the horizontal axis where the region under the graph of the density function, if it were made out of cardboard, would balance.

Example 5   Find the mean age of the US population, using the density function of Example 4.

Solution   The approximate formulas for p are

$$p(t) = \begin{cases} 0.015 & \text{for } 0 \le t \le 40 \\ 0.0263 - 0.00028t & \text{for } 40 < t \le 93.3. \end{cases}$$

Using these formulas, we compute:

$$\text{Mean age} = \int_0^{100} tp(t)dt = \int_0^{40} t(0.015)dt + \int_{40}^{93.3} t(0.0263 - 0.00028t)dt$$

$$= 0.015\frac{t^2}{2}\Big|_0^{40} + 0.0263\frac{t^2}{2}\Big|_{40}^{93.3} - 0.00028\frac{t^3}{3}\Big|_{40}^{93.3} \approx 36 \text{ years}.$$

The mean is shown is Figure F.26..

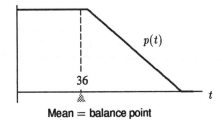

Mean = balance point

Figure F.26: Mean of age distribution

---

[2]Provided all the relevant improper integrals converge.

Normal Distributions

How much rain do you expect will fall in your home town this year? If you live in Anchorage, Alaska, the answer would be something close to 15 inches (including the snow). Of course, you don't expect exactly 15 inches. Some years there will be more than 15 inches, and some years there will be less. Most years, however, the amount of rainfall will be close to 15 inches; only rarely will it be well above or well below 15 inches. What does the density function for the rainfall look like? To answer this question, we look at rainfall data over many years. It lies on a bell-shaped curve which peaks at 15 inches and slopes downward approximately symmetrically on either side. This is an example of a normal distribution.

Normal distributions are frequently used to model real phenomena, from grades on an exam to the number of airline passengers on a particular flight. A normal distribution is characterized by its *mean*, μ, and its *standard deviation*, σ. The mean tells us where the data is clustered: the location of the central peak. The standard deviation tells us how closely the data is clustered around the mean. A small value of σ tells us that the data is close to the mean; a large σ tells us the data is spread out. The formula for a normal distribution is as follows.

A **normal distribution** has a density function of the form

$$p(x) = \frac{1}{\sigma\sqrt{2\pi}} e^{-(x-\mu)^2/(2\sigma^2)},$$

where μ is the mean of the distribution and σ is the standard deviation, with $\sigma > 0$.

The factor of $1/(\sigma\sqrt{2\pi})$ in front of the function is there to make the area under its graph equal to 1. That the factor $\sqrt{2\pi}$ is involved is one of the truly remarkable discoveries of mathematics.

To model the rainfall in Anchorage, we use a normal distribution with $\mu = 15$. The standard deviation can be estimated by looking at the data; we will take it to be 1. (See Figure F.27.)

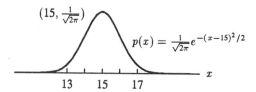

$$p(x) = \frac{1}{\sqrt{2\pi}} e^{-(x-15)^2/2}$$

Figure F.27: Normal distribution with $\mu = 15$ and $\sigma = 1$

In the following example, we will verify that for a normal distribution a certain percentage of the data always lies within a certain number of standard deviations from the mean.

Example 6   For Anchorage's rainfall, use the normal distribution with the density function

$$p(x) = \frac{1}{\sqrt{2\pi}} e^{-(x-15)^2/2},$$

to compute the fraction of the years with rainfall between
(a)   14 and 16 inches,   (b)   13 and 17 inches,   (c)   12 and 18 inches.

Solution   (a)   The fraction of the years with annual rainfall between 14 and 16 inches is $\int_{14}^{16} \frac{1}{\sqrt{2\pi}} e^{-(x-15)^2/2} \, dx$.

Since there is no elementary antiderivative for $e^{-(x-15)^2/2}$, we find the integral numerically. Its value is about 0.68.

$$\begin{array}{l} \text{Fraction of years with rainfall} \\ \text{between 14 and 16 inches} \end{array} = \int_{14}^{16} \frac{1}{\sqrt{2\pi}} e^{-(x-15)^2/2} \, dx \approx 0.68.$$

(b)   Finding the integral numerically again:

$$\begin{array}{l} \text{Fraction of years with rainfall} \\ \text{between 13 and 17 inches} \end{array} = \int_{13}^{17} \frac{1}{\sqrt{2\pi}} e^{-(x-15)^2/2} \, dx \approx 0.95.$$

(c)

$$\begin{array}{l} \text{Fraction of years with rainfall} \\ \text{between 12 and 18 inches} \end{array} = \int_{12}^{18} \frac{1}{\sqrt{2\pi}} e^{-(x-15)^2/2} \, dx \approx 0.997.$$

Since 0.95 is so close to 1, we expect that most of the time the rainfall will be between 13 and 17 inches a year.

Notice that in the preceding example, the standard deviation is 1 inch, so rainfall between 14 and 16 inches a year is within one standard deviation of the mean. Similarly, rainfall between 13 and 17 inches is within 2 standard deviations of the mean, and rainfall between 12 and 18 inches is within three standard deviations of the mean. The fractions of the observations within one, two, and three standard deviations of the mean calculated in the previous example hold for any normal distribution.

Rules of Thumb for Any Normal Distribution

- About 68% of the observations are within one standard deviation of the mean.
- About 95% of the observations are within two standard deviations of the mean.
- Over 99% of the observations are within three standard deviations of the mean.

Problems for Section F

In Problems 1–3, sketch graphs of a density function which could represent the distribution of income through a population with the given characteristics.

1. A large middle class.
2. Small middle and upper classes and many poor people.
3. Small middle class, many poor and many rich people.

4. A large number of people take a standardized test, receiving scores described by the density function p graphed in Figure F.28. Does the density function imply that most people receive a score near 50? Explain why or why not.

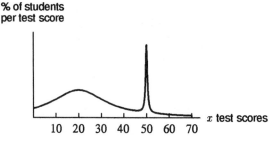

% of students per test score

10 20 30 40 50 60 70 x test scores

Figure F.28: Density function of test scores

5. Figure F.29 shows the distribution of elevation, in miles, across the earth's surface. Positive elevation denotes land above sea level; negative elevation shows land below sea level (i.e., the ocean floor).

   (a) Describe in words the elevation of most of the earth's surface.
   (b) Approximately what fraction of the earth's surface is below sea level?

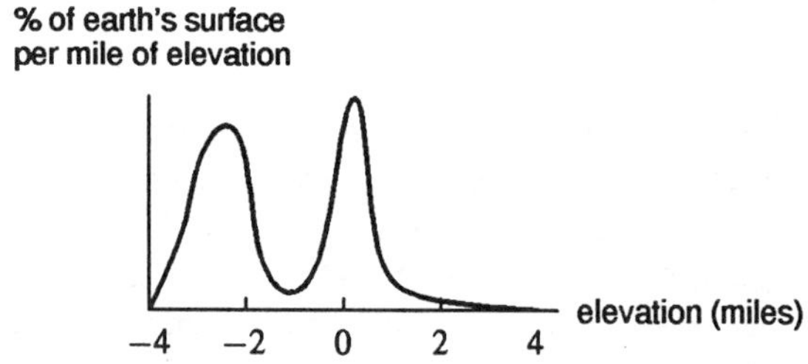

% of earth's surface per mile of elevation

−4   −2   0   2   4   elevation (miles)

Figure F.29

6. Consider a pendulum swinging through a small angle. The x-coordinate of the bob moves between $-a$ and a, as shown in Figure F.30.

   (a) Draw the density function for the location of the x-coordinate of the pendulum bob (i.e., neglect up-and-down motion). In order to do this, imagine a camera taking pictures of the pendulum at random instants. Where is the bob most likely to be found? Least likely? [Hint: Consider the speed of the pendulum at different points in its path. Is the camera more likely to take a photograph of the bob at a point on its path where the bob is moving quickly, or where it is moving slowly?]

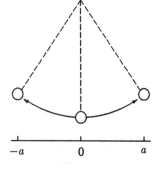

$-a$ 0 a

Figure F.30

   (b) Now sketch the function

   $$f(x) = \begin{cases} \dfrac{1}{\pi\sqrt{a^2 - x^2}} & -a < x < a; \\ 0 & |x| \ge a. \end{cases}$$

   How does this graph compare with the one you drew in part (a)?

(c) Assuming that the function given in part (b) is the density function for the pendulum, what do you expect

$$\int_{-a}^{a} \frac{1}{\pi\sqrt{a^2 - x^2}} \, dx$$

to be? Check this by computing the integral.

(d) Does it seem reasonable, physically speaking, that $f(x)$ "blows up" at a and $-a$? Explain your answer.

7. IQ scores are believed to be normally distributed with mean 100 and standard deviation 15.

(a) Write a formula for the density distribution of IQ scores.

(b) Estimate the fraction of the population with IQ between 115 and 120.

8. Show that the area under the graph of the density function of the normal distribution

$$p(x) = \frac{1}{\sqrt{2\pi}} e^{-(x-15)^2/2}$$

is 1. This function has no elementary antiderivative, so you must do it numerically. Make it clear in your solution what limits of integration you used.

9. (a) Using a calculator or computer, sketch graphs of the density function of the normal distribution

$$p(x) = \frac{1}{\sigma\sqrt{2\pi}} e^{-(x-\mu)^2/(2\sigma^2)}$$

(i) For fixed μ (say, $\mu = 5$) and varying σ (say, $\sigma = 1, 2, 3$).

(ii) For varying μ (say, $\mu = 4, 5, 6$) and fixed σ (say, $\sigma = 1$).

(b) Explain how the graphs confirm that μ is the mean of the distribution and that σ shows how closely the data is clustered around the mean.

10. Let v be the speed, in meters/second, of an oxygen molecule, and let $p(v)$ be the density function of the speed distribution of oxygen molecules at room temperature. Maxwell showed that

$$p(v) = av^2 e^{-mv^2/(2kT)},$$

where $k = 1.4 \times 10^{-23}$ is the Boltzmann constant, T is the temperature in degrees Kelvin (at room temperature, $T = 293$), and $m = 5 \times 10^{-26}$ is the mass of the oxygen molecule in kilograms.

(a) Find the value of a.

(b) Estimate the median and the mean speed. Find the maximum of $p(v)$.

(c) How do your answers in part (b) for the mean and the maximum of $p(v)$ change as T changes?

G REVIEW OF POLAR COORDINATES

Polar coordinates are another way of describing points in an xy-plane. The x- and y-coordinates can be thought of as instructions on how to get to the point. To get to the point $(1, 2)$ go 1 unit horizontally and 2 units vertically. Polar coordinates can be thought of the same way. There is an r-coordinate, which tells you how far to go along the ray extending to the point from the origin, and there is the θ-coordinate, which is an angle, and it tells you the angle the ray makes with the positive x-axis. (See Figure G.31.)

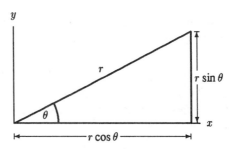

Figure G.31: Polar coordinates

Example 1   Give the polar coordinates of the points $(1,0)$, $(0,1)$, $(-1,0)$, and $(1,1)$.

Solution   To get to the point $(1,0)$, you go 1 unit along the horizontal axis, and there you are. So its r-coordinate is 1 and its θ-coordinate is 0, since you go along the x-axis.

The point $(0,1)$ is also one unit from the origin, so you start out the same way as before, by going 1 unit out along the ray. Then you have to go around the circle of radius 1 through an arc of $\pi/2$ to get to the point $(0,1)$. So $r = 1$ and $\theta = \pi/2$.

The point $(-1,0)$ also has r-coordinate equal to 1, but this time you have to go halfway around the circle to get there, so its θ-coordinate is π.

The point $(1,1)$ is a distance of $\sqrt{2}$ from the origin, and the ray from the origin to it makes an angle of $\pi/4$ with the horizontal ray. Thus, it has r-coordinate equal to $\sqrt{2}$ and θ-coordinate equal to $\pi/4$. See Figure G.32.

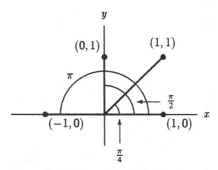

Figure G.32: Four points showing Cartesian
and polar coordinates

Conversion Between Polar and Cartesian Coordinates

Suppose a point has Cartesian coordinates (x, y). Look at Figure G.31. The distance r of the point from the origin is the length of the hypotenuse, which is $\sqrt{x^2 + y^2}$ by Pythagoras' theorem. The angle θ with the positive half of the x-axis satisfies $\tan \theta = y/x$.

On the other hand, if you are given r and θ, then from trigonometry you can see that $x = r \cos \theta$ and $y = r \sin \theta$.

Relation Between Polar and Cartesian Coordinates

$$x = r \cos \theta \qquad r = \sqrt{x^2 + y^2}$$

$$y = r \sin \theta \qquad \tan \theta = y/x.$$

Example 2   Give the Cartesian coordinates of the points with polar coordinates $(2, 3\pi/2)$ and $(2, 1)$.

Solution   Both points have r-coordinate equal to 2, so they are 2 units from the origin. The first one has θ-coordinate equal to $3\pi/2$, which is three quarters of a full revolution, so it is three quarters of the way around the circle of radius 2, at $(0, -2)$.
   The second point has θ-coordinate equal to 1. From the formulas above, we see that

$$x = 2 \cos 1 = 1.0806 \quad \text{and} \quad y = 2 \sin 1 = 1.6830.$$

Problems for Section G

For Problems 1–7, give Cartesian coordinates for the points with the following polar coordinates (r, θ). The angles are measured in radians.

1. $(1, 0)$ 2. $(0, 1)$ 3. $(2, \pi)$ 4. $(\sqrt{2}, 5\pi/4)$

5. $(5, -\pi/6)$ 6. $(3, \pi/2)$ 7. $(1, 1)$

For Problems 8–15, give polar coordinates for the points with the following Cartesian coordinates. Choose $0 \le \theta < 2\pi$.

8. $(1, 0)$ 9. $(0, 2)$ 10. $(1, 1)$ 11. $(-1, 1)$

12. $(-3, -3)$ 13. $(0.2, -0.2)$ 14. $(3, 4)$ 15. $(-3, 1)$

16. Every point in the plane can be represented by some pair of polar coordinates, but are the polar coordinates (r, θ) uniquely determined by the Cartesian coordinates (x, y)? In other words, for each pair of Cartesian coordinates, is there one and only one pair of polar coordinates for that point? Why or why not?

H CHANGE OF VARIABLES

You can often make more sense of a function if you simplify it with a substitution. For example, consider the function in Figure 11.49 on page 28 that gives corn production as a function of climate. The main concern of the people who produced this graph was the effect on corn production of a *change* in the climate, so they presented the function in a slightly different way. Figure H.33 shows the new way.

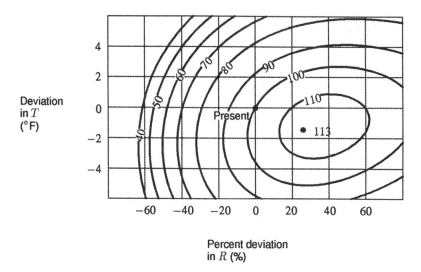

Deviation
in T
(°F)

Percent deviation
in R (%)

Figure H.33: Corn production as a function of climate change

Notice that the point marked "Present" is at the point with coordinates $(15, 76)$ on the old graph, and is at the origin of the new graph, i.e., it is the point with coordinates $(0, 0)$. In the new graph, everything is expressed in terms of deviations from the present values, which is the most sensible way of expressing things when you are trying to understand the problem of the effect of climatic change. Thus, for example, the point on the vertical axis that used to be marked 78 is now marked 2, since it is 2 more than the current value. In addition, on the horizontal axis the rainfall deviations are expressed as as percentages. Thus, since a rainfall of 12 inches would represent a 20% decrease from the current rainfall of 15 inches, the point that used to be marked 12 on this axis is now marked -20.

This is an example of a change of coordinates. The underlying dependence of corn production on rainfall and temperature has not changed, but the coordinates we use to express the rainfall and temperature have changed. In terms of the contour map, this means that the contours stay the same, but the numbers marked on the axes change.

Shifts and Stretches

In the example on corn production above, we made two sorts of changes in our coordinates: first we set things up so that the current temperature and rainfall became the origin of our new coordinate system, and then we scaled the rainfall so that it was expressed as a percentage rather than a number of inches. Let's consider these two sorts of changes separately.

How Do You Change the Origin?

If you are working with coordinates x and y, and you want the point (a, b) to become the origin of your coordinate system, you define new coordinates

$$u = x - a, \quad v = y - b.$$

Then the point that used to have x-coordinate a and y-coordinate b now has u- and v-coordinates both equal to zero, i.e., it is at the origin of your new coordinate system.

Example 1   Choose better coordinates for the equation

$$z = x^2 - 2x + 3 + y^2 + 4y.$$

Solution   If we complete the square on both x and y, we get

$$z = (x - 1)^2 + (y + 2)^2 - 2.$$

This is a parabola-shaped bowl, or paraboloid, whose base is at $x = 1$, $y = -2$, $z = -2$. If we make the change of coordinates

$$u = x - 1 \quad v = y + 2,$$

then the equation becomes

$$z = u^2 + v^2 - 2.$$

The vertex of the paraboloid is now at $u = 0$, $v = 0$.

Rescaling Coordinates

Another common change of coordinates is scaling. We saw an example of this above when we expressed rainfall figures as a percentage of the current amount, rather than in absolute terms.

Scaling coordinates can be very useful in seeing relationships between different things. For example, the van der Waal's gas equation relates the pressure, temperature, and volume of a certain quantity (one mole) of gas. The van der Waal's equation for helium is

$$T = 0.41581\frac{1}{V} - 0.0098546\frac{1}{V^2} - 0.28882P + 12.187VP$$

and the van der Waal's equation for oxygen is

$$T = 16.574\frac{1}{V} - 0.52754\frac{1}{V^2} - 0.3879P + 12.187VP.$$

Here pressure is measured in atmospheres, temperature in degrees Kelvin, and volume in cubic decimeters. The formulas for the two gases are quite different. Figures H.34 and H.35 show the contour maps for these functions.

Notice that although the numbers marked along the sides are different, the contours look quite similar. Each contour map has one contour with a point that is both an inflection point and point where the slope is zero. That point is called the critical point for the gas, and has a special chemical significance. Its P- and V-coordinates are called the critical pressure and volume respectively, and the value of T there is called the critical temperature. The similarity between the contours of the two different gases suggests that if we scale the P, V, and T-coordinates so that the critical values are all equal to 1, then the gases will have the same equation, thus revealing some unifying underlying behavior of these two different gases.

For helium, the critical pressure is 2.2498, the critical volume is 0.0711, and the critical temperature is 5.1984. To make these equal to one in our new coordinate system, we divide the old coordinates by these values. The new coordinates will be denoted by lower case letters. The change of coordinate equations are

$$p = P/2.2498, \quad v = V/0.0711, \quad t = T/5.1984.$$

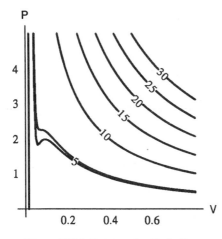

Figure H.34: Gas equation for helium   **Figure H.35**: Gas equation for oxygen

To substitute these into the van der Waals equation, we solve them for the old coordinates:

$$P = 2.2498p, \quad V = 0.0711v, \quad T = 5.1984t.$$

The gas equation for helium becomes

$$5.1984t = 0.41581\frac{1}{0.0711v} - 0.0098546\frac{1}{(0.0711v)^2}$$
$$+ 0.28882(2.2498p) + 12.187(0.0711v)(2.2498p),$$

which simplifies to

$$t = 1.125\frac{1}{v} - 0.375\frac{1}{v^2} - 0.125p + 0.375pv.$$

For oxygen, the critical pressure is 49.717, the critical volume is 0.09549, and the critical temperature is 154.28. Making the change of coordinates

$$p = P/49.717, \quad v = V/0.09549, \quad t = T/154.28,$$

we get the same equation as for helium.

Any change of coordinates that involves multiplying the variables by a constant is called a scaling of the variables. Another situation where you scale the variables is when you change the units of measurement.

I   THE IMPLICIT FUNCTION THEOREM

Consider the equation

$$z^3 - 7yz + 6e^x = 0.$$

Can it be solved for z? Does it determine a function $z = f(x, y)$ of x and y? The test is whether, in principle, the equation determines a table of values for f. Let's try to construct such a table. For instance, to evaluate $z = f(0, 1)$ you must solve for z after setting $x = 0$ and $y = 1$ in the equation. That is, you must solve for z in the equation

$$z^3 - 7z + 6 = 0.$$

As it happens, the equation easily factors, yielding

$$z^3 - 7z + 6 = (z-1)(z-2)(z+3) = 0,$$

so there are three solutions, $z = 1, z = 2$, and $z = -3$. Since a function must take single definite values and we have been given no rule to decide which solution should equal $f(0,1)$, we conclude that the original equation alone does not define z as a function of x and y.

Perhaps the original equation defines three functions f, g, and h of x and y, with, say, $f(0,1) = 1, g(0,1) = 2$, and $h(0,1) = -3$. To test this hypothesis, let's try to evaluate f, g, and h at $(x, y) = (0.02, 1.01)$. You must solve for z after setting $x = 0.02$ and $y = 1.01$. That is, you must solve for z in the equation

$$z^3 - 7.07z + 6e^{0.02} = 0.$$

Again, there are three solutions. They can be found numerically to equal $1.012701, 2.003794$, and -3.016495. Since $(0.02, 1.01)$ is near $(0, 1)$, it would make sense for $f(0.02, 1.01)$ to be near $f(0,1) = 1$, for $g(0.02, 1.01)$ to be near $g(0,1) = 2$, and for $h(0.02, 1.01)$ to be near $h(0,1) = -3$. So we have:

$$f(0.02, 1.01) = 1.012701,$$
$$g(0.02, 1.01) = 2.003794,$$
$$h(0.02, 1.01) = -3.016495.$$

We have found the key. We will say that $f(x, y)$ is defined for (x, y) near $(0, 1)$ to be the solution z of $z^3 - 7yz + 6e^x = 0$ that is near 1, that $g(x, y)$ is defined for (x, y) near $(0, 1)$ to be the solution z of $z^3 - 7yz + 6e^x = 0$ that is near 2, and that $h(x, y)$ is defined for (x, y) near $(0, 1)$ to be the solution z of $z^3 - 7yz + 6e^x = 0$ that is near -3. We will not attempt to define f, g, or h for values of (x, y) that are far from $(0, 1)$.

A brief table of values for g is given in Table I.3.

TABLE I.3: *One Solution of* $z^3 - 7yz + 6e^x = 0$

x \ y	0.98	0.99	1.00	1.01	1.02
-0.02	1.96741	1.99580	2.02312	2.04949	2.07502
-0.01	1.95477	1.98385	2.01177	2.03868	2.06468
0.00	1.94158	1.97142	2.00000	2.02747	2.05398
0.01	1.92778	1.95847	1.98776	2.01586	2.04292
0.02	1.91330	1.94492	1.97501	2.00379	2.03145

It was computed with many applications of Newton's root-finding method. For example, $g(0.02, 1.01)$ is the solution z near 2 of the equation $z^3 - 7.07z + e^x = 0$. To compute it, start with the approximate value $z_0 = 2$, then successively improve the approximation using the formula $z_{n+1} = z - \frac{m(z_n)}{m'(z_n)}$, where $m(z) = z^3 - 7.07z + e^{0.02}$. See TableI.4.

TABLE I.4: *Newton's Method for* $z^3 - 7.07z + e^{0.02} = 0$

n	z_n
1	2.003811757
2	2.003794225
3	2.003794225

It is difficult to compute Table I.3 for $g(x, y)$ because the equation to be solved:

$$z^3 - 7yz + 6e^x = 0$$

is nonlinear. But for the function g we are only interested in values of (x, y, z) near $(0, 1, 2)$, and the equation ought to be approximately linear for (x, y, z) near a single point. Since linear equations are easy to solve, let's replace the nonlinear equation by a linear approximation valid for (x, y, z) near $(0, 1, 2)$.

First find the local linearization of $m(x, y, z) = z^3 - 7yz + 6e^x$ at $(x, y, z) = (0, 1, 2)$. Evaluation of the partial derivatives m_x, m_y, and m_z at $(0, 1, 2)$ is straightforward, and we find that

$$m(x, y, z) \approx 0 + 6x - 14(y - 1) + 5(z - 2)$$

for (x, y, z) near $(0, 1, 2)$. It follows that the solutions $z = g(x, y)$ for (x, y) near $(0, 1)$ of

$$z^3 - 7yz + 6e^x = 0$$

should be close to the solutions of

$$6x - 14(y - 1) + 5(z - 2) = 0.$$

The last equation is linear and easily solved, giving

$$z = -0.8 - 1.2x + 2.8y.$$

We conclude that

$$g(x, y) \approx -0.8 - 1.2x + 2.8y \quad \text{for } (x, y) \text{ near } (0, 1).$$

In fact, this last approximation is the local linearization of g at $(0, 1)$. It has been used to generate easily Table I.5 of approximate values for g, which can be compared with Table I.3 of exact values.

TABLE I.5: *Linear Approximation*

		y				
		0.98	0.99	1.00	1.01	1.02
	-0.02	1.968	1.996	2.024	2.052	2.080
	-0.01	1.956	1.984	2.012	2.040	2.068
x	0.00	1.944	1.972	2.000	2.028	2.056
	0.01	1.932	1.960	1.988	2.016	2.044
	0.02	1.920	1.948	1.976	2.004	2.03

The Implicit Function Principle

The analysis of the previous example can be fairly summarized as follows. A nonlinear equation is hard to solve. Locally it may resemble a linear equation that is easy to solve. The solution of the linear equation will be a good approximation for the solution of the nonlinear equation, at least locally.

More formally, we state the Implicit Function Principle.

The Implicit Function Principle

Suppose that $f(x, y, z)$ is a function, that $f(a, b, c) = 0$, and that $L(x, y, z)$ is the local linearization of f at (a, b, c). If the linear equation,

$$L(x, y, z) = 0,$$

can be solved for z in terms of x and y, then that solution is a close approximation of a function $z = g(x, y)$ that is defined for (x, y) near (a, b) to be the only solution z near c of the equation

$$f(x, y, z) = 0.$$

A little more can be said. The approximation of $g(x, y)$ mentioned by the Implicit Function Principle is actually the local linearization of g at (a, b). Since partial derivatives can be read off from the local linearization, the Implicit Function Principle gives a method of computing $g_x(a, b)$ and $g_y(a, b)$.

Example 1   Use the Implicit Function Principle to investigate solutions of the equation

$$e^z + xyz + x^3y^4 - 2 = 0$$

near the solution $(x, y, z) = (1, 1, 0)$.

Solution   Let $f(x, y, z) = e^z + xyz + x^3y^4 - 2$. Compute:

$$f_x(x, y, z) = yz + 3x^2y^4,$$
$$f_y(x, y, z) = xz + 4x^3y^3,$$
$$f_z(x, y, z) = e^z + xy.$$

Hence,

$$f_x(1, 1, 0) = 3$$
$$f_y(1, 1, 0) = 4$$
$$f_z(1, 1, 0) = 2.$$

Therefore:

$$f(x, y, z) \approx 3(x - 1) + 4(y - 1) + 2(z - 0)$$
$$= -7 + 3x + 4y + 2z \quad \text{for } (x, y, z) \text{ near } (1, 1, 0).$$

The linearization of the equation $f(x, y, z) = 0$ near $(1, 1, 0)$ is

$$-7 + 3x + 4y + 2z = 0$$

which can be solved for z, giving
$$z = 3.5 - 1.5x - 2y.$$

The Implicit Function Principle applies. It asserts that for (x, y) near enough to $(1, 1)$ there is exactly one solution z near 0 of the original (nonlinear) equation

$$e^z + xyz + x^3y^4 - 2 = 0.$$

Moreover, an approximation for the solution $z = g(x, y)$ is given by the local linearization for g near $(1, 1)$:

$$z = g(x, y) \approx 3.5 - 1.5x - 2y.$$

In other words, all solutions of $e^z + xyz + x^3y^4 - 2 = 0$ near $(1, 1, 0)$ are approximated by $(x, y, 3.5 - 1.5x - 2y)$ for (x, y) near $(1, 1)$.

Our final example shows some of the limitations of the Implicit Function Principle.

Example 2   Use the Implicit Function Principle to investigate solutions of the equation

$$x^2 + y^2 + z^2 = 25$$

near the solution $(x, y, z) = (3, 4, 0)$.

Solution   Consider the equivalent equation $f(x, y, z) = 0$, where $f(x, y, z) = x^2 + y^2 + z^2 - 25$. From $df = 2x\, dx + 2y\, dy + 2z\, dz$ we get the local linearization

$$f(x, y, z) \approx 6(x - 3) + 8(y - 4) + 0(z - 0) \quad \text{for } (x, y, z) \text{ near } (3,4,0).$$

The linearization of the equation $f(x, y, z) = 0$ near $(3, 4, 0)$ is

$$6(x - 3) + 8(y - 4) = 0,$$

which cannot be solved for z, because z does not appear in it. Therefore, the Implicit Function Principle does not give an approximation for z as a function of (x, y) near $(3, 4)$.

To see what has happened we turn to geometry. The equation $x^2 + y^2 + z^2 = 25$ describes a sphere of radius 5 centered at the origin. Do values of (x, y) near $(3, 4)$ determine unique values of z near 0 such that (x, y, z) lies on the sphere and hence satisfies the given equation? The answer is no, for two reasons. For points such as $(x, y) = (3.01, 4.01)$ where $x^2 + y^2 > 25$, there is no z at all such that (x, y, z) is on the sphere. For points such as $(2.99, 3.99)$ where $x^2 + y^2$ is slightly less than 25, there are two z values near 0 that solve the equation, namely $z = \sqrt{25 - x^2 - y^2}$ and $z = -\sqrt{25 - x^2 - y^2}$. See Figure I.36.

The Implicit Function Principle could not apply near $(3, 4, 0)$ because the equation $x^2 + y^2 + z^2 = 25$ does not determine definite z values for all (x, y) near $(3, 4)$; and so, there is no function to approximate near $(3, 4)$.

One final comment is in order. The linearization $6(x - 3) + 8(y - 4) = 0$ of the equation $f(x, y, z) = 0$ near $(3, 4, 0)$ is the equation of the tangent plane to the sphere at $(3, 4, 0)$. The absence of z in the equation of the plane indicates that the plane is vertical. See Figure I.37. In some sense it is the fact that the tangent plane is vertical at $(3, 4, 0)$ that makes it possible for the sphere to turn under itself at that point, which is why there can be two different nearby points on the sphere for single values of (x, y) near $(3, 4)$.

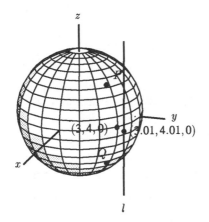

Figure I.36: $x^2 + y^2 + z^2 = 25$

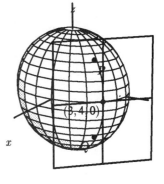

Figure I.37: Vertical Tangent Plane

J NOTES ON KEPLER, NEWTON AND PLANETARY MOTION

If you looked at the stars every night, you would see that they stay in the same positions with respect to each other from night to night, but that during the course of one night the dome of stars rotates slowly about the North Star. If you studied the stars more carefully, however, you would eventually notice some of them moving with respect to the others. If you followed the path of these stars night after night as they wandered amongst the other stars, you might even see their position from one night to the next move forward and then backward. These wanderers are the planets. Five, besides earth, are visible to the naked eye. Ever since people, thousands of years ago, first observed these erratic paths, they have endowed the planets with supernatural powers; for example, astrology is based largely on their positions with respect to the fixed stars. The names by which we know them, Mercury, Venus, Mars, Jupiter, and Saturn, are the names of Ancient Roman gods. People also tried to make sense out of the strange paths, and the mathematical explanation of planetary motion was the first breakthrough human kind made in understanding the natural laws of the universe.

First Steps: Eratosthenes and Copernicus

The first important realization is

> The earth is round, a sphere of radius about 4000 miles.

This was known to some at least as far back as ancient Greece. Eratosthenes gave a reasonable estimate of the radius of the earth by observing the angle of the sun at noon on June 21 at two different locations (see Problem 1).

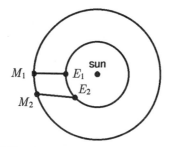 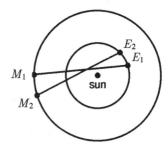

Figure J.38: The retrograde motion of Mars. (M_1 and M_2 indicate Mars on the first and second night; similarly for earth and E_1, E_2)

> The earth spins once every 24 hours around a central axis passing through the north and south poles.

The rotational axis of the earth points at the North Star. Thus, if you watched the North Star throughout the night, the rotation of the earth would make it look as if the dome of stars was spinning slowly around it, completing one full revolution in 24 hours.

The Greek philosopher Aristotle held that the earth remained fixed while all other heavenly bodies moved around it. Indeed, the very idea of the earth spinning seems preposterous. Wouldn't we just be spun right off the planet? In fact, the acceleration you would feel at the equator caused by the earth's spinning is only about 1% that of gravity, about the same as the acceleration you'd feel at the edge of merry-go-round with 30 foot radius that goes around once every 2 minutes (see Problem 2). You could probably crawl on your hands and knees around that merry-go-round and still keep up with it.

In the middle ages Nicolaus Copernicus (1473-1543) proposed the more modern point of view, that the apparent motion of the stars each night is caused by the earth's rotation. Copernicus also contradicted previous theories by placing the sun at the center of the solar system:

> The earth and the planets orbit around the sun. The earth completes one revolution around the sun each 365 1/4 days (approximately). The moon orbits around the earth completing one revolution in about 27.32 days.

In fact Eratosthenes had already proposed this theory in ancient times, because the motion of the earth and planets around the sun explains the apparently complicated motion of the planets. For example, suppose the earth and Mars are on the same side of the sun as shown on the left of Figure J.38. The earth is closer to the sun than Mars and completes one orbit faster. Thus, on consecutive nights, although Mars has moved, you have moved even more, so that Mars appears to have moved backwards. If the earth is moving counterclockwise, then in facing Mars you would have to look back further to your right the second night. On the other hand, if the earth and Mars are on opposite sides of the sun, the second night would find Mars a little to the left of its position the first night, as shown on the right of Figure J.38, and so it would appear to have moved forward. When Mars appears to move backwards, we say it exhibits *retrograde motion*.

Kepler's Laws for Planetary Motion

The orbits of the planets around the sun or the moon around the earth are not circles. The moon's distance to the earth varies from 220,000 to 260,000 miles. Each of the planets, including earth, has a closest and furthest point from the sun. So what are the shapes? In the last half of the 16th century Tycho Brahe accumulated data about the positions of the planets. Johann Kepler (1571-1630) studied the data for years and after some false starts involving platonic solids and the "music of the spheres," he arrived at three laws for planetary motion:

Kepler's Laws

- The orbit of each planet is an ellipse with the sun at the focus. In particular, the plane of the orbit contains the sun.

- As a planet orbits around the sun, the line segment from the sun to the planet sweeps out equal areas in equal times.

- The ratio p^2/d^3 is the same for every planet orbiting around the sun, where p is the period of the orbit (time to complete one revolution) and d is the mean distance of the orbit (average of the shortest and furthest distance from the sun).

It is very important to understand what these laws are saying. In Law I, an ellipse is not just any old squashed circle. It is a specific geometric figure having very special properties. An ellipse is a closed curve in the plane such that the sum of the distances from any point on the curve to two fixed points, called the foci of the ellipse, is constant. If the two foci are located at $(0, -b)$ and $(0, b)$ on the y-axis, then it can be shown (see Problem 3) that the constant sum of distance is $2d$ and that the equation of the ellipse is

$$\frac{x^2}{c^2} + \frac{y^2}{d^2} = 1,$$

where d is the mean distance and $c^2 = d^2 - b^2$. Note as well that the sun is not at the center of the planet's elliptical orbit, but rather at a focus. This is a crucial distinction.

Kepler's Second Law implies that the speed of the planet is not constant. In order to sweep out the same area in one unit of time when the planet is near the sun as it does when it is far from the sun, the planet must move faster when it is near the sun (see Figure J.39).

The third law says that p^2/d^3 is the same for all planets. In particular, this means if you know the period p for a planet, then the mean distance d is determined, and vice versa. Newton later

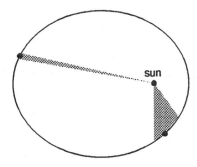

Figure J.39: Equal areas in equal time

showed that the constant value of p^2/d^3 depends on the mass of object about which the planets are orbiting.

Kepler's Laws, impressive as they are, were purely descriptive; they didn't explain the motion of the planets, they merely described it. Newton's great achievement was to find an underlying cause for them.

Newton's Laws of Motion

In 1687, Isaac Newton published *Philosophiae Naturalis Principia Mathematica*. The title is usually shortened to Principia Mathematica and is Latin for "Mathematical Principles of Natural Philosophy."

In *Principia*, Newton developed a theory of motion that explained Kepler's Laws, a theory which placed the concept of force at the center of physics. It also introduced many of the key ideas of calculus, although often in a disguised, geometric form. Newton began with the observation we have made in the previous section: curving motion is an indication of acceleration. He then tried to find a specific law of acceleration that would explain Kepler's Laws, and arrived at his Universal Law of Gravitation, which states that given two objects of mass M and m, the force of attraction between them is proportional to the product of their masses and the inverse square of the distance r separating them:

$$F = GMm/r^2,$$

where G is a universal constant.

In the next section, we will explain Newton's approach. It[3] does not use derivatives or vectors or cross products, but rather similar triangles and geometry.

Newton's First and Second Law of Motion

First Newton defines the mass of a body and observes that this is directly proportional to its weight. He then defines "motion" of a body, which we would call momentum, to be the product of its mass and velocity. This first general law of motion is

Law I

Every body continues in a state of rest, or of uniform motion in a right [straight] line, unless it is compelled to change that state by forces impressed on it.

This seems simple enough but it has profound consequences. In particular, it says a planet can move in an ellipse rather than a straight line only if there is a force acting on it.

Law II

The change of motion is proportional to the motive force impressed; and is made in the direction of the right [straight] line in which that force is impressed.

Notice that the law is also careful to describe the direction of the change: Newton recognized that force and acceleration are vector quantities. In modern terms, this law says that the force vector is proportional to the acceleration vector:

$$\vec{F} = m\vec{a}.$$

---

[3]The version given here is based on the article "Newton and the Transmutation of Force" by Tristan Needham (Math. Monthly 100(1993), 119-137).

Newton's Explanation of Kepler's Second Law

Newton's Second Law and his Universal Law of Gravitation are the keys to Kepler's Second Law. Newton's Second Law says that the acceleration vector of a planet points in the direction of the gravitational force acting on it, and the Law of Gravitation says that the gravitational force points toward the sun. Together they imply that the acceleration vector of a planet orbiting the sun must always point towards the sun. There is a special name for orbital motion that satisfies this condition. We define *centripetal motion about the fixed point A* to be motion where the acceleration is always directed towards A. Newton proved the following statement relating centripetal motion to Kepler's Second Law and the part of the First Law that says the motion lies in a plane, thus providing a dynamical explanation for Kepler's purely descriptive statements.

Newton's First Theorem

Suppose an object is moving in a plane containing the point A in such a way that the line segment from A to the object sweeps out equal areas in equal times. Then the motion is centripetal about A, that is, the acceleration of the object is always directed towards A. Conversely, if the motion of an object is centripetal about the point A, then the object must move in a plane containing A and the line segment from A to the object sweeps out equal areas in equal times.

Newton's Proof

Think of the path of the object as made up of short straight lines which represent the motion over equal short time intervals; the shorter the time interval, the shorter the segments and the more closely they approximate the actual path of motion. Figure J.40 shows two consecutive line segments PQ and QR. If there were no acceleration, the object when it reaches Q would continue in a straight line to S, and the distances PQ and QS would be equal. Instead, the object changes direction and moves to R. Thus, SR indicates the direction of the change in motion, that is. the acceleration at point Q.

First, suppose the motion is in a plane containing A and equal areas are swept out in equal times. Then Figure J.40 lies in a plane and triangles PQA and QRA have the same area. On the other hand, triangles QSA and PQA have the same area, because the bases PQ and QS have equal length and, since P, Q, and S all lie on the same line, the altitudes from A are the same. Thus, triangles QSA and QRA have equal area. The only way this could happen is if SR is parallel to QA (see the right of Figure J.40), because they share the side QA, so the altitudes from S and R to QA must be the same. Therefore, the acceleration at Q is directed towards A.

The entire argument can be reversed to give the converse: if the acceleration is towards A, then equal areas are swept out in equal times.

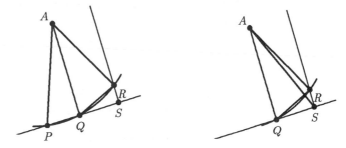

Figure J.40: Centripetal acceleration and equal areas ($PQ = QS$ and RS parallel to AQ)

A Modern Proof

Here is a modern proof using cross products, derivatives and vectors. Consider the quantity $\vec{r} \times \vec{v}$ where $\vec{r}$ represents the vector from the point A to the moving object and $\vec{v}$, as usual, is the velocity vector. Notice that the magnitude of $\vec{r} \times \vec{v}$ is just twice the area of the triangle spanned by $\vec{r}$ and $\vec{v}$ and hence represents the rate at which area is being swept out. Also, the direction of $\vec{r} \times \vec{v}$ is perpendicular to the plane containing $\vec{r} \times \vec{v}$. If we can show that $\vec{r} \times \vec{v}$ is constant, then we can show that area is being swept out at a constant rate and that $\vec{r}$ and $\vec{v}$ always lie in the same plane (so that $\vec{r} \times \vec{v}$ always points in the same direction). After some very messy computations, one can show that cross products satisfy the product law for derivatives. Thus,

$$\frac{d}{dt}(\vec{r} \times \vec{v}) = \frac{d\vec{r}}{dt} \times \vec{v} + \vec{r} \times \frac{d\vec{r}}{dt}$$

Now $\frac{d}{dt}(\vec{r})$ is the rate of change of the position vector and hence is the velocity vector. Of course $\frac{d}{dt}(\vec{v})$ is just the acceleration vector $\vec{a}$. Thus,

$$\frac{d}{dt}(\vec{r} \times \vec{v}) = \vec{v} \times \vec{v} + \vec{r} \times \vec{a}$$

Recall that if two vectors point in the same or opposite directions their cross product is zero and vice versa. Thus, $\vec{v} \times \vec{v} = 0$. Furthermore $\vec{r} \times \vec{a} = 0$ precisely when $\vec{a}$ is the same or opposite direction as $\vec{r}$, that is when the motion is centripetal. Thus, the motion is planar with equal areas swept out in equal time exactly when $\frac{d}{dt}(\vec{r} \times \vec{v}) = 0$ which occurs exactly when $\vec{r}$ and $\vec{a}$ are in the same or opposite directions.

Kepler's First and Third Laws

The Apple and the Moon

The equivalence between Kepler's Second Law and centripetal motion tells us the direction of the acceleration experienced by a planet: it is always toward the sun. But what about its magnitude?

Newton realized there must be a force bending the path of the moon as it circled the earth. He also knew that a falling body, for example an apple from a tree, experienced a constant acceleration of about $g = 32$ ft/sec^2 towards earth (see Problem 9 for how one might measure this). Newton's insight was that the force pulling the apple to the earth extended to the moon and was the same force that accelerated the moon in its orbit. To calculate how this force varies with distance from the center of the earth, one needs only the distance from center to surface, i.e., the radius r of the earth, and the distance R from the center of the earth to the moon. These quantities were known in Newton's time; r is about 4000 miles and R is around 240,000. Therefore, $R = 60r$.

By results in Section 16.3 on page 295, we know that the acceleration of an object moving in a circle of radius R at constant speed v is

$$a = \frac{v^2}{R} = \frac{4\pi^2 R}{p^2},$$

where $p = 2\pi R/v$ is the period.

The period of the moon is 27.32 days. Thus, its acceleration is

$$a = \frac{4\pi^2(240,000 \times 5280)}{(27.32 \times 24 \times 60 \times 60)^2} = 0.00898 \text{ ft/sec}^2.$$

Thus, the ratio of the acceleration of gravity at the surface of the earth to the acceleration of the moon is

$$\frac{g}{a} = \frac{32}{0.00898} \approx 3,560.$$

The ratio of the inverse square of the radii is

$$\frac{1/r^2}{1/R^2} = \frac{R^2}{r^2} = \frac{(60r)^2}{(r)^2} = 60^2 = 3600.$$

It looks like the acceleration is proportional to the inverse square of the distance from the earth. Thus, is Newton's Law of Gravitation born!

There is another way to obtain the Inverse Square Law by using Kepler's Third Law. Suppose all the planets had circular orbits, so the mean distance d is just the radius r of the orbit. Then Kepler's Third Law says

$$p^2 = kr^3,$$

where k is a constant. The acceleration of each planet is

$$a = \frac{4\pi^2 r}{p^2}.$$

Therefore,

$$a = \frac{4\pi^2 r}{kr^3} = \frac{4\pi^2}{k}\left(\frac{1}{r^2}\right).$$

In other words, the acceleration of the planets is proportional to the inverse square of the distance between the planet and the sun.

Centripetal Force from the Center of an Ellipse

Newton considered other laws for centripetal force besides the Inverse Square Law. The simplest is that the magnitude of the force is directly proportional to the distance r from the fixed point A. That is, the acceleration is directed toward A and has magnitude kr rather than k/r^2. What is striking is that the orbits for such a force are also ellipses, only the fixed point A is at the center of the ellipse rather than a focus. This observation plays a key role in one of Newton's proofs of the Inverse Square Law, and it is this proof we will give. First here is the result about kr centripetal force.

Newton's Second Theorem

Suppose an object moves centripetally about the point A such that the orbit is an ellipse with A at the center. Then the acceleration of the object is proportional to its distance r from A. Conversely, suppose an object moves centripetally about the fixed point A in such a way that the acceleration is always proportional to the distance from the object to A. Then the orbit of the object is an ellipse with A at the center. In both cases, the constant of proportionality is $4\pi^2/p^2$, when p is the period.

Proof. Our proof is not Newton's and draws on examples of parametric equations. Consider the motion of period $p = 2\pi/\sqrt{k}$ given by

$$x = a\cos\sqrt{k}t, \quad y = b\sin\sqrt{k}t.$$

The resulting curve, as we have observed, is the ellipse $\frac{x^2}{a^2} + \frac{y^2}{b^2} = 1$, whose center is the origin. The acceleration of this motion is

$$\frac{d^2x}{dt^2} = -ka \cos \sqrt{k}t, \quad \frac{d^2y}{dt^2} = -kb \sin \sqrt{k}t.$$

Thus, the acceleration vector $\vec{a}$ satisfies $\vec{a} = -k(x, y)$. In other words, the acceleration always points in the opposite direction of the position vector (x, y), namely back to the origin. We conclude that the motion is centripetal about the origin. Moreover, the period $p = 2\pi/\sqrt{k}$, so that the constant of proportionalities $k = 4\pi^2/p^2$. By adjusting a, b, and k we can get the motion of an elliptical orbit of any period centered at the origin, and all of the resulting motions are centripetal. By Theorem 1 (the equivalence of centripetal motion and Kepler's Second Law), once you know the orbit and period of a centripetal motion, the motion is completely determined. There is only one way you can sweep out equal areas in equal time and complete one revolution in the presented period. Thus, our given parametric equations must describe all possible centripetal motions about the origin having an ellipse centered at the origin as the orbit. We have already computed the acceleration for these parametric equations:

$$\vec{a} = -k(x, y),$$

which has magnitude, $k\sqrt{x^2 + y^2}$, which is proportional to the distance from the origin, and the constant $k = 4\pi^2/p^2$. Thus, all centripetal motions about the origin whose orbit is an ellipse centered at the origin have an acceleration proportional to the distance from the origin, and the constant of proportionality $k = 4\pi^2/p^2$. Since we can always take the point A in the statement of the theorem as the origin, this proves the first half of the theorem. The converse is simply a matter of solving the second order differential equation

$$\frac{d^2y}{dx^2} = -kx, \quad \frac{d^2y}{dy^2} = -ky.$$

As we know from the harmonic oscillator, the solutions are combinations of $\cos \sqrt{k}t$ and $\sin \sqrt{k}t$. In particular, the orbits are periodic of period $2\pi/\sqrt{k}$ and form closed curves. If one chooses the x-axis to go through the point on the curve farthest from the origin, it is then possible to write the solutions in the form

$$x = a \cos \sqrt{k}t, \quad y = b \sin \sqrt{k}t.$$

Thus, the orbits are ellipses with the origin at the center with period p satisfying $k = 4\pi^2/p^2$.

The Transmutation of Centripetal Forces

Kepler's First Law says that the sun is at the focus of an ellipse, not the center, so we must deal with centripetal motion about a focus. Newton shows us how to change the fixed point of centripetal motion. Specifically, he considered the following problem. Suppose an object moves in a certain orbit centripetally about a fixed point A. How would the magnitude of the acceleration towards A have to change if the object moved in the same orbit but centripetally about a different point B? This "transmutation" of forces sounds very complicated but Newton gives a surprising easy formula for the transmutation.

Suppose we have an object moving centripetally about the point A. Suppose at one instant it is at P and Δt seconds later it is at Q, as shown in Figure J.41. If there were no acceleration the object would have continued out the tangent line at P and would arrive Δt seconds later at some point R. Where is R on the tangent line? The line segment QR leading from where the object would have been, R, to where it actually is Δt seconds later, Q, must point in the direction of the acceleration at

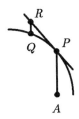

Figure J.41: The acceleration at point P

P. Thus, QR is parallel to AP and so to find R just draw the line parallel to AP through Q and find where it meets the tangent line. The length of QR tells us how far this object has "fallen" towards A because of the acceleration at P. We know for constant acceleration a, that distance traveled in time Δt is $\frac{1}{2}a(\Delta t)^2$. Since in a short time Δt the acceleration is nearly constant, we have that the magnitude of the acceleration at P is

$$a = \frac{2QR}{(\Delta t)^2}.$$

We are a little bit sloppy here and are letting QR stand for the length of the segment QR. Moreover, that equal sign in the equation for the acceleration a really means that as Δt approaches 0 and the point Q gets closer and closer to P, the limit of $2QR/(\Delta t)^2$ is the acceleration at P.

The key idea now is to eliminate Δt in our formula for the acceleration at P. We have to do this because if we change the center of centripetal motion, even if we keep the same orbit, the planet will trace out the orbit at different speeds since the areas swept out from the new point are different from the areas swept out from the old point. This will be the case even if the rate at which area being swept out is the same.

Let k be the constant rate at which area is being swept out. Then k is the area of triangle PQA divided by Δt. Thus, to eliminate Δt we only need to compute the area of PQA. As $\Delta t \to 0$, the line segment PQ lines itself up along the tangent line at P, so the area of PQA behaves in the limit as if its base were PQ and its altitude were the distance h from A to the tangent line. Thus,

$$1/2\frac{PQ \cdot h}{\Delta t} = k \quad \text{so} \quad \Delta t = \frac{1}{2k}(PQ \cdot h)$$

where k is the constant rate at which area is swept out and h is the distance from A to the tangent at P. Therefore, the acceleration at P is

$$a = \frac{2QR}{(\Delta t)^2} = 8k^2\frac{QR}{(PQ \cdot h)^2}$$

To emphasize which parts of the formula depend on the centripetal point A, use the subscript A: write R_A instead of R and h_A instead of h. Then we have that the acceleration a_A for centripetal motion about A:

$$a_A = 8k^2\frac{QR_A}{(PQ \cdot h_A)^2}.$$

Now suppose the same orbit were centripetal about B and that area is being swept out at the same constant rate k. Then the centripetal acceleration a_B about B is

$$a_B = 8k^2\frac{QR_B}{(PQ \cdot h_B)^2}.$$

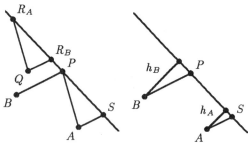

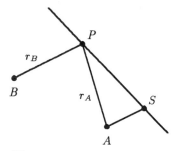

Figure J.42: Similar triangles for QR_A/QR_B and h_A/h_B.

Figure J.43: Transmutation of centripetal acceleration.

We want to determine the relationship between QR_A and QR_B and between h_A and h_B. This is easily done with similar triangles. Draw a line from A parallel to BP and let S be the point where this line meets the tangent through P, as shown on the left of Figure J.42. Then QR_B is parallel to BP which is parallel to AS. Also QR_A is parallel to AP. Thus, the triangles QR_AR_B and APS are similar so

$$\frac{QR_B}{QR_A} = \frac{AS}{AP}.$$

In the same way, by similar triangles as shown on the right of Figure J.42,

$$\frac{h_A}{h_B} = \frac{AS}{BP}.$$

Therefore, if the constant rate k at which area is swept out is the same for both motions, we have that the ratio of the acceleration is

$$\frac{a_B}{a_A} = \left(\frac{QR_B}{QR_A}\right)\left(\frac{PQ \cdot h_A}{PQ \cdot h_B}\right)^2 = \left(\frac{QR_B}{QR_A}\right)\left(\frac{h_A}{h_B}\right)^2 = \left(\frac{AS}{AP}\right)\left(\frac{AS}{BP}\right)^2$$

We have arrived at our transmutation formula. Given a point P on an orbit and points A and B, draw the tangent through P and a line parallel to BP meeting the tangent at S, as shown in Figure J.43. Let $r_A = AP$ and $r_B = BP$. Then if area is swept out at the same rate for centripetal motions about A and B, the acceleration a_A and a_B for these motions satisfies:

$$a_B = a_A \frac{(AS)^3}{r_A r_B^2}$$

The Inverse Square Law and Kepler's First and Third Laws

We are now ready to give Newton's proof that Kepler's First Law implies an Inverse Square Law for the acceleration of gravity. Moreover, the constant of proportionality for that acceleration is $4\pi^2(d^3/p^2)$ when d is the mean distance and p is the period.

Newton's Third Theorem

Suppose an object is moving in an ellipse centripetally about a focus B of an ellipse. Suppose the mean distance of the ellipse is d and the period of the motion is p. If r is the distance from the object to B, then the magnitude of the acceleration satisfies

$$a = \frac{k}{r^2}, \text{ where } k = \frac{4\pi^2 d^3}{p^2}$$

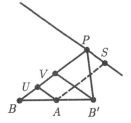

Figure J.44: $AS = UP =$
$(BP + VP)/2 = (BP + BP')/2.$

To see why this is true, suppose the motion were instead centripetal about the center A of the ellipse with the same period p. Then the acceleration a_A toward A, by Newton's Third Theorem, would be directly proportional to the distance r_A from A with constant of proportionality $4\pi^2/p^2$. Transmuting this acceleration to the given motion about B, we have:

$$a = a_B = a_A \frac{(AS)^3}{r_A r_B^2} = \left(\frac{4\pi^2}{p^2} r_A\right) \frac{(AS)^3}{r_A r_B^2} = \frac{4\pi^2}{p^2}(AS)^3 \cdot \frac{1}{r^2}$$

Thus, all we have to do is show that AS is the mean distance d.

Let B' be the other focus of the ellipse. Draw lines parallel to the tangent at P, one line passing through the center A and the other line passing through the focus B'. Suppose these lines meet BP at U and V as shown in Figure J.44.

Then since AS is parallel to BP, we have $AS = UP$. Since A is halfway between B and B' and UA and VB' are parallel, U is halfway between B and V. Thus, $UP = (BP + VP)/2$. Now we use two facts about an ellipse. First, a ray of light emanating from focus B is reflected by the tangent at P to the other focus B'. Thus, the angles made with the tangent by BP and BP' are equal, which makes triangle VPB' an isoceles triangle (remember that VB' is parallel to the tangent). Thus, $VB = B'P$. We have:

$$AS = UP = (BP + VP)/2 = (BP + BP')/2.$$

Next, we use the fact that $BP + BP'$ is constant for an ellipse and equal to twice the mean distance d. Thus, $AS = d$ and the proof is completed.

It follows from Kepler's Laws that every planet orbiting the sun has an acceleration whose magnitude is proportional to the inverse square of the distance r between the planet and the sun, and the constant of proportionality is the same for all planets. Indeed, Kepler's First and Second Laws imply that planetary motion is centripetal about the sun and that the orbits are ellipses with the sun at a focus. Therefore, by Newton's Third Theorem, the magnitude of the acceleration for each planet is proportional to yr^2 and the constant of proportionality is $4\pi^2 d^3/p^2$. By Kepler's Third Law this constant is the same for each planet.

The Dependence of Gravity on Mass

To this point, we have concentrated on acceleration rather than force. The striking point of Kepler's Laws is that there is no dependence at all on the masses of the individual planets. This reminds us of Galileo's discovery that the rate at which a body accelerates in free fall does not depend on the mass of the body. Since force is mass m times acceleration a, there is only one way the mass m can cancel out: the force of gravity between them must be proportional to both m and M. Then the magnitude of the force of gravity must be given by Newton's Universal Law:

$$F = \frac{GMm}{r^2}$$

where G is a constant. For a planet of mass m orbiting the sun of mass M, we have

$$F = ma = GMm/r^2,$$
$$a = GM/r^2.$$

The constant of proportionality is GM. By Newton's Third Theorem, the constant is $4\pi^2 d^3/p^2$. Thus, we have

$$GM = \frac{4\pi^2 d^3}{p^2}$$

In other words, suppose an object orbits around an object of mass M with mean distance d and period p. If we know M, we can then compute the gravitational constant G. Or if we know the gravitational constant G, we can compute the mass M. This applies to planets orbiting the sun, or the moon orbiting the earth, or Jupiter's moon orbiting Jupiter.

Newton's Law of Gravitation Implies Kepler's Laws

We have shown how Kepler's Laws imply Newton's Universal Law of Gravitation. Suppose we instead assume Newton's Law and try to derive Kepler's Laws. It really comes down to this. If you know the acceleration of an object at all times and you know the position and velocity of the object at some initial time, can you then compute the position and velocity at all future times uniquely? Intuitively it seems you should be able to. If you know the acceleration, you can use that to update continually the velocity and with the velocity you can continually update the position. This is, in effect, what Euler's method does to solve a differential equation.

Suppose you are given Newton's Law of Gravitation. Then you know each planet as it orbits the sun experiences centripetal acceleration towards the sun, of magnitude k/r^2. Wait until the planet is at its closest point Q to the sun. Suppose its distance from the sun at that point S and its velocity is v. Consider all possible ellipses with focus at the sun and closest point to the sun at Q. It is possible to find exactly one whose mean distance d determines a period p satisfying $k = 4\pi^2 d^3/p^2$ such that in order to sweep out the entire ellipse in time p the planet must pass through Q with velocity v. Consider now a "fictitious" planet tracing out that ellipse in period p. Then by Newton's Third Theorem, the planet must have the acceleration k/r^2 required by Newton's Law. Since the fictitious planet also has velocity v at position Q, its motion has the same acceleration and initial velocity and position as the actual planet. By our assumption that acceleration together with initial velocity and position determine motion completely, the fictitious and actual motion must be the same. That is, the actual planet travels in the given ellipse with period p and mean distance d satisfying $k = 4\pi^2 d^3/p^2$.

Other Conic Sections

It is possible that an object has such a high velocity that it does not orbit around the sun, but rather passes once through our solar system and then escapes the sun's gravitational attraction. What sort of path does it trace out? It is one of the other possible conic sections, either one branch of a hyperbola or a parabola. The case of the hyperbola is considered in Problem 10.

Problems for Section J

1. Here is how Eratosthenes estimated the circumference, and hence the radius, of the earth. On the longest day of the year, he knew that at Syrene, Egypt, the sun could be seen reflected at the bottom of a deep well; that is, the sun was directly overhead. On the same day, at Alexandria, Egypt, about 500 miles due north of Syrene, the shortest shadow of a 9 foot vertical pole cast by the sun was 1 foot long. Use this information to estimate the circumference of the earth. (See Figure J.45)

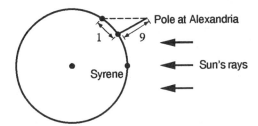

Figure J.45: One foot shadow cast by nine foot pole at Alexandria, while sun directly overhead at Syrene.

2. Compute the acceleration at a point on the equator caused by the earth's rotation. Use feet per second per second as the units. The radius is 4000 miles and the period is, as you know, 24 hours. Compare your answer to the value for gravity, $g = 32$ ft/sec/sec. What velocity would a point at the edge of a merry-go-round of radius 25 feet need in order to achieve the same acceleration? What would the period of the merry-go-round be?

3. Suppose an ellipse has foci at $(0, b)$ and $(0, -b)$ in the xy-plane and that the mean distance to the focus at $(0, b)$ is d. Show that the constant sum of the distances from any point on the ellipse to the two foci is $2d$. Then show that the equation for the ellipse is

$$\frac{x^2}{c^2} + \frac{y^2}{d^2} = 1$$

where d is the mean distance and $c^2 = d^2 - b^2$.

4. Show that there is no single point $x = a$ on the x-axis at which you can place an object of mass $2m$ such that the force exerted by that object on a unit mass placed at x, for all possible x, is the same as that exerted by two objects of mass m, one placed at $x = -1$ and the other at $x = 1$. Show that this is the case even if you restrict the unit mass to possible x such that $|x| > 2$. Show that this is the case even if you allow the single mass to be something other than $2m$.

5. What would happen if you drilled a tunnel from the north to the south pole and dropped a stone into the tunnel? You might expect that the stone might oscillate back and forth from the north pole to the south. On the other hand, if the acceleration of the earth's gravity varies as $1/r^2$, where r is the distance from the center of the earth, then something very strange would happen as the stone passed through the center of the earth. The acceleration there would be infinite, and maybe the stone would be shot right out through the south pole!

   Newton showed that if you are inside a hollow spherical shell, all the forces of gravity exerted by the mass of the shell cancel: there is no force at all. On the other hand, if you are outside the shell, you feel the same force that you would if the entire shell's mass were concentrated at the center of the sphere. Use this to show that if the density of the earth is constant, then the force of gravity at a distance r from the center is directly proportional to r. Use this to show that for our stone dropped into the north-south tunnel, the distance r from the stone to the center of the earth satisfies

$$\frac{d^2r}{dt^2} = -kr$$

What does this imply about the motion of the stone?

6. Suppose a particle moves in the xy-plane so that its acceleration vector $\vec{a}$ always points to the origin and has magnitude proportional to the distance to the origin. Choose the x-axis so that the closest point to the origin on the particle's path is $(a, 0)$. Explain why at that point the velocity vector is perpendicular to the x-axis. Show that with the given x and y coordinates, if we choose to define time $t = 0$ to be the instant when the particle is at $(a, 0)$, then the particle satisfies the differential equations

$$\frac{d^2x}{dt^2} = -kx, \quad \frac{d^2y}{dt^2} = -ky, \quad k > 0$$

with initial conditions $x(0) = a$, $\frac{dx}{dt}(0) = 0$, and $y(0) = 0$, $\frac{dy}{dt}(0) = c$. Here c is the velocity in the y-direction at time $t = 0$. Now show that the solution to these differential equations is

$$x = a \cos \sqrt{k}t, \quad y = b \sin \sqrt{k}t$$

where $b = c/\sqrt{k}$.

7. Experiment on a computer or calculator with centripetal force laws. You will need a program that will plot trajectories (solutions) for systems of differential equations. For example, if you want to look at orbits for the k/r centripetal force law, you need to solve a system with four variables: position variables x and y and velocity variables $u = dx/dt$ and $v = dy/dt$. The system looks like this

$$\frac{dx}{dt} = u, \quad \frac{dy}{dt} = v, \quad \frac{du}{dt} = \frac{-kx}{x^2+y^2}, \quad \frac{dv}{dt} = \frac{-ky}{x^2+y^2}.$$

Check that these equations imply that the acceleration vector $(d^2x/dt^2)\vec{i} + (d^2y/dt^2)\vec{j} = (du/dt)\vec{i} + (dv/dt)\vec{j}$ has the correct direction and magnitude. Then let the computer plot the x and y variables starting from some initial values for x, y, u, and v. Try other laws: k/r^3, kr^2. Are orbits always closed?

8. Show that if A and B are points on one side of a line l, then the shortest path from A to the line l and then back to B, is the path that bounces off of l making equal angles with l coming and going. (Hint: Think of the shortest path from A to the point B' on the other side of l such that l bisects BB'). Use this to show that the path from one focus A of an ellipse to a tangent l then back to the other focus B makes equal angles with l coming into and leaving the point of tangency (Hint: Show that the path from A to the point of tangency to B is the shortest path from A to l to B and apply first part.

9. There are a number of ways to measure the acceleration g of the earth's gravity at the surface of the earth. You would probably think that all you need to do is time how long it takes an object to fall a certain distance. Measuring intervals of time less than a second, however, is

very difficult by hand and most free falls are over in two or three seconds. Galileo slowed the fall by using inclined ramps. A more clever way that does not require accurate measurement of time involves pendulums. A pendulum of length l, when displaced an angle θ from the vertical, experiences an acceleration

$$l\frac{d^2\theta}{dt^2} = -g\sin\theta$$

You might want to try to derive this; it is not hard. When θ is small, $\sin\theta \approx \theta$, and hence we have the equation of the harmonic oscillator

$$\frac{d^2\theta}{dt^2} + \frac{g}{l}\theta = 0,$$

which has solutions of period $p = 2\pi/\sqrt{g/l}$. What is surprising is that the period is independent of the initial displacement angle θ, and thus even as the pendulum swings are dissipated by friction, the period for one full swing stays the same. This fact is the basis for pendulum clocks. Since p is easy to compute and the length l is easy to measure, we have a good way to determine g:

$$\sqrt{g/l} = 2\pi/p \quad \text{so} \quad g = 4\pi^2 l/p^2.$$

Try it. Take a piece of string (as long as you can make it and still swing conveniently) and tie a weight to the string. Then count the total number of complete swings (to and fro) in one minute. Use this to compute p and therefore g.

10. A hyperbola is a curve such that the *difference* of the distances from any point on the curve to two fixed points (called the foci) is constant. The equation for a hyperbola centered at the origin is

$$-\frac{x^2}{c^2} + \frac{y^2}{d^2} = 1$$

where $2d$ is the constant difference in distances from foci at $(0, b)$ and $(0, -b)$ and $c^2 = b^2 - d^2$. Show that

$$x = c(e^{kt} - e^{-kt}), \; y = d(e^{kt} + e^{-kt})$$

satisfies $-x^2/c^2 + y^2/d^2 = 1$ and also

$$\frac{d^2x}{dt^2} = kx, \; \frac{d^2y}{dt^2} = ky.$$

Thus the given motion has an acceleration pointing away from the origin with magnitude proportional to the distance from the origin.

INDEX

Acceleration, 80, 300–302
  circular motion, 301
  linear motion, 301
  vector, 300–302
    components of, 301
    limit definition of, 300
Addition of vectors, 71
  components, 71
  geometric view, 67
  properties, 81
Ampere's law, 352
Approximation
  linear, 126–134
  quadratic, 171–174
Archimedes' Principle, 423
Area
  of surface, 402
  vector, 380
    of parallelogram, 393
Aristotle, 490
Average value of function
  one-variable, 461
  two-variable, 228

Beef consumption, 3
Boltzmann distribution, 479
Boundary
  condition, 165
  of region, 193
  point, 193
Bounded region, 193
Burger's equation, 171

Cartesian coordinates
  conversion to
    cylindrical, 256
    polar, 480
    spherical, 259
    three-dimensional, 10
Catalog of surfaces, 56
Central vector field, 370
Centripetal force, 495, 496
Chain rule, 151–156
  tree diagram for, 153

Change of
  origin, 482
  parameter, 284
  variable, 273–276, 482
Circulation, 347
  and conservative field, 371
  density, 426
  and nonconservative field, 371
  surface, around, 435
Closed region, 193
Cobb-Douglas production function, 33–34
  contour diagram of, 33
  formula for, 34
  returns to scale, 42
Components of vector, 69
  arbitrary direction, 90
  finding, 72
Composite function, 151
Cone, parameterization of, 309
Conservative vector field, 363–366
  definition of, 363
  and circulation, 371
  and gradient field, 363, 365
Constrained optimization, 206–215
  analytical solution
    inequality constraints, 210
  graphical approach, 206
  Lagrange multipliers, 207, 209
  and Lagrangian function, 214
Consumption vector, 83
Contour diagram, 26–34
  and algebraic formula, 30
  Cobb-Douglas, 33
  critical point
    local maximum, 185
    local minimum, 188

    saddle point, 32, 188
  and density, 222
  and differential, 131
  linear function, 46
  and partial derivative, 115
  reading, 2
  and table, 32
Contour line, 27
Contour map, *see* contour diagram
Coordinate
  plane, 12
  axis, 10
Coordinates
  converting Cartesian to
    cylindrical, 256
Coordinates
  Cartesian, three-space, 10
  converting Cartesian to
    polar, 480
  converting Cartesian to
    spherical, 259
  cylindrical, 256–258
  polar, 479
  space-time, 83
  spherical, 259–261
Copernicus, 489
Corn production, 28
Correlation coefficient, 199
Critical point, 184
  discriminant of, 191
  how to find, 185
  local maximum, 185
    contour diagram of, 185
    graph of, 186
  local minimum, 185
    contour diagram of, 188
    graph of, 186
  saddle point, 187
    contour diagram of, 188
    graph of, 188
  and second derivative test, 189, 191
Cross product, 96–102

506

and area, 100
components of, 99, 459
definition of, 96
and determinant, 102
diagram of, 98
and equation of plane, 100
moment, 97
properties of, 98
torque, 97
volume of parallelepiped,
101

Curl
device for measuring, 425
Curl field, 438
Curl of vector field, 425–431
Cartesian coordinates, 429,
445
components of, 428, 447
criterion for gradient field,
437
definition of, 427
formula for, 429
notation for, 430
Curve
complicated, 285
fitting, 198
integral, 333
length of, 299
level, 27
and graph, 30
oriented, 342
parameter, 313
parameterization, 288–292
piece-wise smooth, 343
representation
explicit, 291
implicit, 291
parametric, 291
Cylinder, parameterization of,
306
Cylindrical coordinates, 256–
258
conversion to Cartesian,
256
integration in, 257
volume element, 257
and volume, 258

Definite integral, one-variable,
459–462
definition of, 460
diagram of, 459, 460
and Fundamental Theo-
rem of Calculus, 462
interpretation as
area, 460
average value, 461
density function, 462
total change, 462
and Monte Carlo Method,
247
Definite integral, three-variable,
244–246, 256–261
Definite integral, two-variable,
222–241
definition of, 224
interpretation
average value, 228
density function, 228
volume, 226
and Monte Carlo Method,
249
polar coordinates, 251–
254
Demand function, 219
Density
and definite integral, 222,
228
two-variable
function, 232
Density function, 470, 471
joint, 265
and dependence, 268
and independence, 270
normal distribution, 271
one-variable, 462, 469
properties of, 471
two-variable, 263, 264
and probability, 266
Dependent variable, 2
Derivative
directional, 137–141
higher-order partial, 157
mixed partial, 160
ordinary, 110
partial, 110–124
second-order partial, 157
Determinant, 458

Determinant, 102
Difference quotient
and partial derivative, 112
Differentiability
definition of, 134
and local linearity, 133
and partial derivatives, 134
Differentiable function, 133
Differential, 130
Differential equation
and flow, 336
and vector field, 336
Differential equation, partial,
162–167
boundary condition, 165
heat equation, 163
wave equation
one-dimensional, 167
representation of, 166
traveling, 166
Differential, the, 133
computing, 132
and local linearity, 131
notation, 132
Diffusion equation, 163
Directional derivative, 137–
141
Cartesian coordinates, 140
computing, 137
definition of, 138
and gradient vector, 146
Discriminant, 190, 191
Displacement vector, 66–73
Distance formula, 14
Divergence, 410–415
alternative notation for,
415
Cartesian coordinates, 411,
443
and curl, 438
definition of, 410
free, 411
meaning of, 412
and vector field, 410–415
Divergence Theorem, 418–421,
440
Dot product, 86–90
components, 88
definition of, 86

and equation of plane, 89
formula for, 86
properties of, 88
Double integral, *see* definite
integral, two-variable

Electric field, 392, 417
Eratosthenes, 489
Euler's method, 334
Euler's theorem, 125
Explicit representation of curve,
291
Extremum, *see* also global ex-
tremum, local extremum
Extremum, 184
global, closed bounded
region, 193

Field, *see* also vector field
conservative, 366
curl, 438
electric, 392, 417
force, 325
gravitational, 327, 423
magnetic, 352, 389, 442
Flow
circulating, 348
and differential equation,
336
fluid, 324
heat, 162
through
constant, flat surface,
381
variable, curved surface,
382
with zero circulation, 348
Flow lines
Euler's method, 334
numerical solution, 334
Flux integral, 380–403
and area vector, 380
calculating through
closed surface, 386
constant, flat surface,
381
parameterized surfaces,
401
surface graph, 393

variable, curved surface,
382
definition of, 383
and Divergence Theorem,
418
and orientation, 380
Force
centripetal, 495, 496
gravitational, 80
vector, 80
Fox population, 222
Function
Cobb-Douglas, 33–34
composite, 151
demand, 219
density, 471
one-variable, 469
two-variable, 264
differentiable, 133
differential of, 130
fixing one variable, 4–7,
18, 51
harmonic, 422
joint cost, 218
Lagrangian, 214
linear, 20, 43–48
linear approximation of,
see linearization, local
notation, 2
potential, 364
quadratic, 189
section of, 18–20
smooth, 160
three-variable, 51, 53
level surface of, 53
surface, 56
table of, 52
two-variable, 2
algebraic formula, 3
contour diagram of, 26
graph of, 15–21
surface, 56
Fundamental Theorem of Calculus, 462
line integral, 361, 440
one-variable, 360, 462

Galileo, 499
Gas equation, 483

Gauss' law, 392
Global extremum
closed bounded region,
193
definition of, 184
how to find, 192
Global warming, 152
Gradient search, 199
Gradient vector, 143–147
alternative notation, 145
components, 144
constant, 361
and directional derivative,
146
field, 328
geometric definition of,
143
line integral of, 361
properties of, 144
and rate of change, 147
three-variable, 147
Graph
area under, 460
concave down, 190
concave up, 190
and partial derivative, 114
plane, 44
saddle-shaped, 190
three-space, in, 11
two-variable function, 44
Gravitational field
picture of, 325
Green's theorem, 375
Guitar string, vibrating, 122
Gulf stream, 324, 331

Harmonic functions, 421
Head index, 9
Heat equation, 163
Heated metal plate, 110
Heated room, 115
Helix, 283
Helmholtz's theorem, 433
Higher-order partial derivative,
157
Histogram, 469, 470
Hopf-Cole transformation, 171

Ideal gas equation, 133

508

Implicit Function Theorem, 484
Implicit representation of curve, 291
Independent variable, 2
Inertia, moment of, 279
Instantaneous
    rate of change, 112
    velocity, 297
Integral
    definite, *see* definite integral
    flux, *see* flux integral
    iterated, *see* iterated integral
    line, *see* line integral
    one-variable, *see* definite integral, one-variable
    tables of, 467
    three-variable, *see* definite integral, three-variable
    two-variable, *see* definite integral, two-variable
Integrand, 460
Integration
    Cartesian coordinates, 234, 244, 460
    cylindrical coordinates, 257
    limits of, 460
    non-rectangular region, 237–241
    one-variable, 459
    order of, 237, 241
    polar coordinates, 251–254
    rectangular region, 225
    spherical coordinates, 260
    tables of, 467
    techniques
        tables, 467
Interior
    of region, 193
    point, 193
Intersection
    of curve and surface, 292
    of line and plane, 295
Irrotational vector field, 428
Iterated integral, 232
    and double integral, 234
    graphical view, 232

non-rectangular region, 237–241
numerical view, 234
and triple integral, 244
variable limits, 239

Jacobian, 274
Joint
    cost function, 218
    density function, 265
        and dependence, 268
        and independence, 270

Kepler's Laws, 80, 489–500

Lagrange multipliers, 207, 209
    constrained optimization, 208, 209
    lambda, meaning of, 211, 212
Lagrangian function, 214
Lanchester model, 336
Laplace equation, 169
Law
    Ampere's, 352
    Gauss', 392
    gravitation, 327, 495
    inverse square, 495, 498
    Kepler's, 491, 494
    motion, 492
    Newton's, 492, 495
    planetary motion, 491
Least squares, 198
Left-hand sum, 459
Level
    curve, 27
      and graph, 30
    sets, 27
    surface, 53
Line
    contour, 27
    least squares, 198
    parametric equation for, 290
    regression, 198
Line integral, 342–349
    circulation, 347
    computing, 352–358
    conversion to one-variable

integral, 353
definition of, 342
Fundamental Theorem of Calculus, 361, 440
and gradient vector, 361
interpretation of, 345
notation for, 357
path-independence, 363
properties of, 348
what it tells us, 344
Linear approximation, *see* linearization, local
Linear function, 20, 43–48
    contour diagram of, 46
    equation for, 44
    graph of, 44
    numerical view, 45
    table of, 45
    two-variable, 43
Linearization, local, *see* also local linearity, 126–134
    and differential, 131
    and Implicit Function Principle, 487
    one-variable function, 456
    table, 129
    three-variable or more, 130
    two-variable function, 126, 128
Lissajous figure, 285
Local extremum
    definition of, 184
    gradient search, 199
Local linearity, *see* also linearization, local
    definition of, 133, 134
    and differentiability, 133

Magnetic field, 352, 389, 442
Maximum, *see* global extremum and local extremum
Maxwell's equation, 423
Mean, 474–476
Median, 473
Metal plate, heated, 110
Minimum, *see* global extremum and local extremum
Mixed partial derivative, 160
Moment, 97

Monte Carlo Method, 247–250

Newton, 489–500
    and Kepler's Second Law, 493
    Law of Gravitation, 327, 495, 500
    laws of motion, 492
Nonconservative vector field
    and circulation, 371
Normal distribution, 476
    and standard deviation, 477
Normal vector, 89

One-variable integral, *see* definite integral, one-variable
Open region, 193
Optimization
    constrained, 206–215
    unconstrained, 196–201
Orientation
    curve, 342
    surface, 380
Origin, 10

Parabolic cylinder, 21
Parameter
    change, 284
    curve, 313
    rectangle, 314
        area vector, 397
Parameterization, 282–285, 306–314
    curve, 288–292
        changing, 284
        complicated, 285
        graph of function, 288
    independence of, 357
    line in three-space, of, 290
    and line integral, 357
    plane, of, 308
    surface, 306–314
        of revolution, 309
        using spherical coordinates, 310
    three-dimensional, 283
    using position vector, 289
Parametric equation, *see* parameterization

rameterization
    Lanchester model, 336
Parametric representation of curve, 291
Partial derivative, 110–124
    alternative notation, 112
    computing
        algebraically, 121
        graphically, 115
        numerically, 113
    and contour diagram, 115
    definition of, 112
    and difference quotient, 112
    and differentiability, 134
    function, 123
    and graph, 114
    higher-order, 157
    mixed, 160
    and rate of change, 111–112
    second-order, 157
Partial differential equation, *see* differential equation, partial
Path-independence, 363
Perpendicular vector, 88
Piece-wise smooth curve, 343
Plane, 43
    contour diagram of, 46
    equation for, 44, 89, 100
    graph of, 44
    parameterization of, 308
    points on, 45
    tangent, 127
Planetary motion, 489–500
Point
    boundary, 193
    interior, 193
    sink, 418
    source, 417
Polar coordinates, 479
    area element, 252
    conversion to Cartesian, 480
    integration in, 251–254
Population vector, 83
Position vector, 307, 325
Positive normal, 380

Potential
    energy, 363
    function, 364
Price vector, 83
Probability, 263, 469
    conditional, 268
    and density function, 266, 471
    histogram, 263, 469
    independent variables, 269
Production function
    Cobb-Douglas, 33–34
    general formula, 33
Projection, stereographic, 317
Properties of
    addition and scalar multiplication, 81
    cross product, 98
    dot product, 88
    gradient vector, 144
    line integral, 348
    vector field, 366
Pythagoras' theorem, 14

Quadratic
    approximation, 171–174
    function, 189
        discriminant of, 190
        and second derivative test, 189

Rankine model, 378
Rate of change, 111–112, 137, 143, 147
Rectangular coordinates, *see* Cartesian coordinates
Region
    bounded, 193
    closed, 193
    open, 193
Regression line, 198
Rescaling coordinates, 483
Retrograde motion, 490
Returns to scale, 42
Riemann sum
    one-variable, 459, 460
    three-variable, 244
    two-variable, 224, 236
Right-hand

rule, 96
sum, 459

Saddle, 19
Saddle point, 187
    contour diagram of, 188
    graph of, 188
Scalar multiplication
    geometric definition of,
    68
    and line integral, 348
    properties, 81
Scalar product, 86
Second derivative test, 189,
    191
Second-degree Taylor expan-
    sion, 172, 173
Second-order partial derivative,
    157
    what it tells us, 158
Section of functions, 18–20
Simpson's rule, 251
Smooth
    curve, 343
    function, 160
    surface, 384
Solenoidal vector field, 411
Space-time coordinates, 83
Speed, 299
    and velocity, 78
Sphere
    equation for, 14
    parametrizing, 311
Sphere, parameterization of,
    310
Spherical coordinates, 259–261
    conversion to Cartesian,
    259
    integration in, 260
    parametrizing a sphere,
    311
    volume element, 260
Standard deviation, 476
    of normal distribution, 477
State equation, 133
Steam tables, 129
Stereographic projection, 317
Stokes' Theorem, 434–440
Streamline, 333

Subtraction of vectors, 71
    components, 71
    geometric view, 67
Sum, Riemann, 459, 460
Surface
    of revolution, 309
    area, 402
    orientation, 380
    parameterization, 306–314
    smooth, 384
    and three-variable func-
    tion, 56
    and two-variable function,
    56

Table
    and contour diagram, 32
    and linear function, 45
    reading, 3
    and three-variable func-
    tion, 52
    and two-variable function,
    3
Table of integrals, 467
Tangent
    line and velocity vector,
    299
    plane, 127
Taylor expansion
    second-degree, 172, 173
Theorem
    Divergence, 420, 440
    Euler's, 125
    Fundamental (of Calcu-
    lus), 360
    Green's, 375
    Helmholtz's, 433
    Pythagoras', 14
    Stokes', 434, 440
Three-variable integral, *see* def-
    inite integral, three-variable
Torque, 97
Tree diagrams, 153
Triple integral, *see* definite in-
    tegral, three-variable
Triple product
    volume of parallelepiped,
    101
Two-variable integral, *see* def-

inite integral, two-variable

Unconstrained optimization, 196–
    201
Unit vector, 73

Van der Waal, 483
Variable
    dependent, 2
    independent, 2
Vector, 66
    addition, 71
    area, 380
      of parallelogram, 393
      parameter rectangle, 397
    components, 69, 72
    consumption, 83
    cross product, 96–102
    displacement, 66–73
    dot product, 86–90
    geometric definition of,
    66
    gradient, 143–147
    magnitude, 71
    n-dimensional, 82
    normal, 89
    notation, 82
    perpendicular, 88
    population, 83
    price, 83
    scalar multiplication, 68
    subtraction, 71
    unit, 73
    velocity, 295–300
      and tangent line, 299
    and work, 87
    zero, 71
Vector field, 324–329
    central, 370
    conservative, 363–366
    curl free, 428
    curl of, 425–431
    definition of, 325
    divergence free, 411
    divergence of, 410–415
    flow, 333
    flow line, 331–337
      definition of, 333
    gradient, 328, 360

how to sketch, 325
integral curve, 333
interpreting formula for,
  325
irrotational, 428
properties of, 366
solenoidal, 411
streamline, 333
Vector product, *see* cross product
Velocity
  field, 324

instantaneous, 297
and speed, 78, 299
vector, 295–300
  components of, 297
  definition of, 295
  and tangent line, 299
Vibrating guitar string, 122
Volume, 226, 258
  parallelepiped, 101
Vortex, free, 378, 442

Wave equation, 166

Wave, the, 4
Weather map, 2, 6
Wind-chill factor, 8
Work, 86, 345
  done by a force, 345
  and vector, 87

X-sections, 18

Y-sections, 18

Zero vector, 71